Astrophysics and Space Science Proceedings

For further volumes:
http://www.springer.com/series/7395

Preface

The origin of elements is among the fundamental aspects of our universe; cosmochemistry tries to answer when, how and where the chemical elements arose after hydrogen was created during primordial nucleosynthesis following the Big Bang. However, quantitative answers to these fundamental questions began to emerge only in the late fifties, with the pioneering works of Burbidge, Burbidge, Fowler and Hoyle, and Cameron. Since then there had been significant progress in the understanding of synthesis of elements in stars.

Cosmochemistry, however, remains a fertile area of research, as there remain many outstanding problems. A comprehensive approach to cosmochemistry requires a combination of a number of topics like primordial nucleosynthesis, stellar nucleosynthesis, explosive nucleosynthesis and solar abundance. The Kodai school on 'Synthesis of elements in stars' was organized to provide a glimpse of this exciting area of research to astrophysicists of tomorrow, motivated young students from India and abroad. The lectures are thus aimed at researchers who would like to venture deeper into this exciting arena.

The school drew strength from considerable in-house expertise at IIA in a number of areas critical for the school. A highlight of the school, however, was the faculty participation by a number of leading astrophysicists from different parts of the world.

Following a traditional and inspiring invocation from Upanishad and a brief inaugural function, the school was opened for technical sessions. David Lambert set the tone of the scientific sessions with the lead talk on 'Synthesis of elements in stars: an overview'. The basic properties of nuclei were explained by Arun Mangalam in a series of lectures. The lectures by C Sivaram put the primary issue in cosmochemistry in perspective through a discussion on cosmological nucleosynthesis of light elements. Aruna Goswami discussed some current issues in the present understanding of the Galactic chemical evolution. Gajendra Pandey explained how stellar spectra can be analyzed using 'Curve of growth technique'. Kameswara Rao of IIA talked about the high resolution Echelle spectrograph at VBO, Kavalur and discussed some results obtained from analysis of data acquired using this instrument. These lectures provided the background for the series of lectures by other speakers that followed. Apart from the regular class room lectures, students had ample time for hands-on sessions coordinated by Goswami, Reddy and Pandey.

The book has been organized into three parts to address the major issues in cosmochemistry. Part I of the book deals with stellar structure, nucleosynthesis and evolution of low and intermediate-mass stars. The lectures by Simon Jeffery outline stellar evolution with discussion on the basic equations, elementary solutions and numerical methods. Amanda Karakas's lectures discuss nucleosynthesis of low and intermediate-mass stars covering nucleosynthesis prior to the Asymptotic Giant Branch (AGB) phase, evolution during the AGB, nucleosynthesis during the AGB phase, evolution after the AGB and massive AGB stars. The *slow* neutron-capture process and yields from AGB stars are also discussed in detail by Karakas. The lectures by S Giridhar provide some necessary background on stellar classification.

Part II deals with explosive nucleosynthesis that plays a critical role in cosmochemistry. The lectures by Kamales Kar provide essential background material on weak-interaction rates for stellar evolution, supernovae and r-process nucleosynthesis. He also discusses in detail the solar neutrino problem. Massive stars, their evolution and nuclear reaction rates from the point of view of astronomers and nuclear physicists are discussed by Alak Ray. His lectures also describe the various stages of hydrostatic nuclear fuel burning with illustrative examples of how the reactions are computed. He also discussed core-collapse (thermonuclear vs. core-collapse) and supernovae in brief. The lectures by Marcel Arnould address the phenomena of evolution of massive stars and the concomitant non-explosive and explosive nucleosynthesis. He highlights a number of important problems that are yet unresolved but crucial for our understanding of Galactic chemical evolution. The p-process nucleosynthesis attributed to the production of proton-rich elements, a topic of great importance but yet less explored is also discussed in his lectures.

The third and the final part of the book addresses use of solar system abundances to probe cosmochemistry quantitatively. The lectures by Bruce Fegley address cosmochemistry of the major elements; while the lectures by Katharina Lodders discuss elemental abundances in Solar, meteoritic and outside the solar system.

Cosmochemistry is still an evolving branch of astrophysics, with many challenges. The book is expected to serve as a contemporary reference material for research in cosmochemistry. We would like to take this opportunity to thank all the contributors for making this book a reality.

Bangalore,
April 2009

Aruna Goswami
B. Eswar Reddy

Acknowledgement

This school would not have been possible without the dedicated support of many. We extend our sincere thanks to professor Siraj Hasan, Director, Indian Institute of Astrophysics and professor Vinod Krishan for their all round support for the school.

We are particularly grateful to the school faculty from India and abroad for readily accepting to participate, prepare lecture notes and spend time with the students.

The organization of the school is a collective effort of the coordinators, the convener, the members of the local organizing committee and many others. We are thankful to the administrative department of IIA and the staff members of Kodaikanal Solar observatory for their help and support in various activities of the school.

Contents

List of Contributors

C. Simon Jeffery
Armagh Observatory, College Hill, Armagh ET61 9DG, Northern Ireland
csj@arm.ac.uk

Amanda I. Karakas
Research School of Astronomy & Astrophysics
akarakas@mso.anu.edu.au

Sunetra Giridhar
Indian Institute of Astrophysics, Bangalore 560034, India
giridhar@iiap.res.in

Kamales Kar
Saha Institute of Nuclear Physics, Bidhannagar, Kolkata 700064, India
kamales.kar@saha.ac.in

Alak Ray
Tata Institute of Fundamental Research, Mumbai 400005, India
akr@tifr.res.in

Marcel Arnould
Institut d'Astronomie et d'Astrophysique, Universite' Libre de Bruxelles, CP-226, B-1050 Brussels, Belgium
marnould@astro.ulb.ac.be

Bruce Fegley
Planetary chemistry Laboratory, Department of earth and planetary sciences, Washington University, St. Louis, MO63130-4899, USA
bfegley@wustl.edu

Laura Schaefer
Planetary chemistry Laboratory, Department of earth and planetary sciences, Washington University, St. Louis, MO63130-4899, USA
laura_s@levee.wustl.edu

Katharina Lodders
Planetary chemistry laboratory, Department of earth and planetary sciences and McDonnell centre for the space sciences, Washington University, Campus box, 1169, One Brookings Drive, Saint Louis, MO63130, USA
lodders@wustl.edu

School Faculty

Marcel Arnould
Institut d'Astronomie et d'Astrophysique,
Universite' Libre de Bruxelles, CP-226, B-1050 Brussels, Belgium
marnould@astro.ulb.ac.be

Bruce Fegley
Planetary chemistry Laboratory
Department of earth and planetary sciences, Washington University
St. Louis, MO63130-4899, USA
bfegley@wustl.edu

Sunetra Giridhar
Indian Institute of Astrophysics, Bangalore 560034, India
giridhar@iiap.res.in

Aruna Goswami
Indian Institute of Astrophysics, Bangalore 560034, India
aruna@iiap.res.in

C. Simon Jeffery
Armagh Observatory, College Hill, Armagh ET61 9DG, Northern Ireland
csj@arm.ac.uk

Kamales Kar
Saha Institute of Nuclear Physics, Bidhannagar, Kolkata 700064, India
kamales.kar@saha.ac.in

Amanda I. Karakas
Research School of Astronomy & Astrophysics
akarakas@mso.anu.edu.au

David L. Lambert
McDonald Observatory, University of Texas at austin, Austin
dll@astro.as.utexas.edu

Katharina Lodders
Planetary chemistry laboratory
Department of earth and planetary sciences and McDonnell centre for the space sciences
Washington University, Campus box, 1169, One Brookings Drive, Saint Louis, MO63130, USA
lodders@wustl.edu

Arun Mangalam
Indian Institute of Astrophysics, Bangalore 560034, India
mangalam@iiap.res.in

Gajendra Pandey
Indian Institute of Astrophysics, Bangalore 560034, India
pandey@iiap.res.in

N. Kameswara Rao
Indian Institute of Astrophysics, Bangalore 560034, India
nkrao@iiap.res.in

Alak Ray
Tata Institute of Fundamental Research, Mumbai 400005, India
akr@tifr.res.in

Eswar Reddy
Indian Institute of Astrophysics, Bangalore 560034, India
ereddy@iiap.res.in

C. Sivaram
Indian Institute of Astrophysics, Bangalore 560034, India
sivaram@iiap.res.in

List of Participants

Tiago Mendes de Almeida
Cidade Universitaria, Sao Paulo-SP-Brazil

Sonam Arora
Department of Physics, Punjab University, India

Andrea Borch
Indian Institute of Astrophysics, Bangalore-560034, India

Bhavya B.
Department of Physics, CUSAT, Cochin, Kerala, India

R. S. Keerthi Chandar
Department of Physics, Bharathiar University, Tamil Nadu, India

K. Chandrashekhar
Indian Institute of Astrophysics, Bangalore-560034, India

Sukanta Deb
Department of Physics & Astrophysics, Delhi University, Delhi, India

Nandita Debnath
Department of Physics, Tezpur University, Napam, Sonitpur - 784028, Assam, India

Thubstan Dorje
Indian Astronomical Observatory, Hanle, Leh-Ladakh, India

Krithika Dota
Department of Physics, Mumbai University, Mumbai, India

Koshy George
ISRO, Bangalore - 560017, India

Suruchi Goel
Physical Research Laboratory, Navrangpura, Ahmedabad - 380009, India

Gagan Gupta
Department of Physics, Panjab University, Chandigarh, India

Singh Abhishek Indrajit
University of Mumbai, Kalina Campus , Santa Cruz, Mumbai - 400098, India

Vishal Joshi
Physical Research Laboratory, Navrangpura, Ahmedabad - 380009, India

Devika Kamath
Department of Physics, Christ College, Bangalore - 560029, India

Sreeja S. Kartha
Indian Institute of Astrophysics, Bangalore - 560034, India

Chrisphin Karthick
ARIES, Manora Peak, Nainital - 263129, India

Rajwinder Kaur
Punjab University, Patiala, Punjab, India

Pranav Kumar
Department of Physics & Astrophysics, University of Delhi, Delhi, India

Blesson Mathew
Indian Institute of Astrophysics, Bangalore - 560034, India

Ritesh K. Mishra
Physical Research Laboratory, Navrangpura, Ahmedabad - 380009, India

Rana Nandi
SINP, Theory Division, 1/AF, Bidhannagar, Kolkata - 700064, India

H. S. Nataraj
Indian Institute of Astrophysics, Bangalore - 560034, India

Vinicius Moris Placco
Cidade Universitaria, Sao Paulo-SP-Brazil

Ananta C. Pradhan
Indian Institute of Astrophysics, Bangalore - 560034, India

Yogesh Prasad
Dept. of Physics, H.N.B. Garhwal University
Srinagar Garhwal- 246174, Uttarakhand, India

Ashish Raj
Physical Research Laboratory, Navrangpura, Ahmedabad - 380009, India

N. G. Rudraswami
Physical Research Laboratory, Navrangpura, Ahmedabad - 380009, India

Krishna Prasad Sayamanthula
Indian Institute of Astrophysics, Bangalore - 560034, India

Arul Selvam
Department of Physics, Madurai Kamaraj University, Tamil Nadu, India

S. Sujatha
M P Birla Inst. of Fundamental Research, 43/1 Race Course Road
Bharatiya Vidya Bhavan Campus, Bangalore - 560001, India

Ramya Sethuram
Indian Institute of Astrophysics, Bangalore - 560034, India

Bharat Kumar Yerra
Indian Institute of Astrophysics, Bangalore - 560034, India

Part I

Stellar Structure, Nucleosynthesis and Evolution of Low and Intermediate-mass Stars

Stellar Structure and Evolution: An Introduction

C. Simon Jeffery

Armagh Observatory, College Hill, Armagh BT61 9DG, N. Ireland, UK
csj@arm.ac.uk

Summary. The synthesis of new elements takes place inside stars. How do stars evolve and distribute this creation to the universe at large? This article starts with the observables that the theory of stellar evolution aims to reproduce, and gives a quick overview of what that theory predicts (Sects. 2–3). It presents the equations governing stellar structure and evolution (Sects. 4–6) and the physics of stellar interiors (Sects. 7–9). Approximate and numerical methods for their solution are outlined (Sects. 10–11) and the general results of stellar structure and evolution are discussed (Sects. 12–13). The structure and evolution of horizontal-branch stars, hydrogen-deficient stars and other stellar remnants are also considered (Sects. 14–15).

Keywords: Stars: interiors – Stars: evolution – Stars: horizontal-branch – Stars: AGB and post-AGB – HR and C-M diagrams – Equation of state – Convection – Atomic Processes – Nucleosynthesis

1 Introduction

What are the stars? How do they shine? What are they made of? These questions have challenged mankind ever since he began to explore the world around him and appreciate the awesome splendour of the night sky. Just as challenging are questions about what we ourselves are made of, and where we come from. Only in the last hundred years have we started to find answers that approach a coherent understanding of the universe we inhabit.

Fundamental to understanding the stars are measurements of distance and brightness, colour and constancy. Any theory of what stars are and how they behave must be able to explain these observations. Deeper insight is gained from measurements of chemical composition and the relationships between stars and the interstellar medium. The big story will show how elements are manufactured by nuclear reactions deep inside the stars – nucleosynthesis – and then transported to the stellar surface and into the interstellar medium.

A. Goswami and B.E. Reddy (eds.), *Principles and Perspectives in Cosmochemistry*,
Astrophysics and Space Science Proceedings,
DOI 10.1007/978-3-642-10352-0_1,

Table 1. The Sun

Mass:	$M_\odot = 1.98892(25) \times 10^{30}$ kg
Radius:	$R_\odot = 6.9599(7) \times 10^{8}$ m
Surface gravity:	$g_\odot = 2.7397(5) \times 10^{2}\,\mathrm{m\,s^{-2}}$
Effective temperature:	$T_{\mathrm{eff},\odot} = 5770(6)$ K
Luminosity:	$L_\odot = 3.826(8) \times 10^{26}$ W
Surface hydrogen mass fraction:	$X_\odot = 0.71$
Surface helium mass fraction:	$Y_\odot = 0.265$
Surface metal mass fraction:	$Z_\odot = 0.025$
Age:	$t_\odot \approx 4.567 \times 10^{9}$ y

The object of these lectures is to explain the physics of stellar interiors, to use this physics to make stellar models and hence to understand how stars work and evolve. The lectures will demonstrate how models of stars are constructed, and explain how these models predict stars should evolve.

This article starts by introducing some of the fundamental observational material (Sect. 2), and by providing an early preview of the stellar evolution theory (Sect. 3). Fundamental timescales and the equations of stellar structure and evolution are derived in Sects. 4–6. The micro-physics (equation of state, opacity and nuclear physics) are discussed in Sects. 7–9. Some methods for calculating approximate solutions and full numerical solutions are presented (Sects. 10–11). Subsequent sections deal with the evolution of main-sequence stars (Sect. 12), white dwarfs and supernovae (Sect. 13), horizontal-branch stars (Sect. 14) and hydrogen-deficient stars (Sect. 15).

The text is based on a series of six lectures given at the 2008 Kodai School on Synthesis of the Elements in Stars[1] and on a more extended course given in the University of St Andrews and Trinity College, Dublin over a period of some twenty years. At Kodaikanal, the core material comprised four lectures. Two more advanced lectures covered horizontal-branch stars (Sect. 14) and hydrogen-deficient stars (Sect. 15). Development of the core material was originally drawn from several seminal texts [1, 2, 3, 4, 5, 6, 7, 8, 9]. Section 15 and parts of Sect. 14 are based on [10, 11].

Variables

In considering stellar structure, we will meet a number of quantities. The most important are:

- stellar mass M: often given in solar units $M_\odot$,
- stellar radius R: often given in solar units $R_\odot$,
- stellar luminosity L: often given in solar units $L_\odot$,

[1] Kodaikanal Observatory, Indian Institute of Astrophysics, 2008, April 29 - May 13.

Table 2. Physical Constants: CODATA 2006

speed of light in vacuum	$c = 2.997\,92458 \times 10^{8}$	$\mathrm{m}\,s^{-1}$
gravitational constant	$G = 6.674\,28(67) \times 10^{-11}$	$\mathrm{m^3\,kg^{-1}\,s^{-2}}$
Planck constant	$h = 6.626\,068\,96(33) \times 10^{-34}$	$\mathrm{J}\,s$
electron charge	$e = 1.602\,176\,487(40) \times 10^{-19}$	C
electron mass	$m_\mathrm{e} = 9.109\,382\,15(45) \times 10^{-31}$	kg
proton mass	$m_\mathrm{p} = 1.672\,621\,637(83) \times 10^{-27}$	kg
Avogadro's number	$N_\mathrm{A} = 6.022\,141\,79(30) \times 10^{23}$	$\mathrm{mol^{-1}}$
atomic mass unit	$m_\mathrm{u} \equiv 10^{-3}\,\mathrm{kg\,mol^{-1}}/N_\mathrm{A}$	kg
Boltzmann constant	$k = 1.380\,6504(24) \times 10^{-23}$	$\mathrm{J}\,K^{-1}$
Stefan-Boltzmann constant	$\sigma = 5.670\,400(40) \times 10^{-8}$	$\mathrm{W\,m^{-2}\,K^{-4}}$
Radiation constant	$a \equiv 4\sigma/c$	$[\mathrm{J\,m^{-3}\,K^{-4}}]$
Wien's displacement constants	$b = \lambda_\mathrm{max} T = 2.897\,7685(51) \times 10^{-3}$	m K
	$b' = \nu_\mathrm{max}/T = 5.878\,933(10) \times 10^{10}$	$\mathrm{Hz\,K^{-1}}$

- surface gravity g: a force measured in $\mathrm{m\,s^{-2}}$,
- effective temperature T_eff: usually given in degrees Kelvin.
- mass-fractions of hydrogen X, helium Y and other elements Z, respectively,
- and age t: usually given in years, millions (t_6) or billions of years (t_9).

The flux emitted from the surface of a star is the product of the stellar surface area ($4\pi R^2$) and emissivity per unit area assuming that the surface radiates as a black body:

$$L = 4\pi R^2 \sigma T_\mathrm{eff}^4. \tag{1}$$

where σ is the Stefan-Boltzmann constant. The surface gravity is simply given by

$$g = GM/R^2, \tag{2}$$

where G is the gravitational constant. The conservation equation for chemical composition can be simply written

$$X + Y + Z = 1. \tag{3}$$

In situations where the abundances of individual elements or nuclides are important (*e.g.* nuclear reaction rates), relative abundances by mass will be given as x_i, where i is either the atomic number, or denotes the nuclide in some other distinct way. Values for the Sun are given in Table 1; errors on the last digits are shown in parentheses. Table 2 provides constants used throughout the text and enables many equations to be evaluated.

2 The Hertzsprung–Russell Diagram

The most important correlations amongst stellar properties are contained in a type of diagram developed independently by Ejnar Hertzsprung and Henry

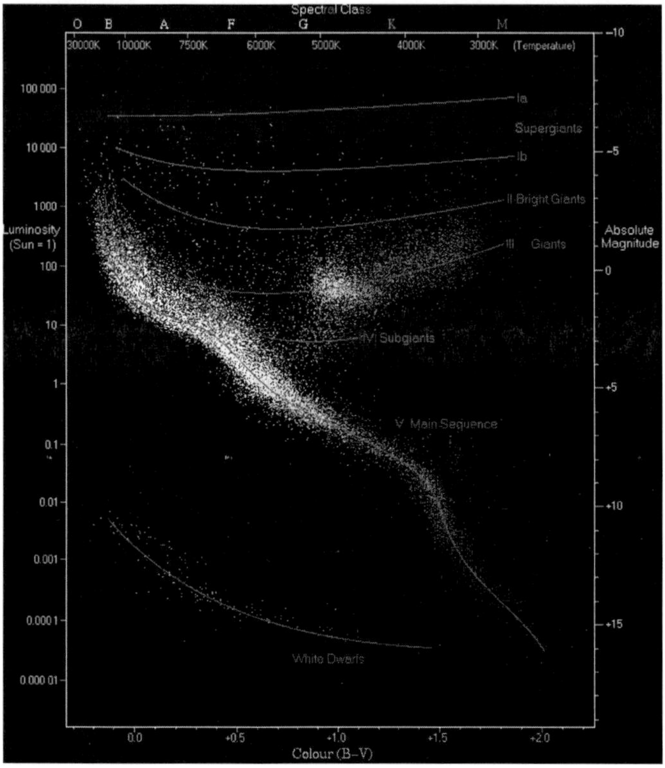

Fig. 1. Hertzsprung-Russell diagram: a plot of luminosity (absolute magnitude) against the colours of the brightest stars ranging from the high-temperature blue-white stars on the left side of the diagram to the low temperature red stars on the right side. Original image by Richard Powell licensed for derivative works and redistribution under the Creative Commons Attribution ShareAlike 2.5 License

Russell [12, 13, 14]. The original Hertzsprung–Russell (HR) diagram showed the distribution of *spectral type and absolute magnitude* (or brightness) for stars with known distances (Fig. 1). The latter is required to convert an apparent brightness (*e.g.* $m_{\rm V}$) to an absolute magnitude ($M_{\rm V}$). The diagram demonstrates that stars do not appear with *any* combination of spectral type and brightness, but fall on well-defined sequences, *e.g.* the main sequence, the giant branch, the white dwarfs, and so on.

A more convenient form is the *colour – magnitude* diagram, in which either an apparent or absolute magnitude is plotted against a photometric colour index, being the ratio of brightness at one wavelength to that at another. Such a diagram is particularly useful for comparing the properties of stars in a cluster, which may be assumed to lie at approximately the same distance. The additional supposition that all stars in a cluster are of the same age has

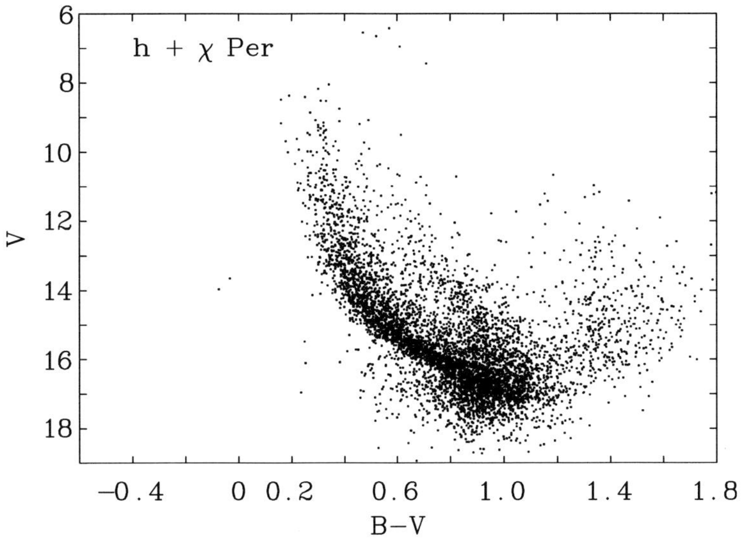

Fig. 2. Colour-magnitude diagram for the young galactic open clusters NGC 869 and NGC 884 (=h and χ Persei; based on [15])

important consequences for understanding stellar evolution – although the supposition may not always be correct!

Wien's displacement law states that there is a relation between the temperature of a black body and the wavelength at which the maximum energy is emitted: $\lambda_{\rm max} = b/T$. From this it was recognised that there should be a connection between the colour of a star and its effective (or surface) temperature. With a theory of stellar atmospheres, the relationship between spectral type, colour index and effective temperature became concrete. In addition, a correction to account for light emitted at unobserved wavelengths allows the apparent visual magnitude to be converted to a bolometric magnitude, and hence to luminosity. The use of an *effective temperature – luminosity* diagram is common in theoretical work (*cf.* Fig. 8). It is important to recognise and understand the connections and differences between all three types of diagram.

2.1 Cluster Diagrams

Figure 1 contains stars of widely varying mass, age and composition, and can only be constructed for stars whose distances can be measured directly. In the case of Galactic and globular clusters, we assume that all stars formed at approximately the same time, from a gas-cloud of roughly uniform composition. Because the cluster members are at the same distance, their relative brightnesses provide an HR diagram in which only the zero-point of the luminosity axis is unknown.

Galactic or open clusters are normally associated with the hydrogen-rich disk of the Galaxy, hence are young with metallicities ($Z \approx 0.01-0.03$) similar to the solar value. Their colour-magnitude diagrams (*cf.* Figs. 2 – 3) typically show:

- Most stars on the main-sequence (MS).
- A turn-off (TO) point somewhere between F and O stars (depends on cluster age, Fig. 3).
- A Hertzsprung gap between stars leaving the MS and Giants (Fig. 2).
- A relatively flat Giant Branch (GB) (Fig. 3).

Globular clusters are associated with the gas-poor halo of the Galaxy, and are very old objects with generally low metallicities ($Z \approx 0.0001-0.01$). Their colour-magnitude diagrams (*cf.* Figs. 4 – 5) typically show:

- Only late-type stars on the MS.
- A turn-off-point around spectral-type G (depends on age and metallicity)
- A continuous sequence from the TO through subgiants up to a steep GB
- A horizontal branch (HB), red in some clusters, blue in others (usually depends on metallicity)
- An asymptotic giant branch (AGB) sequence above the HB and parallel to the GB

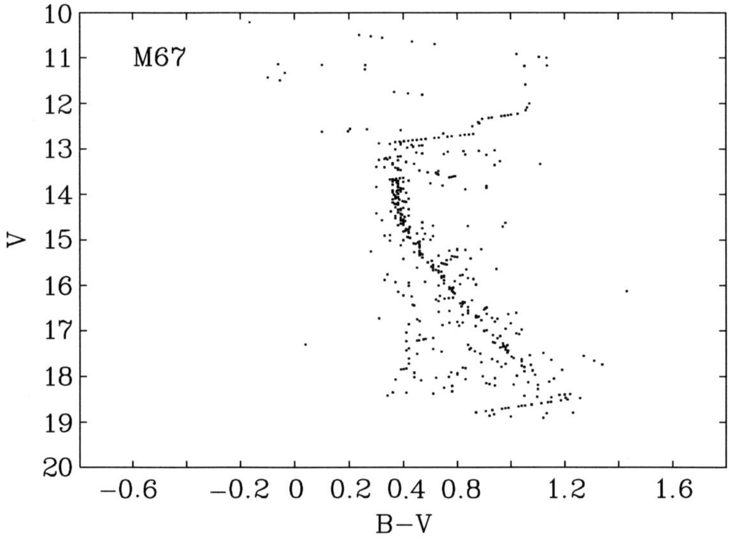

Fig. 3. Colour-magnitude diagram for the old galactic open cluster M67 (based on [16])

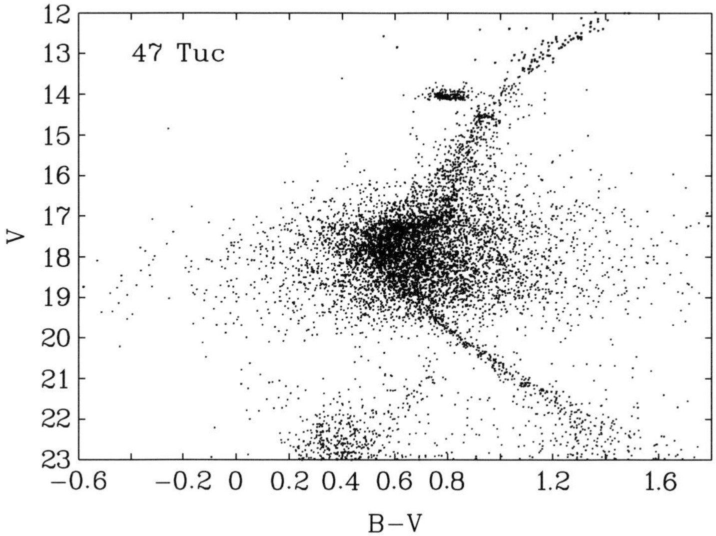

Fig. 4. Colour-magnitude diagram for the metal-rich galactic globular cluster 47 Tuc ($Z = 0.004$, [Fe/H] $= -0.76$; based on [17])

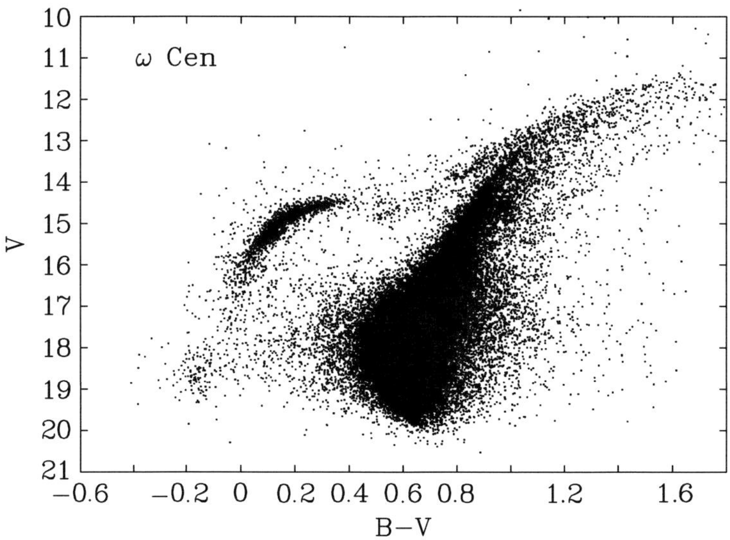

Fig. 5. Colour-magnitude diagram for the moderately metal-poor galactic globular cluster ω Cen ($Z = 0.0006$, [Fe/H] $= -1.62$; based on [18])

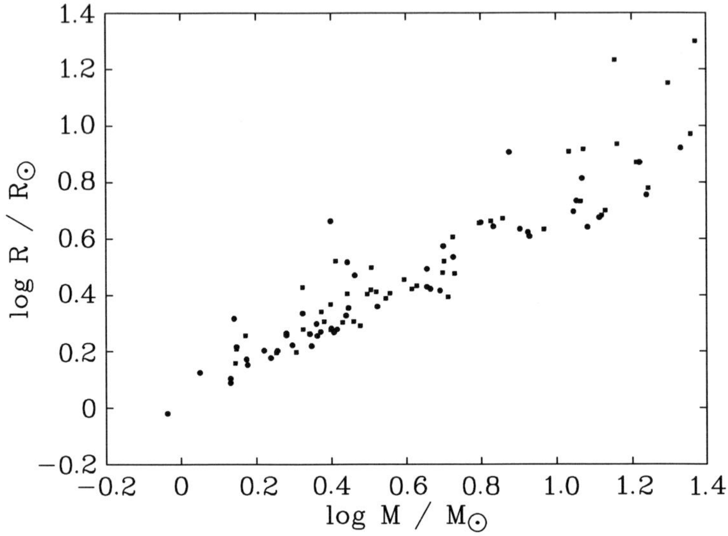

Fig. 6. The mass-radius relation for both components in each of fifty eclipsing binary stars. The primaries and secondaries are shown as squares and circles (based on [19])

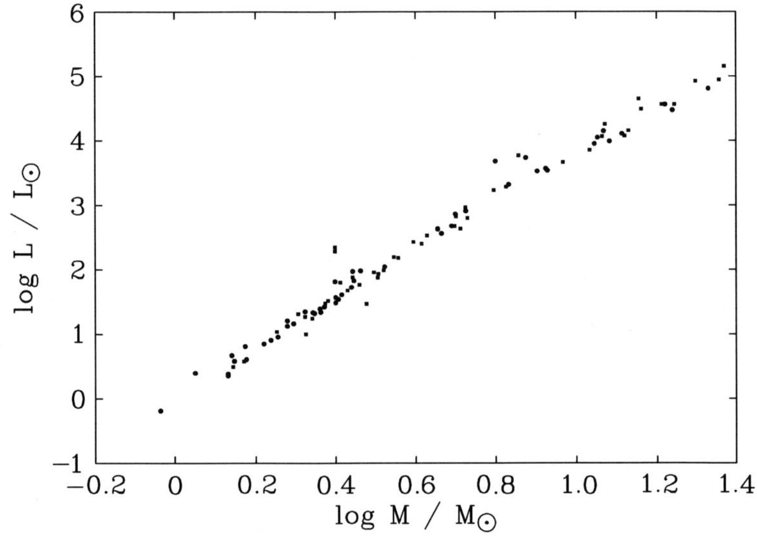

Fig. 7. The mass-luminosity relation for the same stars as in Fig. 6

2.2 The Temperature–Luminosity Relation

The most obvious empirical relation between stellar properties is provided by the main-sequence in the HR diagram, where there exists a direct correlation between stellar luminosity and effective temperature of the form

$$L \propto T_{\text{eff}}^{\alpha} \tag{4}$$

where on average $\alpha \approx 6$, although the value is higher at both the upper and lower ends of the main sequence.

2.3 The Mass–Luminosity and Mass-Radius Relations

Another important empirical relation between stellar properties is provided by eclipsing binary stars, for which direct measurements of mass, radius and luminosity can often be obtained for both components. Plotting $\log R$ and $\log L$ against $\log M$ for main-sequence stars (Figs. 6 and 7), we find straight lines which mean that L follows a relation

$$L \propto M^{\beta}. \tag{5}$$

where on average $\beta \approx 3.8$. Our theory of stellar structure must reproduce these results.

3 Stellar Evolution – A Sneak Preview

We have seen that the luminosity of a star depends on its mass. Since the luminosity L determines the rate at which a star uses up its available fuel, and L goes as $M^{\beta}, \beta > 1$, it is evident that stars of different mass effectively age at different rates. The cluster colour-magnitude diagrams are essentially snapshots of stellar evolution in which all the stars have the same age, but cover a wide range of mass.

We cannot observe the passage of an individual star through the colour-magnitude diagram, but we can attempt to model its evolution from birth to death. A simplified picture of the evolution of two stars is shown in Fig. 8. This diagram provides a convenient way to outline the basic features of stellar evolution. These are common to a majority of stars, with the greatest differences occurring at the extremes of mass.

Star Formation

At the current time in our Galaxy, it appears that star formation takes place in massive interstellar clouds, with dimensions $\approx 10\,\text{pc}$, density $\approx 5 \times 10^{9}\,\text{atoms}\,\text{m}^{-3}$, and temperature $\approx 10\,\text{K}$. The Galaxy is pervaded by a magnetic field aligned approximately parallel to the galactic plane. This magnetic field is strongly tied to the ionized plasma in the interstellar medium.

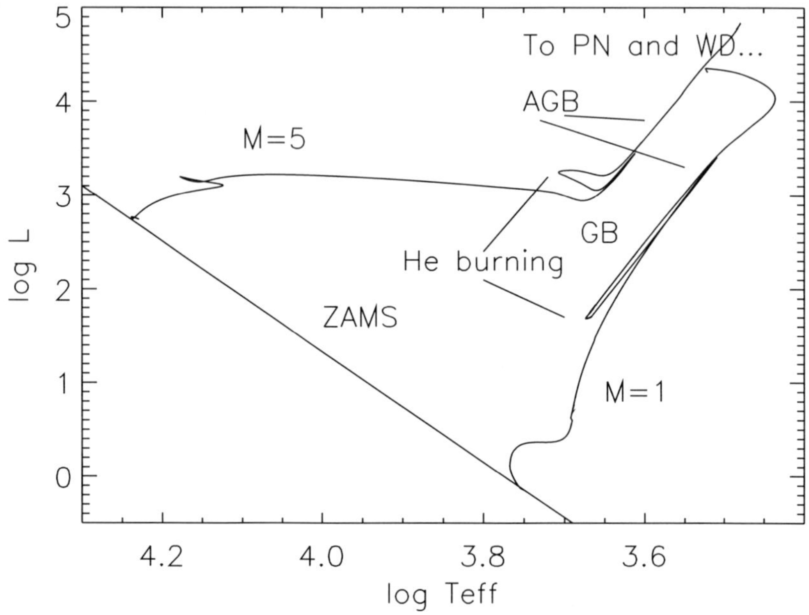

Fig. 8. Schematic $L - T_{\rm eff}$ diagram showing the evolution of 1 and 5 $M_{\odot}$ mass stars. (Based on a figure by J. Lattanzio. Evolution tracks computed by R. G. Izzard.)

Random turbulent process or other small perturbations in the field lead to local potential wells where the ISM can condense, pulling the magnetic field with it, and leading to a Rayleigh-Taylor (or runaway) instability. It is thought that this provides the initial mechanism for the formation of dark clouds of interstellar matter.

A number of processes including refrigeration (to keep the clouds cool) and accretion disks (to dispose of angular momentum) enable these dark clouds to collapse under their own gravity until they are sufficiently dense and hot for nuclear reactions to begin. At this point, we consider a star to have been born.

The Main Sequence

Stars spend over 90% of their lives on the main sequence. Hydrogen is converted to helium in the stellar core (Fig. 9). Since four protons are slightly heavier than one helium nucleus, the excess mass is converted into energy which keeps the core hot, maintains internal pressure and allows the star to shine. During main-sequence burning, the star increases its radius slightly as the hydrogen content in the core drops. More detail will be given in Sect. 12.

The Giant Branch

As hydrogen in the core is depleted, nuclear reactions switch off, and the core contracts. Hydrogen-rich material outside the core is compressed and heated

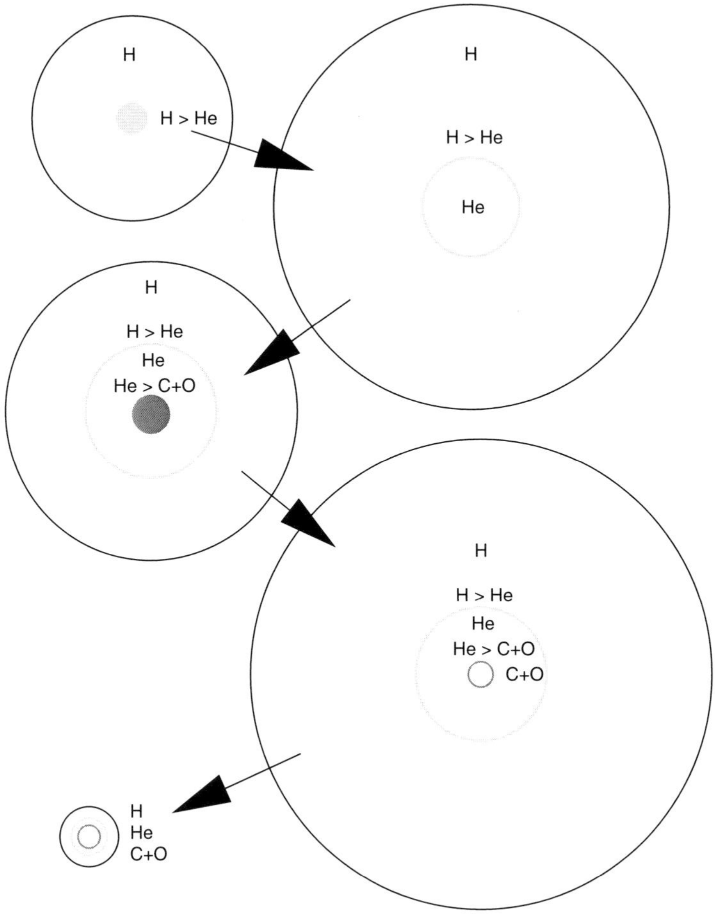

Fig. 9. A simplified view of the internal chemical structure of a star during the major phases of stellar evolution. The panels represent the main-sequence, through the giant-branch, helium-burning, asymptotic-giant branch and white dwarf phases. Filled circles and thick lines represent nuclear-burning regions. Not to scale

by the contraction, so hydrogen-burning shifts to a shell. In low-mass stars ($0.5 - 3M_\odot$), the outer layers of the star expand and the star becomes a red giant. During this phase of evolution, the hydrogen-burning shell adds helium to the core beneath it, which consequently contracts and heats. At the same time, the shell becomes hotter, thinner and more luminous. The hydrogen-rich outer layers (the envelope) expand, and the entire star evolves upwards along the giant branch.

Low-mass Helium Burning

Once the helium core reaches a critical mass, nuclear burning of helium begins, producing carbon and oxygen in a relatively long-lived phase of evolution. The energy derived from helium-burning heats the helium core and forces it to expand. Thus the hydrogen-burning shell may actually get weaker at this point, the *overall* luminosity drops and the star contracts. At this stage, the star may be either blue or red, corresponding to the horizontal-branches in Figs. 4 and 5. We shall learn more about these stars in Sect. 14.

Intermediate-mass Helium Burning

In stars heavier than about $2.3\mathrm{M}_{\odot}$ and with $Z \approx 0.02$, helium burning reactions start before the core becomes compact and before the hydrogen-burning shell gets very thin. Core expansion following helium ignition is therefore mild and the total luminosity does not drop by much. However the radius does become smaller, producing a blue-loop in the Hertzsprung-Russell diagram for the duration of core-helium burning.

Asymptotic Giant-Branch (AGB) Stars

With core-helium exhaustion, the focus of nuclear-burning in low- and intermediate-mass stars shifts to a double-shell structure. Because the He-burning shell is thermally unstable, a process known as thermal-pulsing is established. The helium-burning shell burns faster, and so keeps going out. When sufficient fresh helium has accumulated below the hydrogen-burning, the helium-shell reignites, the intershell region is forced to expand, and hydrogen-burning is extinguished. The interaction between these thermal pulses and convection in different layers of the star makes this one of the most important processes in cosmology; more detail is presented by Karakas (these proceedings).

Of consequence here is that when the hydrogen-burning shell is operating, it is very thin and very powerful, so the star becomes very luminous and very large. The hydrogen envelope is thus consumed from below (by the shell) and blown away from the surface by a strong stellar wind.

Planetary-Nebula Formation and White-Dwarf Cooling

Ultimately, the expansion of the AGB star results in a dynamical instability which causes the expulsion of the outer layers as a planetary nebula. Unable to sustain nuclear reactions, whatever is left of the envelope starts to contract. When it has cooled sufficiently the star becomes a white dwarf: Sect. 13. We will discuss whether the white dwarf stage is the end of the story in Sect. 15.

Massive Stars

Following main-sequence burning, the ultimate evolution of stars more massive than $\approx 8\mathrm{M}_{\odot}$ is quite different. Initial evolution is similar, the star expands rapidly and starts to cross the HR diagram towards the red giant region. For $M \geq 20\,\mathrm{M}_\odot$, core helium ignition occurs at a progressively earlier stage (*i.e.* closer to the main-sequence) as the mass increases. Core-helium burning arrests the redward evolution until core helium is exhausted, and redward evolution resumes. As the inactive carbon-oxygen core contracts, it is sufficiently massive that its temperature increases to the point where carbon-carbon reactions can occur. There follows a rapid sequence of nuclear-burning episodes producing elements with increasing atomic weight. Hydrostatic equilibrium is maintained until the core consists primarily of iron (^{56}Fe), at which point further nuclear reactions require more energy than they release. The stellar core collapses and a supernova explosion follows. This explosion is responsible for the production of quantities of heavy nuclei quite distinct from those created in AGB stars. These processes are discussed in greater detail by Arnould (these proceedings).

Binary Stars

This very simple description applies to those stars which evolve as single stars or as members of a wide binary system which do not interact. It is increasingly clear that a large fraction of stars are born in binary or multiple systems in which two stars exchange material at some point during their evolution. The possibilities of what can happen thereafter are too numerous to be able to cover here, but some of the more bizarre possibilities will be considered later.

4 Stellar Time Scales

We want now to put some physical argument behind our cartoon picture of stellar evolution. It is evident that stars such as the Sun do not change their properties rapidly. It is important to understand and appreciate the relative magnitudes of the timescales on which a star can change.

4.1 Dynamical (Free-Fall) Time

Consider a star as a ball of gas which is held up by the pressure forces within it. How long would the star survive if the pressure forces were removed? The answer is known as the free-fall or dynamical time ($t_{\rm ff}$ or $t_{\rm dyn}$). A simple estimate is given by the time required for a body to fall through a distance of the order R (the stellar radius) under the influence of a (constant) gravitational acceleration equivalent to the surface gravity $g = GM/R^2$ of a star of mass M. From Newton's 2nd law,

$$R = \frac{1}{2} g t_{\rm ff}^2 = \frac{1}{2} \frac{GM}{R^2} t_{\rm ff}^2$$

$$\Rightarrow t_{\rm ff} \approx 2.2 \times 10^3 (R^3/M)^{1/2}\,\mathrm{s}$$

where R and M are in solar units. Writing the mean density as $\rho = (3/4\pi)M/R^3$, we also have:

$$t_{\rm ff} \approx (2\pi G\rho/3)^{-1/2}. \tag{6}$$

$t_{\rm ff}$ is also the characteristic time for a significant departure from hydrostatic equilibrium to alter the state of a star appreciably, the time taken for a body orbiting at the surface of the star to make one complete revolution, and the time for a sound wave to propagate through the star.

4.2 Thermal (Kelvin) Time

Next, consider a star as a ball of hot gas which acts as reservoir of heat energy. How long would it take for this energy to radiate away if it were not replenished? This is the thermal or Kelvin-Helmholtz timescale ($t_{\rm K}$ or $t_{\rm th}$). If the total kinetic (thermal) energy of the star is $E_{\rm kin}$, the timescale is approximately

$$t_{\rm K} = \frac{E_{\rm kin}}{L}. \tag{7}$$

We shall see later that $E_{\rm kin}$ is related to $E_{\rm grav}$, the gravitational binding energy of the star, by the Virial theorem:

$$E_{\rm kin} = -\frac{1}{2}E_{\rm grav}.$$

Since we can write

$$E_{\rm grav} = -\int_0^M \frac{Gm}{r}dm = -q\frac{GM^2}{R} \tag{8}$$

where q is a dimensionless constant of order unity[2], then

$$t_{\rm K} = \frac{q}{2}\frac{GM^2}{LR} \approx 3\times 10^7 \frac{qM^2}{LR}\ {\rm y} \tag{9}$$

where M, L and R are in solar units.

The thermal time is the relaxation time for departure of a star from thermal equilibrium. It is also the time that would be required for a star to contract from infinite dispersion to its present radius if L were to remain constant during its entire contraction.

[2] $q = 3/5$ for a sphere of uniform density, and becomes smaller with increasing central condensation.

4.3 Nuclear Time

It is clear, for example from radio-carbon dating of rocks in the earth's surface, that the solar system must be very much older than the "Kelvin age" of $\approx 3 \times 10^7$ years. It is now taken for granted that the main source of stellar energy comes from nuclear reactions. The fusion of four protons (hydrogen nuclei) to an alpha-particle (helium nucleus) is associated with the release of energy Q, where $Q \approx 26$ MeV. The total available energy is thus

$$E_{\rm nuc} = qM/4m_{\rm p}Q,$$

where again q is a dimensionless factor of order unity representing the fraction of the star available as nuclear fuel. The nuclear time is simply the time taken to radiate this energy

$$t_{\rm nuc} = \frac{E_{\rm nuc}}{L}. \qquad (10)$$

For hydrogen-burning in main-sequence stars,

$$t_{\rm nuc} \approx 1 \times 10^{11} q \frac{M}{L} \text{ y}. \qquad (11)$$

We adopt $q = 0.1$ for Table 3.

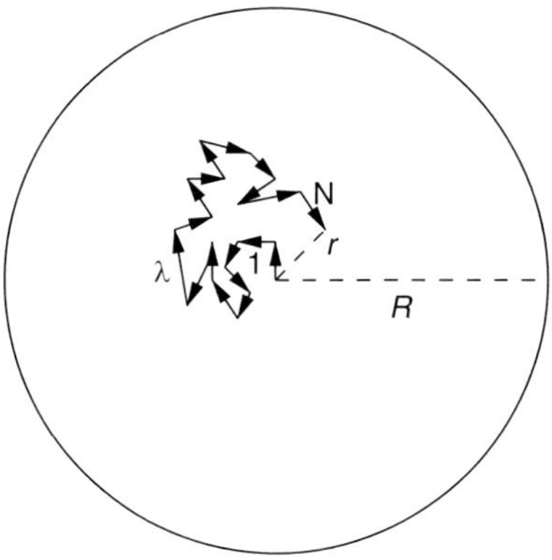

Fig. 10. Diffusion processes in radiative energy transport

4.4 Diffusion Time

Energy released at the centre of a star is used to heat the stellar material and hence oppose gravitational forces, otherwise collapse would be inevitable and hydrostatic equilibrium could not be maintained. Energy liberated as photons cannot therefore leave a star immediately, but must interact with the stellar material by a series of scattering collisions, mainly with electrons. Since scattering is an isotropic process, energy transport is most correctly described by the diffusion equation. Section 5 will examine this further, but essentially the photon-path can be described by a random-walk consisting of N steps, each of length λ. Whilst the total distance travelled by the photon is $N\lambda$, it may be shown that the nett distance travelled is $r = \sqrt{N}\lambda$, because of the scatterings. Thus to escape from the star, the photon must travel a distance R, which it will do in a time

$$t_{\rm diff} \approx \frac{R^2}{\lambda c} \approx 5 \times 10^5 R\,{\rm y} \tag{12}$$

(for solar-type stars) which may be compared with the escape time for non-interacting particles (such as neutrinos):

$$t_{\rm esc} = R/c = 2.3 R\,{\rm s} \tag{13}$$

with R in solar units.

Table 3. Comparative timescales

Star	M	R	L	$T_{\rm eff}$	$t_{\rm esc}$	$t_{\rm ff}$	$t_{\rm diff}$	$t_{\rm kin}$	$t_{\rm nuc}$
	$M_\odot$	$R_\odot$	$L_\odot$	K	s^{-1}	s^{-1}	y^{-1}	y^{-1}	y^{-1}
Sun	1	1	1	5760	2.3	$2.2\,10^{3}$	$1.7\,10^{5}$	$3.0\,10^{7}$	$1.0\,10^{10}$
MS	5	3	450	15342	6.9	$2.6\,10^{4}$	$1.5\,10^{6}$	$5.5\,10^{5}$	$1.1\,10^{8}$
MS	10	5	6 300	22958	12	$7.8\,10^{4}$	$4.3\,10^{6}$	$9.5\,10^{4}$	$1.6\,10^{7}$
MS	0.2	0.3	0.0022	2280	0.69	$1.6\,10^{2}$	$1.5\,10^{4}$	$1.8\,10^{9}$	$9.1\,10^{11}$
RG	1	100	1 500	3600	230	$2.2\,10^{6}$	$1.7\,10^{9}$	$2.0\,10^{2}$	$6.6\,10^{6}$
WD	0.6	0.01	0.001	10243	0.023	$1.7\,10^{0}$	$1.7\,10^{1}$	$1.1\,10^{12}$	
					$t_\odot^{-1}$	$t_\odot^{-1}$	$t_\odot^{-1}$	$t_\odot^{-1}$	$t_\odot^{-1}$
MS	5				3.0	12	$9.0\,10^{0}$	$1.8\,10^{-2}$	$1.1\,10^{-2}$
MS	10				5.0	35	$2.5\,10^{1}$	$3.2\,10^{-3}$	$1.6\,10^{-3}$
MS	0.2				0.3	0.073	$9.0\,10^{-2}$	$6.0\,10^{1}$	$9.1\,10^{1}$
RG	1				100	1 000	$1.0\,10^{4}$	$6.6\,10^{-6}$	$6.6\,10^{-4}$
WD	0.6				0.001	0.000 77	$1.0\,10^{-4}$	$3.6\,10^{4}$	

4.5 Comparative Timescales

Table 3 shows timescales for main-sequence stars of four different masses, a red giant and a white dwarf, using an estimated radius and Eqs. (1), (5), (6), (9), (11), (12), and (13). The upper block shows conventional units of time; the lower block shows timescales relative to that of the Sun. The luminosities for the red giant and white dwarf are also approximations.

5 Equations of Stellar Structure

The simplest picture of a star is that of an isolated body of gas sufficiently massive that the only significant forces are self-gravity and internal pressure. In the simplest case, we can assume spherical symmetry and neglect the influence of rotation, magnetic fields and external gravitational influences (in fact, under these assumptions, equilibrium solutions will enforce spherical symmetry).

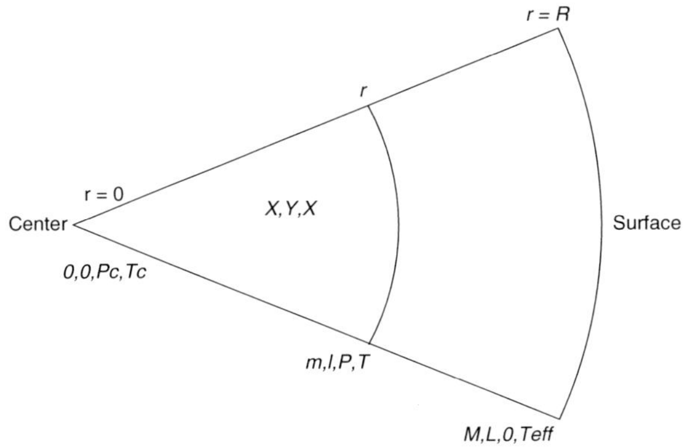

Fig. 11. The variables of stellar structure

Consider a spherical system containing a fluid of mass M and radius R (Fig. 11). The internal structure of this system is described by the distribution of various quantities, including

- r: radius
- $m(r)$: mass of fluid contained within a spherical shell of radius r
- $l(r)$: luminosity or total energy passing through a shell of radius r
- $T(r)$: temperature of the fluid at radius r
- $P(r)$: pressure in the fluid at radius r
- $\rho(r)$: fluid density at radius r

These quantities are expressed as a function of radius r, running from $r = 0$ at the stellar centre to $r = R$ at the stellar surface. However, we could have equally chosen a different independent variable, for example mass, m. In order to keep the notation simple, stellar structure variables will usually be written in short form, *i.e.* $m \equiv m(r), l, P, T, \rho$. The problem of stellar structure is then to determine the run of these quantities throughout the star, for which we require a system of equations.

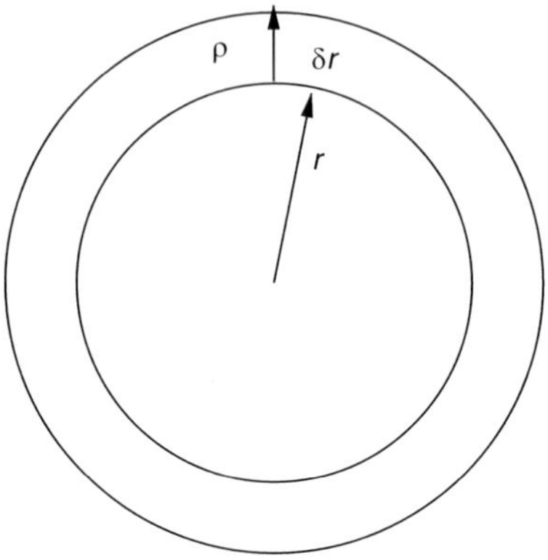

Fig. 12. Derivation of the density equation

5.1 Mass Continuity

Consider a spherical shell of radius r, thickness δr ($\delta r \ll r$) and density ρ (Fig. 12). It has a mass (volume × density)

$$\delta m = 4\pi r^2 \rho \delta r.$$

In the limit $\delta r \to 0$,

$$\frac{dm}{dr} = 4\pi r^2 \rho. \tag{14}$$

The same argument provides the mass of the volume enclosed within the sphere of radius r

$$m = 4\pi \int_0^r r^2 \rho dr.$$

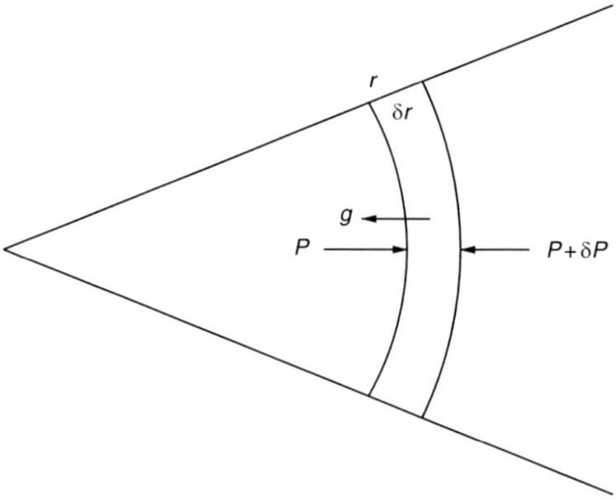

Fig. 13. Gravitational and pressure forces in hydrostatic equilibrium

5.2 Hydrostatic Equilibrium

We must next consider the forces acting at any position within this sphere (Fig. 13). Consider an element of material at radius r. The sphere of radius r acts as a gravitational mass situated at the centre giving rise to an inward gravitational force

$$g = \frac{Gm}{r^2}. \tag{15}$$

If there is a pressure gradient through the sphere, there will be an additional force. If the element has a thickness δr and cross-section δA, then a net force arises if the pressure on the inner and outer surfaces are unequal, so that the inward force is, since $\delta P = (dP/dr)\delta r$,

$$\left[P(r) + \frac{dP}{dr}\delta r - P(r)\right]\delta A = \frac{dP}{dr}\delta r \delta A.$$

But the mass of the volume element is $\delta m = \rho\delta r\delta A$, then the sum of inward forces due to gravity and pressure will be

$$F = \delta m\left(g + \frac{1}{\rho}\frac{dP}{dr}\right) = -\delta m\frac{d^2 r}{dt^2}. \tag{16}$$

For the element considered in Fig. 13 to be in equilibrium, the parenthesis in Eq. (16) must be zero, and by substitution from Eq. (15),

$$\frac{dP}{dr} = -\frac{Gm\rho}{r^2}. \tag{17}$$

This is the equation of hydrostatic equilibrium which, amongst other things, demonstrates that in order to oppose gravity, pressure must increase towards the centre of the star.

5.3 Virial Theorem

A very important result can be obtained when an entire self-gravitating system is in equilibrium, *i.e.* if Eq. (17) is satisfied at all r, since it becomes possible to derive a simple relation between average internal pressure and the gravitational potential energy of the system. This is done by multiplying both sides of Eq. (17) by $4\pi r^3$ and integrating from $r = 0$ to $r = R$ to obtain

$$\int_0^R 4\pi r^3 \frac{dP}{dr} dr = -\int_0^R \frac{Gm}{r} 4\pi r^2 \rho dr.$$

Integrate the lhs by parts ($du = dP/dr.dr$, $dv = 4\pi r^3$), and substitute $dm = 4\pi r^2 \rho dr$

$$\left[4\pi r^3 P\right]_0^R - 3\int_0^R 4\pi r^2 P dr = -\int_0^M \frac{Gm}{r} dm.$$

Since $P(R) = 0$, the first term is zero. Substituting $4\pi r^2 dr = dv$

$$-3\int_0^V P dv = -\int_0^M \frac{Gm}{r} dm.$$

The lhs becomes $-3\langle P\rangle V$ where $\langle P\rangle$ is the volume-averaged pressure and V is the volume of the system, and the rhs is simply the gravitational potential energy of the system E_{grav}. Thus the average pressure needed to support a system with gravitational energy E_{grav} and volume V is given by

$$\langle P\rangle = -\frac{1}{3}\frac{E_{\mathrm{grav}}}{V}. \tag{18}$$

The statement that the average pressure needed to support a self-gravitating system is one third of the stored gravitational energy is called the Virial Theorem. The physical meaning of pressure depends on the system itself, but it can be applied to clusters of galaxies as well as to individual stars. Let us consider two cases for gas in a star, where the equation of state relates the pressure of the gas (P) to the translational kinetic energy of the gas particles (E_{kin}).

Non-Relativistic Gas

For a gas of non-relativistic particles, since $P = nkT = kT/V$ and $E_{\mathrm{kin}} = \frac{3}{2}kT$, then

$$P = \frac{2}{3}E_{\mathrm{kin}}/V. \tag{19}$$

Then for a self-gravitating system of volume V and gravitational energy E_{grav}, the gravitational and kinetic energies are related by

$$2E_{\mathrm{kin}} + E_{\mathrm{grav}} = 0$$

and the total energy of the system, which is the sum of kinetic and gravitational energies, is given by

$$E_{\mathrm{tot}} = E_{\mathrm{kin}} + E_{\mathrm{grav}} = -E_{\mathrm{kin}},$$

and

$$E_{\mathrm{tot}} = E_{\mathrm{grav}}/2. \tag{20}$$

These equations are of fundamental importance. Note the following. If a system is in hydrostatic equilibrium and it is tightly bound, the gas particles have high kinetic energy. They are *hot*. If the system evolves slowly, close to hydrostatic equilibrium, then changes in kinetic and gravitational energies are simply related to changes in the total energy. We shall see some of the consequences later in considering the collapse of gas clouds.

Ultra-Relativistic Gas

For a gas of ultra-relativistic particles, $E_{\rm kin} = 3kT$ and hence

$$P = \frac{1}{3}E_{\rm kin}/V \qquad (21)$$

from which we find

$$E_{\rm kin} + E_{\rm grav} = 0. \qquad (22)$$

Thus hydrostatic equilibrium is possible only if the total energy is zero. As the ultra-relativistic limit is approached, *e.g.* as the gas temperature increases, the binding energy decreases and the system is easily disrupted. This type of instability occurs in stars where the radiation pressure is very high (pressure comes from photons which are, by definition, relativistic) *e.g.* supermassive stars, or where pressure comes from degenerate electrons and the electron temperature become very energetic, *e.g.* massive white dwarfs. Again, we shall examine these later.

Jean's criterion for gas-cloud collapse

An important corollary to the Virial theorem is connected with the condition for gravitational collapse. Stars are believed to form when large clouds of gas and dust contract. However, this contraction can only take place if the internal pressure of the cloud due to the thermal motion of the cloud particles is weaker than the gravitational forces compressing the cloud. Formally, a cloud is *bound* if the total energy is less than zero. Knowing the gravitational binding energy (Eq. (8)) and adopting $q = 1$, and setting the thermal energy per particle to $\frac{3}{2}kT$, the total energy of the cloud is:

$$E_{\rm kin} + E_{\rm grav} = \frac{3}{2}\frac{M}{\mu m_{\rm H}}kT + \frac{GM^2}{R}.$$

For the total system to be bound,

$$E_{\rm kin} + E_{\rm grav} < 0$$

implies

$$M_{\rm cloud} > M_{\rm J} = \frac{3kT}{2G\mu m_{\rm H}}R, \qquad (23)$$

or

$$T_{\rm cloud} < T_{\rm J} = \frac{2}{3}\frac{2G\mu m_{\rm H}}{k}\frac{M}{R}, \qquad (24)$$

or

$$\rho_{\rm cloud} > \rho_{\rm J} = \frac{3}{4\pi M^2}\left(\frac{3kT}{2G\mu m_{\rm H}}\right)^3, \qquad (25)$$

where the critical values are known as the Jeans mass, temperature and density.

Equation (24) illustrates an important consequence of cloud contraction; if the system is to remain bound, *i.e.* if it is to continue to contract and become a star, then the internal temperature of the collapsing cloud or protostar,

$$T_{\rm I} \propto \frac{M}{R}. \tag{26}$$

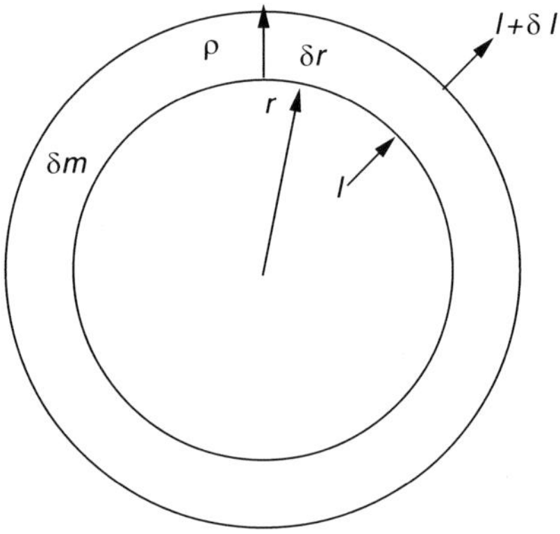

Fig. 14. Derivation of the energy equation

5.4 Energy Conservation

Since energy moves through the star, the consequences for stellar structure imposed by the law of conservation of energy must be considered, whether that energy comes from the release of nuclear or gravitational energy. Consider a volume element $dv = 4\pi r^2 dr$ at radius r (Fig. 14).

The law of conservation of energy states that the total amount of energy leaving a volume element dv must equal the total amount of energy entering that element plus the energy lost or produced within that element.

Assuming spherical symmetry, it is convenient to consider a shell of mass δm. Let l be the amount of energy entering the bottom of shell, and $l + \delta l$ be the amount of energy leaving the top of the shell. Let us denote the energy absorbed or produced within the shell, per unit mass, by ε. Then we can equate energy out to energy in plus energy produced or lost:

$$l + \delta l = l + \varepsilon \delta m,$$

whence

$$\frac{dl}{dm} = \varepsilon \tag{27}$$

or, with Eq. (14),

$$\frac{dl}{dr} = 4\pi r^2 \rho \varepsilon. \tag{28}$$

There are several sources/sinks of energy which we shall consider elsewhere, including:

- nuclear, *e.g.* $4\mathrm{p} + 2\mathrm{e}^- \rightarrow \alpha + 2\mathrm{e}^+ + 2\mathrm{e}^- + 4.768 \times 10^{-29}\,\mathrm{kg} \rightarrow \alpha + 26.72\,\mathrm{MeV}$
- neutrino, *e.g.* $\gamma + \mathrm{e}^- \rightarrow \mathrm{e}^- + \nu + \bar{\nu}$
- nonadiabatic expansion, *i.e.* $dQ = TdS = dE + PdV \neq 0$

5.5 Energy Transport

The fact that there is a temperature difference between the interior and surface of the Sun implies that there must be a temperature gradient and hence a flux of energy. Conservation of energy constrains the energy flux, but the way energy is transported establishes the temperature gradient. There are three major transport mechanisms:

- radiation: generally by photons, but also by neutrinos.
- convection.
- conduction: by electrons in white dwarfs.

Radiative Equilibrium

Energy transported by radiation obeys Fick's law of diffusion:

$$F = -D\frac{daT^4}{dr}$$

where F is the flux density, aT^4 is the radiation energy density[3] and D is a diffusion coefficient. D is related to the *opacity* κ by

$$D = c/\kappa\rho.$$

F must be multiplied by $4\pi r^2$ to obtain the total flux l, whence

$$l = -\frac{4\pi r^2 c}{\kappa\rho}\frac{daT^4}{dr}$$

$$\Rightarrow \frac{dT}{dr} = \frac{3\kappa\rho}{4acT^3}\frac{l}{4\pi r^2}. \tag{29}$$

Equation (29) represents the temperature gradient within the star when the star is in radiative equilibrium. Combining with Eq. (17) (hydrostatic equilibrium) and taking logs, we define a gradient

$$\frac{d\ln T}{d\ln P} = \nabla_{\mathrm{rad}} = \frac{3}{16\pi ac}\frac{\kappa l P}{GmT^4} \tag{30}$$

[3] $a = 4\sigma/c$ is the radiation constant.

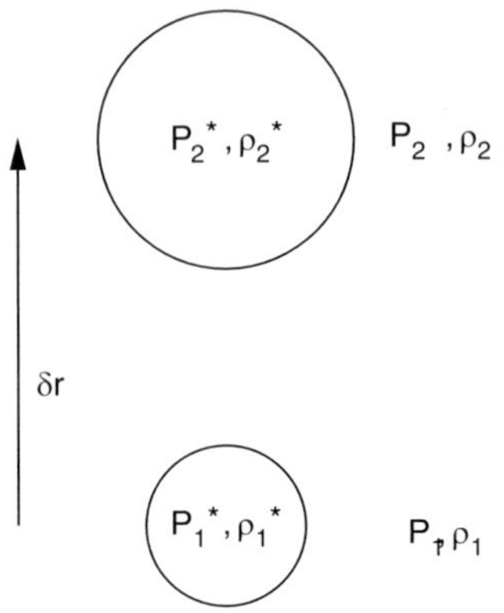

Fig. 15. Convective energy transport

Convective Equilibrium

An element of gas is at some radius r. Consider its upward displacement by a distance dr, allowing it to expand adiabatically until the pressure within is equal to the pressure outside (Fig. 15). Then release the element. If it continues to move upwards, the layer in question is convectively unstable and convective motions will persist.

Let the pressure and density within and without the sphere be denoted by P^*, ρ^*, and P, ρ respectively. Then initially

$$P_1^* = P_1 \quad \text{and} \quad \rho_1^* = \rho_1.$$

After the perturbation, $P_2^* = P_2$ and

$$\rho_2^* = \rho_1^* \left(\frac{P_2^*}{P_1^*}\right)^{1/\gamma},$$

where $P\rho^{-\gamma}$ is a constant and $\gamma = 5/3$ for a highly-ionized gas. For radiative equilibrium, we require $\rho_2^* > \rho_2$ so that the net force (bouyancy+gravitation) is downwards and the element will return to its starting position. Eliminating asterisks and writing $P_1 = P_2 + dP$, we obtain

$$\frac{P}{\rho}\frac{d\rho}{dP} > \frac{1}{\gamma} \tag{31}$$

for radiative equilibrium. This condition is related to the temperature gradient assuming some equation of state (*e.g.* $P = \rho kT/\mu m$) so that Eq. (31) becomes

$$\frac{d\ln T}{d\ln P} \equiv \nabla < \nabla_{\rm ad} \equiv \frac{\gamma - 1}{\gamma}. \tag{32}$$

There are two main circumstances under which Eq. (32) fails:

1. In the centre of main-sequence stars, the radiation flux $l/4\pi r^2$ can become very large, whilst $\kappa\rho$ remains small. Thus the temperature gradient $d\ln T/d\ln P$ required for radiative equilibrium (Eq. (30)) becomes large, and the material becomes convectively unstable. This gives rise to nuclear-driven convective cores in massive stars, and also convective zones in helium-burning stars.
2. In ionisation zones, the adiabatic exponent γ becomes smaller approaching unity, and κ becomes large, so that radiative equilibrium may be violated for small values of the temperature gradient. This gives rise to opacity-driven convection in the envelopes of cool stars.

Mixed Equilibrium

Equations (30) and (32) give the temperature gradient ∇ in radiative and convective equilibrium respectively. Where there are situations when it is clear which equation to use, the situation frequently arises when material may be naturally convective, but the convective efficiency ξ is sufficiently low that radiation carries a substantial fraction of the flux. In these cases, ξ must be derived from a suitable theory of convection.

With this in mind, it is useful to rewrite the equation for the temperature gradient

$$\frac{d\ln T}{d\ln P} = \nabla = (1-\xi)\,\nabla_{rad} + \xi\nabla_{ad} \tag{33}$$

such that:

- $\xi = 0$: radiative equilibrium
- $0< \xi <1$: non-adiabatic convection
- $\xi = 1$: fully adiabatic convection

5.6 The Equations of Stellar Structure

We have now derived the four basic (time-independent) equations of stellar structure. These are mass continuity (Eq. (14)), hydrostatic equilibrium (Eq. (17)), conservation of energy (Eq. (28)), and energy transport (Eq. (33)). These form a set of coupled first order ordinary differential equations relating one independent variable, *e.g.* r, to four dependent variables *i.e.*, m, l, P, T, which uniquely describe the structure of the star, note that any variable could be used as the independent variable. In an Eulerian frame, the spatial coordinate r is the independent variable. For most problems in stellar structure and evolution it is usually more convenient to work in a Lagrangian frame, with mass as the independent variable. Transforming, we obtain:

$$\frac{dr}{dm} = \frac{1}{4\pi r^2\rho},$$

$$\frac{dl}{dm} = \varepsilon,$$

$$\frac{dP}{dm} = -\frac{Gm}{4\pi r^4},$$
$$\frac{dT}{dm} = -\frac{Gm}{4\pi r^4}\frac{T}{P}\nabla.$$

Boundary Conditions

To solve such a system, boundary conditions are required. In the Lagrangian frame, the boundaries are at $m = 0$ (centre) and $m = M$ (surface). In the centre, the enclosed mass and luminosity are defined:

$$r(m = 0) = 0, \tag{34}$$

$$l(m = 0) = 0, \tag{35}$$

and at the surface temperature and pressure are defined:

$$T(m = M) = T_{\text{eff}} = (L/4\pi R^2)^{1/4}, \tag{36}$$

$$P_{\text{gas}}(m = M) = 0. \tag{37}$$

Constitutive Relations

In order to close the system of four first order o.d.e.s and four boundary conditions, additional equations are required to relate the principal state variables to the local micro-physics. Thus ε, ρ, ∇ refer to energy generation, density and energy transport, the last depending on ξ and κ, the convective efficiency and opacity and may be expressed in terms of P, T and the local composition of the stellar material X, Y, Z, or x_i. These relations will be derived more explicitly in Sects. 7–9.

Equation of state:	$\rho = \rho(P, T, x_i)$.
Nuclear energy generation:	$\varepsilon = \varepsilon(\rho, T, x_i)$.
Opacity:	$\kappa = \kappa(\rho, T, x_i)$.
Convective efficiency:	$\xi = \xi(\rho, T, x_i)$.
Energy transport:	$\nabla = \nabla(\rho, T, \xi, \kappa, x_i)$.

Solution

A quantitative physical description of the stellar interior is provided by a solution to these equations, *i.e.* by the run of P, T, m and l as a function of r throughout the domain $0 < r < R$, the radius of the star. These solutions are characterised uniquely by $M = m(R)$, the total mass of the star, and by $x_i(r)$, the run of chemical composition through the star. Various approximate and numerical schemes will be examined in Sects. 10–11.

6 Equations of Stellar Evolution

It will have been seen (Sect. 5) that the equations of stellar structure can be constructed independent of time t. Solutions refer exclusively to stars in complete equilibrium – they are solutions of stellar structure and not of evolution. However, the fact that nuclear reactions change the chemical composition of a region where nuclear energy is produced means that the structure of a star must change in time. There are three ways in which t enters the equations of stellar structure.

6.1 Thermal Expansion (Contraction)

If a region of a star contracts or expands it will release or absorb energy according to the first law of thermodynamics, which states

$$dE = -TdS = -(dU + PdV).$$

Hence the energy equation (Eq. (28)) needs to incorporate additional terms:

$$\frac{dL}{dm} = \varepsilon_{\text{nuc}} - T\frac{dS}{dt}. \tag{38}$$

6.2 Nucleosynthesis

Nuclear reactions produce energy *and* transmute elements from one species to another. A reaction $i(j,k)l$ destroys species i and j and creates species k and l at a rate r_{ij}. The rate of change of abundance of species i will then be given (in mass fraction) by

$$\frac{dx_i}{dt} = -(1 + \delta_{ij})r_{ij}m_{\text{u}} \tag{39}$$

where the bracket allows for more than one nucleus of species i to be destroyed in a reaction, and m_{u} is the mean atomic mass unit. Thus for the p-p chain:

$$\frac{dX}{dt} = -4r_{\text{pp}}m_{\text{u}},$$

$$\frac{dY}{dt} = -\frac{dX}{dt}.$$

6.3 Mixing

Convection provides an efficient transport mechanism for *both* energy *and* matter. Stellar evolution calculations have traditionally assumed that a fully convective region is fully mixed.

However, the student should be aware that, when dealing with timescales which are short compared with the convective turnover time, or with layers that are close to convective boundaries, the assumption of complete mixing may break down. There is currently much interest in three-dimensional modelling of convection regions, in understanding how convection cells behave when they overshoot a convective boundary, and in the behaviour of boundaries between materials of significantly different mean molecular weight (thermohaline mixing) [20].

The consequences of mixing, whatever the mechanism, for stellar evolution calculations is that nucleosynthesis may transform the mixture in part of a convection zone. Alternatively a convective boundary may move to incorporate material of different composition. The new material must then be mixed throughout the convection zone. After a time interval δt, the new composition (x'_i) in the convection zone (cz) will be given by

$$\langle x'_i \rangle = \int_{\mathrm{cz}} \left(x_i(m) + \delta t \frac{dx_i(m)}{dt} \right) dm \Big/ \int_{\mathrm{cz}} dm. \tag{40}$$

7 Equation of State

We adopt the premise that matter inside stars consists of an almost perfect gas. The properties of a gas are often referred to as *state variables.* The macroscopic properties of a gas are described completely by three quantities, *i.e.* any three of *pressure, density, temperature, entropy,* and *number density.*

An equation of state (EOS) describes relations between these properties, *e.g.* pressure, density and temperature. The form of the equation of state depends on whether the fluid particles can be treated as classical or non-classical, relativistic or non-relativistic.

7.1 Gas Laws

In 1662 Robert Boyle[4] performed a series of experiments which showed that gas volume varies inversely with pressure:

$$PV = \text{constant}.$$

In 1787, the French physicist Jacques Charles found that oxygen, nitrogen, hydrogen, carbon dioxide and air expand to the same extent over the same 80 degree change in temperature. In 1802, Joseph Louis Gay-Lussac published results of similar experiments indicating a linear relationship between volume and temperature:

$$V/T = \text{constant}.$$

At about the same time (1801), John Dalton observed that the total pressure exerted by a gaseous mixture is equal to the sum of the partial pressures of each individual component:

[4] born at Lismore Castle, Munster, Ireland, as the seventh son and fourteenth child of Richard Boyle, the "Great Earl of Cork"

$$P_{\text{total}} = P_1 + P_2 + \ldots + P_n.$$

In 1834, Emile Clapeyron combined Boyle's and Charles' laws into the first statement of the ideal gas law initially formulated as

$$PV = R(T_{\text{C}} + 267)$$

where the temperature is expressed in degrees Celsius. With the constant being improved to 273.15 and temperature being measured in degrees Kelvin (after the Belfast born William Thompson),

$$PV = RT_{\text{K}},$$

where $R = N_{\text{A}}k$ is the ideal gas constant.

7.2 Pressure

There are important differences between fluid deep within a star and a laboratory gas. First, stellar material is ionized, *i.e.* it is a plasma, which allows greater compression and interatomic distances as small as $\approx 10^{-15}$ m to be achieved, compared with $\approx 10^{-10}$ m for neutral gases. Second, it is in thermodynamic equilibrium with radiation; the radiation intensity is governed by Planck's law. Third, the fluid particles may be non-classical and/or relativistic. Therefore the effects of quantum mechanics and special relativity must be considered.

Pressure due to a classical or quantum gas is physically expressed by collisions between the gas particles and the gas boundary – thus it is directly related to the kinetic energy of the particles. Consider a box of dimension D containing N particles, each having velocity $\mathbf{v}$ and momentum $\mathbf{p} = m\mathbf{v}$. Assuming that collisions with the boundaries are elastic, each collision imparts momentum $2p_z$, and occurs at a rate $v_z/2D$. The rate of transfer of momentum to unit area is then $p_z v_z/D^3$, and averaging over all particles,

$$P = \frac{N}{D^3}\langle p_z v_z\rangle = n\langle p_z v_z\rangle.$$

If the gas is isotropic,

$$\langle p_x v_x\rangle = \langle p_y v_y\rangle = \langle p_z v_z\rangle = \langle \mathbf{p}\cdot\mathbf{v}\rangle/3$$

$$\Rightarrow P = \frac{n}{3}\langle \mathbf{p}\cdot\mathbf{v}\rangle. \tag{41}$$

In general, the relation between the energy, momentum and velocity of a particle is

$$\varepsilon_{\text{p}}^2 = p^2c^2 + m^2c^4, \quad v = pc^2/\varepsilon_{\text{p}},$$

In the non-relativistic limit: $p \ll mc$, so $\varepsilon_{\text{p}} = mc^2 + p^2/2m$ and $v = p/m$. In this case, $\mathbf{p}\cdot\mathbf{v} = mv^2$ and

$$P = \frac{n}{3}mv^2 = \frac{2}{3}n(\frac{1}{2}mv^2) = \frac{2}{3}U_{\rm kin} \tag{42}$$

where $U_{\rm kin}$ is the translational kinetic energy *density* of the particles. In the ultra-relativistic limit: $p \gg mc$, so $\varepsilon_{\rm p} = pc$ and $v = c$. Hence $\mathbf{p} \cdot \mathbf{v} = pc^2$ and

$$P = \frac{1}{3}n(pc) = \frac{1}{3}U_{\rm kin}. \tag{43}$$

7.3 The Classical Ideal Gas

From kinetic theory, the pressure of an ideal gas can have the form

$$P_{\rm gas} = nkT \tag{44}$$

where n is the number density of particles, and k is Boltzmann's constant. With Eq. (42), we obtain for a non-relativistic ideal gas the mean energy per particle

$$E_{\rm kin} = U_{\rm kin}/n = \frac{3}{2}kT. \tag{45}$$

The number density is related to the mass density ρ via the mean particle weight (sometimes mean molecular weight) μ

$$n = \frac{\rho}{m_{\rm u}\mu}.$$

7.4 Mean Mass per Particle

The mean mass per particle is essentially a reduced mass. That is, for a mixture of elements with atomic mass a_k, and mass fraction x_k, the *mean mass per atom* is

$$\mu_{\rm a}^{-1} = \sum_k \frac{x_k}{a_k}.$$

By contrast, the *mean mass per particle* depends on the number of free electrons per atom. Assuming that the material is fully ionized, each atom having z_k electrons,

$$\mu^{-1} = \sum_k \frac{x_k(1+z_k)}{a_k}. \tag{46}$$

It is often useful to know the *mean mass per electron*, which can be computed similarly or from the identity

$$\mu^{-1} = \mu_{\rm e}^{-1} + \mu_{\rm a}^{-1}.$$

Consider some examples:

- neutral hydrogen: $\mu = \mu_{\rm a} = 1$.
- ionized hydrogen: $\mu = \mu_{\rm a}/2 = 1/2$.

- a mix of neutral hydrogen and helium with mass fractions X and Y:

$$\mu^{-1} = \mu_{\rm a}^{-1} = \frac{X}{1} + \frac{Y}{4}.$$

- a fully-ionized hydrogen-helium-metal mix (X,Y,Z):

$$\mu^{-1} = \frac{2X}{1} + \frac{3Y}{4} + \frac{\langle z+1\rangle Z}{\langle a\rangle}.$$

Assuming that for metals, $\langle a\rangle \approx 2\langle z\rangle$ and $\langle z\rangle \gg 1$, and rearranging

$$\mu = \left(2X + \frac{3Y}{4} + \frac{Z}{2}\right)^{-1}.$$

If the gas is partially ionized, which is the case close to the stellar surface, then it is necessary to solve the Saha equation to determine the distribution of ions for every element.

7.5 Degenerate Electron Gas

The equation of state described in Sect. 7.3 is valid when the gas can be described as *classical*, *i.e.* when the average separation of particles is large compared with the de Broglie wavelength[5]. As the average separation becomes small, the lightest particles will experience a breakdown of classical physics. This occurs when the particle density n significantly exceeds the critical "quantum concentration"

$$n_{\rm Q} = \left(\frac{2\pi mkT}{h^2}\right)^{3/2}$$

(in the non-relativistic limit), which is another way of stating the requirement that the average occupation of any quantum state should be small. In the regime where quantum effects dominate $n_{\rm e} \gg n_{\rm Q}$, which is equivalent to the condition that

$$kT \ll \frac{h^2 n_{\rm e}^{2/3}}{2\pi m_{\rm e}},$$

[5] de Broglie wavelength: a particle of mass m moving with a velocity v will under suitable conditions exhibit the characteristics of a wave with wavelength

$$\lambda_{\rm B} = h/mv = h/p.$$

Compton wavelength: when the particle is moving relativistically, the de Broglie wavelength may be written

$$\lambda_{\rm C} = h/mc = 2.4 \times 10^{-12}\,{\rm m}$$

whence it may be seen that a quantum gas is a cold gas. The standard of coldness is set by the density, not the temperature. A cold gas is described as degenerate because the particles occupy the lowest possible energy states and (for electrons), obey the Pauli-exclusion principal. The energy of the most energetic electrons in a cold electron gas is known as the Fermi energy ($\varepsilon_{\rm F}$); the momentum of these particles is the Fermi momentum $p_{\rm F}$, which has the unique property[6]

$$p_{\rm F} = \left(\frac{3n_{\rm e}}{8\pi}\right)^{1/3} h.$$

The equation of state for a degenerate gas is obtained by evaluating the internal energy. Substituting $\varepsilon_{\rm p} = mc^2 + p^2/2m$ into an integral over quantum states gives for the non-relativistic case $p_{\rm F} \ll m_{\rm e}c$

$$U = \left(m_{\rm e}c^2 + \frac{3p_{\rm F}^2}{10m_{\rm e}}\right).$$

Since for an ideal gas $P_{\rm NR} = (2/3)U_{\rm kin}$ (Eq. (42)), and substituting for $p_{\rm F}$ and $n_{\rm e} = \rho/m_{\rm u}\mu_{\rm e}$,

$$P_{\rm e} = \frac{n_{\rm e}p_{\rm F}^2}{5m_{\rm e}} = K_{\rm NR}\left(\frac{\rho}{m_{\rm u}\mu_{\rm e}}\right)^{5/3}, \quad K_{\rm NR} = \frac{h^2}{5m_{\rm e}}\left(\frac{3}{8\pi}\right)^{2/3}. \tag{47}$$

Similarly, for an ultra-relativistic degenerate electron gas, $\varepsilon_{\rm p} = pc$ gives

$$U = \frac{3}{4}p_{\rm F}c$$

and from $P_{\rm UR} = (1/3)U_{\rm kin}$,

$$P_{\rm e} = \frac{n_{\rm e}p_{\rm F}c}{4} = K_{\rm UR}\left(\frac{\rho}{m_{\rm u}\mu_{\rm e}}\right)^{4/3}, \quad K_{\rm UR} = \frac{hc}{4}\left(\frac{3}{8\pi}\right)^{1/3}. \tag{48}$$

7.6 Photons

Thermal radiation may be characterized as a photon gas, consisting of zero-mass bosons with zero chemical potential. Without proof, the photon number density may be written [6, Sect. 2.3],

$$n = bT^3, \quad b = 2.4048\pi\left(\frac{k}{hc}\right)^3 = 2.03. \times 10^7\,{\rm K^{-3}\,m^{-3}}$$

The internal energy density is

---

6 The Fermi momentum implies that the de Broglie wavelength of the most energetic electrons in a degenerate gas is comparable with $n_{\rm e}^{-1/3}$, the average distance between electrons.

$$U_r = aT^4, \quad a = \frac{8}{15}\frac{\pi^5 k^4}{(hc)^3} = 7.565 \times 10^{-16}\,\mathrm{J\,K^{-4}\,m^{-3}} \tag{49}$$

and the pressure due to this radiation is

$$P_r = U/3 = aT^4/3. \tag{50}$$

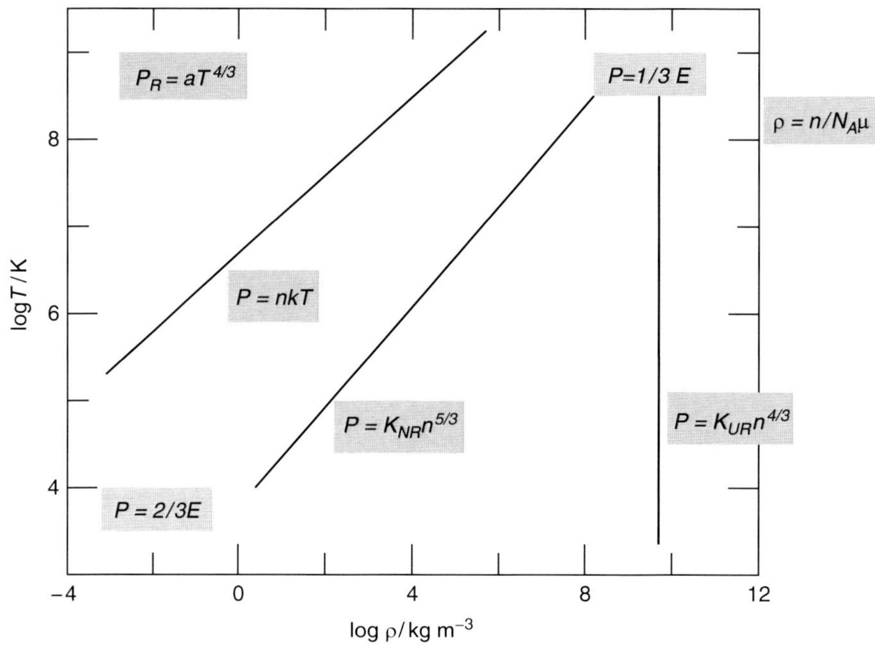

Fig. 16. A map of density-temperature space illustrating the different regimes for the equation of state, including the classical, relativistic, and degenerate limits

7.7 Total Pressure

The total pressure is the sum of the gas and radiation pressures, and the gas pressure can be written as the sum of ion and electron pressures:

$$P_t = P_g + P_r = P_i + P_e + P_r$$

Stability

Recall that the internal temperature of a star is proportional to its gravitational potential energy, and hence $T_I \approx M/R$ (Eq. (26)). Meanwhile the particle density $n \approx M/R^3$. Thus if the particles form a classical gas, the ratio of gas to radiation pressure

$$\frac{P_\mathrm{r}}{P_\mathrm{g}} = \frac{aT_\mathrm{I}^4/3}{(n_\mathrm{e}+n_\mathrm{i})kT_\mathrm{I}} \propto \frac{T_\mathrm{I}^3}{n_\mathrm{e}+n_\mathrm{i}} \propto \frac{M^3/R^3}{M/R^3} \propto M^2.$$

For increasing mass, radiation pressure becomes increasingly important.

Which equation?

The exact form of the EOS to use depends on whether the gas particles are classical or degenerate (depends on T/n), and on whether they are relativistic or non-relativistic (related to n or T). Figure 16 illustrates which EOS to apply in the temperature-density space typically found in stars.

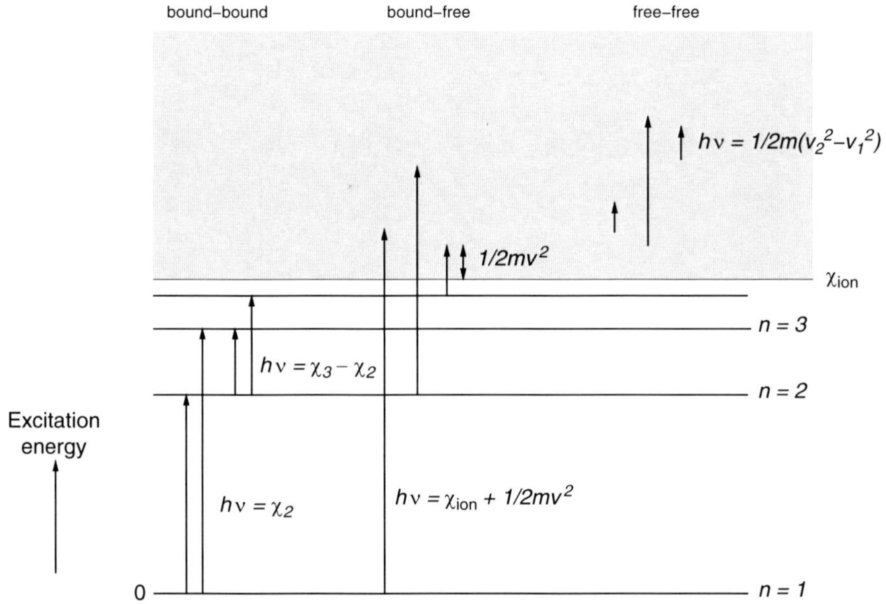

Fig. 17. The principle photon-electron interactions contributing to radiative opacity in stellar interiors

8 Stellar Opacity

The equation for energy transfer (Eq. (33)) includes a term κ, representing the opacity of stellar material to the flow of energy by radiation or conduction. Roughly speaking, it represents the ability of stellar material to absorb radiation. It is equivalent to the inverse of its heat conductivity. In order to evaluate the total radiative opacity, it is necessary to consider all of the microscopic processes whereby radiation at a specific frequency ν is absorbed, and how these processes can be combined as a single macroscopic quantity.

There are four basic processes involved in the interaction of photons with atoms.

1. bound-bound absorption – bb
2. bound-free absorption – bf
3. free-free absorption – ff
4. electron scattering – es

The first three represent true absorption because in each case a photon is absorbed. The fourth only changes the direction of a photon. The contribution from thermal conduction by electrons must also be considered.

In discussing the absorption processes, consider the energy level diagram for a simple atom (Fig. 17).

8.1 Bound-Bound Absorption

The bound energy levels of an ion lie at discrete excitation energies $\chi_{\rm exc}$ above the lowest-lying level (or ground state). Bound electrons may only occupy these levels, and when moving between them absorb (or emit) photons at a precise frequency (spectral line) given by

$$E = h\nu_{12} = \chi_2 - \chi_1.$$

If a photon of frequency ν_{12} collides with an ion where an electron occupies level 1, that electron may be excited into level 2, with a probability B_{12}. This transition probability (or oscillator strength) is effectively the bound-bound absorption coefficient for a single line, $a_{\rm bb}(\nu_{12}) = B_{12}$. When multiplied by the occupation number of each level 1, and summed over transitions between all levels in all ions, the total monochromatic bound-bound opacity is obtained

$$\kappa_{\rm bb}(\nu) = \sum_{ions} \sum_{1} N_1 \sum_{2} a_{12}(\nu). \tag{51}$$

Bound-bound absorption is very important in stellar atmospheres and stellar envelopes, where many millions of lines, particularly from iron-group elements, have been shown to have a profound influence on stellar models. The frequencies which dominate the opacity must correspond roughly to the peak of the Planck function, *i.e.* $\nu_{\rm max}/T = b'$ (Wien's law again). Most atomic transitions have wavelengths longer than 100 Å, so for $T > 5 \times 10^6$ K, bound-bound absorption becomes less important.

8.2 Bound-Free Absorption

At some excitation energy χ_{ion}, an electron is no longer bound to the parent ion and the levels which it may occupy are no longer discrete but form a continuous sequence or *continuum* of states. A continuous spectrum of absorption can be found *only* if the final state corresponds to a free electron. If the initial state is bound, then we have bound-free absorption, or photoionization. The kinetic energy of a free electron is simply $E = 1/2m_{\mathrm{e}}v^2$. The energy involved in a bound-free transition from level n is

$$E = \chi_{\mathrm{ion}} - \chi_n + 1/2m_{\mathrm{e}}v^2.$$

If a photon has energy $E = h\nu > \chi_{\mathrm{ion}} - \chi_n$, it will be absorbed with a probability $a_{\mathrm{bf}}(n,\nu)$ and will ionize the parent atom or ion. Since most elements in stellar interiors are highly ionized, we may introduce an approximation for the bound-free absorption coefficient for a hydrogenic atom in level n:

$$a_{\mathrm{bf}}(n,\nu) = \frac{64\pi^4 m_e e^{10} Z'^4 g_{\mathrm{bf}}}{3\sqrt{3}ch^6 n^5 \nu^3} \tag{52}$$

where g_{bf} is the Gaunt factor – a quantum mechanical correction factor, Z' is the effective charge of the nucleus and e, m_{e}, c, and h have their usual meaning. For hydrogen, $g_{\mathrm{bf}} \approx 1$, $Z' \approx 1$. For other atoms, g_{bf} can be quite large and in most cases is not known. Under this approximation:

$$\chi_{\mathrm{ion}} - \chi_n = \frac{2\pi^2 m_{\mathrm{e}} e^4}{h^2} \frac{Z'^2}{n^2}.$$

Summing over all ions and all levels, the total bound-free absorption coefficient becomes

$$\kappa_{\mathrm{bf}}(\nu) = \sum_{\mathrm{ions}} \sum_{n} N_n a_{\mathrm{bf}}(n,\nu) \tag{53}$$

where N_n represents the total number of ions per unit volume in a given quantum state with quantum number n. Note that $a_{\mathrm{bf}}(n,\nu) \propto \nu^{-3}$, and that Wien's law demands $\nu_{\mathrm{max}} = b'T$. Hence Eq. (52) implies $\kappa \propto T^{-3}$.

8.3 Free-Free Absorption

In the case that both initial and final states are unbound, an electron can absorb a photon of frequency ν_{ff} to move to a state with energy given by $h\nu_{\mathrm{ff}} = 1/2m_{\mathrm{e}}(v_2^2 - v_1^2)$. The free-free absorption coefficient for one atom and one free electron per unit volume can be written in the hydrogenic approximation as

$$a_{\mathrm{ff}}(\nu) = \frac{4\pi e^6 Z'^2 g_{\mathrm{ff}}}{3\sqrt{3}chm_{\mathrm{e}}^2 v \nu^3} \tag{54}$$

where v is the initial electron velocity and $g_{\rm ff}$ is the Gaunt factor for free-free absorption. The total absorption coefficient must be found for a sum similar to Eq. (53), by averaging over electron velocities,

$$\kappa_{\rm ff}(\nu) = \sum\nolimits_{\rm ions} N_{\rm ions} \int_{v} a_{\rm ff}(\nu) n_{\rm e}(v) dv \tag{55}$$

normally assuming a Maxwellian velocity distribution $n_{\rm e}(v)$, such that $\langle v \rangle \geq \surd(\pi kT/m_{\rm e})$.

8.4 Electron Scattering

Classically, scattering may be pictured as an elastic collision between two particles. If $h\nu \ll mc^2$, where m is the mass of the scatterer and ν the frequency of the photon, the scatterer is scarcely moved by the collision and the frequency of the photon will not alter (This approximation only breaks down for stars with very high central temperatures). Although a photon is not absorbed, scattering slows down the rate of energy escape by continually changing the photon direction.

Scattering is independent of frequency but does depend to some extent on density and the degree of ionisation. In practice, for $T > 100\,000$ K where hydrogen and helium are fully ionized, the density dependence is negligible but the composition is not, since this determines the number of electrons per ion. The absorption coefficient per electron is

$$\sigma_{\rm e} = 8\pi e^4/3c^4 m_{\rm e}^2 \tag{56}$$

and per unit volume

$$\kappa_{\rm es} = \frac{\sigma_{\rm e} n_{\rm e}}{\rho} = \frac{4\pi}{3} \frac{e^4 m_{\rm p}^2}{c^4 m_{\rm e}^2}(1+X) = 0.020(1+X)\,{\rm m^2\,kg^{-1}}. \tag{57}$$

For obvious reasons, this mechanism dominates at high temperatures, and hence in fully-ionized stellar cores. It makes a negligible contribution to the total absorption at low temperatures, except in the absence of other absorbers such as neutral hydrogen.

8.5 Total Absorption Coefficient

The total monochromatic absorption coefficient is given by the sum of the contributions from all four processes:

$$\kappa(\nu) = \kappa_{\rm bb}(\nu) + \kappa_{\rm bf}(\nu) + \kappa_{\rm ff}(\nu) + \kappa_{\rm es} \tag{58}$$

The above has completely ignored stimulated emission. It may be shown that this effect can be accurately corrected by using a reduced absorption coefficient

$$\kappa^*(\nu) = \kappa(\nu)(1 - e^{(-h\nu/kT)})$$

The Rosseland Mean Opacity

Stellar opacity calculations must consider all atoms and ions and, for stellar structure calculations, provide an appropriate average integrated over all frequencies.

To compute this average, it is not sufficient simply to weight every frequency equally since the contribution of any one transition must reflect the relative radiative flux at that frequency. Stellar structure calculations most commonly use a weighted average

$$\frac{1}{\kappa} = \int_0^\infty \frac{w_\nu}{\kappa_\nu} d\nu \bigg/ \int_0^\infty w_\nu d\nu \ .$$

in which the derivative of the Planck function maximises the opacity contribution where the flux is strongest:

$$\frac{1}{\kappa_{\rm Ross}} = \int_0^\infty \frac{dB_\nu}{dT} \frac{d\nu}{\kappa_\nu^*} \bigg/ \int_0^\infty \frac{dB_\nu}{dT} d\nu \ . \tag{59}$$

$\kappa_{\rm Ross}$ is known as the Rosseland mean opacity, and $B_\nu(T)$ represents the energy density for black body radiation:

$$B(\nu, T) = \frac{2h\nu^3}{c^2} \frac{1}{e^{h\nu/kT} - 1}. \tag{60}$$

Approximate Values

Since a vast number of data are required to calculate the contribution to the Rosseland mean from many radiative processes, the calculation of stellar opacities is a long and involved process. Also, since the total opacity (Rosseland mean) is an harmonic mean, the opacity must be recalculated for every chemical mixture; thus

$$\kappa = \kappa(\rho, T, x_i)$$

where x_i represents the abundances of elements $i, i = 1, .., N$. For modern stellar structure calculations, detailed tables of precalculated opacities are usually used. In the case of high ionization, the Saha equation gives $N_{i,n}$, the average number of electrons which are bound in the $n^{\rm th}$ state as:

$$N_{i,n} = n_{\rm e} n^2 \frac{h^3}{2(2\pi m_{\rm e} kT)^{3/2}} e^{\chi_n/kT} .$$

$n_{\rm e}$ is the number density of free electrons and n^2 is the statistical weight of the $n^{\rm th}$ state. Substituting this and Eq. (52) into Eq. (53) and rearranging [1, p. 67] gives Kramers' Law of Opacity

$$\kappa_{\rm bf} = 4.34 \times 10^{24} \frac{\bar{g}}{t} Z(1 + X) \frac{\rho}{T^{3.5}} \ {\rm m^2 \, kg^{-1}} \tag{61}$$

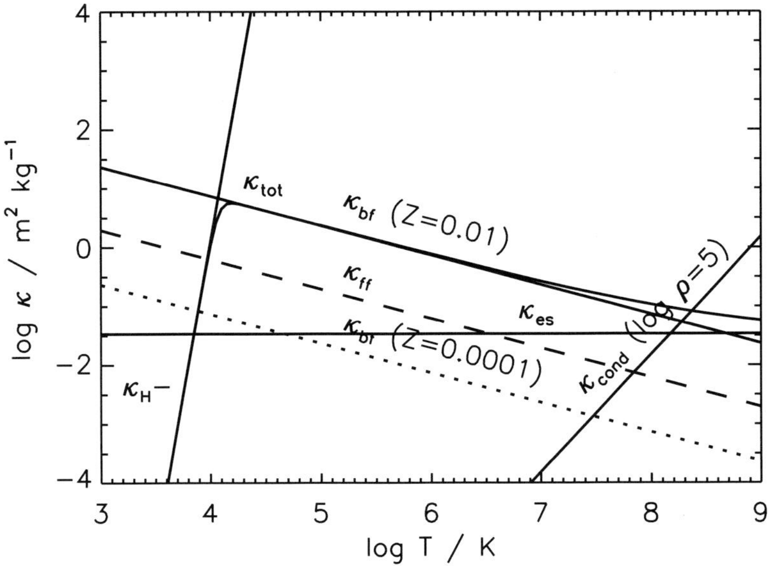

Fig. 18. Approximate opacities given by Eqs. (57), (61), (62), (63) and (65) for $\rho = T_6^3$ and $X = 0.7$, except $\kappa_{\rm cond}$ which is illustrated for a carbon-oxygen mixture. $\kappa_{\rm tot}$ represents the appropriate sum of $\kappa_{\rm H^-}$, $\kappa_{\rm bf}$ and $\kappa_{\rm es}$

(see Fig. 18). A similar derivation for free-free transitions gives

$$\kappa_{\rm ff} = 3.68 \times 10^{21} \bar{g_{\rm ff}} (X+Y)(1+X) \frac{\rho}{T^{3.5}} \,{\rm m^2\,kg^{-1}}. \tag{62}$$

At high temperatures, we have already seen that electron scattering dominates:

$$\kappa = \kappa_{\rm es} = 0.02(1+X)\,{\rm m^2\,kg^{-1}}.$$

At low temperatures, the high ionization approximation breaks down and the opacity is dominated by a steeply rising contribution due to the dissociation of H^- [7, p. 164]:

$$\kappa_{\rm H^-} \approx 2.5 \times 10^{-32} (Z/0.02) \rho^{1/2} T^9 \,{\rm m^2\,kg^{-1}}. \tag{63}$$

The Sun

In order to put some numbers on these formulae, consider the solar interior. In the centre, $\rho_c \approx 10^5\,{\rm kg\,m^{-3}}$ and $\kappa_{\rm c} \approx 10^{-1}\,{\rm m^2\,kg^{-1}}$. Therefore $\kappa\rho \approx 10^4\,{\rm m^{-1}}$ and the photon mean-free path $\langle\lambda\rangle \approx 10^{-4}\,{\rm m}$. At lower temperatures, *e.g.*, where $\rho \approx 10^3\,{\rm kg\,m^{-3}}$, $\kappa \approx 10\,{\rm m^2\,kg^{-1}}$. Thus $\kappa\rho \approx 10^4\,{\rm m^{-1}}$ also. The mean temperature gradient in the Sun $dT/dr \approx 2 \times 10^{-2}\,{\rm K\,m^{-1}}$, and therefore the temperature difference between photon emission and absorption $\Delta T \approx 2 \times 10^{-6}$ K. This demonstrates that radiation and matter in deep stellar interiors is in true thermodynamic equilibrium.

8.6 Electron Conduction

In very dense stellar material, the mean free path of the photon becomes so small that it is no longer the most efficient carrier of energy. Thermal conduction by electrons becomes the dominant transport mechanism, controlled by a conductive opacity $\kappa_{\rm cond}$ in much the same way as conduction by photons is controlled by the radiative opacity. From the diffusion approximation we obtain

$$\kappa_{\rm cond} = \frac{4acT^3}{3D_e\rho}, \tag{64}$$

where D_e is a diffusion coefficient for electrons with the general form $D_e \approx c_{\rm V} v_{\rm e} \lambda/3$. It is relatively straightforward to obtain an approximate expression for D_e, whence

$$\kappa_{\rm cond} \approx 4 \times 10^{-9} \frac{\mu_e^2}{\mu_I} Z_c^2 \left(\frac{T}{\rho}\right)^2 \,\mathrm{m^2\,kg^{-1}} \tag{65}$$

where $Z_{\rm c}$ is the ion charge. Note how, in contrast to radiative opacities, the conductive opacity is inversely proportional to density – heat travels *more* easily in dense media. The total opacity to energy transport by the combined effects of conduction and radiation is also an harmonic sum,

$$\kappa^{-1} = \kappa_{\rm rad}^{-1} + \kappa_{\rm cond}^{-1}.$$

9 Thermonuclear Physics

After the discovery that atoms have masses which are integral multiples of the mass of the hydrogen atom, it was realised that, if a suitable mechanism could be found, all atoms could be created from the fusion of hydrogen. The problem is that the electrostatic force implies a strong repulsion between atomic nuclei, which all carry positive electric charge. Typically, the e-s potential energy for 2 protons, separated by 2 proton radii ($\approx 10^{-15}$ m) is

$$E_{\rm pot} = e^2/4\pi\varepsilon_0 r \approx 3 \times 10^{-13}\,\mathrm{J}. \tag{66}$$

The average kinetic energy of a proton, at 10^7 K is

$$E_{\rm kin} = 3/2kT \approx 2 \times 10^{-16}\,\mathrm{J} \tag{67}$$

which is insufficient to overcome the electrostatic potential barrier. So while Eddington argued that the interiors of stars were likely sites for the synthesis of the elements, antagonists (including James Jeans) pointed out the energetics were against it. Eddington's rejoinder was that *"We do not argue with the critic who urges that stars are not hot enough for this process; we tell him to go and find a hotter place."*

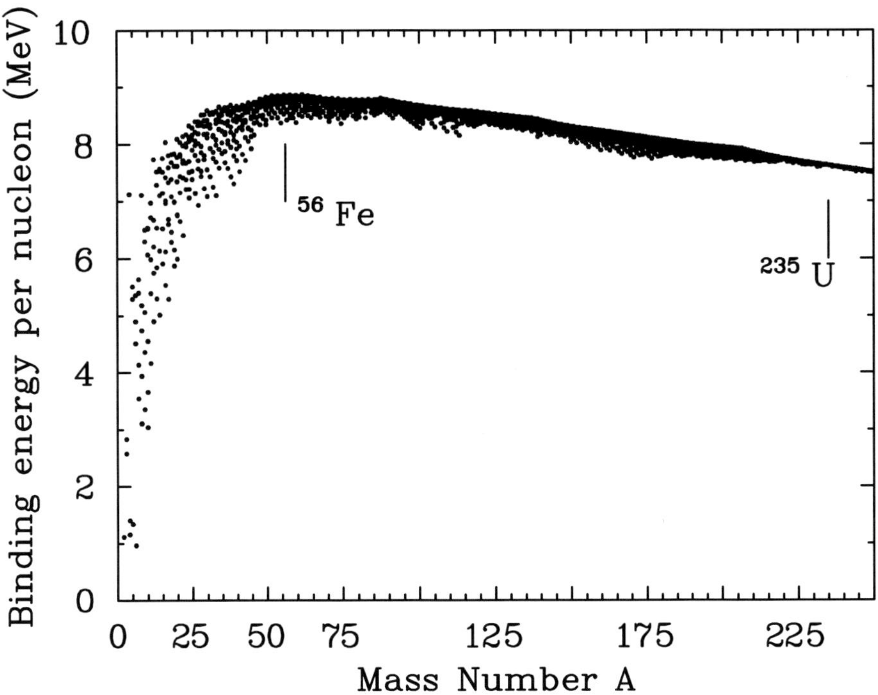

Fig. 19. Binding energy Q per nucleon as a function of atomic mass

9.1 Fusion

Consider the interaction of two protons: the p-p reaction. The fundamental forces which control the interaction can be expressed in four forms: (Table 4).

Table 4. Forces governing the interaction of two protons

force	source	range	nuclear reactions
gravitational	mass	$1/r^2$	no
electrostatic	charge	$1/r^2$	yes
weak nuclear	baryon-lepton	$1/r^w : w \gg 2$	some
strong nuclear	baryon-baryon	$1/r^s : s \gg 2$	yes

For the p-p reaction the electrostatic and strong forces are significant. Since $E_{\rm kin} \ll E_{\rm pot,max}$, classical physics states that, although the resultant 2p nucleus would be energetically favourable, the two protons cannot approach one another to within a separation r_1. However, quantum mechanics describes the proton as a wavefunction ψ given by the solution of the Schrödinger equation

$$\frac{\partial^2\psi}{\partial r^2} + \frac{2m}{\hbar^2}\left(E_{\rm kin} - E_{\rm pot}\right)\psi = 0 \tag{68}$$

For $r > r_1$ and $r < r_2$, $(E_{\rm kin} - E_{\rm pot})$ is positive and ψ is real:

$$r > r_1 : \psi \propto \sin kr : k = 2m/h^2(E_{\rm kin} - E_{\rm pot})$$

$$r_2 < r < r_1 : \psi \propto e^{-kr}$$

$$r < r_2 : \psi \propto \sigma \sin kr$$

where σ represents the probability of barrier penetration. There is therefore a finite probability of the proton *tunnelling* through to a position r_2 and combining with the target proton. The tunnelling effect also allows alpha- and beta-decay processes to occur, whereby a particle can *escape* from the potential well in the atomic nucleus if it has sufficient kinetic energy.

Formally, the solution of Eq. (68) can be used to obtain an expression for the penetration probability and hence a reaction cross-section σ.

In the case of the p-p reaction, if the tunnelling operation is successful, an unstable nuclide consisting of 2 protons is created. What can happen next is that either the inverse reaction occurs (one proton escapes from the nucleus) or else one proton quickly releases a positron to remove excess electric charge, and a neutrino to conserve momentum and lepton number, and becomes a neutron thus forming a deuterium nucleus.

Reaction Notation

We introduce the notation

$$i_1(i_2, o_3)o_4$$

where i are the input particles and o are the output particles and where i_1 and o_4 are the principal participants in the reaction. Both i_2 and o_3 may be absent, or may represent one or more particles. Thus the p-p reaction may be written

$${}^1{\rm H}({\rm p}, \beta^+\nu){}^2{\rm H},$$

and neutron decay may be written

$${\rm n}(\beta^-, \bar{\nu}).$$

Binding Energy

All atomic nuclei consist of Z protons and N neutrons. The total rest mass of the energy of the individual particles is always greater than the rest mass energy of the nucleus. The deficit represents the binding energy of the nucleus

$$Q(Z,N) = (Zm_{\rm p} + Nm_{\rm n} - m(Z,N))c^2 \tag{69}$$

which is also the nett amount of energy released during the construction of a given nuclide. Q is shown for stable nuclides as a function of atomic mass (A = Z+N) in Fig. 19.

For any given reaction, we are interested in two quantities

1. the reaction rate: this is related to the penetration probability and the energy distribution of the particles
2. the energy released: this is (approximately) the change in binding energy per nucleon

In the latter case, we also need to know how much of the energy is available for heating the stellar material (photons, electrons, positrons) and how much is 'lost' in the form of neutrinos.

9.2 Reaction Rates

Penetration Probability

The penetration probability takes the form of a cross-section σ. In its simplest form, the incoming wavefunction takes the form of a decaying exponential ($\psi(r) \propto e^{-kr}$), so that the penetration probability

$$\sigma \propto (e^{-kr})^2.$$

Considering the reaction constant k and writing the equation in terms of the relative energy E of the nuclei and the Gamow energy (a measure of the magnitude of the Coulomb barrier)

$$E_{\rm G} = (\pi\alpha Z_i Z_j)^2 2\mu c^2$$

where $\alpha = e^2/4\pi\varepsilon_0 hc \approx 1/137$ is the fine structure constant and μ is the reduced mass, the penetration probability may be written [6, pp. 102–104]:

$$\sigma \approx \frac{S(E)}{E} \exp\left(-(E_{\rm G}/E)^{1/2}\right) \tag{70}$$

where $S(E)$ is determined by the nuclear physics.

Conversion to a reaction rate (reactions per second per gram) requires an average over particle velocities and number densities:

$$\langle\sigma v\rangle = \frac{\int_0^\infty \sigma v N(v) dv}{\int_0^\infty N(v) dv} \tag{71}$$

$$r_{ij} = n_i n_j \langle\sigma v\rangle/\rho. \tag{72}$$

Velocity Distribution

Assuming a Maxwellian velocity distribution

$$N(v)dv = \left(\frac{\mu}{2\pi kT}\right)^{3/2} \exp\left(-\frac{\mu v^2}{2kT}\right) 4\pi v^2 dv.$$

Writing $E = \frac{1}{2}\mu v^2$ and substituting for $N(v)$ into Eq. (71)

$$\langle\sigma v\rangle = \left(\frac{8}{\pi\mu}\right)^{1/2} \left(\frac{1}{kT}\right)^{3/2} \int_0^\infty E\sigma e^{-E/kt} dE.$$

Reaction Rate

Now substituting for σ and multiplying by number densities:

$$r_{ij} = \frac{n_i n_j}{\rho}\left(\frac{8}{\pi\mu}\right)^{1/2}\left(\frac{1}{kT}\right)^{3/2}\int_0^\infty S(E)\exp\left(-\frac{E}{kt}-\left(\frac{E_{\rm G}}{E}\right)^{1/2}\right)dE. \quad (73)$$

The product of the two exponentials (velocity distribution times penetration probability) has a maximum when the energy

$$E_0 = (E_{\rm G}/4)^{1/3}(kT)^{2/3}. \quad (74)$$

Fusion principally occurs when the relative particle energy E lies in a narrow energy range around E_0 known as the fusion window, given by:

$$\Delta = \frac{4}{3^{1/2}2^{1/3}}E_{\rm G}^{1/6}(kT)^{5/6}. \quad (75)$$

Simplifying Eq. (73) and extracting key terms,

$$r_{ij} \propto n_i n_j S(E_0)\exp\left(-3\frac{E_{\rm G}}{4kT}\right)^{1/3},$$

it may be seen that in a mixture with many species present, fusion will favour reactions with small Coulomb barriers and high number densities.

Energy Yield

If Q_{ij} is the energy released per reaction ij, the total energy released is

$$\varepsilon_{ij}(\rho, T, n_i, n_j) = r_{ij}Q_{ij}. \quad (76)$$

9.3 Reaction Networks

Hydrogen-burning (in particular) does not proceed via a single reaction, but through a chain or cycle involving several different nuclear reactions. In this case, the reaction rate of the network is governed by the rate of the slowest reaction in the network, but the energy released per product nucleon will be that for the entire cycle, thus:

$$\varepsilon_{\rm cycle} \approx r_{\rm slowest}Q_{\rm cycle}.$$

The principle reaction cycles during main-sequence burning are the proton-proton (p-p) chains and the CNO cycles.

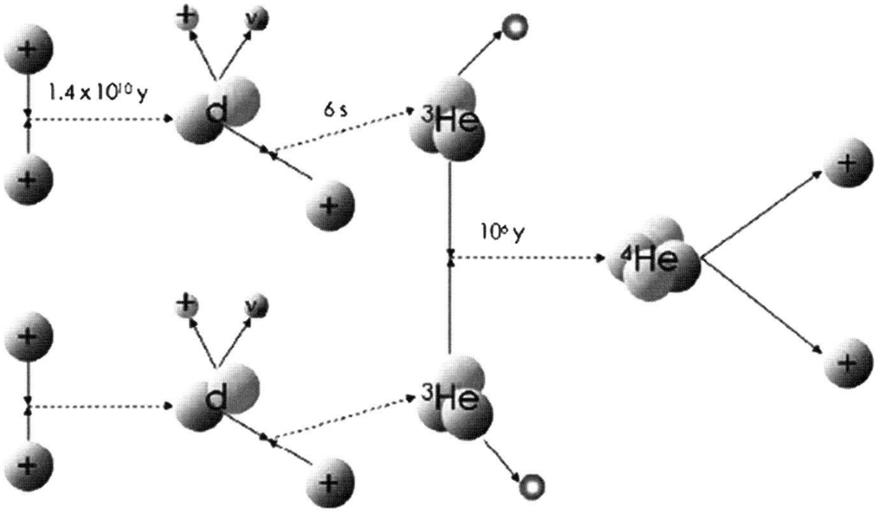

Fig. 20. The p-p I chain, with relative lifetimes for each of principal reactions

Table 5. The p-p chains. The bottom two lines show energy yield per ^{4}He nucleus produced and the relative total yield over the main-sequence lifetime of a solar-type star

p-p I	p-p II	p-p III
^{1}H(p,$\beta^+\nu$)^{2}H		
^{2}H(p,γ)^{3}He		
^{3}He(^{3}He,2p)^{4}He	^{3}He(α,γ)^{7}Be	
	^{7}Be(β^-, ν)^{7}Li	^{7}Be(p,γ)^{8}B
	^{7}Li(p,α)^{4}He	^{8}B$\rightarrow^8$Be* + β^+ + ν
		^{8}Be* $\rightarrow$ 2 ^{4}He
26.2 MeV	25.7 MeV	19.1 MeV
85%	15%	0.02%

p-p chains

The simplest p-p chain begins with the p-p reaction already discussed. The rest mass of the proton is 1.00728 u, where u is the atomic mass unit, 1/12$^{\rm th}$ the mass of a ^{12}C atom. The fusion of two protons, including a positron decay, yields a deuterium nucleus with rest mass 2.01355 u. The energy excess is thus

$$E_{\rm excess} = \delta mc^2 = (2m_{\rm p} - m_{\rm d})c^2 = 1.502 \times 10^{-13}\,{\rm J}.$$

This is the slowest reaction in any of the p-p chains with a rate given by

$$r_{\rm pp} \propto x_{\rm H}^2(\rho/T_6^{2/3})\exp(25.44 - 33.81/T_6^{1/3}).$$

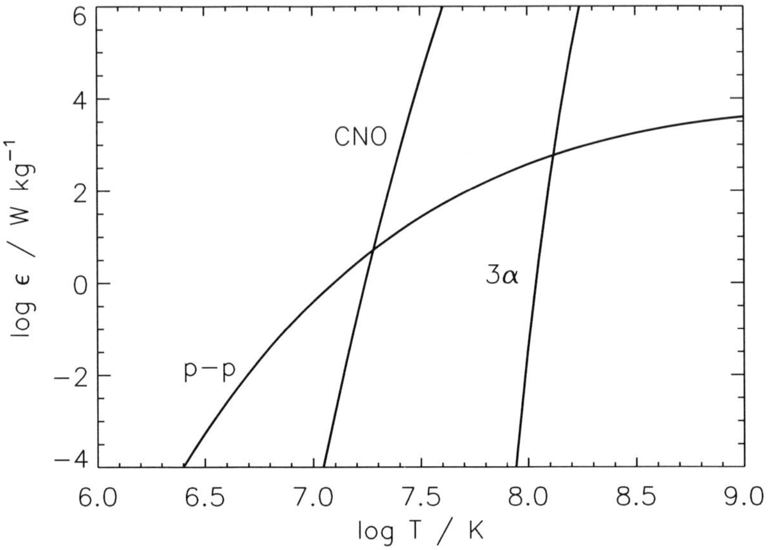

Fig. 21. Energy generation rates as a function of temperature for the p-p ($X = 0.7$), CNO ($X = 0.7, Z = 0.02$) and 3α ($Y = 0.98$) processes, all for $\rho = 10^5\,\mathrm{kg\,m^{-3}}$

Fusion with an additional proton creates a ^{3}He nucleus and a photon in a very prompt reaction[7], and the energy excess has increased to

$$E_{\mathrm{excess}} = \delta mc^2 = (3m_{\mathrm{p}} - m_{^3\mathrm{He}})c^2 = 8.697 \times 10^{-13}\,\mathrm{J}.$$

The slower fusion of two ^{3}He nuclei finally yields a ^{4}He nucleus, two protons and a nett energy excess

$$E_{\mathrm{excess}} = \delta mc^2 = (4m_{\mathrm{p}} - m_{^4\mathrm{He}})c^2 = 3.956 \times 10^{-12}\,\mathrm{J} = 24.7\,\mathrm{MeV}.$$

The total energy available to the star is less than this because neutrinos produced by the p-p reaction and in the p-p II and p-p III branches immediately remove a fraction of the energy yield from the star.

As the hydrogen abundance drops and the ^{3}He and ^{4}He abundances increase, additional reactions (Table 5) come into play and eventually the p-p II and p-p III chains dominate over p-p I.

Two additional points should be noted about the p-p burning: (1) p-p II results in the destruction of ^{7}Li and (2) the neutrinos produced in each reaction have quite distinct energy signatures.

In a region not far below where the p-p and CNO reactions are equally important, the p-p energy generation rate can be expressed as [21]:

[7] This is the fastest hydrogen-burning reaction and, since ^{2}H is significantly abundant in seawater and can be readily isolated, is the reaction most likely to used for terrestrial power generation by nuclear fusion.

$$\varepsilon_{\rm pp} = \frac{2.4 x_{\rm H}^2 \rho}{T_9^{2/3}} \exp(-3.380/T_9^{1/3})\,{\rm W\,kg^{-1}}. \tag{77}$$

Table 6. The CNO cycles

CN	NO	NOF	OF
$^{12}C\ (p,\gamma)^{13}N$			
$^{13}N \rightarrow ^{13}C + \beta^+ + \nu$			
$^{13}C\ (p,\gamma)^{14}N$			
$^{14}N\ (p,\gamma)^{15}O$	$^{14}N\ (p,\gamma)^{15}O$		
$^{15}O \rightarrow ^{15}N + \beta^+ + \nu$	$^{15}O \rightarrow ^{15}N + \beta^+ + \nu$		
$^{15}N\ (p,\alpha)^{12}C$	$^{15}N\ (p,\gamma)^{16}O$	$^{15}N\ (p,\gamma)^{16}O$	
	$^{16}O\ (p,\gamma)^{17}F$	$^{16}O\ (p,\gamma)^{17}F$	$^{16}O\ (p,\gamma)^{17}F$
	$^{17}F \rightarrow ^{17}O + \beta^+ + \nu$	$^{17}F \rightarrow ^{17}O + \beta^+ + \nu$	$^{17}F \rightarrow ^{17}O + \beta^+ + \nu$
	$^{17}O\ (p,\alpha)^{14}N$	$^{17}O\ (p,\gamma)^{18}F$	$^{17}O\ (p,\gamma)^{18}F$
		$^{18}F \rightarrow ^{18}O + \beta^+ + \nu$	$^{18}F \rightarrow ^{18}O + \beta^+ + \nu$
		$^{18}O\ (p,\alpha)^{15}N$	$^{18}O\ (p,\gamma)^{19}F$
			$^{19}F\ (p,\alpha)^{16}O$
(^{12}C destroyed)	(^{16}O destroyed)		

CNO cycles

The second major hydrogen-burning network is a catalytic process involving carbon, nitrogen and oxygen and which becomes important at temperatures above $\approx 1.5 \times 10^7$ K. While nearly all nuclei will react with protons at some temperature, the results are only interesting from the point of view of long-lived nuclear burning if the supply of nuclei matches the amount of reactant. In the case of the CNO cycles, the capture of a proton by a ^{12}C nucleus to form a ^{13}N nucleus is followed by further proton captures and β^+ decays. The fourth proton-capture (by ^{15}N) may give ^{16}O, but at low temperatures is more likely to give an α particle and a ^{12}C nucleus, replenishing the supply of ^{12}C so that the cycle may continue (Table 6).

At higher temperatures, additional catalytic cycles come into operation starting, successively, with ^{14}N, ^{15}N and ^{16}O. The energy produced is $\approx$25.02 MeV, independent of the cycle, but the rate is cycle dependent, since it is controlled by the slowest reaction in each cycle. In the case of the CN and the NO cycles, the slowest reaction is $^{14}N(p,\gamma)^{15}O$. When the cycle is operating in equilibrium, the local abundance of participating elements other than H, He and ^{14}N will drop to a low value; ^{12}C and ^{16}O are effectively converted to ^{14}N by the CN and NO cycles respectively.

The overall reaction rate is determined by the $^{14}N(p,\gamma)^{15}O$ reaction,

$$r_{\rm CN} \propto \frac{1}{T_9^{2/3}} \exp(-15.23/T_9^{1/3})$$

where $T_9 = T/10^9$. If $x'_{\rm N14} \approx x_{\rm N14} + x_{\rm C12}$ (CN cycle) or $x'_{\rm N14} \approx x_{\rm N14} + x_{\rm C12} + x_{\rm O16}$ (CN + NO cycles), then the overall reaction rate in the region where p-p and CNO reactions are equally important takes the form

$$\varepsilon_{\rm CNO} \approx \frac{3.2 \times 10^{21} x_{\rm H} x'_{\rm N14} \rho}{T_9^{2/3}} \exp(-15.23/T_9^{1/3})\,{\rm W\,kg^{-1}}. \qquad (78)$$

Note that in addition to helium nuclei, thermonuclear fusion through either p-p chains or CN cycles produces positrons, neutrinos, and high-energy photons (γ rays). The positrons annihilate with electrons to give photons (and some neutrinos). The photon energy is used to maintain the temperature gradient through the star via the energy transfer equation. The neutrino energy leaves the star immediately since the neutrino opacity is negligible.

The energy produced by the p-p chains and CN and CNO cycles is plotted as function of temperature in Fig. 21. For more accurate work, equilibrium abundances of the reactants should be computed and the most up-to-date values for the reaction rates should be used [22, 21, 23, 24].

3α and α-capture reactions

When hydrogen is completely converted to helium by either of the above processes, the next most energetically favourable reaction is the combustion of helium:
^{4}He $(\alpha, \gamma)^8$Be – 92 keV
^{8}Be $(\alpha, \gamma)^{12}$C* – 288 keV
^{12}C$^* \rightarrow ^{12}$C + 2γ + 7.65 MeV

This reaction has a very high threshold, and the first two stages are endothermic. After the initial α capture, the ^{8}Be nucleus has a high probability of decaying to two α particles. Only if a high energy α particle collides with the ^{8}Be nucleus within 10^{-16} s will it form an excited ^{12}C nucleus. In nearly all cases this ^{12}C* nucleus will decay back to a ^{4}He nucleus and a ^{8}Be nucleus. However if the forward reaction is sufficiently rapid, a small pool of ^{12}C* nuclei will accumulate, some of which can decay by photon cascade (or positron-electron pair emission) to give a stable ^{12}C nucleus and 7.27 MeV. The energy released by this reaction is [23]:

$$\varepsilon_{3\alpha} = 5.1 \times 10^4 \frac{x_{He}^3 \rho^2}{T_9^3} \exp(-4.4027/T_9)\,{\rm W\,kg^{-1}}. \qquad (79)$$

Note the extremely steep temperature gradient for this reaction (Fig. 21), and compare it with that for the CNO and p-p reactions.

When the ^{4}He abundance is reduced by conversion to ^{12}C, further α captures become more frequent:
^{12}C (α, γ) ^{16}O + 7.16 MeV,
^{16}O (α, γ) ^{20}Ne + 4.73 MeV,

^{20}Ne (α, γ) ^{24}Mg + 9.32 MeV
Meanwhile, any ^{14}N left over from the CNO-process will have been destroyed by an additional set of α captures:
^{14}N (α, γ) ^{18}F + 4.42 MeV,
^{18}F (, $\beta^+\nu$) ^{18}O,
^{18}O (α, γ) ^{22}Ne + 9.67 MeV

Other reactions

When hydrogen and helium are completely exhausted, heavier elements will burn at successively higher temperatures. Since the most abundant products of helium burning are ^{12}C and ^{16}O, important reactions include:

$T_9 \approx$0.5-1:
^{12}C + ^{12}C $\rightarrow^{23}$Na + p + 2.24 MeV (56%)
^{12}C + ^{12}C $\rightarrow^{20}$Ne + α + 4.62 MeV (44%)

T_9 >1:
^{16}O + ^{16}O $\rightarrow^{31}$P + p + 7.68 MeV (61%)
^{16}O + ^{16}O $\rightarrow^{28}$Si + α + 9.59 MeV (21%)
^{16}O + ^{16}O $\rightarrow^{31}$Si + n + 1.5 MeV (18%)

T_9 >3:
^{28}Si "burning" – at these temperature, photodisintegration leads to a soup of protons, neutrons and heavy nuclides in which an enormous number of possible reactions are possible.

9.4 Nucleosynthesis of elements

As the hydrogen-burning reaction chains suggest, a complex web of nuclear reactions will occur at the same time. Yields are determined by the branching ratios and rates of individual reactions. The chemical composition of this thermonuclear soup can be obtained, at least for steady-state burning, by setting up a system of simultaneous equations that model the entire reaction network and solving for the equilibrium abundances of each chemical species.

Even in steady-state nuclear burning, some species are destroyed and others created, so the chemical composition of a star will change. If we know the reaction rate r_{ijk} for a reaction in which species i and j are destroyed and species k is produced, then the rates of change of species i (or j) and k are simply related by

$$\frac{dn_i}{dt}|_{ijk} = -(1+\delta_{ij})r_{ijk}\frac{dn_k}{dt}|_{ijk} = -(\frac{dn_i}{dt} + \frac{dn_j}{dt}) \tag{80}$$

where $n_{i,j,k}$ are the relative abundances by number.

These time-dependent equations represent the fundamental equations of *stellar evolution*. A succession of nuclear reactions (H, He, C and subsequent burning) represents the mechanism whereby all the chemical species have been created. Many exotic elements can only be produced in core-collapse supernova explosions, often through the capture of fast neutrons (r-process). Others are produced by the capture of slow neutrons in the envelopes of low- and intermediate-mass stars (s-process). The basic framework for the theory is elegantly illustrated in [25, Fig. I.2]. Such processes are described in more detail elsewhere in these proceedings.

9.5 Neutrinos

Neutrinos are produced as electron (or positron) decay or capture products in nearly all thermonuclear reaction networks. The neutrinos carry energy away from the stellar interior because their interaction with the stellar material is very small (but not zero). To give some idea of the scale, a typical neutrino capture cross section is given by $\sigma_\nu \approx 10^{-48}\varepsilon_\nu^2\,\mathrm{m}^2$ where ε_ν is the neutrino energy in MeV. Then the mean free path is $\lambda = 1/n\sigma_\nu \approx 10^{18}\varepsilon_\nu^{-2}/\rho$ m, where n is the number density of targets with a mean molecular weight near unity. For ordinary stars λ is very large, but in the cores of supernovae, with $\rho \approx 10^{17}\,\mathrm{kg\,m^{-3}}$ and $\varepsilon_\nu \approx 20\,\mathrm{MeV}$, then $\lambda \approx 25\,\mathrm{m}$ may be possible. The consequences for energy transport and hence the physics of the supernova core are dramatic.

Usually the neutrino losses are modest. Measurements of the solar neutrino flux have been used to test models of stellar structure. However the *neutrino luminosity* can be crucial during some stages of evolution – and can lead to a negative flux gradient! This becomes particularly severe in the late stages of stellar collapse when conditions arise under which large numbers of neutrinos can be created. A very large burst of neutrinos is produced in a supernova explosion, with flux approximately equal to the total photon flux. Such a large burst was observed from SN 1987A, despite the latter being in another galaxy.

In addition to the nuclear decay/capture neutrinos, neutrinos can be produced in a number of other ways which can become important at some stages in stellar evolution.

Urca process

The Urca process is essentially an electron capture followed by β-decay, with the emission of a neutrino and anti-neutrino pair, *e.g.*

$$^{35}\mathrm{Cl}(\beta^-,\nu)^{35}\mathrm{S} \rightarrow ^{35}\mathrm{S}(,\beta^-\bar{\nu})^{35}\mathrm{Cl}$$

The energy carried by the $\nu\bar{\nu}$ pair is lost to the star, but the process is unlikely to be important except in very evolved cores with high densities and low temperatures.

Photo-Neutrino Process

Electron scattering by photons may alternatively result in pair neutrino production, with a nett energy loss as shown:

$$\gamma + \beta^- \rightarrow \beta^- + \gamma \rightarrow \beta^- + \bar{\nu} + \nu$$

$$\varepsilon_\nu \approx 10^3 T_9^8 \,\mathrm{W\,kg^{-1}}$$

This can be produced by ordinary bremsstrahlung and is an important energy loss mechanism in hot white dwarfs.

Pair-Neutrino Process

Similarly:

$$\gamma + \gamma \leftrightarrow \beta^- + \beta^+ \leftrightarrow \bar{\nu} + \nu$$

$$\varepsilon_\nu \approx 10^{14.7} \frac{T_9^3}{\rho} \exp(-11.88/T_9) \,\mathrm{W\,kg^{-1}}$$

Plasma-Neutrino Process

In a dense plasma electromagnetic waves can be quantized and behave like relativistic Bose particles with finite mass. These "plasmons" can decay into either $\beta^- + \beta^+$ or $\bar{\nu} + \nu$ pairs.

10 Approximate Solutions

Although general solutions of the stellar structure equations require numerical techniques, approximate models provide valuable insight.

10.1 Polytropic Gas Spheres

Rearranging Eq. (17)

$$m = -\frac{dP}{dr}\frac{r^2}{\rho G} \tag{81}$$

differentiating and substituting Eq. (14)

$$\frac{dm}{dr} = 4\pi r^2 \rho = -\frac{d}{dr}\left(\frac{dP}{dr}\frac{r^2}{\rho G}\right)$$

$$\Rightarrow \frac{1}{r^2}\frac{d}{dr}\left(\frac{r^2}{\rho}\frac{dP}{dr}\right) = -4\pi G \rho \tag{82}$$

Suppose the equation of state takes the polytropic form

$$P = K\rho^\gamma \tag{83}$$

where K is a constant and $\gamma = (n+1)/n$. n is said to be the polytropic index. Then Eq. (82) becomes Poisson's equation

$$\frac{1}{r^2}\frac{d}{dr}\left(\frac{r^2}{\gamma}K\rho^{\gamma-1}\frac{d\rho}{dr}\right) = -4\pi G\rho \tag{84}$$

with boundary conditions $\rho = \rho_c$ and $d\rho/dr = 0$ at $r = 0$.

The resulting solutions are called polytropes, and are encountered in connection with:

1. adiabatic changes ($\gamma = c_P/c_V$) applicable to stars in adiabatic convective equilibrium.
2. equation of state of degenerate matter $\gamma = 5/3$ (NR), $\gamma = 4/3$ (UR), applicable in white dwarfs.
3. models where P_r/P =constant.

By writing

$$y = (\rho/\rho_c)^{-n}$$

and

$$x = \left(\frac{4\pi G \rho_c^{(2-\gamma)}}{(n+1)K} \right)^{1/2} r,$$

Eq. (84) may be rewritten in the more convenient form:

$$\frac{1}{x^2} \frac{d}{dx} \left(x^2 \frac{dy}{dx} \right) + y^n = 0 \tag{85}$$

which is known as the Lane-Emden equation, with boundary conditions $y(0) = 1$ and $(dy/dx)_{x=0} = 0$.

Polytropic models have played a very important role in the development of stellar structure theory, but although popular before the computer age, still involve the solution of quite messy differential equations.

10.2 Clayton Models

A simpler approach to solving the structure equations is to guess a form for the density profile $\rho(r)$ and to use this as a starting point for an approximate solution. The equations can be tackled sequentially, again starting with Eqs. (14) and (17) to obtain the pressure structure. An equation of state provides the temperature structure. Introducing an opacity $\kappa(\rho, T)$, Eq. (33) can be used to obtain the power flow, which can be compared with the power flow obtained from the nuclear power density $\varepsilon(\rho, T)$ and Eq. (28). For approximate solutions, these last will bear little similarity.

Pressure

Clayton (1986) proposed a simple parameterisation of the pressure profile P. The pressure gradient is directly constrained by Eq. (17) (hydrostatic equilibrium):

$$\frac{dP}{dr} = -\frac{Gm\rho}{r^2}.$$

Near the centre, r is small and $m \approx 4\pi r^3 \rho_c/3$, hence

$$\frac{dP}{dr} = -\frac{4\pi}{3} G \rho_c^2 r.$$

Near the surface $m \approx M$, hence

$$\frac{dP}{dr} = -\frac{GM\rho}{r^2}.$$

Thus the pressure gradient must be zero at the centre, initially varying linearly with radius but again approaching zero as ρ/r^2 decreases towards the surface. The essence of the Clayton model is to guess a simple form for $P(r)$ which takes these into account:

$$\frac{dP}{dr} = -\frac{4\pi}{3} G\rho_c^2 r \, \exp\left(-r^2/a^2\right) \tag{86}$$

where a is a length parameter to be specified. Integrating, with $P(R) = 0$, gives

$$P = \frac{2\pi}{3} G\rho_c^2 a^2 \left[\exp\left(-r^2/a^2\right) - \exp\left(-R^2/a^2\right)\right] \tag{87}$$

defining a family of stellar models parameterised by ρ_c, a and R. By definition the minimum in dP/dr occurs at a distance $r = a/\sqrt{2}$ from the centre of the star.

Temperature and Density

Corresponding expressions for temperature and density can then be obtained. From hydrostatic equilibrium

$$Gmdm = -4\pi r^4 dP$$

and integrating:

$$\frac{Gm^2}{2} = -4\pi \int_0^r r'^4 \frac{dP}{dr'} dr'.$$

Substituting for the pressure gradient:

$$m = (4/3)\pi a^3 \rho_c \Phi(x) \tag{88}$$

where $x = r/a$ and

$$\Phi^2(x) = 6 \int_0^x x'^5 \exp\left(-x'^2\right) dx' = 6 - 3\left(x^4 + 2x^2 + 2\right) \exp\left(-x^2\right). \tag{89}$$

Then the density can be found directly (mass equation)

$$\rho = \frac{1}{4\pi r^2} \frac{dm}{dr} = \rho_c \left[\frac{x^3 \exp\left(-x^2\right)}{\Phi}\right] \tag{90}$$

and from the equation of state for an ideal classical gas (Eq. (41))

$$T = \frac{P\mu m_{\mathrm{u}}}{\rho k} \tag{91}$$

where $\mu = 2/(1 + 3X + Y/2)$. At small r, these quantities will be reasonably reliable.

The exponential term in Eq. (90) may be expanded as a power series to obtain approximate expressions valid for small r:

$$\rho = \rho_c \left(1 - \frac{5r^2}{8a^2} + \ldots\right),$$

$$T = T_c \left(1 - \frac{3r^2}{8a^2} + \ldots\right).$$

High ρ_c Models

The model is further simplified when $\rho_c \gg \langle\rho\rangle$. Then $a \ll R$ and terms proportional to $\exp(-a^2/R^2)$ can be neglected. This is reasonable when applied to the Sun, where $a = R_\odot/5.4$. With small a, Eq. (88) gives

$$M = m(R) = \frac{4\pi\rho_c a^3}{3}\Phi(R/a) \approx \frac{4\pi\rho_c a^3\sqrt{6}}{3} \tag{92}$$

and the average density is about $\sqrt{6}(a/R)^3\rho_c$.

It can then be shown that the density at $r = a$ is $0.53\rho_c$ and that a sphere of radius a contains $0.28M$. Substituting $r = 0$ into Eq. (87),

$$P_c \approx \frac{2\pi}{3}G\rho_c^2 a^2$$

and eliminating a with Eq. (92) gives

$$P_c \approx (\pi/36)^{1/3} G M^{2/3} \rho_c^{4/3} \approx 0.44 G M^{2/3} \rho_c^{4/3} \tag{93}$$

which is expected to be valid for any homogeneous star where the mass is concentrated towards the centre. Other simple models (*e.g.* polytropes) give similar results.

Modelling the Sun

The simple model described above is characterised by ρ_c, a and R. Equation (92) provides a in terms of M and ρ_c. If we take $M = \mathrm{M}_\odot$, $R = \mathrm{R}_\odot$, and $\rho_c = 9 \times 10^4\,\mathrm{kg\,m^{-3}}$ from a standard solar model, then $a = \mathrm{R}_\odot/5.4$. Equations (87), (90) and (91) then provide the variation in pressure, temperature and density inside the Sun (Fig. 22). The variation of these solutions with mass can be investigated quite simply.

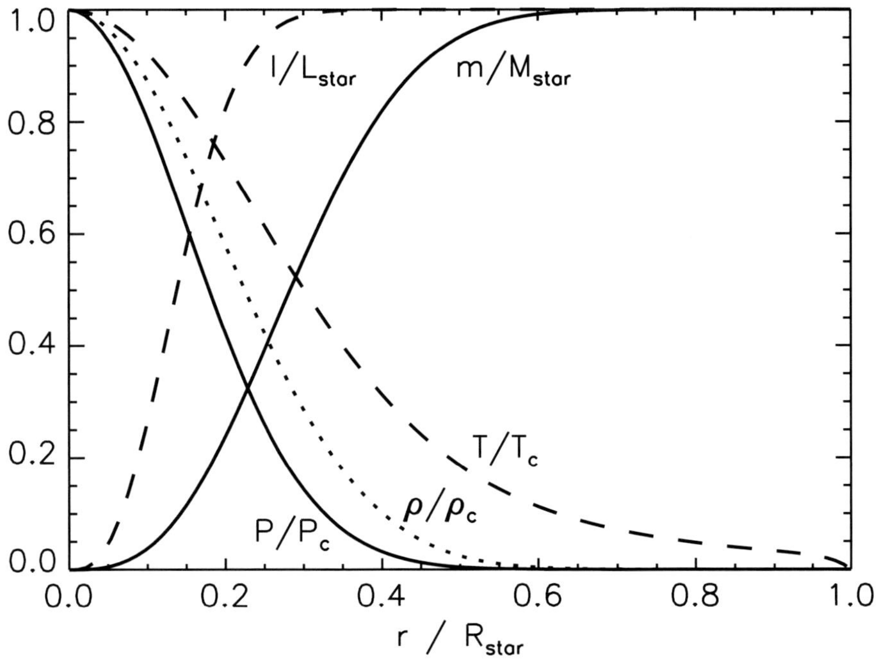

Fig. 22. Distribution of P, m, ρ, T, and l with r in the Sun as given by the approximations (87), (88), (90), and (91) and by integrating Eq. (28) with Eqs. (77) and (78)

10.3 Minimum Mass of a Star

We have found expressions for the central temperature of a star. This must be greater than the ignition temperature for the hydrogen fusion if a star is to result. For a perfect gas, the pressure given by Eq. (93) must be the same as that given by the equation of state. Equating the pressures gives

$$kT_{\rm c} \approx (\pi/36)^{1/3} G\mu m_{\rm u} M^{2/3} \rho_{\rm c}^{1/3}$$

Since $T_{\rm c} \propto \rho_c^{1/3}$, the temperature of a contracting gas cloud continues to rise as it contracts until either nuclear reactions are initiated or electron degeneracy prevents further contraction. The condition for stardom is that the density at which electron degeneracy prevents further contraction corresponds to a temperature greater than that required for the ignition of nuclear reactions.

An estimate of this critical density/temperature is obtained by assuming that the electrons are degenerate and the ions are classical and, to simplify the algebra, that the star is entirely composed of hydrogen. The central pressure is given by

$$P_{\rm c} = K_{\rm NR} n_{\rm e}^{5/3} + n_{\rm i} kT_{\rm c}$$

where $n_{\rm e} = n_{\rm i} = \rho_{\rm c}/m_{\rm u}$

$$P_c = K_{NR}(\rho_c/m_u)^{5/3} + (\rho_c/m_u)kT_c$$

This must satisfy hydrostatic equilibrium (Eq. (93)) and so, eliminating P_c,

$$kT_c \approx (\pi/36)^{1/3} G m_u M^{2/3} \rho_c^{1/3} - K_{NR}(\rho_c/m_u)^{2/3}$$

which has the form

$$kT_c = A\rho_c^{1/3} - B\rho_c^{2/3},$$

where kT_c reaches a maximum of $A^2/4B$ at a density of $(A/2B)^3$. Substituting, we find

$$(kT_c)_{max} \approx (\pi/36)^{1/2} G^2 M^{4/3} m_u^{8/3}/4K_{NR}$$

and

$$M_{min} \approx (36/\pi)^{1/2}(4K_{NR}/G^2 m_u^{8/3})^{3/4}(kT_{ign})^{3/4}$$

For $T_{ign} \approx 1.5 \times 10^6$ K, $M_{min} \approx 0.1\,M_\odot$.

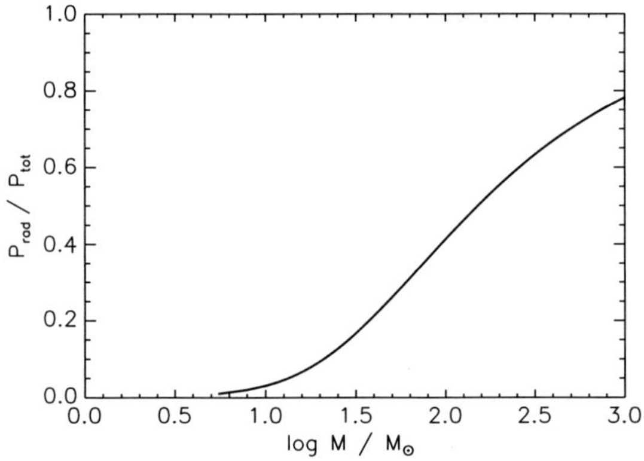

Fig. 23. The ratio of mean radiation to total pressure as a function of mass

10.4 Maximum Mass of a Star

Pressure in the stellar core is the sum of gas pressure $P_g = \rho_c kT_c/m_u\mu$ and radiation pressure $P_r = aT_c^4/3$. Denoting $P_g = \beta P_c$ and $P_r = (1-\beta)P_c$, T can be eliminated to give

$$P_c = \left[\frac{3}{a}\frac{(1-\beta)}{\beta^4}\right]^{1/3}\left[\frac{k\rho_c}{m_u\mu}\right]^{4/3} \tag{94}$$

Equating this to the pressure from Eq. (93) to satisfy hydrostatic equilibrium yields

$$\left[\frac{\pi}{36}\right]^{1/3} GM^{2/3} = \left[\frac{3}{a}\frac{(1-\beta)}{\beta^4}\right]^{1/3}\left[\frac{k}{m_u\mu}\right]^{4/3} \tag{95}$$

This equation can be used to plot (1–β) (the radiation pressure fraction) as a function of mass (Fig. 23). When this quantity is more than ≈ 0.5, radiation pressure leads to strong stellar winds and dramatic mass loss from the stellar surface.

11 Methods for Numerical Solution

Recall that the problem is to find the structure of a star with given mass M and chemical composition $x_i(m), i = 1, 2, \ldots$ by solving the system of equations described in Sect. 5, together with microphysics such as that described in Sects. 7–9.

In general, there is no analytic solution although approximate solutions can be found. In practice, it is necessary to find a numerical solution. If all four boundary conditions were given, say, at the centre, this would require a straightforward numerical integration from the centre to the surface. This is obviously not the case. There are two methods commonly used to deal with the problem.

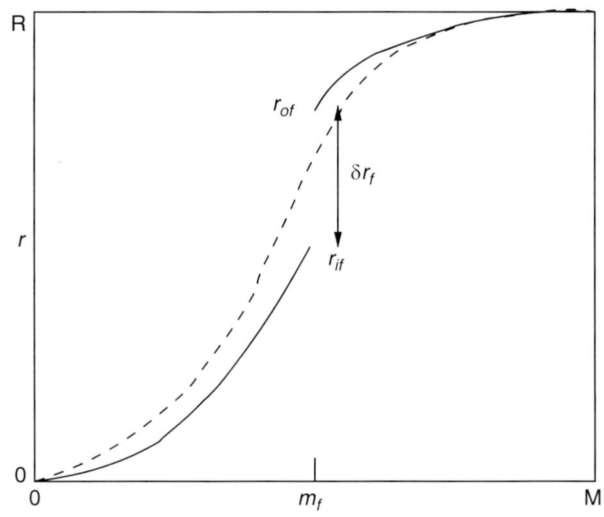

Fig. 24. Schematic of solution for one variable using the shooting method. The trial solution is shown as a solid line, the desired solution as a dashed line

11.1 Shooting Method

The principle of this method divides the star into two parts, an inner and outer part, for which separate solutions will be obtained by estimating a set

of additional boundary conditions. These are then adjusted until they match one another. The procedure works as follows.
Inward solution. At $m = M$ we have $P_\mathrm{s} = 0$, $T_\mathrm{s} = T_\mathrm{eff}$[8]. Estimate R and L. Integrate inwards to the same point m_f, where we obtain P_if, T_if, l_if, and r_if.
Outward solution. At $m = 0$ we have $r_\mathrm{c} = 0$, $l_\mathrm{c} = 0$. Estimate P_c and T_c. Integrate outwards to some fitting mass m_f, where we obtain P_of, T_of, l_of and r_of.

For any arbitrary starting values of $R, L, P_\mathrm{c}, T_\mathrm{c}$ we have, in general, $\delta P_\mathrm{f} = P_\mathrm{if} - P_\mathrm{of} \neq 0$, $\delta T_\mathrm{f} = T_\mathrm{if} - T_\mathrm{of} \neq 0$, $\delta l_\mathrm{f} = l_\mathrm{if} - l_\mathrm{of} \neq 0$, and $\delta r_\mathrm{f} = r_\mathrm{if} - r_\mathrm{of} \neq 0$. The goal is therefore to improve these starting values iteratively until $\delta P_\mathrm{f} = \delta T_\mathrm{f} = \delta l_\mathrm{f} = \delta r_\mathrm{f} = 0$.

Repeat the inward solution with $R + \delta R$, L and again with R, $L + \delta L$, and repeat the outward solution with $P_\mathrm{c} + \delta P_\mathrm{c}$, T_c, and with P_c, $T_\mathrm{c} + \delta T_\mathrm{c}$.

From the resulting changes in r_{if}, r_{of}, etc., form the derivatives $\partial \delta r_\mathrm{f} / \partial R$, $\partial \delta r_\mathrm{f} / \partial L$, $\partial \delta r_\mathrm{f} / \partial P_\mathrm{c}$, etc. Then corrections to R, L, P_c, T_c, may be found by solving

$$\begin{pmatrix} \frac{\partial \delta r_f}{\partial R} & \frac{\partial \delta r_f}{\partial L} & \frac{\partial \delta r_f}{\partial P_c} & \frac{\partial \delta r_f}{\partial T_c} \\ \frac{\partial \delta l_f}{\partial R} & \frac{\partial \delta l_f}{\partial L} & \frac{\partial \delta l_f}{\partial P_c} & \frac{\partial \delta l_f}{\partial T_c} \\ \frac{\partial \delta p_f}{\partial R} & \frac{\partial \delta p_f}{\partial L} & \frac{\partial \delta p_f}{\partial P_c} & \frac{\partial \delta p_f}{\partial T_c} \\ \frac{\partial \delta t_f}{\partial R} & \frac{\partial \delta t_f}{\partial L} & \frac{\partial \delta t_f}{\partial P_c} & \frac{\partial \delta t_f}{\partial T_c} \end{pmatrix} \begin{pmatrix} \delta R \\ \delta L \\ \delta P_c \\ \delta T_c \end{pmatrix} = \begin{pmatrix} \delta r_f \\ \delta l_f \\ \delta p_f \\ \delta t_f \end{pmatrix} \tag{96}$$

for δR, δL, δP_c, and δT_c. These corrections are applied and new estimates are used to repeat the entire process, iterating until convergence is obtained, *i.e.* by satisfying some criteria such as $\delta P_f < \varepsilon$, $\delta T_f < \varepsilon$, $\delta l_f < \varepsilon$, and $\delta r_f < \varepsilon$.

The result is a set of values for R, L, P_c, T_c and solutions for r, l, P, and T on the interval $m = [0, M]$.

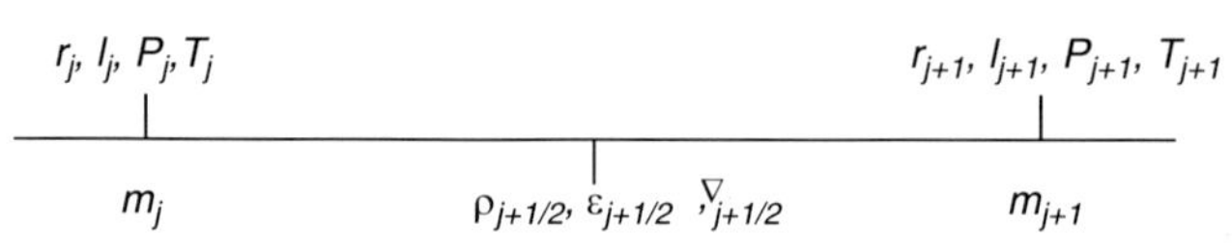

Fig. 25. Discrete variables used in difference equations

11.2 Difference Method

An alternative approach can be obtained by writing the o.d.es for stellar structure as difference equations. To do this, consider the star divided into J discrete shells, numbered $0, \ldots, j, \ldots, J$ from the center to the surface. The mass interior to shell j is denoted m_j, with $m_0 = 0$ and $m_J = M$ (Fig. 25). Replacing other continuous variables by the discrete variables r_j, l_j, P_j, T_j, we

[8] In practice it is better to use a non-zero starting condition such as $P_\mathrm{s} = (-2/3)g/\kappa$

$$
\mathbf{A}=\begin{pmatrix}
\frac{\partial B_1}{\partial r_1} & \frac{\partial B_2}{\partial r_1} & \cdots & \cdots & \cdots & \cdots & \cdots & \cdots & \cdots & \cdots & \cdots & 0 \\
\frac{\partial B_1}{\partial l_1} & \frac{\partial B_2}{\partial l_1} & \cdots & \cdots & \cdots & \cdots & \cdots & \cdots & \cdots & \cdots & \cdots & \vdots \\
\vdots & & \ddots & & & & & & 0 & & & \vdots \\
0 & & & \frac{\partial G_{4,j-1}}{\partial T_{j-1}} & \frac{\partial G_{1,j-1}}{\partial T_j} & \frac{\partial G_{2,j-1}}{\partial T_j} & \frac{\partial G_{3,j-1}}{\partial T_j} & \frac{\partial G_{4,j-1}}{\partial T_j} & 0 & 0 & & \vdots \\
0 & & 0 & 0 & \frac{\partial G_{1,j}}{\partial r_j} & \frac{\partial G_{2,j}}{\partial r_j} & \frac{\partial G_{3,j}}{\partial r_j} & \frac{\partial G_{4,j}}{\partial r_j} & \frac{\partial G_{1,j}}{\partial r_{j+1}} & & & \vdots \\
\vdots & & & 0 & \frac{\partial G_{1,j}}{\partial l_j} & \frac{\partial G_{2,j}}{\partial l_j} & \frac{\partial G_{3,j}}{\partial l_j} & \frac{\partial G_{4,j}}{\partial l_j} & \frac{\partial G_{1,j}}{\partial l_{j+1}} & & & \vdots \\
\vdots & & & 0 & \frac{\partial G_{1,j}}{\partial P_j} & \frac{\partial G_{2,j}}{\partial P_j} & \frac{\partial G_{3,j}}{\partial P_j} & \frac{\partial G_{4,j}}{\partial P_j} & \frac{\partial G_{1,j}}{\partial P_{j+1}} & & & \vdots \\
\vdots & & & 0 & \frac{\partial G_{1,j}}{\partial T_j} & \frac{\partial G_{2,j}}{\partial T_j} & \frac{\partial G_{3,j}}{\partial T_j} & \frac{\partial G_{4,j}}{\partial T_j} & \frac{\partial G_{1,j}}{\partial T_{j+1}} & & & \vdots \\
\vdots & & & 0 & 0 & 0 & 0 & 0 & \frac{\partial G_{1,j+1}}{\partial T_{j+1}} & & & \vdots \\
\vdots & & & & & & & 0 & & \ddots & & \vdots \\
\vdots & & & & & & & & & \cdots & \frac{\partial C_3}{\partial P_J} & \frac{\partial C_4}{\partial P_J} \\
0 & \cdots & \cdots & \cdots & \cdots & \cdots & \cdots & \cdots & \cdots & \cdots & \frac{\partial C_3}{\partial T_J} & \frac{\partial C_4}{\partial T_J}
\end{pmatrix}
$$

$$
\mathbf{x}=\begin{pmatrix} \delta r_1 \\ \delta l_1 \\ \vdots \\ \delta T_{j-1} \\ \delta r_j \\ \delta l_j \\ \delta P_j \\ \delta T_j \\ \delta r_{j+1} \\ \vdots \\ \delta P_J \\ \delta T_J \end{pmatrix} \quad \text{and} \quad \mathbf{b}=\begin{pmatrix} -B_1 \\ -B_2 \\ \vdots \\ -G_{4,j-1} \\ -G_{1,j} \\ -G_{2,j} \\ -G_{3,j} \\ -G_{4,j} \\ -G_{1,j+1} \\ \vdots \\ -C_3 \\ -C_4 \end{pmatrix}
$$

Fig. 26. Block structure for difference equations

arrive at a system containing $4(J+1)$ variables. Since r_0, l_0, P_J and T_J are given by the boundary conditions, we have $4J$ unknowns.

The ordinary differential equations can be discretized and Eq. (14) can be written for example, as:

$$
\frac{r_{j+1}-r_j}{m_{j+1}-m_j} = \left(\frac{1}{4\pi r^2 \rho}\right)_{j+1/2} .
$$

Since there are 4 such equations for J shells, we have a system of $4J$ non-linear simultaneous equations in $4J$ unknowns. Equation (14) can be rewritten in the form:

$$G_{1,j} \equiv (r_{j+1} - r_j) - (m_{j+1} - m_j)\frac{1}{4\pi r^2_{j+1/2}\rho_{j+1/2}} = 0; quad j = 1, \ldots, J-1.$$

Likewise Eqs. (17), (28) and (33) become $G_{2,j}, G_{3,j}, G_{4,j} = 0$, boundary conditions (34)–(37) become $B_1 = 0$, $B_2 = 0$, $C_3 = 0$, $C_4 = 0$. Hence we have a system of non-linear equations of the form:

$$E_k(y_1, y_2, \ldots, y_{4J}) = 0; \quad k = 1, 4J. \tag{97}$$

Supposing that there exists an approximate solution:

$$y_1^{(0)}, y_2^{(0)}, \ldots, y_{4J}^{(0)},$$

then the system can be linearized using, for example, a Newton-Raphson iteration. In general, an approximate solution $y_i^{(0)}$ will not satisfy the system Eq. (97). The goal is therefore to find a set of corrections δy_i such that

$$y_i^{(s+1)} = y_i^{(s)} + \delta y_i; \quad i = 1, 4J,$$

by solving the block-diagonal system of equations

$$\sum_{i=1}^{4J} \left(\frac{\partial E_k}{\partial y_i}\right)^{(s)} \delta y_i = -E^{(s)}; \quad k = 1, 4J,$$

$$\equiv \quad \mathbf{A} \cdot \mathbf{x} = \mathbf{b}$$

where $\mathbf{A}$, $\mathbf{x}$ and $\mathbf{b}$ are matrices (Fig. 26).

This system may be solved using a substitution scheme known as the Henyey method [26], the corrections are applied to the approximate solution, and then iterated to convergence [27].

12 Stellar Evolution

12.1 Pre-Main-Sequence Evolution

Before discussing the structure and evolution of main-sequence stars, it is helpful to make some remarks about how stars reach the main sequence in the first place.

It is well established observationally that stars form in massive interstellar dust clouds. If the Jeans condition is satisfied (cf. Eq. (25)), then parts of the cloud will start to collapse, leading to the release of gravitational energy. Initially, the star will be very luminous but cool, and may be identified with "Hot Cores" in giant molecular cloud complexes. Low-mass protostars are fully convective and the family of solutions lies on the left side of the HR diagram on a path known as the Hayashi track. The contracting protostar evolves down this track, heating in the process.

Once the core temperature is high enough (hydrogen and helium must be fully ionized) the opacity ($\kappa_{\rm bf}+\kappa_{\rm es}$) fall sufficiently for a radiative core to form. At this point the star contracts at constant L towards the main-sequence, and the convective envelope shrinks. By this stage, the stars have mostly emerged from their dusty cocoon and become visible, low-mass stars as T Tauri stars and intermediate mass stars as Herbig Ae/Be stars.

As contraction continues, the core continues to heat until $T_{\rm c}$ exceeds the critical value for the ignition of deuterium burning (the second reaction in the p-p I chain). Providing the star is sufficiently massive, this will be followed by the p-p reaction and/or the CNO-cycle and full hydrogen burning will be established. At this point the star has reached the "zero-age main sequence" (ZAMS).

12.2 The Zero-Age Main Sequence

The location of stars along the ZAMS is a function of mass and composition. For radiation dominated stars

$$L \propto M^3/\kappa.$$

The mass-luminosity relation averaged over the whole main sequence (Sect. 2.3) is empirically

$$L \propto M^{3.5},$$

but local variations reflect changes in the dominant physics.

At low mass, $M/{\rm M}_\odot < 0.4$, the star is fully convective due to the high opacity of the incompletely ionized gas, and

$$L \propto 0.23 M^{2.3}.$$

With increasing mass $0.4 < M/{\rm M}_\odot < 7$, the star first develops a radiative core. By about $1.4\,{\rm M}_\odot$, the (opacity-driven) convective envelope has disappeared. Depending on metallicity, a nuclear-driven convective core will develop. In this interval,

$$L \propto M^{4.75}.$$

For $M/{\rm M}_\odot > 7$, the star maintains a radiative envelope and increasingly massive convective core. For the most massive stars, the convective core can comprise over 50% of the mass of the star. In this mass range,

$$L \propto M^3.$$

Nuclear Energy

The principal nuclear energy generation also depends on the mass and composition. For Population I ZAMS stars ($Z \approx 0.02$), energy produced by the p-p chain and CNO cycle is approximately equal for $M \approx 1.9 M_{\odot}$. An increasing contribution from the CNO cycle will develop during main-sequence burning as the central hydrogen concentration drops.

For Population II stars ($Z < 0.001$) the p-p chain is generally dominant at low masses ($M < 2 M_{\odot}$), though the CNO-cycle will continue to be important at high masses.

Convection

Convection occurs in the core when the CNO cycle contributes more than $\approx 20\%$ of the total energy output, since $\varepsilon_{\rm CN} \approx T^{17}$ implies that the temperature gradient in the core is very steep. The fact that stars in the $1 - 1.4\,M_{\odot}$ range can develop CNO burning during main-sequence evolution also implies that such stars (including the Sun) may develop a convective core.

The significance of a convective core is twofold. In the first place, convection mixes proton-rich material from a region much larger than that which is actually burning, thus increasing the total fuel available and extending the potential main-sequence lifetime by a comparable amount.

Second, the shape of the evolution track is altered. In a star with a radiative core nuclear burning shifts smoothly from the hydrogen-depleted core to a thick hydrogen-poor shell and then to a thin hydrogen-rich shell. The transition from main-sequence to post main-sequence evolution is comparatively smooth (Fig. 8). In a star with a convective core, nuclear burning switches off abruptly as the entire convective core is exhausted. This precipitates a minor collapse of the depleted core before ignition of the hydrogen shell, and produces a "hook" in the evolution track at the end of main-sequence evolution.

Convective envelopes are found when the hydrogen ionization zone is sufficiently deep inside the star which occurs for $T_{\rm eff} < 8\,000\,{\rm K}$ and $M < 1.4\,M_{\odot}$. The mass in the convective envelope increases with decreasing M until the entire star becomes convective at $M \leq 0.2 - 0.3\,M_{\odot}$.

Opacity

The composition of a main-sequence stars has a direct effect on both the luminosity and effective temperature by changing the mean molecular weight and the opacity. Since the effects are not linear, we state simply that, as a rough guide, the core opacity controls the luminosity (by modifying the temperature gradient and hence the rate at which energy leaves the star) and the envelope opacity controls the radius (or $T_{\rm eff}$).

We can deduce what will happen to the star during main-sequence evolution by considering the changes in chemical composition in the core. As hydrogen is consumed, X decreases and Y increases. This has two consequences. κ is reduced (Eqs. (53), (57)) and μ is increased.

From hydrostatic equilibrium, increasing μ requires that ρ_c must increase. From the Virial theorem it then follows that T_c must also increase, and this leads to an increase in ε_c and L. So a main-sequence star will become more luminous as it depletes its hydrogen core.

Distribution of Mass and Luminosity

The nuclear energy generation, the energy transport and the opacity also have consequences for the distribution of mass $m(r)$ and luminosity $l(m)$ within the star. The first indicates how centrally condensed the star is. One measure is the fraction of the total mass located within 50% of the radius (or 12% of the volume). Stars with small convective cores are more centrally condensed than those with radiative cores so that a $2\,M_\odot$ star has 95% of its total mass inside this volume. For a $0.4\,M_\odot$ star, the figure is 60%, and for a $10\,M_\odot$ star, it is 90%.

The temperature exponent of the energy production rate determines where the power is generated. Consequently the CNO-cycle generate 90% of the total luminosity within the innermost 15% of the mass. With a "softer" temperature exponent, p-p burning extends this figure up to about 30%.

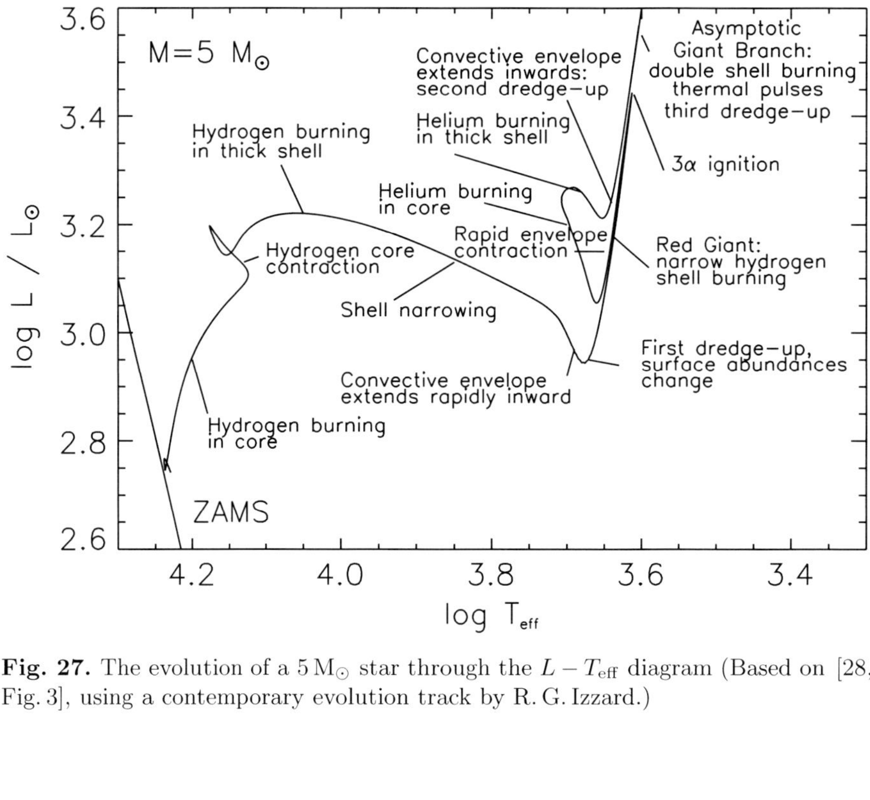

Fig. 27. The evolution of a $5\,M_\odot$ star through the $L - T_{\text{eff}}$ diagram (Based on [28, Fig. 3], using a contemporary evolution track by R. G. Izzard.)

12.3 Evolution of a $5\,M_\odot$ Star

Having examined the variation in stellar *structure* along the main sequence, we turn our attention to the *evolution* of a star of given mass ([29], Fig. 27). We have chosen a mass of $5\,M_\odot$ because it is relatively simple, avoiding the high mass-loss associated with more massive stars and the convective envelope associated with low-mass main-sequence stars.

During the main hydrogen-core burning phase, hydrogen is converted to helium via the CNO-cycle. Consequently the star has a convective core, but this decreases in mass from 20% to 8% of the total mass during main-sequence burning. At the same time, the luminosity increases as hydrogen is depleted (see above) all on a nuclear timescale $t_{\rm nuc} \approx 6 \times 10^7$ years.

When the core hydrogen abundance drops to $X_{\rm c} < 0.5$, any further increases in $T_{\rm c}$ fail to compensate for the drop in energy generation and the whole star starts to contract. As nuclear reactions are extinguished, the convective core vanishes, but only when $X_{\rm c} \ll 0.01$. Much of the star's luminosity now comes from the release of gravitational energy, with the core contracting on a thermal timescale $t_{\rm K} \approx 2 \times 10^6$ years.

When $X_{\rm c} \approx 0$, approximately the tip of the "hook" in Fig. 27, the finite temperature gradient through the star removes energy from the core faster than $\varepsilon_{\rm nuc}$ so that the core is forced to contract and cool. Since the greatest cooling is at the centre, the core becomes isothermal on a timescale $t \approx 10^5$ years. The mass of the, now, helium core is $\approx 7\%$ of the total mass. Outside this core, hydrogen-rich material continues to fall inwards, heats, and eventually ignites nuclear reactions. Once ignited, heat from the thick hydrogen-burning shell forces the outer layers of the star to expand.

The base of the hydrogen-burning shell now adds helium to the helium core, which increases in mass. When the core mass $M_{\rm c} \approx 0.1 M_\star$ (the Schönberg-Chandrasekhar limit[9]) the core contracts very rapidly. This contraction and consequent heating feeds back directly into the hydrogen-burning shell, so that the luminosity of the shell ($L_{\rm sh}$) increases rapidly. This does not result in an immediate increase in L (the stellar luminosity), the stellar envelope absorbs much of the excess energy, heats and expands, so that R increases rapidly. Actually, this expansion absorbs so much energy that L falls until the star starts to become convective and take over as the dominant transport process in the envelope. As a consequence, the star crosses from the main sequence to the giant branch very quickly ($\approx 10^6$ years), producing the Hertzsprung gap in HR diagrams for young open clusters.

With the effective temperature now less than 5 000 K, the surface layers become increasingly convective. Whilst at the base of the *giant branch*, the

[9] The Schönberg-Chandrasekhar limit [30] represents a limit above which an isothermal non-degenerate core can support the remainder of the star. Its value is $\approx 0.1 - 0.15 M_\star$ and is valid for stars with total mass between 2 and $6\,M_\odot$. Less massive stars develop an electron-degenerate core and more massive stars ignite helium before this value is reached.

surface convection zone extends deep enough to reach material which has been partially CNO-cycled. In a process known as first dredge-up, nitrogen and helium-rich material dilutes the composition of the previously pristine surface. Deeper in the star the hydrogen-burning shell becomes thinner, hotter and continues to add mass to the core. As the M_c increases, so does L_{sh} and R. Since at $M = 0.5\,M_\odot$, the core is not degenerate, it contracts and heats until, when $T_c \approx 10^8\,K$, 3α reactions are ignited.

Following core-helium ignition, the core expands (R_{sh} increases), the hydrogen-burning shell cool, and both L_{sh} and L drop, as the star approaches a long-lived phase of core-helium burning. With a smaller total radius, the effective temperature also increases and the surface layers become fully radiative during this phase. However, the high temperature exponent of the 3α reaction means that a convection zone develops in the core. In contrast to the CNO-cycle driven convection zone, the convective core mass *increases* during most of core helium-burning, so that by the time $Y_c = 0$, there is relatively massive helium-poor layer surrounding a central carbon-oxygen core.

At this point, helium-burning shifts initially to a thick shell surrounding the inert inner core, causing the outer (helium) core to contract and the hydrogen-burning shell to move inwards and heat. In a re-run of the approach to the giant branch, the increasing power of the hydrogen-burning shell forces the star to expand, cool, and develop a deep convective envelope for the second time, again dredging freshly-processed helium to the surface. Helium shell-burning turns out to be unstable; as the star evolves up the *asymptotic giant branch*, the dominant energy source continues to be the hydrogen shell, with a luminosity which goes [31, 32] roughly as:

$$L_{sh} \approx 10^4 M_c^\delta; \qquad \delta \approx 2, M_c \approx 0.86.$$

As the mass of intershell zone (*i.e.* the helium layer between the hydrogen shell and the carbon-oxygen core) increases to around $0.004\,M_\odot$, the temperature at its base increases to the point when 3α reactions are re-ignited. There follows a complex interaction in which a 3α-driven convection zone is driven upwards into the intershell, the intershell expands and suppresses the hydrogen-burning shell. Briefly, the helium-burning shell adds about $0.001\,M_\odot$ carbon ash to the core beneath, but is itself extinguished after less than 5 y. This allows the intershell to contract, the hydrogen-shell to heat, surface convection to dredge-up new material from the top of the intershell, and the hydrogen-shell to start building up the mass of the intershell until after an interval of some 5000 y, this sequence of *thermal pulses* is repeated.

With increasing luminosity, the star which started with a mass of some $5\,M_\odot$ will lose most of its mass via a stellar wind. Just as the surface layers are being removed above, the hydrogen-envelope is being burnt at an ever-increasing rate from beneath until there is insufficient mass to sustain the hydrogen-burning shell. With a mass of some $0.9\,M_\odot$, electron-degeneracy in the carbon-oxygen core (Sect. 13.1) will prevent it contracting and heating to the point at which nuclear carbon burning becomes possible (Sect. 9.3). At this point the star will effectively die, contracting to become a white dwarf (Sect. 15.6).

12.4 Evolution at Other Masses

The parameter which defines the evolution of a star more than any other is mass. Clearly (Sect. 4), the more massive a star, the more luminous and the more quickly it evolves. There is not enough space here to discuss the individual evolution of stars up and down the main-sequence and at different metallicities. However there are some important classifications to make.

Very high-mass stars with initial masses greater than $\approx 20\,\mathrm{M}_\odot$ will ignite helium ignition while the star is still a blue supergiant. Their extremely high luminosities lead to a radiation-driven stellar wind and substantial mass-loss. The result is likely a Wolf-Rayet star, followed by a core-collapse supernova.

High-mass stars have a final mass greater than the Chandrasekhar mass (Sect. 13.1). In these stars, nuclear burning of carbon, oxygen and heavier elements will be followed by collapse as a type II supernova (Sect. 13.2). This is generally thought to occur for stars with initial masses greater than $\approx 8\,\mathrm{M}_\odot$.

Intermediate-mass stars with initial masses roughly between 3 and $8\,\mathrm{M}_\odot$ ignite helium in a non-degenerate core, thus avoiding both the helium core flash and the core-collapse supernova fate. These stars will all evolve to the asymptotic-giant branch phase and subsequently to become CO white dwarfs.

In *low-mass stars* with initial masses less than about $3\,\mathrm{M}_\odot$, helium-ignition occurs in a degenerate core – usually when the helium-core mass has reached about $0.48\,\mathrm{M}_\odot$. The star subsequently evolves as a horizontal-branch star (Sect. 14). With a sufficiently massive hydrogen envelope (the exact value depends strongly on metallicity), it will evolve up the asymptotic giant branch after core helium exhaustion. Otherwise it will evolve directly to the white-dwarf cooling track.

Very low-mass stars with initial masses below about $0.5\,\mathrm{M}_\odot$ may never ignite helium. Being nearly fully convective, they have a large reservoir of nuclear fuel compared with their luminosity, so their nuclear lifetimes are long compared with the age of the Universe. Ultimately they will fade away as degenerate helium white dwarfs.

13 Stellar Remnants

13.1 White Dwarfs

When a star runs out of nuclear fuel it contracts until either the temperature becomes high enough to ignite another nuclear fuel source or until electron-degeneracy provides sufficient pressure to restore hydrostatic equilibrium. Recall the hydrostatic equilibrium and mass continuity equations:

$$dP/dr = -\rho G m/r^2, \qquad dm/dr = 4\pi r^2 \rho. \tag{98}$$

In the completely degenerate regime, ρ is independent of T, $P = K_{\rm NR}(\rho/m_{\rm u}\mu_{\rm e})^{5/3}$ and we have two o.d.e.s with two b.c.s (cf. the Lane-Emden equation). Hence hydrostatic equilibrium alone determines R and ρ as a function of mass.

Write average values in terms of the central pressure $P_{\rm c}$, M and R:

$$\langle dP/dr \rangle = -P_{\rm c}/R,$$

$$\langle \rho \rangle = \frac{3}{4\pi} M/R^3 \tag{99}$$

and

$$\langle m/r^2 \rangle \approx M/R^2.$$

Then, from Eq. (98),

$$\langle dP/dr \rangle = -\frac{P_{\rm c}}{R} = -\frac{3}{4\pi}\frac{M}{R^3}\frac{GM}{R^2}$$

$$\Rightarrow P_{\rm c} = \frac{3}{4\pi}\frac{GM^2}{R^4}, \tag{100}$$

and substituting $\langle \rho \rangle$ into the equation of state:

$$P_{\rm e} = K_{\rm NR}\left(\frac{\rho}{m_{\rm u}\mu_{\rm e}}\right)^{5/3} = \frac{M^{5/3}}{R^5} K_{\rm NR}\left(\frac{3}{4\pi}\frac{1}{m_{\rm u}\mu_{\rm e}}\right)^{5/3}. \tag{101}$$

Eliminating P between Eqs. (100) and (101), and rearranging for R:

$$R = \frac{K_{\rm NR}}{G m_{\rm u}^{5/3}}\left(\frac{3}{4\pi}\right)^{2/3} M^{-1/3}\mu_{\rm e}^{-5/3} \approx 1400(M/{\rm M}_\odot)^{-1/3}(\mu_{\rm e}/2)^{-5/3}\,{\rm km}. \tag{102}$$

For increasing mass, the white dwarf radius decreases and density increases.

Chandrasekhar Mass Limit

As the mass of a white dwarf is increased, so the density increases and the non-relativistic approximation will break down ($M > 0.2{\rm M}_\odot$). Eventually the equation of state must be replaced by the ultra-relativistic version $P_{\rm e} \propto (\rho/\mu_{\rm e})^{4/3}$. If we consider the case where the two equations of state give the same pressure, we can solve for the density where this transition occurs. Equating pressures from the UR equation of state and Eq. (100):

$$P_{\rm e} = K_{\rm UR}\left(\frac{\rho}{m_{\rm u}\mu_{\rm e}}\right)^{4/3} = \frac{3}{4\pi}\frac{GM^2}{R^4}$$

Expressing the density as above (Eq. (99)), then

$$K_{\rm UR}\left(\frac{1}{m_{\rm u}\mu_{\rm e}}\right)^{4/3}\left(\frac{3}{4\pi}\frac{M}{R^3}\right)^{4/3} = \frac{3}{4\pi}\frac{GM^2}{R^4}$$

$$\Rightarrow K_{\rm UR}\left(\frac{1}{m_{\rm u}\mu_{\rm e}}\right)^{4/3}\left(\frac{3}{4\pi}\right)^{1/3}\frac{M^{4/3}}{R^4} = \frac{GM^2}{R^4}$$

from which R cancels and hence the central pressure becomes indeterminate. Thus a white dwarf can be supported in hydrostatic equilibrium entirely by pressure due to degenerate electrons *provided* those electrons are not fully relativistic. On exceeding the relativistic limit, the radius and density become indeterminate and the star cannot support itself. This occurs when the mass exceeds the Chandrasekhar limit

$$M > M_{\rm Ch} \equiv \left(\frac{K_{\rm UR}}{G}\right)^{3/2}\left(\frac{3}{4\pi}\right)^{1/2}\left(\frac{1}{m_{\rm u}\mu_{\rm e}}\right)^2 \approx 1.4\left(\frac{\mu_{\rm e}}{2}\right)^{-2}{\rm M}_\odot. \quad (103)$$

13.2 Type II Supernovae

When the mass of a star exceeds $M_{\rm Ch}$, and following exhaustion of all available nuclear fuel, electron-degeneracy cannot provide sufficient pressure to maintain hydrostatic equilibrium and the star will continue to contract. In the core, some electrons will combine with heavy nuclei, since they must find vacant energy states. However the main result is that the temperature will rise (Virial theorem!). Several outcomes are possible.

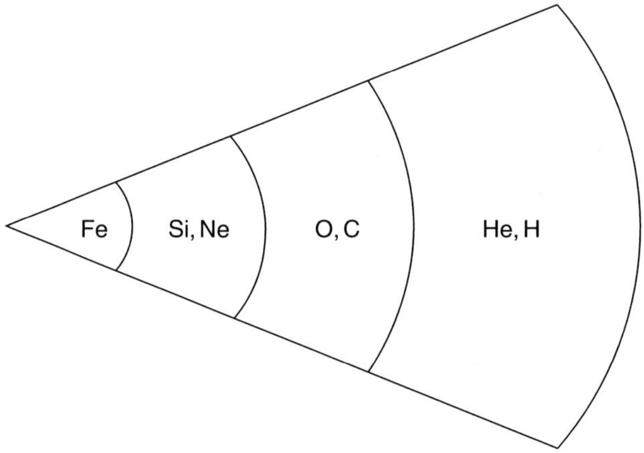

Fig. 28. Schematic of chemical layering in a massive star immediately prior to core collapse

Observations show that stars explode as supernovae, releasing $\approx 10^{43}$ J in a very short time. This requires the detonation of $\approx 0.1\,{\rm M}_\odot$ of material via proton captures onto light elements at temperatures of $\approx 10^9$ K. This can *only* occur if material rich in the necessary material collapses catastrophically.

Consider a star with fairly massive ($> 3M_\odot$) CO core. Gravitational contraction raises T so that reactions such as

$$2\,^{16}\mathrm{O} \rightarrow^{28} \mathrm{Si} +^4 \mathrm{He} \Rightarrow 2\,^{28}\mathrm{Si} \rightarrow^{56} \mathrm{Fe}$$

may occur, creating a star with approximately the chemical structure illustrated in Fig. 28.

Conditions for explosive burning can only occur if the core collapses at a speed $\gg v_\mathrm{s}$, the local sound speed. In a star with an iron core, $v_s \approx 10^7 \mathrm{m\,s^{-1}}$, which implies a collapse timescale ≤ 3 s – essentially a dynamical timescale. Any slower and the star can reorganize itself without an explosion, so a lot of energy ($\approx 10^{14}\,\mathrm{J\,kg^{-1}}$) must be removed rapidly.

A number of processes, possibly, acting in series may provide this very efficient refrigeration. The first process (Fig. 29) [33] involves the photodissociation of iron:

$$^{56}\mathrm{Fe} + \gamma(2.22\mathrm{MeV}) \rightarrow 12\,^4\mathrm{He} + 4\mathrm{n},$$

absorbing $\approx 2 \times 10^{14}\,\mathrm{J\,kg^{-1}}$. As T_c rises and exceeds a critical value ($\approx 2 \times 10^9$ K), photo-dissociation occurs, absorbing the photo-dissociation energy and temporarily halting the T_c increase until photo-dissociation is complete. It is possible that this process alone could trigger a catastrophic collapse of the core, followed by rapid infall, heating and ignition of proton-rich envelope material. The resulting explosion could correspond to a type II supernova outburst.

Additional processes occur when temperatures exceed 10^9 K which contribute to the physics of a supernova explosion. First, when $T_\mathrm{c} \approx 2 \times 10^9$ K, photodissociation of ^{4}He occurs

$$^4\mathrm{He} + \gamma(26\mathrm{MeV}) \rightarrow 2\mathrm{p} + 2\mathrm{n},$$

and a similar refrigeration process is repeated ($\approx 3 \times 10^{16}\,\mathrm{J\,kg^{-1}}$).

By the time, $T_\mathrm{c} \approx 5 \times 10^8$ K, the neutrino luminosity is similar in magnitude to the photon luminosity. At higher temperatures the URCA process is very efficient:

$$\mathrm{e}^+ + \mathrm{n} \rightarrow \mathrm{p} + \bar{\nu}(+10\mathrm{MeV}),$$

$$\mathrm{e}^- + \mathrm{p} \rightarrow \mathrm{n} + \nu(+10\mathrm{MeV}).$$

The neutrons and protons freed by ^{4}He dissociation will dramatically increase the neutrino flux – each neutrino removing 10 MeV from the core and thus providing a second very efficient refrigeration and contraction agent.

Subsequently, when $T_\mathrm{c} \approx 4 \times 10^{10}$ K and $\rho_\mathrm{c} \approx 3 \times 10^{14}\,\mathrm{kg\,m^{-3}}$, the neutrino opacity ($\kappa_\nu$) becomes non-zero. Neutrinos start to be reabsorbed (the reverse URCA process) and the temperature will start to increase very sharply.

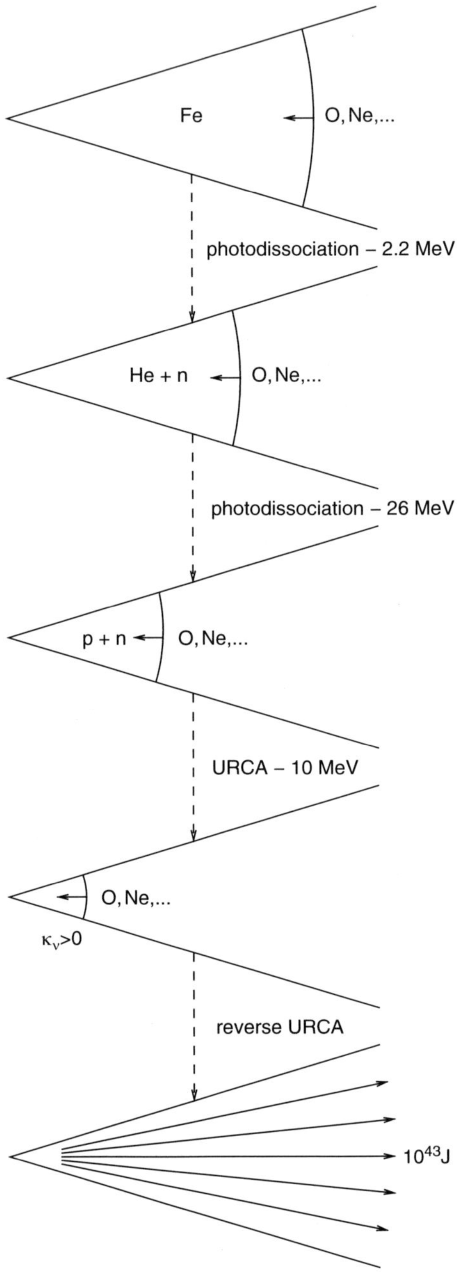

Fig. 29. Illustration of a possible sequence of processes during core collapse leading to a type II supernova explosion. See text for more details

Finally, when $T_c \approx 10^{11}$ K and $\rho_c \approx 10^{16}\,\mathrm{kg\,m^{-3}}$, the contractions are halted by the material becoming neutron-degenerate and incompressible. By now, the outer layers of the star are effectively in free fall, and converting gravitational potential into kinetic energy. As this infalling material collides with the hard incompressible core, a mixture of light elements and protons is rapidly heated to $T \approx 10^{11}$ K, resulting in a runaway thermonuclear detonation.

It is evident that the omission of rotation (in particular) from this simple picture oversimplifies the true physics of a supernova explosion, but our description demonstrates how the timescale for the sequence must be extremely short, ≈ 0.1 s for the entire core collapse and ignition. It also demonstrates how much power can be generated by the explosion, and that most (99%) of it is carried by *neutrinos* rather than photons.

13.3 Neutron Stars

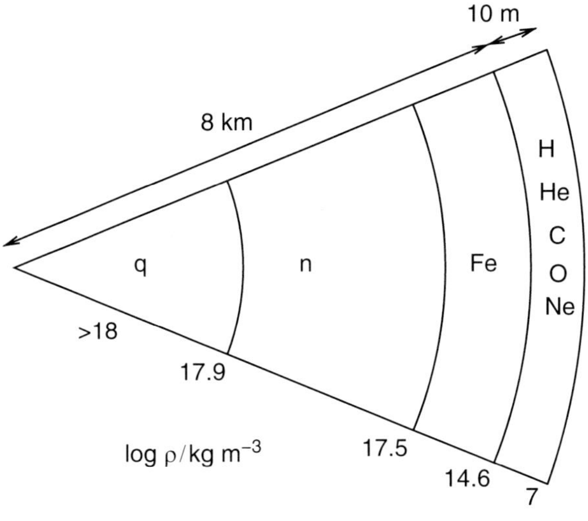

Fig. 30. Schematic structure of a neutron star, showing core made up of quarks and hyperons inside a degenerate neutron layer inside an iron-rich mantle. Approximate scales and densities are also shown

The debris from a type II supernova explosion includes ejecta rich in heavy elements which come from reaction of light elements with the mantle of the iron core and a residual star which is made of neutron-rich material (Fig. 30). A neutron star is believed to have a dense crust made of normal matter, but including many heavy elements. At deeper layers, more neutron-rich (and normally unstable) nuclei are expected and below that, layers containing free electrons and free neutrons. In the core itself, neutrons and possibly more exotic elementary particles including hyperons and free quarks will dominate. The equation of state for neutron-degenerate matter in stars is not yet known, so that the theoretical radius for a 1.5 $M_\odot$ neutron star could be between 10 and 15 km.

14 Horizontal Branch Stars

We have demonstrated that there are two long-lived phases of nuclear-driven stellar evolution. On the main-sequence, stars burn hydrogen in their cores. Core-helium burning manifests in more diverse ways, depending on the mass of the star. In the most massive stars, core helium ignition occurs while the star is still a blue supergiant; the result is likely a Wolf-Rayet star. For stars in the 8 – 20 $M_\odot$ range, core helium ignition occurs as the star is crossing the HR diagram, and will produce a yellow supergiant. After completing helium-burning, such a star will start to burn carbon, then silicon and ultimately become a supernova.

Stars less massive than about 8 $M_\odot$ will avoid the supernova fate. In the 3 – 8 $M_\odot$ range, helium ignition occurs in a non-degenerate core, so is not explosive. Core helium burning is associated with a blueward loop through the Cepheid instability strip, after which the star develops a double shell structure and becomes an asymptotic giant branch star.

The helium flash

At lower masses ($< 2M_\odot$), a star develops an electron-degenerate helium core as it ascends the giant branch. This core does not increase in temperature as the pressure increases ($P \propto \rho^{5/3}$) so will not ignite helium so easily. The hydrogen-burning shell adds processed helium to the core, and itself get progressively hotter and thinner as the mass of the remaining hydrogen envelope is reduced. In principle, the core should be isothermal with the hydrogen-burning shell ($\approx 10^8$ K). In practice, pair-production of neutrinos in the dense e^--degenerate core removes heat and keeps the center of the core cooler than the outer regions.

When the hottest part of the core reaches a critical temperature ($\approx 10^8$ K), a 3α reaction becomes increasingly likely. This occurs when the core mass reaches about 0.48 $M_\odot$. Energy is liberated and provides immediate local heating. Since the local equation of state is insensitive to temperature, runaway heating occurs and sustained helium-burning reactions begin. These stabilise once the temperature has risen sufficiently to lift the local degeneracy and the helium-burning flame has propagated into the centre of the star.

The details of the helium flash are still poorly understood. It is clear that very short timescales are involved, and that nuclear-driven convection mixes some layers of the star. Realistic 3D models of the helium-flash are now becoming possible [34]. It is not clear how much of the star is mixed, or whether the stellar surface reflects the violence of the helium ignition in any way. It seems unlikely since the initially degenerate core has a very high heat capacity; helium burning will force the core to expand, and this will quench (but not extinguish) the hydrogen-burning shell, so the overall star will contract.

Since the mass of the core at helium ignition depends little on prior conditions, the dimensions of the star during stable core helium burning are primarily a product of the mass ($M_{\rm H}$) and composition ($Y_{\rm e}$, $Z_{\rm e}$) of the hydrogen envelope. The envelope composition should reflect the initial composition (Y, Z), but dredge-up of either helium or nitrogen during first giant branch evolution could be significant.

Low-mass helium-core burning stars are identified with horizontal branch (HB) stars in globular clusters.

14.1 Horizontal Branch Stars in Clusters and the Field

In the HR diagram for Hipparcos stars (Fig. 1), the giant branch will be seen to contain a dense clump of stars centered at $B - V = 1, M_{\rm bol} = 2$. Clump stars are essentially low-mass helium-core burning stars belonging primarily to Population I, *i.e.* they are metal-rich and have significant hydrogen envelopes.

In metal-poor environments, such as globular clusters and field stars at high galactic latitudes, these stars appear on a horizontal strip extending blueward of the giant branch. There is a clear correlation between position on this horizontal branch and cluster metallicity. Compare, for example, the horizontal branches in ω Cen ([Fe/H]= −1.62: Fig. 5) and 47 Tuc ([Fe/H]= −0.76: Fig. 4).

The Second-Parameter Problem

While the location of stars on the HB is normally a function of the cluster metallicity (Figs. 4 & 5), this is not always true as Figs. 31 & 32 show. Other cluster pairs with similar metallicity but quite different horizontal-branch morphologies include M2 and M3, both with metallicity [Fe/H]= −1.6, and Arp 2 and Rupprecht 106 with [Fe/H]= −1.7.

There must be another factor at work in these cases – sometimes known as the second-parameter problem for globular clusters. The resolution may have something to do with cluster age, and hence the degree of mixing that stars currently on the HB experienced during first dredge-up, since this correlates with total mass of the star leaving the main-sequence. Other suggestions include cluster density, which may influence the average rotation rate, the degree of mixing and the helium core mass.

14.2 Theoretical Models for Horizontal Branch Stars

Since HB stars are in a long-lived phase of evolution, the time- dependent terms in the stellar structure equations are negligible and equilibrium solutions

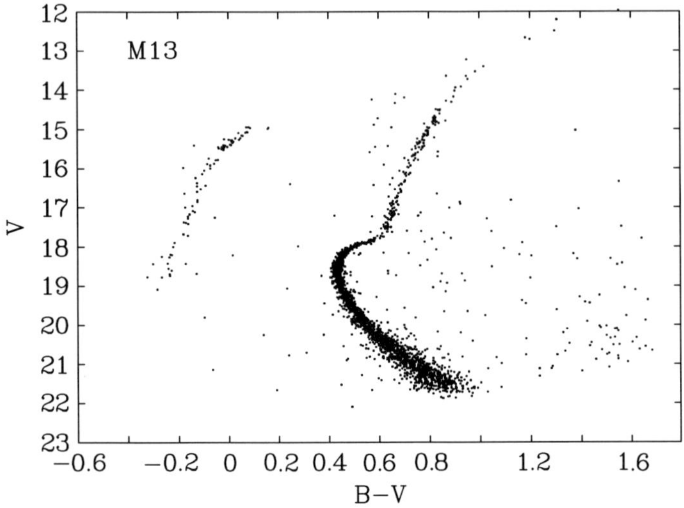

Fig. 31. Colour-magnitude diagram for the metal-poor galactic globular cluster M13 ($Z = 0.0007$, [Fe/H] $= -1.54$; based on [35])

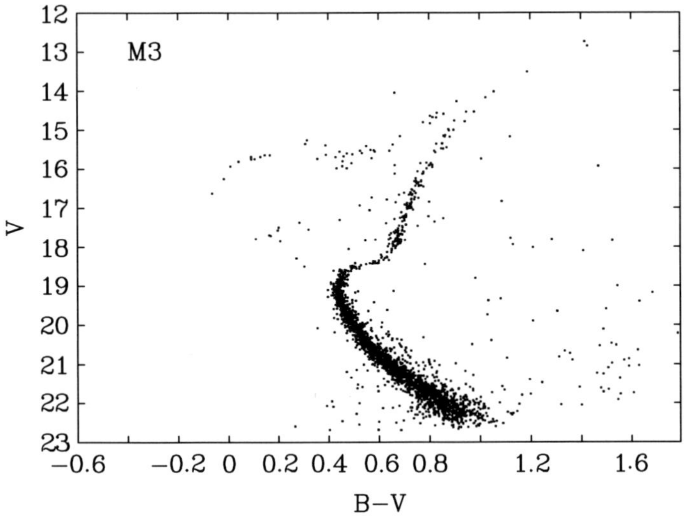

Fig. 32. Colour-magnitude diagram for the metal-poor galactic globular cluster M3, its red horizontal-branch illustrating the 2nd parameter problem ($Z = 0.0007$, [Fe/H] $= -1.57$; based on [35]

provide an excellent approximation to the overall structure. Moreover, the chemical structure of a zero-age HB star is quite simple, having a hydrogen-rich envelope of mass $M_{\rm H}$ surrounding a helium-rich core of mass $M_{\rm c}$. If it is further assumed that both core and envelope are homogeneous with $Y_{\rm c} = 1-Z$ and $Y_{\rm e} = Y, Z_{\rm e} = Z$ (Y, Z being the initial composition), then a sequence of models may be computed for a given $M_{\rm c}, Y, Z$ and a range of $M_{\rm H}$.

In practice, it is necessary to consider how the envelope may have been modified by convective dredge-up while the star was on the giant branch. Thus $Y_{\rm e} > Y$ though dredge-up of helium, and the relative CNO concentrations within Z may have been altered. Both should correlate with $M_{\rm i}$, the initial mass of the star. Another consideration is the extent and location of core carbon enrichment $C_{\rm c}$ during the core helium flash. An estimate of this can be derived by computing the carbon required to provide the energy required to lift the core electron degeneracy, and the position of the temperature maximum immediately prior to core helium ignition. For models with very small $M_{\rm H}$ it is necessary to consider carefully the composition across the core/envelope boundary. For large $M_{\rm H}$, this should initially correspond to the profile of the H-burning shell in the precursor red giant.

In summary, HB models are defined by:
$M_{\rm c}$: Helium core mass,
M : Total mass (or envelope mass $M_{\rm H} = M - M_{\rm c}$),
$Y_{\rm e}$: Envelope helium abundance,
$Z_{\rm e}$: Envelope metallicity, and
$C_{\rm c}$: Core carbon abundance and distribution

Models of HB stars as a function of $M_{\rm H}$ not surprisingly form a roughly horizontal sequence in the $L - T_{\rm eff}$ diagram. The helium core provides a luminosity of some $L \propto M_{\rm c}^{\beta}$ largely irrespective of $M_{\rm H}$[10].

At low mass, the hydrogen envelope is essentially inert, simply providing a high opacity envelope which makes the star larger than it would be otherwise. Once $M_{\rm H}$ exceeds some threshold, the base of the envelope becomes sufficiently hot to sustain hydrogen burning by the CNO cycle, prompting both an increase in radius *and* luminosity. Since increasing radius means decreasing $T_{\rm eff}$, the models become cooler until they merge with the Giant Branch.

Models of HB stars have been computed by many researchers, including [10]. The definitive set in common usage [36] demonstrate the locus of the HB as a function of both $M_{\rm c}$ and $Y_{\rm e}$.

The MS – HB mass discrepancy

If the metallicity of an HB star is known, and a reasonable estimate of the helium abundance can be made, then the luminosity and effective temperature allow a realistic deduction of the total mass. In most globular clusters, this gives answers of $\approx 0.6{\rm M}_{\odot}$. In the same clusters it is also possible to estimate

[10] In fact, setting $M_{\rm H} = 0$ leads us to the helium main sequence, which is equivalent to (and roughly parallel with) the hydrogen main sequence.

the mass of stars at the Turn-Off point, *i.e.*, those stars which are just ending their main-sequence evolution. Invariably, these turn-off stars are more massive than the HB stars in the same cluster. Since the current HB stars must have left the MS before the current epoch and therefore have been more massive than the current turn-off stars, substantial mass must be lost either while the star is on the giant-branch or at helium ignition.

The discovery of substantial stellar winds in red giants appears to account for most of the discrepancy. In some cases, these winds appear able to remove almost all of the envelope hydrogen, though theory suggests that the hydrogen shell will be quenched and the star will evolve to become a helium white dwarf if the envelope mass drops below $M_{\rm H} \leq 0.01{\rm M}_{\odot}$ before the core ignites helium [37,38,39].

14.3 Evolution of Horizontal Branch Stars

For classical HB stars containing a strong hydrogen-burning shell, initial evolution is defined by a balance between He-core and H-shell. As core helium drops and the mean molecular weight increases, the core temperature rises to maintain hydrostatic equilibrium, so the helium core becomes more luminous and expands. The core expansion forces the hydrogen-shell outwards, so its temperature and luminosity initially drop, and the overall luminosity stays roughly constant. However, the total radius (and hence effective temperature) drops because of the weakening of the H-burning shell, so the star evolves horizontally to the blue in the HR diagram. Towards the end of core helium-burning, this process reverses, the core-expansion is reduced and power returns to the H-burning shell, causing the star to evolve back to the red and to higher luminosity. Complete core helium exhaustion is followed by core contraction and hydrogen-shell thinning which moves the star towards the asymptotic giant branch. As what is now a carbon-oxygen core contracts it becomes electron-degenerate. The increasingly powerful H-burning shell adds helium to the layer between the CO core and the H-envelope (the intershell). When sufficient helium has accumulated in the intershell, it ignites in an unstable process (*i.e.* nearly explosively) which leads to intershell expansion, thermal pulses and other interesting phenomena (see Karakas: these proceedings).

Semiconvection

Core-helium burning is always associated with a nuclear-driven convective core. Because the temperature-sensitivity of the 3-α process is so steep, the mass of the convective core increases with falling central helium abundance. Assuming conventional convection theory the outer edge of the convective core, with a molecular weight increasing due to its increasing carbon abundance, will drive into unburnt helium with a lower mean molecular weight. This will create a sharp discontinuity in molecular weight at the convective-core boundary. While the upward buoyancy of a convective element in the carbon-rich core approaches equilibrium close to the boundary, it would remain unstable were it to penetrate above the boundary. This has led to considerable discussion about how to treat convection (or semiconvection) at the horizontal-branch convective-core boundary. The overall result of including some form of convective core extension is to increase the extent of the convective core and hence the lifetimes of horizontal branch stars by some 20%.

14.4 Extreme Horizontal Branch Stars

Extreme horizontal-branch (EHB) stars are horizontal-branch stars with a hydrogen envelope whose mass is insufficient to sustain hydrogen burning;*i.e.* the H-burning shell is inert. They can be thought of as helium main-sequence stars with a very thin hydrogen envelope ($0.0001 \ll M_{\rm H}/{\rm M}_\odot \ll 0.01$).

The evolution of an EHB star is dominated by that of the core; the very thin H-envelope plays a passive rôle. Thus the total luminosity increases during evolution and the star follows a nearly vertical track in the HR diagram.

Following core-helium exhaustion, there being no H-burning shell, the mass of the helium-rich intershell remains fixed. A few mild helium-shell ignitions may occur[11], but the star is ultimately doomed not to reach the asymptotic giant branch. Instead it will contract directly to become a hybrid He/CO white dwarf.

Faint Blue Stars in the Galactic Disk

The observant reader will realise that the existence of an EHB star with $M_{\rm H} \ll 0.01{\rm M}_\odot$ is inconsistent with the minimum hydrogen mass anticipated assuming conventional mass loss from a red giant ($M_{\rm H} \geq 0.01{\rm M}_\odot$). Thus the existence of real EHB stars would appear to be impossible assuming standard single-star evolution.

On the other hand, EHB stars have been identified with a very important class of stars known as subdwarf B (sdB) stars. These were originally identified as faint blue stars in the galactic field. Not being associated with spiral arms or OB associations, they had to be older than typical O and B stars. Measurement of their surface gravities showed them to lie below the upper main-sequence, and it was realised that they must be low-mass stars, and identified with the extreme blue end of the horizontal branch [40]. Their distribution in the Galactic plane is consistent with a relatively old population – they occur up to high galactic latitudes and most probably belong to the thick disk.

[11] the existence of so-called "breathing pulses" towards the end of horizontal-branch evolution may be a numerical artefact due to incomplete treatment of core convection

Extreme Horizontal Branch Stars in Clusters

Extremely-blue horizontal-branch stars were subsequently discovered in a number of globular clusters, notably NGC 6752 [41, 42], NGC 2808 [43] and ω Cen [44, 45]. Notably more difficult to execute than comparable studies in the field, high-resolution spectroscopy demonstrated that cluster EHB stars and field sdB stars are comparable. The advantage of cluster stars is that they provide a sample at the same distance and, ostensibly, of the same age.

Both field and cluster sdB stars are predominantly hydrogen-rich. That is, their surface helium abundances have been suppressed below typical values for B stars by factors of 10 – 100. This can be explained by the high effective temperature and surface gravity of these stars, which makes chemical diffusion an important transport process, mediated by gravitational settling and radiative levitation. In the case of helium, gravity normally wins and it sinks out of sight beneath the hydrogen. However, there also exists a small fraction ($\leq 5\%$) of sdB (and subdwarf O) stars with very high overabundances of helium, such as can only occur if nuclear-processed material is exposed on the stellar surface. These He-rich stars may be more prevalent in some globular clusters than in the field (ω Cen: [46]).

14.5 The Origin of EHB Stars

EHB and sdB stars pose challenges for stellar evolution theory in how to account (a) for their very low hydrogen-envelope masses and (b) for the range of surface helium abundances. Specifically, what mechanism removes almost the entire hydrogen-envelope from a red-giant star? How does it still achieve helium ignition? And how should we explain the existence of helium-rich sdB stars?

sdB Binaries

Some answers to these questions were provided by discoveries that many sdB stars are members of binary systems. Observationally, three (or four) distinct groups are identified [47]:
I. sdB stars with single spectra showing no radial velocity changes (30-35%),
II. sdB stars with composite spectra (usually sdB+G) showing small or no velocity variations. Orbital periods are thought to range from tens of days to several years (15-20%),
III. sdB stars with single spectra showing large velocity variations with periods of hours to days (45-50%).
The last group may be further divided into:
IIIA. those having a white dwarf or other compact companion,
IIIB. those having a low-mass main-sequence star companion.
The presence of a second star and a short-period orbit in group III require a prior interaction between a red giant and its companion in order to explain

why the present-day orbit is smaller than the radius of the giant that birthed the sdB star. Assuming that the existence (or otherwise) of a companion contributes to the formation of the sdB star, it does appear that nature can make the latter in at least four different ways.

Formation Channels

The problem of whether an EHB star can ignite helium at all has been addressed [39, 48, 49]. Models of red giants were computed which include an enhanced mass-loss rate near to the tip of the giant branch, *i.e.* when the core mass is slightly less than the critical value for 3-α ignition. In these models, the hydrogen shell is extinguished and the star starts to evolve to the blue, *i.e.* contraction towards the white dwarf sequence. Depending on conditions when the star left the giant branch, these models show an off-center ignition of helium either during blueward evolution or after the star reaches the white dwarf cooling track. This off-center ignition creates a helium-burning shell which raises the local electron degeneracy and burns its way into the core via a series of mild helium shell flashes. Meanwhile the star evolves through a series of loops to a position on the extreme horizontal branch. The overall conclusion is that if a star reaches within 0.25 magnitudes of the RGB tip, helium ignition *will* occur. The final location on the zero-age horizontal branch only depends on the hydrogen envelope mass.

The question raised by this result is what mechanism can produce an enhanced mass-loss rate close to the red-giant tip? Two have been proposed that involve binary star evolution [50, 51].

Common Envelope Ejection

In the first binary-star formation channel, the sdB progenitor is a giant that fills its Roche lobe (the gravitational equipotential surface surrounding both stars that meets at a single point of contact between them). If it does this near the tip of the giant branch, and if mass ratio of the progenitor mass to that of its companion is larger than ≈ 1.2, then there will be mass transfer on a dynamical timescale (Eq. (6)) from the more massive star. The reason is that as mass is transferred, the mass ratio changes and (in this case) the orbit will shrink. The ratio of the massive star radius to the Roche lobe radius and the rate of mass transfer will both increase in a runaway process. Meanwhile the secondary star cannot assimilate this new material quickly enough, which will fill up and spill over the Roche surface to form a *common envelope* around *both* stars.

With two stars orbiting one another inside this common envelope, friction between the stars and envelope, which will not be rotating at the same rate, will cause the two stars to spiral towards each other. Orbital energy will be converted to heat until sufficient has been lost to eject the envelope [52]. The result is a much closer binary (orbital period typically between 0.1 and 10 d) consisting of the red-giant core and the secondary. If the core ignites helium (see above) it can evolve to become an sdB/EHB star.

The secondary star may be either a main-sequence star or a white dwarf. In the first case, the sdB star was originally the more massive star in the binary system, since it has evolved to reach the GB first. The less massive star must have $M \leq 0.4\mathrm{M}_\odot$ in order for mass transfer to be unstable, and is therefore likely an M-type star.

If the secondary is a white dwarf, then the sdB star must originally have been the less massive star in the binary system. A previous phase of stable mass transfer must have transferred the envelope of the original primary to the secondary, allowing the core to become a helium white dwarf and reversing the mass ratio in the binary system. The common envelope phase occurs when the new primary evolves to become a red giant, and the resulting system contains an sdB star and a white dwarf.

One challenge encountered by this scenario is that the mass predicted for white dwarf companions in short-period sdB binaries should be $< 0.5\mathrm{M}_\odot$. Recent observations have demonstrated several sdB+WD binary systems in which substantially higher masses are required for the unseen companion. In some cases these masses exceed the Chandrasekhar mass limit for white dwarfs (M_{Ch}: Sect. 13) and are a continuing challenge for stellar evolution theory [53].

Stable Roche Lobe Overflow

In the case that the mass ratio is ≤ 1.2, a similar binary to that considered above will experience *stable* mass transfer [54]. With a lower initial mass ratio, the exchange of angular momentum will lead the binary to widen and the rate of transfer to remain low enough for matter to be assimilated comfortably by the companion. The result will be an sdB star with a main-sequence or sub-giant companion of increased mass and an orbital period of tens to hundreds of days.

White Dwarf Mergers

An alternative channel for the formation of sdB stars considers the ultimate fate of binary systems which are so evolved as to comprise two white dwarfs in a close binary orbit. This predicates a binary system which has passed through several stages of mass transfer (whether common-envelope or stable Roche-lobe overflow) to leave these remnants.

In such a long-lived short-period binary, the emission of gravitational waves can remove orbital energy from the system so that over a long time the orbit will shrink until the stars are in contact. The merger timescale is given by

$$t_{\mathrm{merge}} = 10^7 (P/\mathrm{h})^{8/3} \mu^{-1} (M/\mathrm{M}_\odot)^{-2/3}\,\mathrm{yr} \tag{104}$$

where P is the orbital period (in hours) and μ is the reduced mass of the binary [55, Sect. 110], [56]. It will be seen that binaries with periods $\leq 2\,\mathrm{h}$ can potentially come into contact within a realistic time frame.

Upon contact, the less massive white dwarf will have a larger radius (Sect. 13) and will fill its Roche lobe first. Upon transferring a small amount of mass, the white dwarf radius will increase at a rate faster than the binary orbit will widen, and at a rate much faster than matter can be accreted by the companion. This leads to a runaway (or dynamical) disruption of the less massive white dwarf which will form a hot disk in Keplerian orbit around the other more massive white dwarf. Simulations of the process show that the disruption of the white dwarf will take only a few orbits, $P_{\rm orb} \approx 3\,{\rm m}$ at the time of contact [57, 58, 59].

What happens next is a matter of some conjecture. Heating of the disk may initiate some prompt nucleosynthesis [59, 60]. Viscosity in the dense disk will drive angular momentum to the outer edge of the disk, and mass will therefore fall inwards [61]. The result will be a high entropy envelope surrounding the relatively unperturbed massive white dwarf. The envelope will cool and material at its base will be assimilated onto the surface of the underlying core. The next step depends on the initial white dwarfs.

When the mass of the helium-layer accreted into the low-entropy core is sufficiently high, helium ignition will occur at the boundary between the non-degenerate and degenerate material. The star will expand on the short thermal timescale of the envelope ($\approx 10^3\,{\rm y}$) to become a yellow supergiant – a stage which will be described in more detail in Sect. 15.7.

If both white dwarfs were helium white dwarfs [62], the helium-burning layer will have unprocessed helium beneath. Heating from the nuclear reactions will lift the degeneracy of the top layers of the helium core and allow the helium-burning shell to burn into the center in a series of mild pulses. At this point the star should resemble an extreme horizontal branch star. Although the surface layers will be predominantly composed of helium from the disrupted white dwarf, there should be sufficient hydrogen to diffuse to the surface (helium will sink) to form the helium-poor atmosphere usually observed in sdB stars. Of course, this channel will always lead to the formation of a *single* EHB star.

Other formation channels

Recently, another channel has been proposed for forming single sdB stars [63] following the merger of a red giant with a lower-mass main-sequence or brown-dwarf companion during a common-envelope phase of evolution. This route avoids the surface composition problem, but depends on an epoch of enhanced mass loss following spin-up by the merger. In this respect it bears some similarity to the single-star evolution channel, and provides a mechanism for the enhanced mass-loss near the tip of the giant branch. However, the expected contribution is too small to explain all observed single sdB stars.

Another merger model suggests that some post-common-envelope sdB binaries could share the white dwarf merger fate [64]. The idea is that the post-sdB star, which is a hybrid white dwarf containing a partially-burnt CO core with an unburnt but degenerate helium layer, merges with its less-massive helium white dwarf companion. The new star is a helium shell-burning star, but the CO core is of sufficiently low mass that the star's evolution and radius is more strongly determined by the total mass of helium. Both merger routes might equally account for the significant number of hydrogen-poor sdB and sdO stars discussed at greater length in Sect. 15, and which are also located around the extreme blue end of the horizontal branch [65, 66].

Work has also focused on mixing processes associated with violent burning events, including extra mixing at convective boundaries, diffusive mixing and mass loss. For example, it has been proposed that deep mixing driven by a He-flash convection zone can engulf a H-rich envelope and lead to the formation of He-rich sdO stars [67, 68].

14.6 Extreme Horizontal Branch Stars in other Galaxies

The significance of EHB stars in a cosmological context may be recognised from resolved observations of nearby stellar systems and from the integrated flux distributions of more distant systems [69]. Significant numbers of EHB stars in globular clusters, including ω Cen and NGC 6752, have been noted. Meanwhile the nearby spiral galaxy M31 shows a large EHB population very similar to that of ω Cen.

Like globular clusters, giant elliptical galaxies are old star systems which should principally comprise low-mass main-sequence stars, red giants and white dwarfs, and no young blue massive stars; they are known as "old red dead" galaxies. Therefore it was a surprise to discover several of these galaxies show an upturn in the flux distribution at wavelengths shorter than 2000 Å [70, 71]. By creating various models for the population of these galaxies, it has been demonstrated (as originally conjectured) that this ultraviolet excess is due to a substantial population of EHB stars [72, 73].

Such results are significant for studies of stellar evolution. EHB stars are an important diagnostic of age in old populations – it should be possible to use the ultraviolet upturn as an age discriminator for distant galaxies. However, this will depend on a clear understanding of the formation channels for EHB stars at different epochs and metallicities in galactic evolution. Population synthesis calculations which predict relative numbers of stars and predict emergent spectra for galaxies of different ages and metallicities must adopt the best possible physics to achieve this [74].

15 Late Stages of Stellar Evolution: Hydrogen-Deficient Stars

As the section on the evolution of horizontal-branch stars has shown, the story of stellar evolution is not a simple progression from formation to death as a white dwarf or supernova. The late stages of evolution are frequently marked by stars which exhibit remarkable properties – one of the most significant being a near-complete absence of surface hydrogen [11].

The science of hydrogen-deficient stars began when Williamina Fleming noted that "the spectrum of υ Sgr is remarkable since the hydrogen lines are very faint and of the same intensity as the additional dark lines" [75]. Fifteen years later observed Hγ was found to be completely absent from the spectrum of R CrB [76].

These early observations pointed to something unusual in the spectra of a variety of stars, but there was reluctance to accept (or even suggest) that hydrogen might be deficient [77,78,79]. It was not until the 1930's that quantitative spectral analyses forced the conclusion that in R CrB [80] and υ Sgr [81,82] "somehow, a very substantial amount of hydrogen had been lost".

Forty years ago, very few hydrogen-deficient stars were known [114,115], and the distinctions between them were poorly understood [116]. With the benefit of hindsight and on-line databases, it is now easier to explore their history and identify the early landmarks (Table 7). Meanwhile, large-scale spectroscopic surveys have produced a torrent of new stars with hydrogen-deficient spectra (Table 8). Broadly, these hydrogen-deficient stars may be divided into massive stars, low-mass supergiants, hot subdwarfs and white dwarfs.

15.1 Population I and Massive Hydrogen-Deficient Stars

Wolf-Rayet Stars.

Wolf-Rayet (WR) stars were identified as peculiar from the bright bands seen on the continuous spectra of three "small" stars in Cygnus [83] which in retrospect were identified as emission lines due to ionised atoms including helium. By 1894, some 55 WR stars were known, most of which had been discovered by Fleming [84]. Remarkably, there was debate as to whether these stars were either pre-main sequence or highly-evolved stars through the 1960's, while disagreement on whether they were H-deficient persisted into the 1980's.

WR stars are found solely in spiral arms, OB associations and young clusters, and hence are associated with massive star evolution. They can be clearly divided into two sequences, one showing nitrogen-rich spectra, the other being carbon-rich. Numerically, some 230 WR stars are known in the Galaxy, 159 having $m_V < 15$ [118]. Hydrogen has been detected in about half of those analyzed in detail. They are not exclusive to metal-rich environments since about 100 have been found in the Large Magellanic Cloud, and a dozen in the Small Magellanic Cloud.

The strong emission lines provide a clue that WR stars represent a normal stage in the evolution of massive stars. Very high luminosities lead to high radiation pressure and hence to very strong stellar winds. As the star evolves towards and beyond the end of main-sequence hydrogen burning, high mass-loss from the stellar surface exposes first nitrogen-rich material (CNO-processed

Table 7. Various classes of hydrogen-deficient star, with representative prototypes and discovery paper

Prototype	Class	No.	Discovery
Population I			
V1679 Cyg	Wolf-Rayet	≈ 230	[83]
υ Sgr	H-deficient binary	5	[84]
σ Ori E	Intermediate helium B	≈ 30	[85]
SN 1983N, 1984L	SN Ib	≈ 80	[86]
SN 1987M	SN Ic	≈ 120	[87]
Low-mass supergiants			
R CrB	R CrB	≈ 50	[88, 76]
HD 182040	H-deficient carbon	5	[89, 90]
HD 124448	Extreme helium B	17	[91]
MV Sgr	Hot RCrB	4	[92, 93]
FG Sge	Born-again	3	[94]
Hot subdwarfs			
PG1544+488	sdOD / He-sdB	≈ 50	[95, 96]
BD+75°325	compact He-sdO	≈ 50	[97]
BD+37°1977	low-g He-sdO	5	[98]
Central stars of planetary nebulae			
BD+30°3639	[WC]	≈ 50	[99, 100]
A66 30	Of-WR(C)	2	[101]
PG1159–035	O(C) ≡ PG1159	≈ 40	[102]
K 1-27	O(He)	4	[103]
White dwarfs			
HZ 21	DO	≈ 50	[104]
L 930-80	DB	≈ 400	[105]
HZ 43	DC	≈ 360	[106]
	DQ	≈ 120	
	DZ	≈ 80	
AM CVn	AM CVn binary	21	[107]

helium in which carbon and oxygen have been converted to nitrogen) and then carbon-rich material (3α-processed material). These give rise in turn to the WN and WC subclasses.

He-rich B Stars or Intermediate Helium Stars.

σ Orionis is a tightly bound group of young OB stars, with a total mass of some 100 $M_\odot$ or more. Berger [85] noted the spectrum of star E to show a series of excessively strong lines of neutral helium, extending to a continuum discontinuity on the ultraviolet side of the Balmer series limit. Subsequent investigation has shown σ Ori E and similar stars to be chemically peculiar

Table 8. Major surveys yielding hydrogen-deficient stars

	Survey	Reference
HZ		[106]
F	Feige	[108]
FB	Faint Blue Stars	[40]
PG	Palomar-Green	[95]
BPS	HK Objective-Prism	[109]
HS	Hamburg-Schmidt	[110]
HE	Hamburg-ESO	[111]
EC	Edinburgh-Cape	[112]
SDSS	Sloan Digital Sky	[113]

main-sequence B stars. Some 24 are catalogued [114], which may be further divided into fast and slow rotators. Measurements of the helium/hydrogen ratio show that the apparent helium abundances vary with a period of 1–10 d. In σ Ori E, the helium anomaly is associated with a dipole magnetic field of some 10^4 G inclined $\approx 90°$ to the rotation axis; the magnetic caps are metal-poor. It is generally agreed that the surface abundance distribution is governed by diffusion, where both radiative and magnetic fields act selectively to concentrate particular elements in the line-forming region at specific locations on the stellar surface *e.g.* [119].

H-Deficient or υ Sgr Binaries.

Despite an unusual spectrum [84], it was large radial velocity variations that really drew attention to υ Sgr [120]. The star remained unique until [121] recognized its similarity to HD 30353 = KS Per. Two similar stars (V426 Car and V1037 Sco) were discovered during a survey of OB+ stars [122]. BI Lyn was recognized as the fifth "hydrogen-deficient binary" when it was accidentally observed in a survey of subdwarf B binaries [123]. Remarkably, V426 Car had first been noted as a faint star with a spectrum "of the fifth type" (meaning it shows emission lines) in the former constellation of *Argo* in 1892 [124].

The chief characteristics of the class are strong helium lines on a metallic spectrum, and radial velocity variations of several tens of km s^{-1}. The orbital periods range from $\approx 50 - 360$ d. Most of these hydrogen-deficient binaries show emission at Hα and Hβ and also evidence of pulsations.

From high-resolution ultraviolet spectra, [125] showed that υ Sgr is a double-lined system, and hence that the primary has a mass $\approx 3\,M_\odot$ and luminosity $\approx 10^5\,L_\odot$. With systemic radial velocities close to those for circular orbits around the galactic centre, and with all these stars lying less than 200 pc from the galactic plane, the hydrogen-deficient binaries are clearly population I stars, probably with $M > M_{\rm Ch}$, and are excellent candidates for the progenitors of SN Ib/c [126, 125].

Early models suggest at least two phases of mass transfer necessary to explain the very low surface hydrogen abundances *e.g.* [127]. The current best model for the evolution of υ Sgr [9] is that it began as a $10+3\,M_{\odot}$, 150 d binary in which the envelope was blown to infinity, with little change of orbit, as the more massive star approached both Roche Lobe overflow and the Cepheid instability strip simultaneously.

H-Deficient or Type Ib and Ic Supernovae.

The Type II supernovae discussed in Sect. 13 all contain hydrogen in their ejecta and are universally attributed to the explosions of massive stars. Type I supernovae (SN I), by contrast, come in a greater variety. Two classes show no evidence of hydrogen absorption lines and include those which lack the ionized-silicon absorption line (SN Ib) and those which further lack helium lines (SN Ic). As SN Ib age, their spectra show stronger features of helium than SN Ia, until eventually they contain emission lines due to light elements (e.g. oxygen, calcium and magnesium), while SN Ia spectra become dominated by iron emission [128].

The crucial distinction between supernovae of types Ia and Ib/c is that the latter occur near star formation sites and have never been observed in giant elliptical galaxies. Consequently they are associated with core-collapse explosions in massive or intermediate mass stars which have already lost their hydrogen envelopes. Possible candidates include Wolf-Rayet stars and hydrogen-deficient binaries [126, 129].

15.2 Low-Mass Hydrogen-Deficient Supergiants

The remaining classes of hydrogen-deficient star are associated with very late stages of evolution, either low-mass stars approaching the horizontal-branch, or more massive stars which have already past the asymptotic-giant branch and are evolving to become white dwarfs, or yet older stars which have been white dwarfs and somehow found a second life. The first to consider are the low-mass supergiants.

R CrB stars, including H-def C stars.

The R CrB stars are exceptional in that the class is defined by their dramatic light variability rather than by a spectrum. Characterised by spectacular dimmings of five or more magnitudes in the course of days, followed by a slow recovery to maximum light, the R CrB phenomenon is primarily a signature of surface activity, rather than chemical peculiarity. Many discoveries were made from inspection of photographic sky patrol plates *e.g.* [130]. Patrols searching for microlensing events and planetary transits continue to discover new R CrB stars, particularly in the Magellanic Clouds [131].

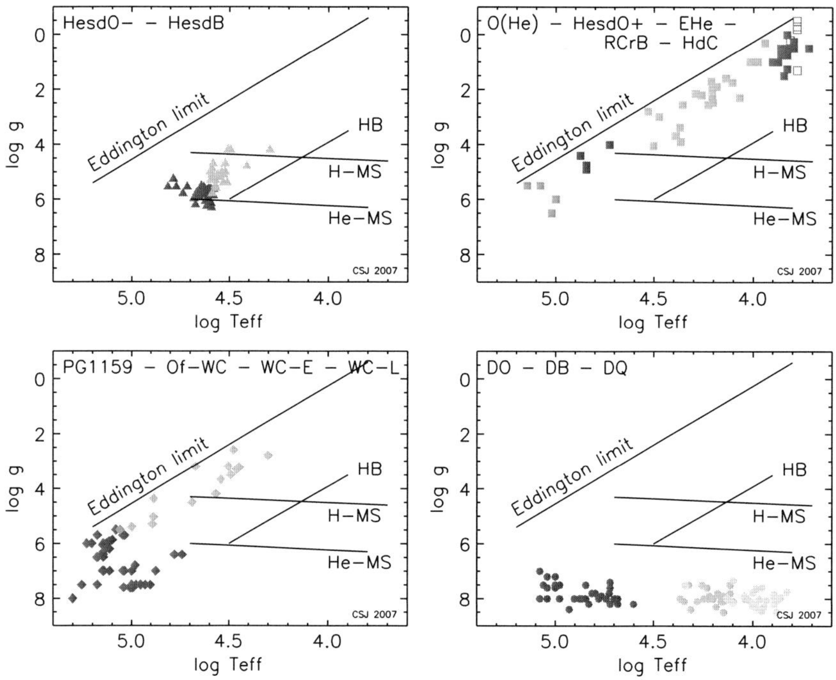

Fig. 33. $g - T_{\rm eff}$ diagram for several classes of low-mass hydrogen-deficient stars described in the text. Different panels identify contemporary evolutionary *proposals*, not necessarily definitive or exclusive *sequences*. In the electronic version, classes are separated by colour. Data have been collated from a wide variety of sources

The majority of R CrB stars would be of spectral types F and G, except that they are missing Balmer lines and the molecular CH bands are replaced by the Swan bands of C_2, indicating an absence of hydrogen and a superabundance of carbon. Approximately 40 are known in the Galaxy, with some 17 in the Large Magellanic Clouds. The overall distribution is similar to that of the Galactic bulge or of an old thick disk. Assuming $M_{\rm V} \approx -5$ based on Magellanic Cloud R CrB stars, it is estimated there may be up to 400 R CrB stars in the Galaxy.

In addition to their irregular fadings thought to be due to directed mass-ejections, R CrB stars show low-amplitude variability with periods of 30 – 60 d and associated with pulsations. All show a substantial infrared excess attributed to a warm dust shell, while a few show evidence of a more extended nebula. It will be seen that these stars are extraordinarily rich in exotic behaviour [132].

While R CrB stars were discovered because of their dramatic variability, it was fortuitous that HD 182040 was chosen for the Harvard spectral type R0, since both the absence of hydrogen lines and CH bands was subsequently noticed with ease [89, 90]. A further three stars with similar spectra were subsequently confirmed to be severely hydrogen deficient [133,134]. Since these stars are spectroscopically similar to the R CrB stars, possibly showing small-amplitude variations [135] but missing the deep minima and infrared excesses, the two groups have been collectively identified hydrogen-deficient carbon stars [134]. Whether the absence of mass-ejection episodes is characteristic of a lower luminosity, a lower carbon abundance, and/or an earlier evolutionary age, has yet to be established.

Extreme helium stars and hot R CrB stars.

In a paper of thirteen lines, Popper [91] announced that spectra of the B2 star HD 124448 (obtained at an elevation of 13°) "show no hydrogen lines in either emission or absorption, although helium lines are sharp and strong". Lines of O II and C II were also present. Popper also remarked that "faint early B-type stars so far from the galactic plane are rare". In the Henry Draper catalogue, only HD 168476 (B5p) and HD 160641 (O9.5Iap) share properties with HD 124448. Subsequent surveys of luminous blue stars have identified a total of 17 *bona fide* extreme helium stars [115] – essentially spectral-types from late O to A with weak or absent hydrogen lines, strong, sharp neutral helium lines, strong C II and N II. Being low-mass supergiants, their high luminosities mean that galactic surveys are essentially complete for these stars – those remaining to be discovered will be highly reddened and on the far side of the Galactic centre. Some, but not all show low-amplitude pulsations.

The kinematics of the R CrB, cool HdC and EHe stars are all similar, being those of an old stellar population strongly concentrated towards the Galactic Centre. They do not share the galactic rotation [136,137]. UW Cen and U Aqr may be in halo orbits. Such a distribution is consistent with a wide range of metallicity.

Mention should be made of four "hot R CrB" stars. MV Sgr was discovered photometrically with cooler R CrB stars [92] and shows the typical light curve and an infrared excess. It has a weak emission spectrum dominated by double-peaked Fe II lines, and an absorption spectrum very similar to HD 124448 [93]. The group is quite heterogeneous [138], with MV Sgr and DY Cen possibly representing evolved R CrB stars, and V348 Sgr and HV 2671 (in the LMC) being something altogether different.

"Born-again" stars.

Three hydrogen-deficient stars stand out because they have been observed to evolve from one corner of the HR diagram to the other, and back, within a human lifetime. FG Sge came to attention [94] as a new blue star as it brightened from being a hot white dwarf in the late 19th century to become a cool yellow supergiant in the late 20th century [139]. In even more spectacular fashion, V4334 Sgr (Sakurai's object) [140, 141] brightened from a faint blue star with $m_j = 21$ ($V > 15$ in 1994) to reach $V = 11$ in late 1995 and to become a yellow supergiant in 1996 [142]. Embedding itself in a thick dust cloud it had virtually disappeared by 1999, but radio measurements suggest that the star has now begun a blueward evolution, with $T_{\rm eff} \approx 12\,000$ K in 2006 [143]. Meanwhile the very hot central star of the planetary nebula A 58 had exhibited a nova-like outburst in 1919 [144]. Examination of historical material *and* exploration of the current nebula suggest that this star followed a rapid "red-giant excursion" similar to that of V4334 Sgr [145]. In all three cases, the photosphere is hydrogen-deficient. V4334 Sgr appears to have lost its hydrogen during its redward evolution [146]; when FG Sge and V605 Aql became hydrogen-deficient is less clear [147, 148].

These stars demonstrate that somehow an extraordinary explosion can occur inside a hot white dwarf and that this is not necessarily a rare event. With three such "born-again" stars observed in a century, a significant fraction of stars that become white dwarfs must go through this phase.

15.3 Hydrogen-Deficient Subdwarfs.

The discovery that certain faint blue star-like objects (quasars) might be used to probe the farthest reaches of the universe [149] created an industry in blue sky surveys (Table 8). Luckily for stellar science, these surveys are strongly contaminated by all sorts of evolved *stars*, including white dwarfs, hot subdwarfs and cataclysmic variables. Many of these are hydrogen-deficient, including the PG1159 stars, and various classes of helium-rich subdwarf and white dwarf.

The designations sdO and sdB were introduced for stars with spectra resembling MK types O and B but with broadened lines of H, He I and He II [104]. The term sdOB introduced for sdB stars which show He II 4686 [150] has been falling out of use in recent years. The hydrogen-dominated sdB stars have been discussed already (Sect. 14.4).

With higher quality data, greater discrimination was achieved [95] by introducing classes sdB-O (some He I 4471) as a subclass of sdB, and dividing sdO (with relative numbers) into sdOA (49: dominated by H absorption with some He II), sdOB (87: dominated by He I and He II, generally some H), sdOC (50: primarily He II, maybe some H, a trace of He I 4471) and sdOD (30: "pure" He I, weak or absent H and He II 4686). The sdOA stars are not important here. The sdOB and sdOC stars have been referred to elsewhere as helium-rich sdO or He-sdO stars, while sdOD stars are also known as helium-rich sdB or He-sdB stars. Note that two definitions for sdOB are not equivalent, and the definition for sdOD is identical to that for the extreme helium B stars [151].

He-sdB stars.

Little attention was paid to the stars classified sdOD in the PG survey for several years. The prototype PG 1544+488 was discussed in a poster paper for which only an abstract was published [96]. JL 87 was the first to be reported in detail [152]. Subsequent observations demonstrate that the class is quite heterogenous, with a wide range of hydrogen/helium ratios. The biggest surprise was that PG 1544+488 turns out to be a double-lined spectroscopic binary comprising two helium-rich subdwarfs [153]. A decade ago, some 50 He-sdB stars were known from the FB, PG, BPS, and HS surveys [115]. The HE and Sloan surveys have added a further 5 definite and 11 possible He-sdB stars, while better spectroscopy has shown several of the original 50 would be better described as He-sdO.

Compact He-sdO stars.

The sdO classification causes some confusion because class members span a range of 2 dex in surface gravity (and hence in luminosity-to-mass ratio). PG classifications do not demonstrate this, but a finer scheme is more successful [154]. Higher-gravity stars lie close to the extreme horizontal-branch / helium main-sequence. A link with the He-sdB stars seems likely (Fig. 33).

An early description referred to these stars as "narrow-line O-type subdwarfs", including BD+75°325, HD 127493, HZ 44 and BD+25°4655 [97]. Like the He-sdB stars, there is considerable diversity as regards binarity and abundance. The stars can be divided (from classification spectra) into those which are carbon-rich, and those which are not. For the most part, all are nitrogen rich. Several occur in composite spectrum binaries, and one [155, HE 0301-3039] is a double He-sdO system similar to the double He-sdB; others are apparently single. With substantial numbers of both helium-enriched and helium-deficient compact sdO stars now identified and analysed spectroscopically [156], an explanation of their origin is tantalizingly close – but not fully formed.

Low-gravity He-sdO stars.

The first indication that some He-sdO stars might have a much lower surface gravity was obtained from spectroscopy of BD+37°1977 [98], which was both the hottest star in the original list observed by the TD1A ultraviolet satellite and clearly more luminous than BD+75°325 (see above). The group remains sparse, with only five established members, including BD+37°1977, BD+37°442, LSE 159, LSE 259 and LSE 263, though the boundaries with other groups (*e.g.* EHe stars, compact He-sdO) are ill-defined. These stars are important; if R CrB and Extreme Helium stars contract to the white dwarf sequence, at some point they should look like these low-gravity He-sdO stars (Fig. 33). Although we know five excellent examples brighter than $V = 12$, fainter counterparts do not appear in surveys more recent than [157].

15.4 Central Stars of Planetary Nebulae

Planetary nebulae attract attention partly because they are photogenic, partly because of their extreme physics and partly because they harbour a very hot compact star at their center. Their spectra are diverse and can be clearly divided into hydrogen-rich and hydrogen-deficient sequences, roughly in the ratio 70:40 [158]. Both sequences can be subdivided, the latter as follows.

[WC] stars.

As in massive, Population I, WC stars, the spectra of [WC] stars are dominated by broad emission from He I, II, C II - IV, N II, III and other light ions. The group extends from early to very late types. Effective temperatures are difficult to determine for stars with optically thick winds, but equivalent measurements indicate surface temperatures ranging from 22 000 K (V348 Sgr: [WC12]) to 140 000 K (Sanduleak 3: [WC3]).

Of-WR(C) stars.

The spectrum is dominated by strong C emission lines. He II 4686 is a strong emission line, narrow in Of(C) and broad in the Of-WR(C). He II 4200, 4541 are in absorption. The two established Of-WR(C) stars are Abell 30 and Abell 78, both of which also show H-deficient material in the inner parts of their nebulae.

O(He) stars.

Intermediate between the [WC] stars and the PG1159/O(C) stars, the O(He) stars show He II, absorption and C IV, N V and O VI emission. Four are known [159]. Two have planetary nebulae (K 1-27 and LoTr 4) and two, so far, do not. The prototype K 1-27 was recognised from a survey of planetary nebula central stars because of the strong ratio of the He II 4686 to [O III 4959], which marked it out as "the most extreme example of the class of old nebulae with very high excitations" [103]. Pulsations have not been detected [160].

O(C) / PG1159 stars.

According to [158], the spectra of O(C) planetary nebula central stars are dominated by an absorption line spectrum of C. During the course of the PG survey, a small group of eight stars showed spectra dominated by helium, but distinguished by a broad absorption from He II 4686, C IV 4658 and possible other C and N contributors [95]. The O(C) and PG1159 spectral classes are essentially equivalent, and it is evident that stars of similar spectral type are to be found both with and *without* planetary nebulae. Stars of both O(He) and O(C) classes appear with and without nebulae in roughly the ratio 1:1.

The PG1159 classification has grown to include some stars showing sharp emission lines, some having less broad absorption and one showing no helium [161, H 1504+65]. The group had attracted immediate attention because a fraction, including the prototype PG 1159-035, showed complex pulsations [102]. It subsequently became clear that these stars were amongst the hottest known, effective temperatures in excess of 100 000 K were reliably established. H 1504+065 is not only very hot ($\approx$ 200 000 K), its surface is deficient in hydrogen and helium, and dominated by a carbon/oxygen mixture with exceptionally high neon and magnesium [162, 163].

Measurements of effective temperature and surface gravity place these stars around the knee where a contracting post-asymptotic giant-branch star becomes fully degenerate and begins to cool down along the white dwarf cooling sequence. Several hydrogen-deficient spectral classes correspond to this evolutionary domain, including the hottest [WC] and Of-WR(C)stars, and the DO white dwarfs (Fig. 33), demanding precision in measurement and raising important questions about stellar evolution [164].

15.5 White Dwarfs

H-deficient white dwarfs.

It is no surprise that Fleming was involved in the discovery of white dwarfs – she had described the spectrum of 40 Eri B as being of type A, in apparent contradiction to its faint absolute magnitude [165]. A difficulty in identifying the first *hydrogen-deficient* white dwarf is linked to the development of a robust classification scheme. The "D" designation was introduced [166], with the suggestion to use DC for continuous spectra, followed by DA, DF, DG, ... DM. To this would be added a two-digit number indicating the estimated surface temperature in thousands of degrees. Five stars were identified as "DBn" [106], but the "Bn" appears to derive from analogy with MK types rather than adherence to Luyten's suggestion, since the DB0 spectrum of HZ 43 appears "nearly continuous". The first DB white dwarfs in the modern sense appear to be L 1573-31 and L 930-80 [105], with no hydrogen lines and He I "outstanding in intensity, ... probably stronger than in any known stellar absorption spectrum". By this time, Luyten had adopted yet another notation to subdivide the DAs and DBs.

As the first example of a DC white dwarf, HZ 43 is remarkable in other ways, and especially since space observatories were established to explore the ultraviolet spectrum and beyond. Early UV observations pointed to a temperature $>$ 100 000 K [167]. Within a decade, X-ray observations had discovered the DO/PG1159 star H 1504+65 [161] – where we have already been.

The class of very hot DO white dwarfs with He II stronger than He I was established [104] with the "interesting star" HZ 21 [104]. In the PG survey [95], the classification [95] is very broad and a composite of four subgroups defined by [168], including DAO (which show hydrogen), a "cool" DOc group, which shows He I and He II, a "hot" DOh group, which shows only He II, and the PG1159 group described above. The overlap between DO and PG1159 can be confusing if the definitions are not used with care.

If the continuous spectra of DC white dwarfs lack both hydrogen and helium, then it is fair to suspect that they are composed of elements with low opacity at high temperatures. Cooler white dwarfs lacking both H and He include the DQ white dwarfs (C I or C_2 absorption) and DZ white dwarfs (Ca II, Mg II, ...).

Far more complete descriptions of the classification and statistics of white dwarfs will be found in the McCook et al. catalogue [169] and its online supplements. These introduce a system of half-integer subclasses linked to the inverse effective temperature ($10 \times 5040/T_{\rm eff}$). Whether a spectral classification should depend on an analysis remains contentious – but it is clearly necessary to distinguish the huge spectral range exhibited within each white dwarf class (DQ, for example). As sky surveys go deeper, there is no shortage of new white dwarfs to discover, with 9316 confirmed in SDSS Data Release 4 [170, 171].

AM CVn binaries.

AM CVn and its ilk stand amongst the most exotic of binary stars. The white dwarf HZ 29 (=AM CVn) differed from normal helium-rich white (DB) dwarfs since the helium lines appear shallow and almost double [107]. These stars remained enigmatic through the 1980's, until 4m telescopes and high-efficiency spectrographs showed them to be interacting white dwarfs with periods in the 5 – 65 min range, *i.e.* double stars separated by tens of earth radii. A wide variety of phenomena are seen; photometric variations with periods different to the orbital period, spectra ranging from broad absorption line systems (AM CVn) to those dominated by He I and He II emission (ES Cet). They include the shortest-period binary known (RX J0806.3+1527=HM Cnc: $P_{\rm orb} = 321\,{\rm s}$) [172], which would fit inside Saturn. Systems appear to show either a high or low state, as in conventional cataclysmic variables, except that in a double-degenerate binary, the transferred material impacts directly onto the stellar surface, with the "disk" more likely being debris filling the space between the accretor's surface and the Roche lobe. Over 20 systems are now known, thanks most recently to the SDSS, with a space density upwards of $10^{-6}{\rm pc}^{-3}$ [173]. The question arises whether these very compact binary stars will eventually merge, and what the products might be: a gamma-ray burst or an R CrB star perhaps? Meanwhile, it seems likely they will dominate the gravitational wave background to be measured by LISA in the 0.3 – 1 mHz range [174].

15.6 Post-AGB evolution

Such a diverse collection of exotic stars places severe demands on the theory of stellar evolution. In the chapter by Karakas in this volumn, the evolution of low-mass stars up to the asymptotic-giant branch has been discussed more or less in detail. In the next few paragraphs, we briefly outline some of the possibilities for "post-AGB" evolution.

"Normal" post-AGB evolution

Conventional theory posits that once the hydrogen-envelope mass has dropped below some critical limit, the thermally-pulsing AGB star will develop a superwind that expels the remaining outer layers. These layers initially form an expanding dusty shell or envelope around the star, obscuring it from view at visible wavelengths. The shell might be called a proto-planetary nebula and may be visible at infrared wavelengths. Since the hydrogen-burning shell in the AGB remnant has been extinguished, the star contracts and heats. As the dust-shell expands, the central star will become visible, initially as an F- or A- star. The contraction proceeds on a thermal timescale (Eq. (9)).

The more massive post-AGB stars ($M \geq 0.6\mathrm{M}_{\odot}$) with higher luminosity contract more quickly ($L \geq 10^4\,\mathrm{L}_{\odot}, t_{\mathrm{K}} \approx 10\,000\,\mathrm{y}$). When their surface temperature exceeds 30 000 K, surface hydrogen is fully ionized, the Lyman continuum opacity drops and ultraviolet photons are able to ionize the much-expanded shell, which now becomes visible as a planetary nebula. Schematically we might represent the evolution as follows:

$$\mathrm{AGB} \rightarrow \mathrm{protoPN} \rightarrow \mathrm{PN} \rightarrow \mathrm{WD}.$$

Less massive post-AGB stars contract more slowly ($L \leq 10^4\,\mathrm{L}_{\odot}$, $t_{\mathrm{K}} \approx 100\,000\,\mathrm{y}$) so that the shell will have dispersed completely before the stellar remnant is hot enough to ionize it. These central stars may appear first as low-gravity F-stars, and later as sdO stars. Their evolution from spectral-type F- to sdO passes through spectral-types A and B; in the absence of an infrared dust shell, such stars are difficult to distinguish from main-sequence stars of similar spectral-type, except that their rotation velocities will be small. In this case:

$$\mathrm{AGB} \rightarrow \mathrm{pAGB} \rightarrow \mathrm{sdO} \rightarrow \mathrm{WD}.$$

The late thermal pulse

In some cases the last helium shell flash will not occur while the star is on the asymptotic-giant branch [175, 176]. If the mass of the intershell is sufficiently high after the superwind phase and the onset of envelope contraction, residual heating can re-ignite the helium shell while the star is contracting. Since the hydrogen-envelope has very little mass and is relatively dense, the expansion of the intershell will not be absorbed by the hydrogen envelope and the result will be an immediate expansion of the star back towards the giant branch. This will occur on the mean thermal timescale of the envelope $\approx 2 \times 10^2$ and $10^3\,\mathrm{y}$, depending on the phase of the thermal pulse cycle at which the star originally left the AGB and hence on precisely when shell ignition occurs. Following ignition, the shell will drive a convection zone through the intershell region, but it is not currently thought that there will be contact between H- and He-rich material.

Once the star has fully expanded to become a yellow or red supergiant, a deep surface convection zone will develop which will dredge helium- and carbon-rich material to the surface, and will completely dilute what is left of the hydrogen envelope. As the star contracts for a second time, its surface will be helium- and carbon-rich. Any planetary nebula ejected at the tip of the AGB phase may still be ionized, since the nebula recombination time is long compared with the thermal timescale of the envelope. Additional helium-rich material may also have been ejected, or be visible as a stellar wind.

As the energy dissipated in such a *late thermal pulse* is radiated away from the star, the envelope contracts and the surface heats. The star will be very luminous, possibly super-Eddington, driving a helium- and carbon-rich stellar wind visible as optical emission lines, giving rise to [WC-L] and [WC-E] type spectra. The surface heating will eventually ionize the inner portions of the nebula expelled when the star returned to the giant phase. As the star further contracts its spectrum will start to resemble that of the PG1159 stars before cooling down the white dwarf sequence for a second time, possibly as a DB white dwarf [177, 178, 164]:

$$\begin{aligned} \mathrm{AGB} \rightarrow \mathrm{PN} \rightarrow \mathrm{LTP} \rightarrow \mathrm{YSG} \rightarrow [\mathrm{WC-L}] \rightarrow [\mathrm{WC-E}] \\ \rightarrow \quad \mathrm{O(C)} \rightarrow \mathrm{PG1159} \rightarrow \mathrm{DB/DQ}. \end{aligned}$$

The very late thermal pulse

The possibility that the last thermal pulse could occur even later, *i.e.* when the star has reached the white dwarf cooling sequence and when the envelope is even more compressed [176], leads to the possibility of an even more violent helium-shell flash and expansion on a timescale of $\leq 10\,\mathrm{y}$. In this case, theory suggests that flash-driven convection can mix material all the way through the He-rich intershell and into the H-rich surface. The surface should appear to become hydrogen-poor and carbon-rich either immediately following the shell flash, or very soon after the surface convection starts to develop [179, 180]. The rapid expansion and prompt change in surface composition observed in Sakurai's object are thought to be due to such a *very late thermal pulse*. The near ballistic expansion expels a dusty envelope, so that the star itself remains invisible for much of the time it spends as a yellow supergiant. Due to the low envelope mass, subsequent contraction would also be more rapid, taking $\leq 100\,\mathrm{y}$ to return to the white dwarf sequence. The combination of extreme flash mixing and subsequent dredge-up could result in a surface dominated by carbon and oxygen, such as that seen in the hot DO white dwarf H1504+65 and the hot DQ white dwarfs [181]:

$$\mathrm{AGB} \rightarrow \mathrm{PN} \rightarrow \mathrm{WD} \rightarrow \mathrm{VLTP} \rightarrow \mathrm{YSG} \rightarrow [\mathrm{WC-E}] \rightarrow \mathrm{H1504+65} \rightarrow \mathrm{DQ}.$$

15.7 Double-Degenerate Mergers

It has become clear that the R CrB and EHe stars probably cannot be formed as a result of a late thermal pulse; in particular their surface carbon abundances are too low [164,60]. It has already been seen (Sect. 14.5) how a binary system comprising two white dwarfs could decay and merge within a Hubble time. If the resulting star re-ignites helium to produce a yellow supergiant [182], the immediate result would be as follows.

He+He white dwarf mergers

The evolution of a star formed from the merger of two helium white dwarfs considered a $0.4\,M_\odot$ helium white dwarf accreting helium at approximately half the Eddington accretion rate ($10^{-5} M_\odot\,yr^{-1}$ [62]). Helium ignites at the core-envelope boundary when $0.067\,M_\odot$ has been accreted and stars expands to become a yellow supergiant in $\approx 10^3$ yr. The accretion is switched off artificially once a pre-selected final mass has been reached, whereupon the star evolves towards the helium main sequence as previously described (Sect. 14.5).

The significance for EHes is that the surface layers of the product should consist primarily of CNO-processed helium, with some contamination by residual hydrogen from the surface layers of the original white dwarfs. The models can reproduce the observed, mass, radius, luminosity, pulsation period (and $\dot{P}$), and surface composition for the EHe star V652 Her with remarkable success [183, 184]. The models also anticipate a population of helium- and nitrogen-rich subdwarf O and B stars:

$$\mathrm{He+He\,WD} \rightarrow \mathrm{DD} \rightarrow \mathrm{EHe} \rightarrow \mathrm{He-sdB} \rightarrow \mathrm{He-sdO} \rightarrow \mathrm{DB}.$$

CO+He white dwarf mergers

In considering the merger of a helium white dwarf with a 0.5 or $0.6\,M_\odot$ CO white dwarf [185], it was assumed (as for the He+He merger) that the CO white dwarf accretes matter from the debris disk at $10^{-5}\,M_\odot\,yr^{-1}$. Helium ignition occurs once sufficient material has been accreted, forcing the star to expand. Helium accretion is switched off at a pre-assigned final mass. In this case, the helium-burning shell is stable, and maintains the star as a supergiant until the helium envelope has been burnt through, leaving a CO core of 0.7 – $0.9\,M_\odot$ as chosen. The core being more massive, the luminosity of the helium-burning shell will be substantially greater than in the case of the He+He merger. At this point, the remaining envelope cannot support itself and the star contracts on a thermal timescale at constant luminosity until it becomes a white dwarf.

This model is successful in accounting for many of the observed properties of more luminous EHes, and also RCrB stars [186]. The absence of any EHe stars in a binary, the observed radii, masses, and surface gravities, observed contraction rates, surface abundances and number densities are all in satisfactory agreement with evolution models for CO+He WD mergers which result in star with a final mass of $0.7 - 0.9\,M_\odot$. Presumably such an EHe continues to contract to become a white dwarf. There is no indication that a nebula exists around any EHes, except possibly MV Sgr. Stars that may represent the descendants of EHes include the low-gravity helium-rich sdO stars and the O(He) stars [187]. After that, these stars should become relatively high-mass CO white dwarfs, *i.e.*

$$\mathrm{CO + HeWD \rightarrow DD \rightarrow RCB \rightarrow EHe \rightarrow HesdO^{+} \rightarrow O(He) \rightarrow DB.}$$

16 Conclusion

These lecture notes have attempted to outline the fundamental theory of stellar structure and evolution from first principles. They have tried to demonstrate the principles by which mainstream stars work. They have also introduced some of modern theory which attempts to explain how various exotic stars were formed. To do so within six lectures and within the constraints of a volume such as this, much useful material has had to be omitted. Many interesting analytic approximations have had to be overlooked, including homology relations and specific solutions of the Lane-Emden equation. With hindsight, discussion of the microphysics has been less than rigorous. The omission of stellar winds, rotation, non-standard mixing and magnetic fields is no longer satisfactory. We have not addressed the physics of very massive or of low-mass stars. The former have a direct impact on nucleosynthesis in the galaxy. The latter are statistically very significant. Neither have we discussed the general consequences of binary star evolution, supernovae explosions, or the evolution of entire stellar populations on nucleosynthesis yields. Fortunately, some of these topics are discussed by other contributors to these proceedings. For others, we are obliged to leave the reader to make their own exploration in the full knowledge that this is a rewarding and fulfilling endeavour.

Acknowledgments

The author is indebted to the organisers of the Kodai School for the invitation to participate, financial support and indefatigable hospitality. He is also grateful for their patience while these extended notes were being prepared for publication. Drs Chris Tout and Amanda Karakas made invaluable contributions to the preparation of the text. In addition to the texts cited, extensive use was made of the SIMBAD database, operated at CDS, Strasbourg, France, and of NASA's Astrophysics Data System.

References

1. M. Schwarzschild, *Structure and evolution of the stars*, Princeton University Press (1958)

2. R. J. Tayler, *The stars: Their structure and evolution*, London: Wykeham (1970)
3. I. S. Shklovskii, *Stars, their birth, life and death*, San Francisco: W. H. Freeman, p13 (1978)
4. R. kippenhahn, A. Weigert, *Stellar Structure and Evolution*, Springer-Verlag: Berlin Heidelberg New York (1990)
5. E. Böhm-Vitense, *Introduction to stellar astrophysics, Vol. 3 - Stellar structure and evolution*, Cambridge University Press (1992)
6. A. C. Phillips, *The physics of stars*, Chichester, UK: Wiley (1994)
7. C. J. Hansen, S. D. Kawaler, *Stellar interiors. Physical Principles, Structure, and Evolution*, Springer-Verlag: Berlin Heidelberg New York (1994)
8. D. Prialnik, *An introduction to the Theory of Stellar Structure and Evolution*, Cambridge University Press (2000)
9. P. Eggleton, *Evolutionary Processes in Binary and Multiple Stars*, Cambridge University Press (2006)
10. C. S. Jeffery, *Studies in Stellar Structure and Evolution*, PhD Thesis, University of St. Andrews (1982)
11. C. S. Jeffery, in *Hydrogen-Deficient Stars*, Astronomical Society of the Pacific Conference Series Vol. **391**, p3 (2008)
12. E. Hertzsprung, *Publikationen des Astrophysikalischen Observatoriums zu Potsdam*, 63 (1911)
13. H. N. Russell, Popular Astronomy **22**, 275 (1914)
14. H. N. Russell, Popular Astronomy **22**, 331 (1914)
15. S. C. Keller, E. K. Grebel, G. J. Miller, K. M. Yoss, AJ **122**, 248 (2001)
16. K. G. Stassun, M. Van den Berg, R. D. Mathieu, F. Verbunt, A&A **382**, 899 (2002)
17. J. E. Hesser, W. E. Harris, D. A. Vandenberg, J. W. B. Allwrite, P. Shott, P.B. Stetson, PASP **99**, 739 (1987)
18. S. C. Rey, Y. W. Lee, C. H. Ree, J. M. Joo, Y. J. Sohn, A. R. Walker, AJ **127**, 958 (2004)
19. K. F. Khalliulin, Master's Thesis, Sternberg Astronomical Institute (1996)
20. P. P. Eggleton, D. S. P. Dearborn, J. C. Lattanzio, ApJ **677**, 581 (2008)
21. W. A. Fowler, G. R. Caughlan, B. A. Zimmerman, ARA&A **13**, 69 (1975)
22. W. A. Fowler, G. R. Caughlan, B. A. Zimmerman, ARA&A **5**, 525 (1967)
23. M. J. Harris, W. A. Fowler, G. R. Caughlan, B. A. Zimmerman, ARA&A **21**, 165 (1983)
24. M. Aikawa, M. Arnould, S. Goriely, A. Jorisson, K. Takahashi, A&A **441**, 1195 (2005)
25. E. M. Burbidge, G. R. Burbidge, W. A. Fowler, F. Hoyle, Reviews of Modern Physics **29**, 547, (1957)
26. L. G. Henyey, L. Wilets, K.H. Böhm, R Lelevior, R. D. Levee, ApJ **129**, 628 (1959)
27. R. Kippenhahn, A. Weigert, E. Hofmeister, Meth. Comp. Phys. **7**, 53 (1967)
28. I. Iben, Jr., Q. J. R. Astron. Soc. **26**, 1 (1985)
29. I. J. Iben, ApJ **143**, 483 (1966)
30. M. Schönberg and S. Chandrasekhar, ApJ **96**, 161, (1942)
31. U.Uus, Nauk. Inf. **17**, 32 (1970)

32. C. S. Jeffery, MNRAS **235**, 1287 (1988)
33. F. Hoyle, W. A. Fowler, ApJ **132**, 565 (1960)
34. D. S. P. Dearborn, J. C. Lattanzio, P. P. Eggleton, ApJ **639**, 405, (2006)
35. S. C. Rey, S. J. Yoon, Y. W. Lee, B. Chaboyer, A. Sarajedini, AJ **122**, 3219 (2001)
36. B. Dorman, R. T. Rood, R. W. O'Connell, ApJ **419**, 596, (1993)
37. P. Demarque, J. G. Mengel, ApJ **164**, 317 (1971)
38. A. V. Sweigart, J. G. Mengel, P. Demarque, A&A **30**, 13 (1974)
39. M. Castellani, V. Castellani, ApJ **407**, 649, (1993)
40. J. L. Greenstein, A. I. Sargent, ApJS **28**, 157 (1974)
41. V. Calo, V. Castellani, J. Danziger, R. Gilmozzi, R. D. Cannon, P. W. Hill, A. Boksenberg, MNRAS **222**, 55 (1986)
42. U Heber, A&A **155**, 33 (1986)
43. C. Sosin, B. Dorman, S. G. Djorgovski, G. Piotto, R. M. Rich, I. R. King et al., ApJ **480**, L35 (1997)
44. J. H. Whitney, R. W. O'Connell, R. T. Rood, B. Dorman, W. B. Landsman, K. P. Cheng et al., AJ **108**, 1350 (1994)
45. J. H. Whitney, R. T. Rood, R. W. O'Connell, N. L. D'Cruz, B. Dorman, W. B. Landsman et al., ApJ **495**, 281 (1998)
46. S. Moehler, S. Dreizler, T. Lanz, G. Bono, A. V. Sweigart, A. Calamida et al., in *Hot Subdwarf Stars and Related Objects, Astronomical Society of the Pacific Conference Series*, vol. **392**, ed. by U. Heber, C. S. Jeffery, R. Napiwotzki, p47 (2008)
47. R. A. Saffer, E. M. Green, T. Bowers, in *12th European Workshop on White Dwarfs, Astronomical Society of the Pacific Conference Series*, vol. **226**, ed. by J. L. Provencal, H. L. Shipman, J. MacDonald, S. Goodchild, p408 (2001)
48. N. L. D'Cruz, B. Dorman, R. T. Rood, R. W. O'Connell, ApJ **466**, 359 (1996)
49. T. M. Brown, A. V. Sweigart, T. Lanz, W. B. Landsman, I. Hubeny, ApJ **562**, 368 (2001)
50. Z. Han, P. Podsiadlowski, P. F. L. Maxted, T. R. Marsh, N. Ivanova, MNRAS **336**, 449 (2002)
51. Z. Han, P. Podsiadlowski, P. F. L. Maxted, T. R. Marsh, MNRAS **341**, 669 (2003)
52. B. Paczynski, *Comments on Astrophysics* **6**, 95 (1976)
53. S. Geier, C. Karl, H. Edelmann, U. Heber, R. Napiwotzki, Astronomische Nachrichten **328**, 708 (2007)
54. J. G. Mengel, J. Norris, P. G. Gross, ApJ **204**, 488 (1976)
55. L. Landau, E. M. Lifshitz, *The classical theory of fields* , Pergamon Press, Oxford (1958)
56. T. R. Marsh, V. S. Dhillon, S. R. Duck, MNRAS **275**, 828 (1995)
57. W. Benz, A. G. W. Cameron, W. H. Press, R. L. Bowers, ApJ **348**, 647 (1990)
58. L. Segretain, G. Chabrier, R. Mochkovitch, ApJ **481**, 355, (1997)
59. J. Guerrero, E. García-Berro, J. Isern, A&A **413**, 257 (2004)
60. G.C. Claton, T.R. Geballe, F. Herwig, C. Fryer, M. Asplund, ApJ **662**, 1220 (2007)
61. D. Lynden-Bell, J. E. Pringle, MNRAS **168**, 603 (1974)

62. H. Saio, C. S. Jeffery, MNRAS **313**, 671 (2000)
63. M. Politano, R. E. Taam, M. van der Sluys, B. Willems, ApJ **687**, L99 (2008)
64. S. Justham, P. Podsiadlowski, Z. Han, On the origin of the single and binary helium-rich sdO stars
65. A. Ahmad, C. S. Jeffery in *14th European Workshop on White Dwarfs, Astronomical Society of the Pacific Conference Series*, vol. **334**, ed. by D. Koester, S. Mochler, p291 (2005)
66. R. Napiwotzki in *Hot subdwarf star and related objects, Astronomical Society of the Pacific Conference Series*, vol **392**, ed. by U. Heber, C. S. Jeffery, R. Napiwotzki, p139 (2008)
67. A. V. Sweigart, in *The Third Conference on Faint Blue Stars* ed. by A. G. D. Philip, J. Liebert, R. Saffer, D. S. Hayes, p3 (1997)
68. M. M. Miller Bertolami, L. G. Althaus, K. Unglaub, A. Weiss, A&A **491**, 253 (2008)
69. T. M. Brown, New Astronomy Review **49**, 474 (2005)
70. A. D. Code, G. A. Welch, ApJ **228**, 95 (1979)
71. F. Bertola, M. Capaccioli, A. V. Holm, J. B. Oke, ApJ **237**, L65 (1980)
72. T. M. Brown, C. W. Bowers, R. A. Kimble, A. V. Sweigart, H. C. Fergyson, ApJ **532**, 308 (2000)
73. S. Yi, P. Demarque, A. J. Oemler, ApJ **492**, 480 (1998)
74. P. Podsiadlowski, Z. Han, A. E. Lynas-Gray, D. Brown, in *Hot Subdwarf Stars and Related Objects, Astronomical Society of the Pacific Conference Series*, Vol. **392**, ed. by U. Heber, C. S. Jeffery, R. Napiwotzki, p15 (2008)
75. W. P. Fleming, Astron. Nachr. **126**, 165 (1891)
76. H. Ludendorff, Astron. Nachr. **173**, 1 (1907)
77. A. H. Joy, M. H. Humason, PASP **35**, 325 (1923)
78. J. S. Plaskett, Publications of the Dominion Astrophysical Observatory Victoria **4**, 1 (1926)
79. C. H. Payne, Stellar Atmospheres; a Contribution to the Observational Study of High Temperature in the Reversing Layers of Stars. PhD thesis, Radcliffe College (1925)
80. L. Berman, ApJ **81**, 369 (1935)
81. O. Struve, F. Sherman, ApJ **91**, 428 (1940)
82. J. L. Greenstein, ApJ **91**, 438 (1940)
83. C. Wolf, G. Rayet, Comptes Rendus **65**, 292 (1867)
84. M. Fleming, Astron. Nachr. **128**, 403 (1891)
85. J. Berger, Compt. rend. séances Acad. Sci. **242**, 2300 (1956)
86. J. H. Elias, K. Matthews, G. Neugebauer, S. E. Persson, ApJ **296**, 379 (1985)
87. A. V. Filippenko, A. C. Porter, W. L. W. Sargent, AJ **100**, 1575 (1990)
88. E. Pigott, H. C. Englefield, Philosophical Transactions Series I **87**, 133 (1797)
89. R. H. Curtiss, Publications of Michigan Observatory **2**, 182 (1916)
90. W. C. Rufus, Publications of Michigan Observatory **3**, 257 (1923)
91. D. M. Popper, PASP **54**, 160 (1942)
92. I. E. Woods, Harvard College Observatory Bulletin **855**, 22 (1928)
93. G. H. Herbig, ApJ **140**, 1317 (1964)
94. C. Hoffmeister, Astron. Nachr. **274**, 176 (1944)

95. R. F. Green, M. Schmidt, J. Liebert, ApJS **61**, 305 (1986)
96. U. Heber, S. Dreizler, K. de Boer, S. Moehler, T. Richtler, Astron Ges., Abstr. Ser. **1**, 16 (1988)
97. J. Greenstein, G. Münch, Carnegie Inst. Washington Year Book **53**, 11 (1953)
98. S. C Wolff, C. A. Pilachowski, R. D. Wolstencroft, ApJ **194**, L83 (1974)
99. C. S. Beals, Trans. IAU **6**, 248 (1938)
100. L. F. Smith, L. H. Aller, ApJ **157**, 1245 (1969)
101. M. Cohen, H. S. Hudson, S. L. Odell, W. A. Stein, MNRAS **181**, 233 (1977)
102. J. T. McGraw, J. Liebert, S. G. Starrfield, R. Green, in *IAU Colloq. 53: White Dwarfs and variable Degenerate Stars*, ed. by H.M. van Horn, V. weidemann, p377 (1979)
103. K. G. Henize, A. P. Fairall, PASP **93**, 435 (1981)
104. J. L. Greenstein, ApJ **144**, 496, (1966)
105. W. J. Luyten, ApJ **116**, 283, (1952)
106. M. L. Humason, F. Zwicky, ApJ **105**, 85 (1947)
107. J. L. Greenstein, M. S. Matthews, ApJ **126**, 14 (1957)
108. J. Feige, ApJ **128**, 267, (1958)
109. T. C. Beers, S. P. Doinidis, K. E. Griffin, G. W. Preston, S. A. Shectman, AJ **103**, 267 (1992)
110. H. J. Hagen, D. Groote, D. Engels, D. Reimers, A&AS **111**, 195 (1995)
111. L. Wisotzki, T. Koehler, D. Groote, D. Reimers, A&AS **115**, 227 (1996)
112. R. S. Stobie, D. Kilkenny, D. O'Donoghue, A Chen, C. Koen, D. H. Morgan et al., MNRAS **287**, 848 (1997)
113. C. Stoughton, R. H. Lupton, M. Bernardi, M. R. Blanton, S. Burles, F. J. Castander et al., AJ **123**, 485 (2002)
114. J. S. Drilling, P. W. Hill in *IAU Colloq. 87: Hydrogen Deficient Stars and Related Objects, Astrophysics and Space Science Library*, vol. **128**, ed. by K. Hunger, D. Schönberner, N. Kameswara Rao, p 499 (1986)
115. C. S. Jeffery, U. Heber, P. W. Hill, S. Dreizler, J. S. Drilling, W. A. Lawson et al. in *Hydrogen Deficient Stars, Astronomical Society of the Pacific Conference Series*, vol **96**, ed. by C. S. Jeffery, U. Heber p471 (1996)
116. A. S. Dinger, Ap&SS **6**, 118 (1970)
117. W. W. Campbell, A&A **13**, 448 (1894)
118. K. A. van der Hucht, New Astronomy Review **45**, 135 (2001)
119. K. Hunger, D. Groote, A&A **351**, 554 (1999)
120. W. W. Campbell, ApJ **10**, 241 (1899)
121. W. P. Bidelman, ApJ **111**, 333 (1950)
122. J. S. Drilling, ApJ **242**, L43 (1980)
123. C. S. Jeffery, R. Aznar Cuadrado, A&A **378**, 936 (2001)
124. W. P. Fleming, A&A **11**, 765 (1892)
125. R. E. Dudley, C. S. Jeffery, MNRAS **247**, 400 (1990)
126. A. Uomoto, ApJ **310**, L35 (1986)
127. D. Schönberner, J.S. Drilling, ApJ **268**, 225 (1983)
128. A. V. Filippenko, ARA&A **35**, 309, (1997)
129. J. C Wheeler, R. P. Harkness, Reports on Progress in Physics **53**, 1467 (1990)

130. D. Hoffleit, AJ **64**, 241 (1959)
131. C. Alcock, R. A. Allsman, D. R. Alves, T. S. Axelrod, A. Becker, D.P. Bennett et al., ApJ **470**, 583 (1996)
132. G. C. Clayton, PASP **108**, 225 (1996)
133. W. P. Bidelman, ApJ **117**, 25 (1953)
134. B. Warner, MNRAS **137**, 119 (1967)
135. W. A. Lawson, P. L. Cottrell, MNRAS **285**, 266 (1997)
136. C. S. Jeffery, J. S. Drilling, U. Heber, MNRAS **226**, 317 (1987)
137. P. L. Cottrell, W. A. Lawson, PASA **15**, 179 (1998)
138. O. De Marco, G. C. Clayton, F. Herwig, D. L. Pollacco, J. S. Clark, D. Kilkenny, AJ **123**, 3387 (2002)
139. A. M. van Genderen, A. Gautschy, A&A **294**, 453 (1995)
140. S. Nakano, Y. Sakurai, M. Hazen, R. H. McNaught, S. Benetti, H. W. Duerbeck et al., IAU Circ. No. **6322**, 1 (1996)
141. H. W. Duerbeck, S. Benetti, ApJ **468**, L111 (1996)
142. H. W. Duerbeck, W. Liller, C. Sterken, S. Benetti, A. M. van Genderen, J. Arts et al., AJ **119**, 2360 (2000)
143. P. A. M. van Hoof, M. Hajduk, A. A. Zijlstra, F. Herwig, A. Evans, G.C. van de Steene et al., A&A **471**, L9 (2007)
144. K. Lundmark, PASP **33**, 314 (1921)
145. G. C Clayton, O. De Marco, AJ **114**, 2679 (1997)
146. M. Asplund, D. L. Lambert, T. Kipper, D. Pollacco, M. D. Shetrone, A&A **343**, 507 (1999)
147. C. S. Jeffery, D. Schönberner, A&A **459**, 885 (2006)
148. G. C Clayton, F. Kerber, N. Pirzkal, O. De Marco, P. A. Crowther, J. M. Fedrow, ApJ **646**, L69 (2006)
149. A. Sandage, ApJ **141**, 1560 (1965)
150. B. Baschek, J. Norris, ApJ **199**, 694 (1975)
151. J. S. Drilling, in *Hydrogen Deficient Stars, Astronomical Society of the Pacific Conference Series*, vol. **96** ed. by C. S. Jeffery, U. Heber, p461 (1996)
152. H. Schulz, G. Wegner, U. Heber, PASP **103**, 435 (1991)
153. A. Ahmad, C. S. Jeffery, A&A **413**, 323 (2004)
154. J. S. Drilling, S. Moehler, C. S. Jeffery, U. Heber, R. Napiwotzki, in *The Garrison Festschrift,* ed. by R. O. Gray, C. J. Corbally, A. G. D. Philip, p27 (2003)
155. T. Lisker, U. Heber, R. Napiwotzki, N. Christlieb, D. Reimers, D. Homeier, Ap&SS **291**, 351 (2004)
156. A. Stroeer, U. Heber, T. Lisker, R. Napiwotzki, S. Dreizler, N. Christlieb, D. Reimers, A&A **462**, 269 (2007)
157. J. S. Drilling, ApJ **270**, L13 (1983)
158. R. H. Mendez, in *Evolution of Stars: the Photospheric Abundances Connection, IAU Symposium*, vol. **145** ed. by G. Michaud, A. V. Tutukov, p375 (1991)
159. T. Rauch, S. Dreizler, B. Wolff, A&A **338**, 651 (1998)
160. R. Ciardullo, H. E. Bond, AJ **111**, 2332 (1996)
161. J. A. Nousek, H. L. Shipman, J. B. Holberg, J. Liebert, S. H. Pravdo, N.E. White, P. Giommi, ApJ **309**, 230 (1986)
162. K. Werner, A&A **251**, 147 (1991)

163. K. Werner, T. Rauch, M. A. Barstow, J. W. Kruk, A&A **421**, 1169 (2004)
164. K. Werner, F. Herwig, PASP **118**, 183 (2006)
165. E. L. Schatzman, *White dwarfs*, Amsterdam, North-Holland Pub. Co.; New York, Interscience Publishers (1958)
166. W. J. Luyten, ApJ **101**, 131 (1945)
167. B. Margon, J. Liebert, M. Lampton, H. Spinrad, S. Bowyer, G. Gatewood, ApJ **209**, 525 (1976)
168. F. Wesemael, R. F. Green, J. Liebert, ApJS **58**, 379 (1985)
169. G. P. McCook, E. M. Sion, ApJS **121**, 1 (1999)
170. D. J. Eisenstein, J. Liebert, D. Koester, S. J. Kleinmann, A. Nitta, P. S. Smith et al., AJ **132**, 676 (2006)
171. D. J. Eisenstein, J. Liebert, H. C. Harris, S. J. Kleinmann, A. Nitta, N. Silvestri et al., ApJS **167**, 40 (2006)
172. G. Ramsay, P. Hakala, M. Cropper, MNRAS **332**, L2 (2002)
173. G. H. A. Roelofs, P. J. Groot, G. F. Benedict, B.E. McArthur, D. Steeghs, L. Morales-Rueda et al., ApJ **666**, 1174 (2007)
174. G. Nelemans, in *The Astrophysics of Cataclysmic Variables and Related Objects, Astronomical Society of the Pacific Conference Series*, vol **330**, ed. by J. M. Hameury, J. P. Lasota, p27 (2005)
175. M. Y. Fujimoto, PASJ **29**, 331 (1977)
176. D. Schönberner, A&A **79**, 108 (1979)
177. K. Werner, U. Heber, A&A **247**, 476 (1991)
178. C.S. Jeffery, CCP7 Newsletter **21**, 5 (1994)
179. F. Herwig, T. Blöcker, N. Langer, T. Driebe, A&A **349**, L5 (1999)
180. F. Herwig, ApJ **554**, L71 (2001)
181. P. Dufour, G. Fontaine, J. Liebert, G. D. Schmidt, N. Behara, ApJ **683**, 978 (1008)
182. R. F. Webbink, ApJ **277**, 355 (1984)
183. C. S. Jeffery, P. W. Hill, U. Heber, A&A **346**, 491 (1999)
184. C. S. Jeffery, V. M. Woolf, D. L. Pollacco, A&A **376**, 497 (2001)
185. H. Saio, C. S. Jeffery, MNRAS **333**, 121 (2002)
186. G. C. Clayton, in *Hydrogen-deficient Stars, Astronomical Society of the Pacific Conference Series*, vol **391**, ed. by A. Werner, T. Rauch, p17 (2008)
187. C. S. Jeffery, in *Hydrogen-deficient Stars, Astronomical Society of the Pacific Conference Series*, vol **391**, ed. by A. Werner, T. Rauch, p53 (2008)

Nucleosynthesis of Low and Intermediate-mass Stars

Amanda I. Karakas

Research School of Astronomy & Astrophysics, Mount Stromlo Observatory, Weston Creek ACT 2611, Australia akarakas@mso.anu.edu.au

Summary. The asymptotic giant branch (AGB) is the last nuclear-burning phase for low and intermediate-mass stars with initial masses between about 0.8 to 8 solar masses. The AGB phase of evolution is very short, comprising less than 1 per cent of the main-sequence lifetime, nevertheless, it is on the AGB that the richest nucleosynthesis occurs for this mass range. The nucleosynthesis is driven by thermal instabilities of the helium-burning shell, the products of which are dredged to the stellar surface by recurrent mixing episodes. Hot bottom burning occurs in the most massive AGB stars, and this also alters the surface composition. I review the evolution and nucleosynthesis from the main sequence through to the tip of the AGB. The nucleosynthesis that occurs during the AGB is explored in detail, including a discussion of the effects of hot bottom burning. I finish with a brief review of the *slow*-neutron-capture process that produces elements heavier than iron.

Keywords: stars: AGB and post-AGB stars — planetary nebulae: general — nuclear reactions, nucleosynthesis, abundances

1 Introduction

All stars in the mass range $0.8M_\odot$ to $\sim 8M_\odot$, including our own Sun, will become *asymptotic giant branch* (AGB) stars (see Herwig [1] for a recent review). The AGB is the last nuclear burning phase of stellar evolution for this mass range, and is brief, lasting much less than 1% of the time spent during core hydrogen (H) burning on the main sequence. AGB stars are evolved objects and are found in the high-luminosity, low temperature region of the Hertzsprung-Russell (HR) diagram. AGB stars have evolved through core H and helium (He) burning and are now sustained against gravitational collapse by alternate H-shell and He-shell burning. The He-burning shell is thermally unstable and flashes or pulses every 10^4 years, depending on the stellar mass. AGB stars are often observed to be chemically different from their less evolved counterparts, and show enrichments in carbon and heavy elements synthesized

A. Goswami and B.E. Reddy (eds.), *Principles and Perspectives in Cosmochemistry*, Astrophysics and Space Science Proceedings,
DOI 10.1007/978-3-642-10352-0_2,

by the *slow* neutron-capture process (the s-process). AGB stars are observed with spectral classes M, MS, S, SC and C, where C-type stars are *carbon-rich* and have C/O ratios exceeding unity. AGB stars are observed to be long-period Mira, semi-regular or irregular variables with periods ≥ 100 days. Many AGB stars are also observed to be losing mass rapidly (typical mass-loss rates $\dot{M} \sim 10^{-5} M_{\odot}$ year^{-1}) through slow, dense outflows (velocities ~ 10 km/second). After the ejection of the envelope, the AGB phase is terminated and the star quickly evolves (in $\approx 10^5$ years) to become a post-AGB star, before finally ending its life as a dense white dwarf. For a review of the post-AGB and planetary nebulae phase of evolution, we refer to [2] and [3].

The range of bolometric luminosities and masses of AGB stars can be estimated from the large sample of long-period variables observed in the Large Magellanic Cloud (LMC). The luminosities range from $M_{\rm bol} \approx -3.8$ to a maximum of ≈ -7.1, that corresponds to current masses between about $\approx 1 M_{\odot}$ to $8 M_{\odot}$ [4]. In Figure 1 we show the colour-magnitude diagram (CMD) for the globular cluster M3. A CMD is just a type of HR diagram in which we use colour index rather than spectral class on the horizontal axis; and use the apparent visual magnitude, V, for the vertical axis. While most of the stars in Figure 1 are located on the main sequence, there are distinct bands to the top right of the diagram and an almost horizontal band above the main sequence. Each of these bands, including the main sequence, represents a distinct phase in the life of low to intermediate-mass stars. In the luminous red part of the diagram (top-right corner) there are a few AGB stars. Very few of the AGB stars in globular clusters are more luminous than the tip of the first giant branch which occurs at a bolometric luminosity of $M_{\rm bol} \approx -3.6$ [5].

AGB stars are particularly important because they are a significant site of nucleosynthesis. The nucleosynthesis during the AGB leads to the production of carbon, nitrogen, fluorine and heavy elements such as barium and lead. Recurrent mixing episodes bring the freshly synthesized material from the core to the envelope, and strong stellar winds ensure that this material is expelled into the interstellar medium (ISM). For these reasons, AGB stars are major factories for the production of the elements in the Universe [7, 1]. Knowledge of AGB evolution and nucleosynthesis is vital to obtain an estimate of the contribution of low and intermediate-mass stars to the chemical evolution of galaxies and stellar systems such as M3. For example, a third of the carbon in our Galaxy today is estimated to have been produced in AGB stars [8].

In these lecture notes, we are primarily interested in the changes to the surface composition of low and intermediate mass stars during the AGB. Before we get to the AGB stage, it is necessary that we first examine the evolution and nucleosynthesis that occurs prior to the AGB and we cover this material in Sect. 3. AGB evolution is discussed next in Sect. 4, and the nucleosynthesis in Sect. 5. We briefly review the s-process in Sect. 6. We begin with some important preliminaries.

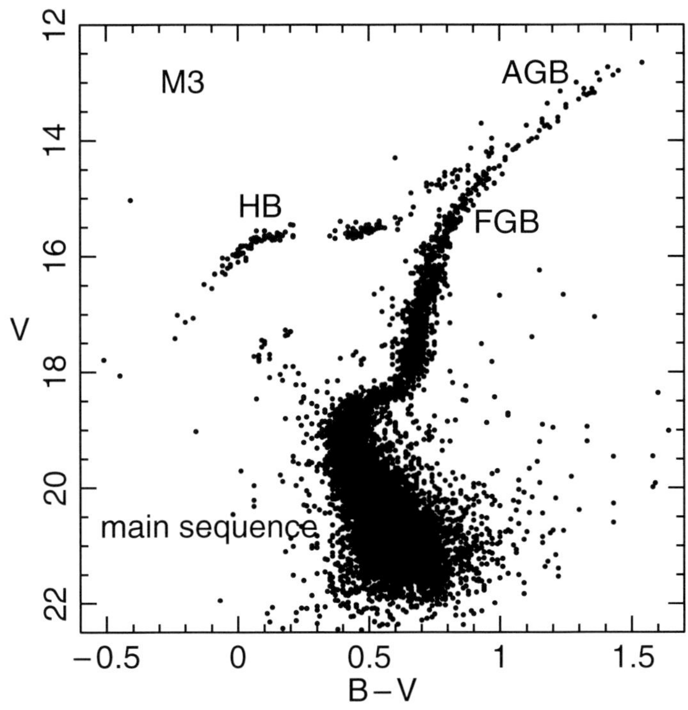

Fig. 1. Colour-magnitude diagram for M3, using data from [6]. The approximate positions of the main sequence, first giant branch (FGB), horizontal branch (HB), and asymptotic giant branch (AGB) are labelled

2 Some preliminaries

In the following lecture notes, it is assumed that the reader has an understanding of the basics of stellar evolution and nucleosynthesis. We refer to other lectures given at the Kodai School, along with the following textbooks: Clayton [9], Rolfs & Rodney [10], and Iliadis [11] cover all aspects of nuclear astrophysics. Stellar interiors and evolution are covered by Hansen, Kawaler, & Trimble [12]. The chemical evolution of galaxies and stellar systems are discussed by Pagel [13], along with an overview of stellar evolution and nuclear astrophysics. The topic of AGB stars is covered in detail in [14]. Lugaro [15] reviews stardust in meteorites, with an emphasis on pre-solar grains that originated in AGB stars.

Before we begin, it is important to make a few definitions. In the following we will refer to *low-mass stars* as stars with initial masses between 0.8 to $\sim 2.25 M_{\odot}$, and *intermediate-mass stars* as stars with an initial masses between 2.25 and $\sim 8 M_{\odot}$. Stars in the range $0.8 M_{\odot}$ to $8 M_{\odot}$ evolve through central hydrogen and helium burning and enter the AGB with an electron-degenerate carbon oxygen (C-O) core. The cores of these stars do not reach the $\approx$ 800 million K temperatures required for carbon burning. Stars with initial masses between 9–11$M_\odot$ may also become AGB stars after core carbon burning; these objects are referred to as *super-AGB stars*. Stars more massive than about 11$M_\odot$ evolve through the central carbon, neon, oxygen and silicon burning stages and end their lives as core collapse supernovae. We do not discuss these massive stars and refer the reader to the lecture notes by Arnould, see also Woosley, Heger, & Weaver [16] and references therein for details.

The dividing mass which separates low and intermediate mass is not arbitrarily chosen but corresponds to the transition from degenerate to non-degenerate core helium ignition (which is at about 2.25$M_\odot$). The lower mass limit of 0.8$M_\odot$ is the minimum mass required to ignite helium and evolve through central helium burning. The upper mass limit of 8$M_\odot$ is the maximum mass which avoids core carbon ignition. The mass ranges above are valid for stars with a composition similar to that of our Sun, which is composed of approximately 70% hydrogen by mass, 28% helium (or more precisely, ^{4}He), and about 2% *metals*. The metal content is referred to as the global metallicity, Z, and is comprised of about half ^{16}O. Using the solar abundances of Anders & Grevesse [17], we obtain a global solar metallicity of $Z = 0.019$. The revised solar abundances of Asplund, Grevesse & Sauval [18] has reduced the global metallicity to $Z \approx 0.015$, primarily because the oxygen abundance of the Sun has been reduced by about a factor of ~ 2 [19].

In real stars, where it is impossible to measure the composition of every element, the Fe/H ratio has been taken to represent the global metallicity of the star due to its ease of measurement. It is usually given as a logarithmic ratio relative to the Fe/H ratio in the Sun, that is,

$$[\mathrm{Fe/H}] = \log_{10}(\mathrm{Fe/H})_{\mathrm{star}} - \log_{10}(\mathrm{Fe/H})_\odot, \quad (1)$$

where [Fe/H] = 0 corresponds to a star with the same iron abundance as in our Sun; a star with [Fe/H] = −1 has an atmospheric composition that has 10 times less iron than in our Sun. Metal-poor stars are defined as those with [Fe/H] < 0, and metal-rich stars with [Fe/H] > 0. In metal-poor stars, the definitions given above are shifted to lower mass, where for example, the minimum mass for degenerate He ignition is about $\approx 1.75M_\odot$ at $Z = 0.0001$. Ancient very metal-poor stars with [Fe/H] ≤ -2 have been found in the halo of our Milky Way Galaxy [20], where the most metal-poor stars discovered have [Fe/H] < −5 [21]. Stars with [Fe/H] < −3 behave very differently to their metal-rich counterparts, however, we will be discussing the evolution and nucleosynthesis of solar-like stars with $Z \approx Z_\odot$ or [Fe/H] ≈ 0, unless otherwise indicated.

The distribution of stellar birth masses is governed by the *initial mass function*. The initial mass function of the local solar neighbourhood [22, 23] indicates that the majority ($\sim$ 90%) of stars have masses less than 0.8$M_\odot$, and that the remainder of the stars (about 10%) have masses between about 0.8 to 8$M_\odot$. Massive stars comprise much less than 1% of all stars. From the perspective of the origin of the elements in the Universe, it is instructive

Table 1. Stellar lifetimes for stars of initial masses between 0.8 to $25M_\odot$. The lifetimes for the 25 and $15M_\odot$ stars are from Woosley et al. [16]; lifetimes for the lower mass stars are from Karakas & Lattanzio [24]. Lifetimes are given in Myr, which is units of 10^6 years, or in Gyr, which is units of 10^9 years

Initial mass ($M_\odot$)	Main-sequence lifetime	Total stellar lifetime
25	6.7 Myr	7.5 Myr
15	11 Myr	13 Myr
5	78 Myr	102 Myr
2	0.87 Gyr	1.2 Gyr
1	9.2 Gyr	12 Gyr
0.8	20 Gyr	32 Gyr

to examine the *stellar lifetimes* of stars in each of these mass groupings. The most important lifetime is the time a star spends on the main sequence, fusing hydrogen to helium in the core. This is because it is the longest lived phase of stellar evolution and therefore representative of the entire nuclear burning lifetime of a star. In Table 1 we show the main sequence and total stellar lifetime for a selection of initial stellar masses.

Stars progressively enrich the interstellar medium with the products of stellar nucleosynthesis. This occurs because massive stars explode as supernovae, releasing vast quantities of processed material, whereas lower mass stars like our Sun slowly lose their outer envelopes by stellar winds. From Table 1 it is clear that stars more massive than about $8M_\odot$ evolve and die very quickly, in under 20 Myr, short by even geological standards. Lower mass stars, on the other hand, evolve much more slowly, with a star like our Sun spending about 10 Gyr on the main sequence. Stars less massive than $\approx 0.8M_\odot$ will spend so long on the main sequence that they will not have had a chance to evolve and release their processed material into the interstellar medium. These very-low mass stars have not yet contributed to the chemical enrichment of our Galaxy.

The stars that evolve through the asymptotic giant branch phase are the 0.8 to $8M_\odot$ objects with relatively long lives. Initial mass function considerations ensure that there are very many of them in the Galaxy, and this means that they contribute in a significant way to the origin and evolution of elements. Their contribution will be quite different to massive stars that explode as supernovae, owing to the different time scales of their contribution. However, as an example of their importance to the composition of the Galaxy, it has been estimated that up to 90% of the dust in the Galaxy today originated in the outflows from AGB stars [25]. If we want to answer the question "Where did the elements come from?", it is essential that AGB stars be studied alongside their more massive cousins!

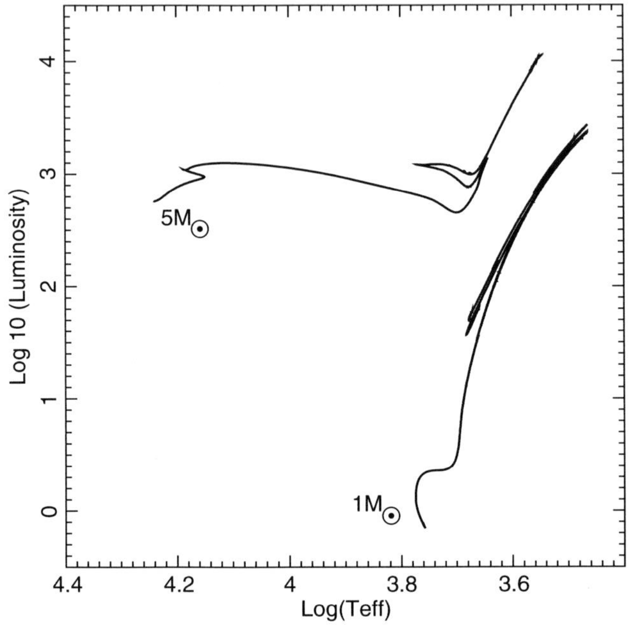

Fig. 2. Hertzsprung-Russell diagram for a $1M_{\odot}$ and a $5M_{\odot}$ star of solar composition. Evolutionary tracks have been plotted from the zero-age main sequence to the asymptotic giant branch. The location of the main sequence is noted by the position of the label. The x-axis is the effective temperature of the model star, shown in units of $\log_{10} T_{\rm eff}$, whereas the y-axis is the luminosity, in units of $\log_{10}(L/L_{\odot})$

3 Evolution and nucleosynthesis prior to the AGB

All stars begin their nuclear-burning life on the main sequence, burning hydrogen to helium in their cores. The majority of a star's nuclear-burning life is spent on the main sequence, which is why we find most of the stars observed in M3 (Figure 1) in this phase of evolution. The stars in M3 typically have masses between about 0.18 to $0.8M_{\odot}$ [26], which means that their main-sequence lifetimes are exceedingly long ($\geq$10 Gyr). In the following, we describe the evolution of stars of masses $1M_{\odot}$ and $5M_{\odot}$ of solar composition (taken as $Z = 0.02$ in this case), and note that the $0.8M_{\odot}$ stars of M3 will undergo the same evolution. One main difference is that all of the 1 and $5M_{\odot}$ stars have long evolved away in old globular clusters such as M3. The $1M_{\odot}$ star spans the range of evolutionary behaviours observed for low-mass stars, which go through the core helium flash and do not experience the second dredge-up. The $5M_{\odot}$ is representative of intermediate-mass stars which do experience the second dredge-up.

In Figure 2 we show the theoretical evolutionary tracks of a 1 and $5M_{\odot}$ star on a HR diagram. In this figure, and in what follows we show model results from Karakas & Lattanzio [24] and Karakas [27]. These models were computed using the Monash version of the Mount Stromlo Stellar Structure code, and we refer to [24] and references therein for details. It is important to remember that the results presented here do depend on the details of the stellar models, with different codes sometimes giving different results. For example, the inclusion of core overshoot during the main sequence will lower the upper mass limit from $\sim 8M_{\odot}$ to $\sim 5-6M_{\odot}$ of a star that enters the AGB with a C-O core (e.g., [28]).

From a quick inspection of Figure 2 it is clear that the $5M_{\odot}$ is much brighter on the main sequence than the $1M_{\odot}$ star by almost three orders of magnitude. The location of the *first giant branch* (also commonly referred to as the "red giant branch") is in the upper right of the diagram. A $1M_{\odot}$ red-giant star will be cooler ($T_{\rm eff} \approx 3,100$ K) than the 5 ($\sim 4,500$ K), although it has a similar luminosity ($\log_{10}(L/{\rm L}_{\odot}) \approx 3.2$). The luminosity of the $5M_{\odot}$ is much higher during the AGB phase. The other main difference is the blue-ward extension of the core helium burning track for the $5M_{\odot}$ model, which is not at all apparent for the lower mass star. Let's now look at the evolution of the $1M_{\odot}$ star in more detail.

3.1 The Evolution of a $1M_{\odot}$ Star

Figure 3 shows a schematic HR diagram for a $1M_{\odot}$ star [29, 30]. Core hydrogen burning occurs radiatively, with the central temperature and density increasing with the mean molecular weight (points 1–3). The temperature during core H burning is $\approx 15 \times 10^{6}$ K, which means that the main energy generation reactions are the pp chains. At core H exhaustion, the hydrogen profile is shown in inset (a) in Figure 3, and the star now leaves the main sequence. Following this, the star crosses the Hertzsprung Gap (points 5–7), and nuclear burning is established in a shell surrounding the contracting ^{4}He core. Inset (b) shows the advance of the H-shell during this evolution. Simultaneously, the outer layers expand and as a consequence become convective, due to an increase in the opacity. When the star reaches the Hayashi limit (point 7), convection has extended all the way from the surface to the deep interior, and the star is now on the first giant branch (FGB). The star is very big (up to $\sim$ 200 times the radius on the main sequence) but most of the mass in the core is within a small fraction of the total radius. A consequence of this is that the outer layers are only tenuously held onto the star and can be lost through an outflow of gas called a stellar wind. About $\sim$30% of a star's total mass can be lost on the FGB, depending upon the length of time a star spends in this phase of evolution.

During the star's ascent of the giant branch, the convective envelope moves inward, mixing the outer layers with internal matter that has experienced partial H-burning, shown in inset (c) of Figure 3. This mixing event is known as

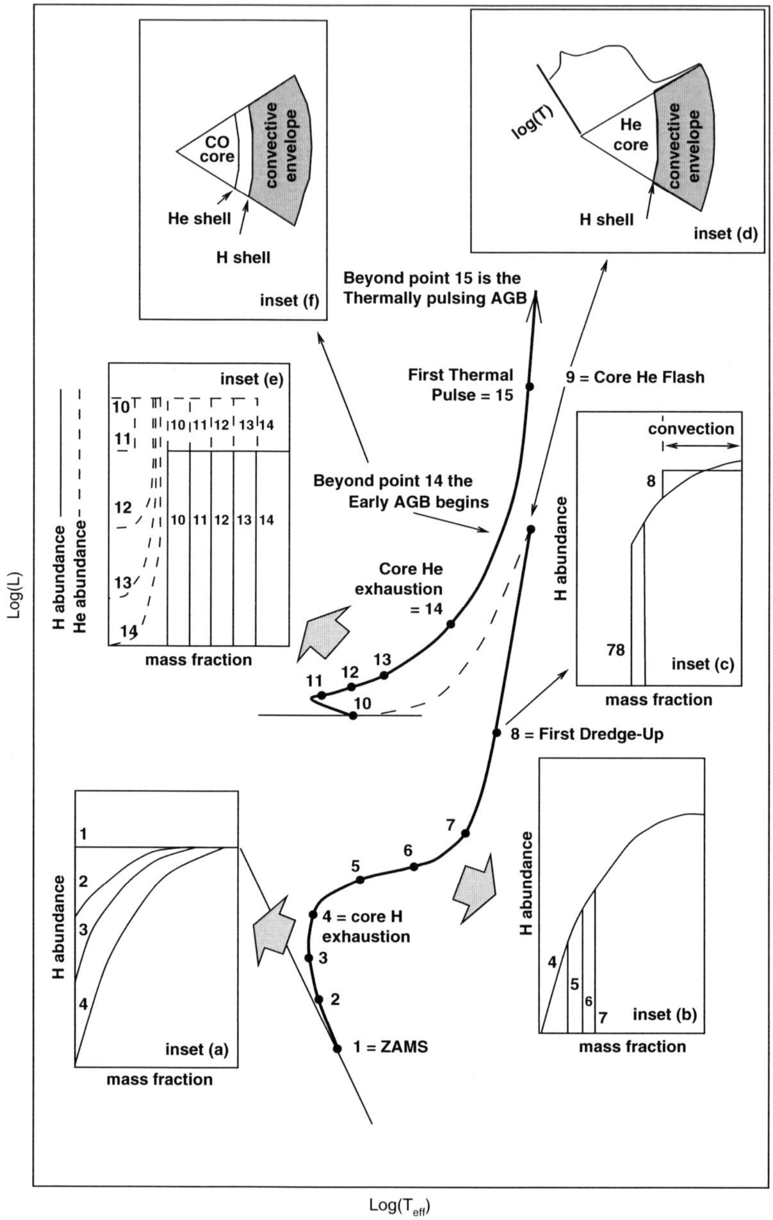

Fig. 3. Hertzsprung-Russell diagram for a $1M_{\odot}$ star of solar composition, from the zero-age main sequence to the tip of the asymptotic giant branch. Figure provided by J. C. Lattanzio

the first dredge-up (FDU) and affects the surface composition of low to intermediate mass stars. Note that the region reached by convection was partially processed by hydrogen burning during the main sequence phase, but is external to the location of the H-shell. The composition of this material is still mostly hydrogen but with some added ^{4}He together with the products of CN cycling. The main surface abundance changes are an increase in the ^{4}He abundance by $\sim 5\%$, a decrease in the ^{12}C abundance by about 20%, and an increase in the ^{14}N and ^{13}C abundances by 30% and a factor of ~ 2, respectively. The number ratio of ^{12}C/^{13}C drops from its initial value to lie between 18 and 30 [31,32]. The FDU leaves behind a sharp composition discontinuity exterior to the position of the H-burning shell. In the lowest-mass stars the relatively long lifetime on the first giant branch allows the H-shell to reach the composition discontinuity and erase it. This does not occur in intermediate mass stars, which leave the first giant branch before the H-shell can erase the discontinuity.

During the ascent of the FGB, the He core continues to contract and heat and becomes electron degenerate. Neutrino energy losses from the centre cause the temperature maximum to move outward, as shown in inset (d) of Figure 3 and in Figure 4. The FGB lifetime is terminated when the necessary temperatures for central helium ignition are reached, at about 100 million K. At this temperature the triple alpha reactions are ignited at the point of maximum temperature but with a degenerate equation of state. The temperature and density are essentially decoupled and this leads to a violent helium ignition that is referred to as the *core helium flash* (point 9). During a core helium flash, the surface luminosity does not change greatly but the helium luminosity may reach up to $10^{10} L_\odot$ during the first flash. The helium-burning luminosity oscillates a few times after the main He-flash, before settling down to quiescent core He-burning (see Figure 4). During the flash, about 3% of the ^{4}He in the core is converted into ^{12}C, however it seems unlikely that there is any mixing between the core and envelope in solar metallicity stars. There are indications that such mixing takes place in extremely low-metallicity stars with [Fe/H] ≤ -3 (e.g., [33,34,35]).

The maximum initial mass for the core He-flash to occur, $M_{\rm flash}$, is about $2.25 M_\odot$ at $Z = 0.02$ using the models of Karakas, Lattanzio, & Pols [36] that include no convective overshoot. Models with convective overshoot during core H burning find that the upper mass limit occurs at lower mass, where $M_{\rm flash} \leq 1.6 M_\odot$ [37]. Note that $M_{\rm flash}$ is also reduced in lower metallicity models. For a star with an initial mass of $1 M_\odot$, the first giant branch is the next longest phase of evolution after the main sequence, and the star will typically spend ≥ 2.8 Gyr there. This is also why the FGB of M3 is the second most populated branch in Figure 1.

Following the core helium flash, the star quickly ($\sim 10^6$ years) moves to the Horizontal Branch (HB), where it burns ^{4}He in a convective core, and hydrogen in a shell (that provides most of the luminosity). This corresponds to points 10–13 in Figure 3. The coulomb repulsion is larger for He than for H,

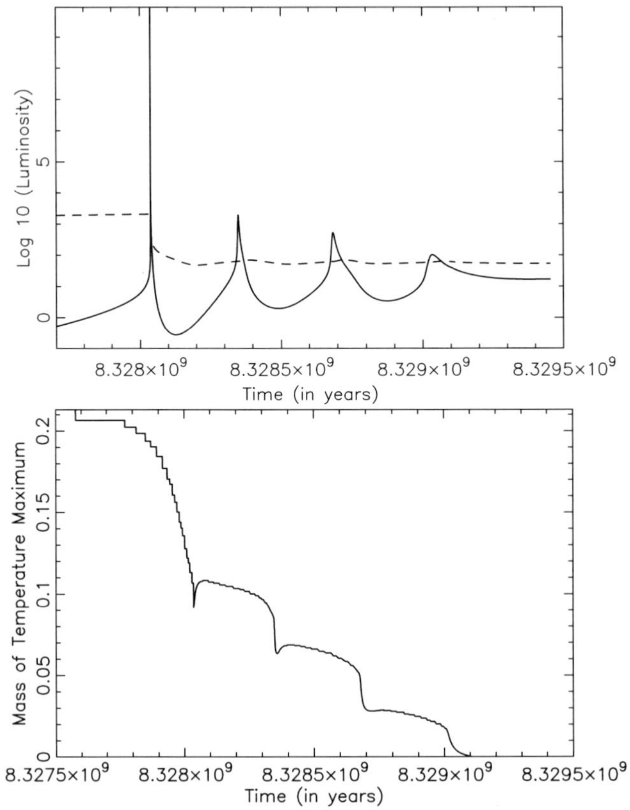

Fig. 4. (*Top panel*) radiated (dashed line) and He-burning (solid line) luminosity during the core helium flash for the $1M_{\odot}$, $Z = 0.004$ model. (*Bottom panel*) mass of the temperature maximum as a function of time. At the flash peak, the maximum temperature is $0.2M_{\odot}$ from the centre of the star. The evolution of a $1M_{\odot}$, $Z = 0.02$ star would be very similar although somewhat less extreme owing to the higher initial metallicity

hence more energy is required by the triple alpha process to maintain the star in hydrostatic equilibrium. This is coupled with the fact that less energy is released by the triple alpha process per reaction than during hydrogen burning. For example, the reaction ^{8}Be + ^{4}He $\rightarrow$ ^{12}C releases 7.65 MeV of energy per reaction, compared to $\sim$ 26 MeV per $4p \rightarrow$ ^{4}He reaction. The overall result is that the core helium burning lifetime is about 100 Myrs, compared to 10 Gyrs during the main sequence.

Helium burning increases the fraction of ^{12}C, which in turn increases the abundance of ^{16}O from the reaction ^{12}C$(\alpha,\gamma)^{16}$O. It is during core helium burning that the star first experiences *semiconvection*: the outer layers of the convective core become stable to the Schwarzschild convection criteria but unstable to that of Ledoux, see [38], [39] and [12] for a detailed description of this phenomenon. The semiconvection causes the composition profile to adjust itself to produce convective neutrality with the resulting profiles as shown in inset (e) of Figure 3. Note that semiconvection is different to core overshooting, which extends the convective border beyond the formal boundary set by the Schwarzschild criteria. If overshooting is included the core hydrogen and helium burning lifetimes are longer, because more fuel is added to the core, along with increases in the mass of the H-exhausted core at the beginning of the AGB (e.g., [40, 41, 42]).

Following core helium exhaustion (point 14) the star ascends the giant branch for the second time, and this is known as the asymptotic giant branch. The properties of the C-O core left after core He-burning depend on a number of factors including the numerical treatment of convection and the inclusion of convective overshoot. The uncertain rate of the $^{12}C(\alpha,\gamma)^{16}O$ reaction also helps to determine the final C/O ratio and composition of the core [43, 44]. As the star ascends the AGB, the core becomes electron degenerate, and the star's energy output is provided by the He-burning shell (which lies immediately above the C-O core), along with the H-burning shell. The structure is shown in inset (f) in Figure 3, and schematically in Figure 10. The AGB is divided into two regimes: the early-AGB, prior to the first instability of the He-burning shell, and the thermally-pulsing AGB (TP-AGB) beyond this (point 15 in Figure 3).

3.2 The Evolution of a $5M_\odot$ Star

The evolution of a $5M_\odot$ is qualitatively similar to the $1M_\odot$ discussed above. The main difference during the main sequence is that the higher temperatures in the core cause CNO cycling to be the main source of H fusion. The higher temperature dependence of these reactions ($T \propto 10^{17}$ compared to 10^4 for the pp chains) results in the formation of a convective core. The main-sequence evolution of the $5M_\odot$ star corresponds to points 1–4 in Figure 5. Following core H exhaustion, the core contracts and H burning is established in a shell as the star crosses the Hertzsprung Gap, points 5–7 and inset (b) of Figure 5. As the star ascends the first giant branch, the inward movement of the convective envelope (point 8) reaches regions where partial hydrogen burning occurred during the main sequence. Thus the products of CN cycling (^{14}N and ^{13}C) are mixed to the surface in much the same way as for lower mass stars; see also inset (c) of Figure 5. The depth of the first dredge-up relative to the total mass of the star is less for the $5M_\odot$ than the $1M_\odot$, a point we come back to in Sect. 3.3.

For these more massive stars, the contracting He cores do not become electron-degenerate during the ascent of the FGB and the ignition of ^4He occurs in the centre under non-degenerate conditions. The star then settles down to a period of central He-burning that lasts about 20 Myrs. The core He-burning timescale is much shorter than the 90 Myrs spent on the main sequence. Intermediate-mass stars spend longer during core He-burning than

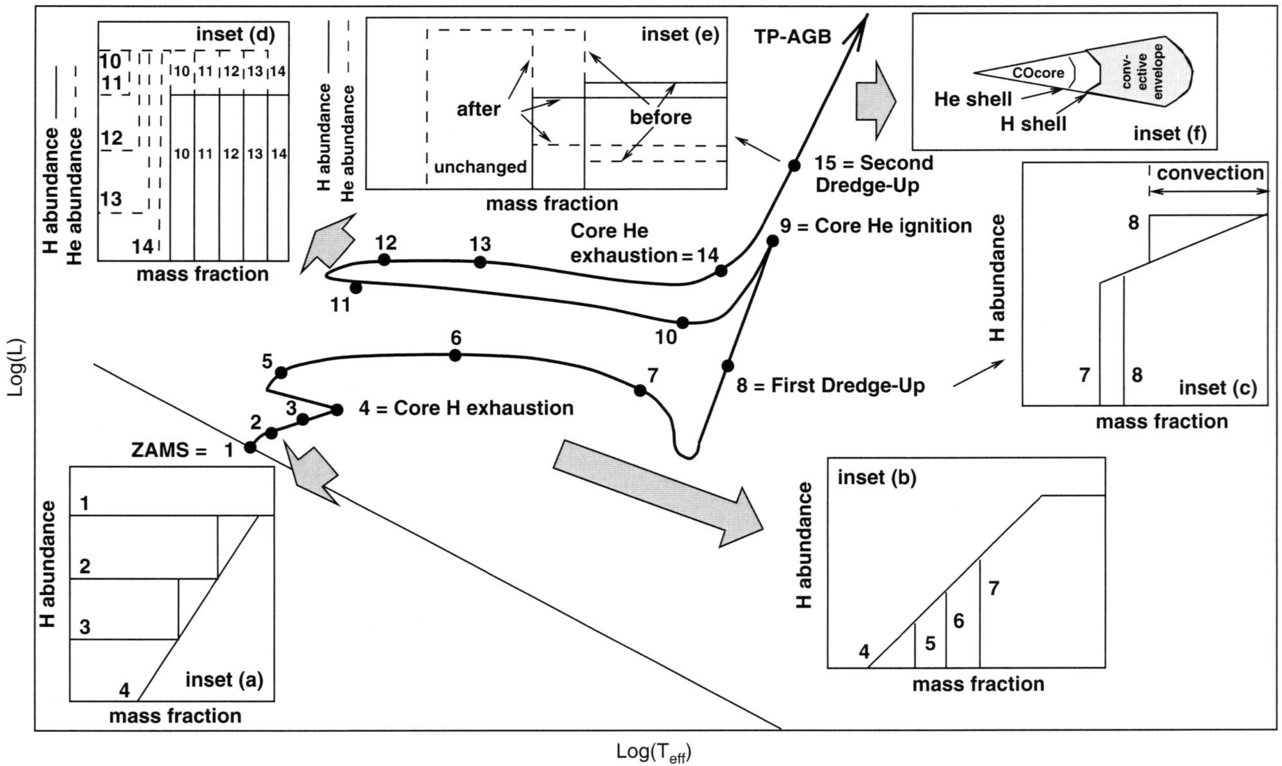

Fig. 5. Hertzsprung-Russell diagram for a $5M_{\odot}$ star of solar composition, from the zero-age main sequence to the tip of the asymptotic giant branch. Figure provided by J. C. Lattanzio

on the FGB (20 compared to 3 Myrs). This is different to lower-mass stars that have very long FGB lifetimes. These lifetimes affect the relative ratio of main sequence to FGB and core He-burning stars that are visible in any cluster.

The extent of the blue loops seen in the HR diagram in Figure 2 is determined by the competition between the two burning sources: central He-burning and H-shell burning. It is during the blue-ward excursion across the HR diagram that the star may cross the instability strip and be observed as a Cepheid variable (points 10–14). Following core He exhaustion, the strong expansion of the star caused by the structural re-adjustment to He-shell burning causes the H-shell to be extinguished as the star begins the ascent of the asymptotic giant branch. With the entropy barrier of the H-shell gone, the convective envelope moves inward and reaches regions where complete H burning had taken place (during the main sequence), resulting in an increase in the surface composition of ^{4}He and ^{14}N. In the process the mass of the H-exhausted core is reduced, because while ^{4}He is mixed outward, H is mixed inward by the convective envelope, see inset (e) in Figure 5. This inward movement of the convective envelope and reduction of the H-exhausted core mass during the early-AGB phase is known as the *second dredge-up* (SDU). There is a critical minimum mass below which the SDU does not occur ($\sim 4.5 M_{\odot}$ at $Z = 0.02$); see Sect. 3.3 and Figure 6. Following the SDU, the H-shell is re-ignited and the first He-shell instability soon follows. The star is now on the TP-AGB, where the structure is qualitatively the same for all masses.

3.3 The First and Second Dredge-up

Prior to the TP-AGB, the surface composition of a star will be altered by the first and possibly second dredge-up events. In this section we look at these changes in more detail. In Figure 6 we show the innermost mass layer reached by the convective envelope during the first (solid lines) and second (dashed lines) dredge-up as a function of the stellar mass. The black lines represent the $Z = 0.02$ models, the red lines the $Z = 0.008$ models, and the blue lines the $Z = 0.004$ models. The FDU does not extend very far inward in the most massive models, depending on Z, but extends further inward during the SDU (Figure 6). The depth reached by the SDU is approximately the same for all the 5 and $6 M_{\odot}$ models, regardless of the initial metallicity. From Figure 6 we predict that the largest changes to the surface composition from the FDU are seen for stars of $\approx 2.5 M_{\odot}$, and the largest changes from the SDU for the most massive stars that experience it ($6 M_{\odot}$ in the case of the models shown in the figure).

In Figure 7 we illustrate the changes made to the surface abundance ratios of ^{12}C/^{13}C, ^{14}N/^{15}N, ^{16}O/^{17}O, and ^{16}O/^{18}O for the $Z = 0.02$ models after the FDU and SDU mixing events. All isotopic ratios are given as number ratios. The ^{12}C/^{13}C ratio shows a noticeable decline after the FDU, from the initial value of 90 to values between 20 and 30 [31, 32]. The decrease in the

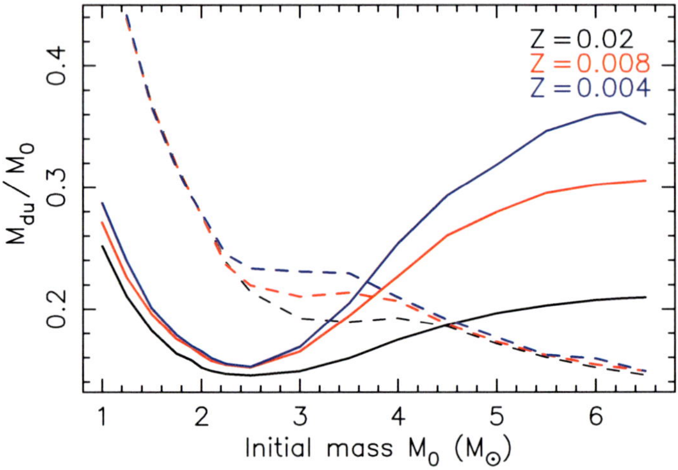

Fig. 6. Innermost mass layer ($M_{\rm du}/M_0$) reached by the convective envelope during the first (solid lines) and second dredge-up (dashed lines) as a function of the stellar mass, M_0

carbon isotopic ratio is caused by the convective envelope moving inward to a position interior to the ^{13}C-rich region left behind after core H burning. In Figure 8 we show the composition as a function of the stellar mass (in $M_\odot$) at the end of core H-burning for the $1M_\odot$ and $3M_\odot$, $Z = 0.02$ models. The abundances are in number fraction, Y, where the mass fraction $X = YA$, and A is the atomic mass. By definition, the mass fraction of all species sums such that $\sum X = 1$. We will use the number fraction Y throughout the text. The position of the maximum inward penetration of the convective envelope is also noted. For all models, hydrogen burning leaves behind a region rich in ^{13}C that is engulfed by the convective envelope during the FDU.

Theoretical predictions of the ^{12}C/^{13}C ratio after the FDU for intermediate-mass stars are in good agreement with the observations, to within $\sim 25\%$ [45, 31, 46, 32]. The situation is the opposite for low-mass stars where the predicted trend, which has the ^{12}C/^{13}C ratio increasing with decreasing mass, has been found to be inconsistent with the observations e.g., [47]. The same is true for Population II giants in globular clusters, where the deviation between theory and observation is more extreme [48, 49, 50, 51]. The same is true for the C/N ratio, which is also observed to be lower in red-giant stars than predicted by standard stellar evolution theory [31]. The observations of low ^{12}C/^{13}C and C/N ratios have been interpreted as evidence for *extra mixing* taking place between the base of the convective envelope and the H-shell. Observations indicate that the conflict between theory and observation does not arise until after the deepest first dredge-up, hence this mixing most likely takes place after the H-shell has erased the composition discontinuity left by the FDU event. The physical mechanism that causes the extra mixing is still

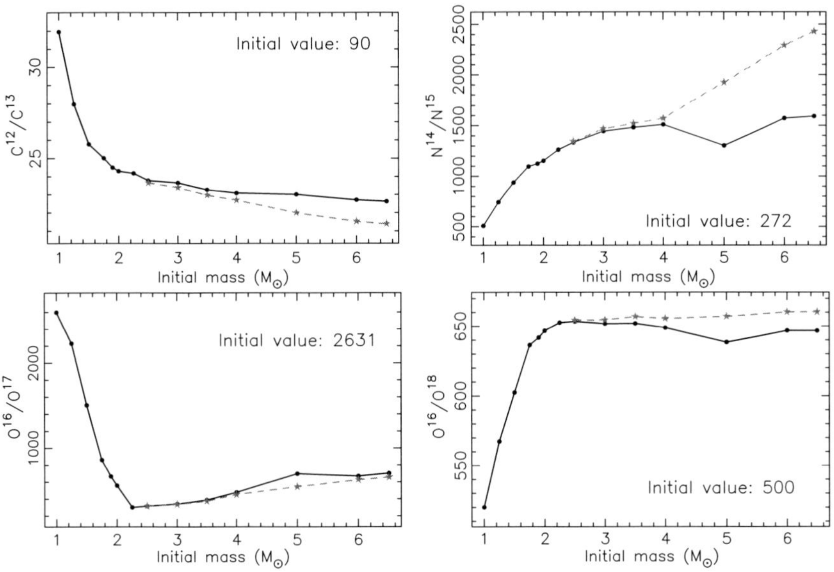

Fig. 7. The surface abundance ratios (by number) of $^{12}C/^{13}C$, $^{14}N/^{15}N$, $^{16}O/^{17}O$, and $^{16}O/^{18}O$ as a function of the stellar mass for the $Z = 0.02$ models. The solid line (with points) show the ratios after the FDU and the dashed red line (with stars) show the ratios after the SDU. The initial ratio is indicated on each panel

uncertain, with meridional circulation caused by rotation [52] and thermohaline mixing suggested as possible candidates [53, 54, 55]. Most algorithms used in deep mixing models are parameterised in some way to reproduce the observed data [56, 57, 32, 58]. For example, Boothroyd & Sackmann [32] take material from the base of the convective envelope into a region just hot enough for some nuclear processing, and then transport the material back to the envelope. The temperature and depth to which the material is mixed, along with the amount of material in the circulation current, is parametrized to obtain the observed abundances.

From Figure 7 and 8 we see that the $^{14}N/^{15}N$ ratio increases dramatically in all models after the FDU and SDU events, regardless of the initial mass or metallicity. For a $2M_\odot$ star, the increase is about a factor of 4, whereas this increases to about a factor of 6 for 3–$4M_\odot$ stars. The composition profiles (Figure 8) demonstrate the mechanism for the large increase in the $^{14}N/^{15}N$ ratio. The increase can be attributed to the production of ^{14}N at the expense of ^{12}C but also to a reduction in the abundance of ^{15}N from CNO cycling. The equilibrium value of the $^{14}N/^{15}N$ ratio is 2.5×10^4 [9], much higher than the $^{14}N/^{15}N$ ratios seen in stellar models after either the FDU or SDU. The maximum inward extent of the convective envelope reaches the first ^{14}N peak

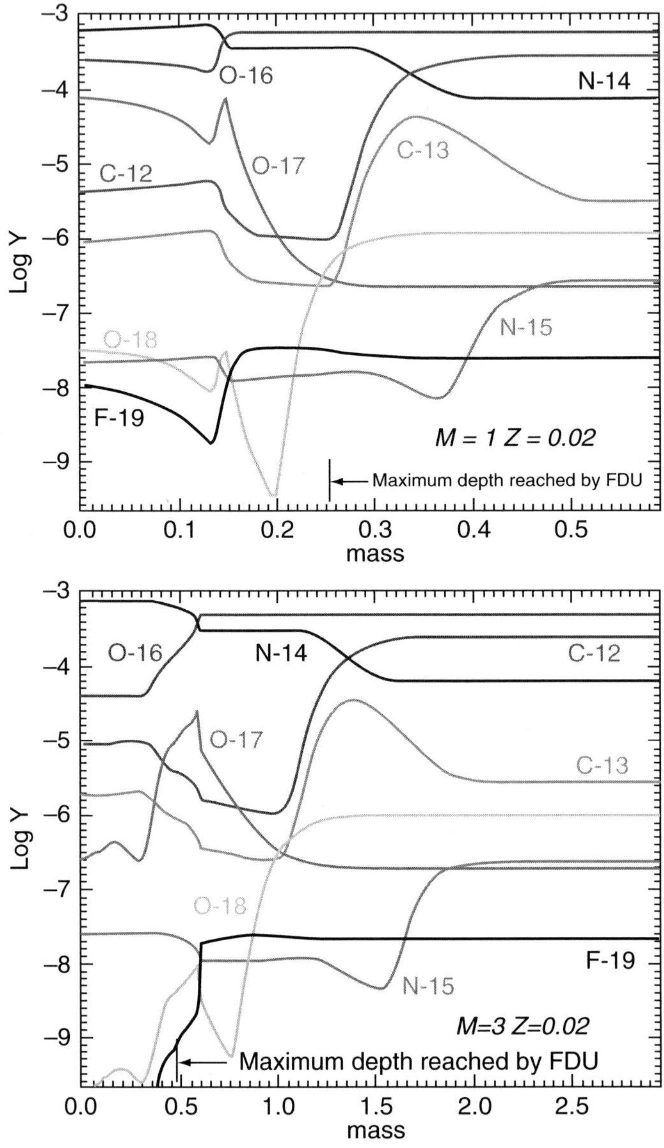

Fig. 8. The composition profile for the C, N, and O isotopes as a function of the interior mass for the $1M_{\odot}$ and $3M_{\odot}$ $Z = 0.02$ models. The unit on the y-axis is the logarithm of the number fraction, Y, where the mass fraction is given by $X = YA$, and A is the atomic mass. The composition profile is a *snap-shot* of the interior composition of the star at an instant in time, in this case at the end of core hydrogen exhaustion

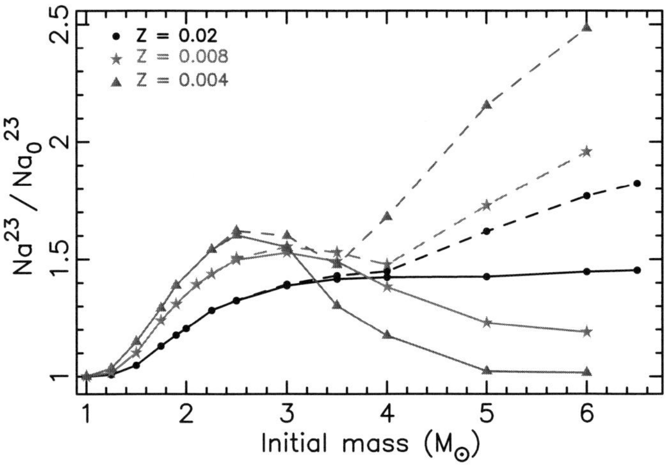

Fig. 9. The fractional changes to the surface ^{23}Na abundances after the FDU (solid lines) and SDU (dashed lines)

but only in models with the deepest FDU, or for those models that experience the SDU, does the convective envelope reach the second ^{14}N plateau caused by ON cycling. Hence, these stars are predicted to have the highest ^{14}N/^{15}N ratios.

Figure 7 shows that the oxygen isotopes experience only small changes from the first and second dredge up, with an increase in ^{17}O, and depletions in ^{18}O, depending on the initial mass and metallicity. Figure 8 shows that the convective envelope may move into the region depleted in ^{18}O and enriched in ^{17}O, but leaves the ^{16}O abundance essentially unchanged. For elements heavier than oxygen, the first and second dredge-up affects the surface abundances of ^{7}Li, ^{19}F, ^{23}Na, and to a lesser extent the Ne, Mg and Al isotopes. Lithium is easily destroyed in stellar atmospheres at relatively low temperatures and is depleted by burning during the pre-main sequence phase. Hence the FDU results in strong surface depletions of this light element. Fluorine is also depleted at the surface in all but the lowest mass models, with maximum depletions on the order of $\sim 20\%$. In Figure 9 we show the fractional change to the surface ^{23}Na abundance as a result of the FDU and SDU. For elements heavier than O, this element shows the largest change, with increases of up to a factor of 2 for the most massive models after the SDU.

4 Evolution during the AGB

We now focus on the thermally-pulsing AGB phase of evolution, which alters the surface abundances of the models in two distinct and important ways. The first is through the operation of the *third dredge-up*, which can occur periodically after each thermal pulse (TP) and is the mechanism for turning

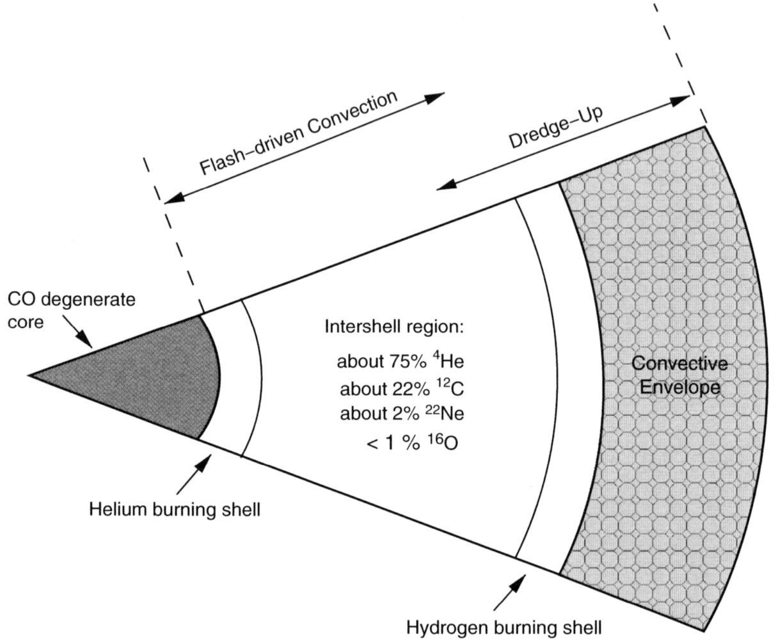

Fig. 10. Schematic structure of an AGB star showing the degenerate C-O core surrounded by a He-burning shell above the core, and a H-burning shell below the deep convective envelope. The burning shells are separated by an intershell region rich in helium ($\sim 75\%$) and carbon ($\sim 22\%$), with some oxygen and ^{22}Ne. From [36]

(single) stars into carbon-rich stars, where the C/O ratio of the envelope exceeds unity. The second mechanism is *hot bottom burning* (HBB). In this section we focus on the structure and evolution of AGB stars, with a discussion of the nucleosynthesis in Sect. 5.

There are many reviews on AGB evolution including Iben [59], Frost & Lattanzio [29], Wood [60], Busso, Gallino, & Wasserburg [7], and Herwig [1]. Here we briefly review the main features. Following core He-exhaustion, the structure of the star is shown schematically in Figure 10. The H-exhausted core of the star, composed primarily of carbon and oxygen, begins to contract. At the same time the outer layers expand and the star ascends the giant branch for the second time. He-burning is ignited in a thin layer around the degenerate C-O core but instabilities quickly set in, due to the thinness of the burning shell. At this stage the star is said to have entered the thermally-pulsing AGB (TP-AGB) phase of evolution. This stage of evolution is characterized by relatively long periods of quiescent H-shell burning, known as the interpulse phase, interrupted by instabilities of the He-shell. The ashes of the H-shell increase the thickness of the hydrogen depleted region, until eventually the next thermonuclear runaway or thermal pulse. During a TP, the He-shell burns fiercely, producing up to $\sim 10^8 L_\odot$ for a few hundred years. This enormous

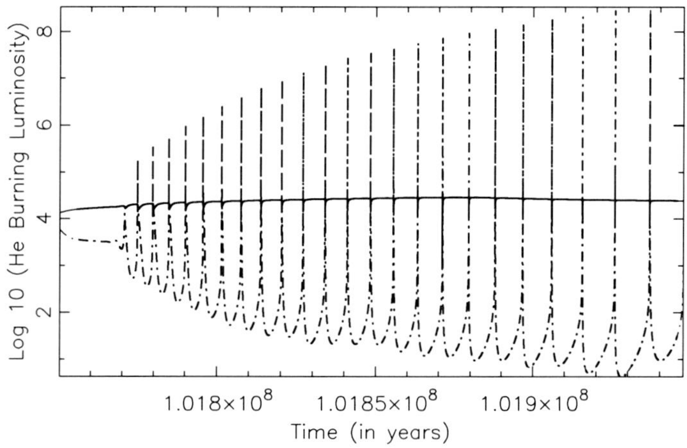

Fig. 11. Temporal evolution of the radiated (solid line) and He-burning (dashed line) luminosities during the TP-AGB for the $5M_{\odot}$, $Z = 0.02$ model. The AGB life cycle is shown by the repetition of the He-shell flashes (24 in total), followed in each case by much longer periods of H-shell burning. The interpulse period of this star is $\approx 7,000$ years

quantity of energy does not reach the surface but goes into expanding the outer layers of the star, causing the H-shell to become extinguished. In Figure 11 we show the He-shell and radiated luminosity from a $5M_{\odot}$, $Z = 0.02$ model during the TP-AGB phase. The energy from the TPs does not noticeably affect the radiated luminosity, which stays approximately constant at about $20{,}000L_{\odot}$.

The energy produced by the flash powers a convective region in the He-shell, which has the effect of homogenizing abundances in this region. In Figure 12 we show the convective regions in the intershell during the first 5 TPs of a $1.9M_{\odot}$, $Z = 0.008$ model. The teardrop-shaped green pockets represent flash-driven convection. Following a TP the star expands, this is shown by the outward movement of the base of the convective envelope, at $\approx 0.56M_{\odot}$. When the flash dies down, the outer convective envelope moves inward in mass to the region mixed by intershell convection. This phase is known as the third dredge-up (TDU), and can occur after each thermal pulse. Note that there are no TDU events visible in Figure 12, where the base of the convective envelope is seen to move inward but does not penetrate the flash-driven convective region. The result of the third dredge-up is to mix the products of partial He-burning (mostly ^{4}He and ^{12}C, see Figure 10) to the surface. This is the process which turns (single) stars into carbon-rich stars. Following dredge-up, the star contracts and the H-burning shell is re-ignited and the star enters a new interpulse phase. The cycle of *interpulse–thermal pulse–dredge-up* may occur many times on the AGB, depending on the initial mass and composition, as well as on the mass-loss rate.

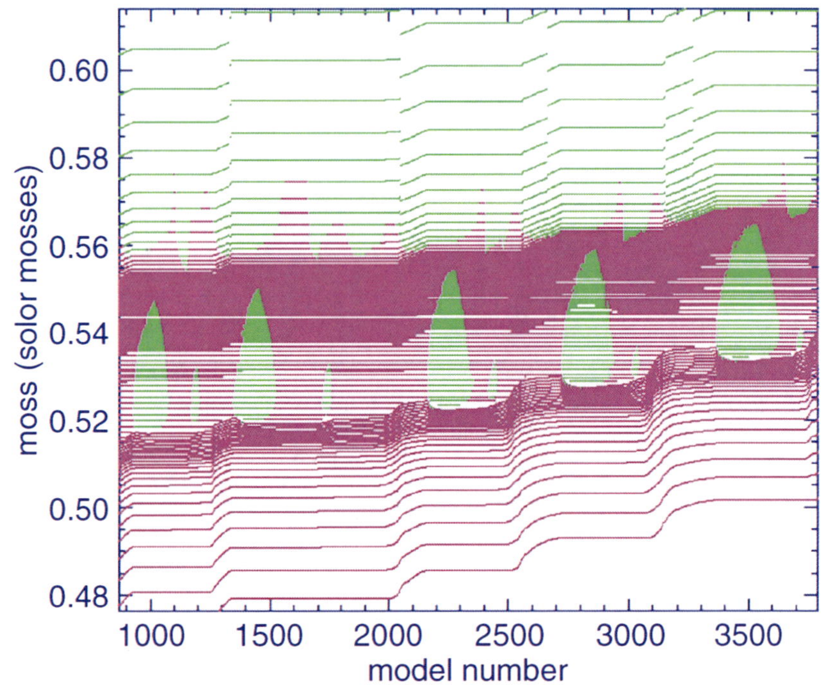

Fig. 12. Convective regions for the $1.9M_\odot$, $Z = 0.008$ model during the first five thermal pulses. The He-intershell extends from about $0.52M_\odot$ to $0.56M_\odot$. The teardrop-shaped pockets correspond to the flash-driven convective region that extends over most of the He-intershell. These have the effect of homogenizing the abundances within the He-intershell. Convective regions are shaded in green and radiative zones in magenta. The x-axis is nucleosynthesis time-step number, which is a proxy for time. For this model, the duration of the convective zones are about 250 years

In the most massive AGB models, the convective envelope can dip into the top of the H-shell, resulting in nuclear burning at the base of the convective envelope. This phenomena is known as hot bottom burning (HBB) and can also dramatically alter the surface composition. It has been known for some time that intermediate mass stars over about $5M_\odot$ develop deep convective envelopes with very high temperatures at the base, allowing for nuclear burning and some energy generation [61]. The observational evidence that HBB is occurring in massive AGB stars came from the lack of luminous carbon-rich stars in the Magellanic Clouds [4]. Many of these luminous, O-rich stars were later found to be rich in lithium [62,63,64]. The first detailed calculations were not made until the early 1990's by Blöcker & Schönberner [65] and Lattanzio [66]. Blöcker & Schönberner found that the linear core-mass-luminosity relation first proposed by Paczynski [67] does not apply to stellar models with HBB. Boothroyd, Sackmann, & Ahern [68] found that HBB prevents the

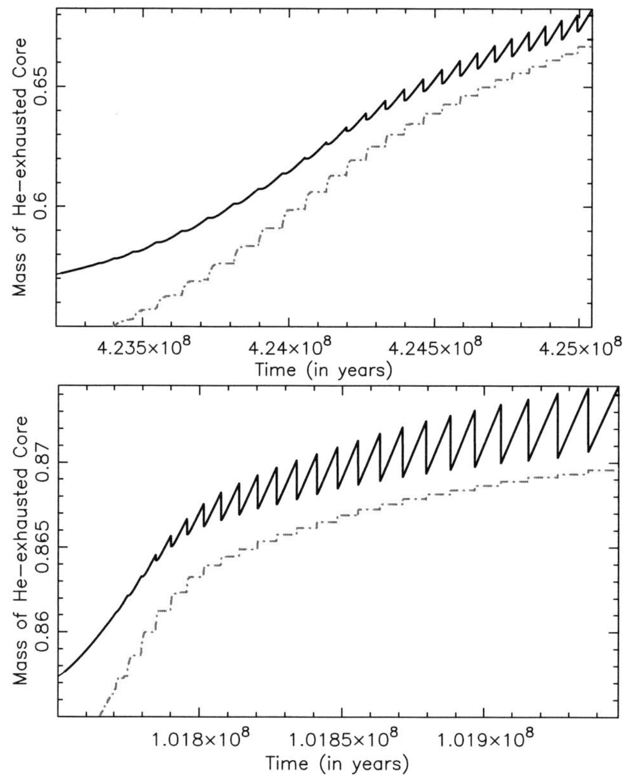

Fig. 13. The mass of the hydrogen-exhausted core (solid line) and helium-exhausted core (dashed line) as a function of time for the $3M_{\odot}$, $Z = 0.02$ model (top) and $5M_{\odot}$, $Z = 0.02$ model (lower)

formation of a carbon-rich atmosphere by burning ^{12}C into ^{14}N and provided a mechanism for the lack of bright C-stars in the LMC.

In summary, the AGB evolutionary cycle can be broken down into four distinct phases:

1. **The on-phase** or thermal pulse, which is when the He-shell burns brightly producing up to 100 million $L_{\odot}$ in a short time ($\sim 10^2$ years). The energy drives a convective zone in the He-intershell.
2. **The power-down phase** when the He-shell dies down. The enormous amount of energy from the TP drives an expansion of the whole star and the H-shell is pushed to cooler regions and is extinguished.
3. **The third dredge-up** phase, which is when the convective envelope moves inwards into regions previously mixed by the flash-driven convection. Carbon and other He-burning products are mixed to the stellar surface.

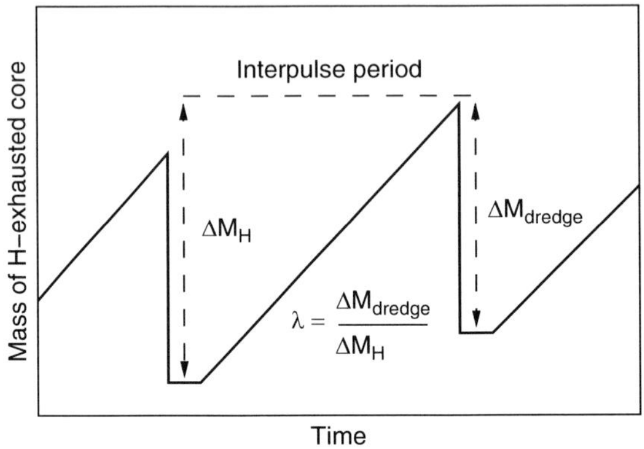

Fig. 14. The definition of λ, shown schematically, where the x–axis represents time and the y–axis represents the mass of the H-exhausted core

4. **The interpulse phase**, which is when the whole star contracts after the TDU; the H-shell is re-ignited and provides most of the surface luminosity for the next 10^4 years or so.

In Figure 13 we show the evolution of the H and He-exhausted cores for the $3M_{\odot}$ and $5M_{\odot}$, $Z = 0.02$ models. These models both show third dredge-up. For the $5M_{\odot}$ model, deep third dredge-up starts after the fifth TP, this corresponds to a decrease in the H-exhausted core mass. The $3M_{\odot}$ model, on the other hand, does not experience much dredge up until the $\sim 13^{\text{th}}$ thermal pulse. One striking difference between the 3 and $5M_{\odot}$ models is the mass of the He-intershell, which is much smaller by about a factor of five in the $5M_{\odot}$ model. This mass difference has important consequences for the production of heavy elements in low-mass AGB stars. This is because at a given dredge-up efficiency, more matter is mixed into the envelope, at a smaller dilution, in lower mass stars than in intermediate-mass AGB stars. We can quantify this by defining λ, which is a parameter used to measure the depth, or efficiency of the third dredge-up. If there is dredge-up, then a fraction of the H-exhausted core will be mixed into the envelope

$$\lambda = \frac{\Delta M_{\text{dredge}}}{\Delta M_{\text{core}}}. \tag{2}$$

where λ is the third dredge-up efficiency parameter, ΔM_{dredge} is the mass mixed into the envelope, and ΔM_{core} is the amount by which the He-intershell increased over the previous interpulse phase; see Figure 14.

The value of λ depends on physical parameters such as the core mass and metallicity of the star. Exactly how λ depends on these quantities is unknown. From the above definition it should be clear that when $\lambda = 1$, the core mass does not grow but remains constant. In Figure 15 we show the efficiency of the TDU as a function of the core mass for two $3M_{\odot}$, $Z = 0.02$ models, computed with two different stellar evolutionary codes. The top panel shows results from Karakas et al. [36] whereas the lower panel shows results from Straniero et al. [69]. The behaviour of λ with core mass is quite different, with the Karakas model [36] showing more efficient third dredge-up at the same core mass.

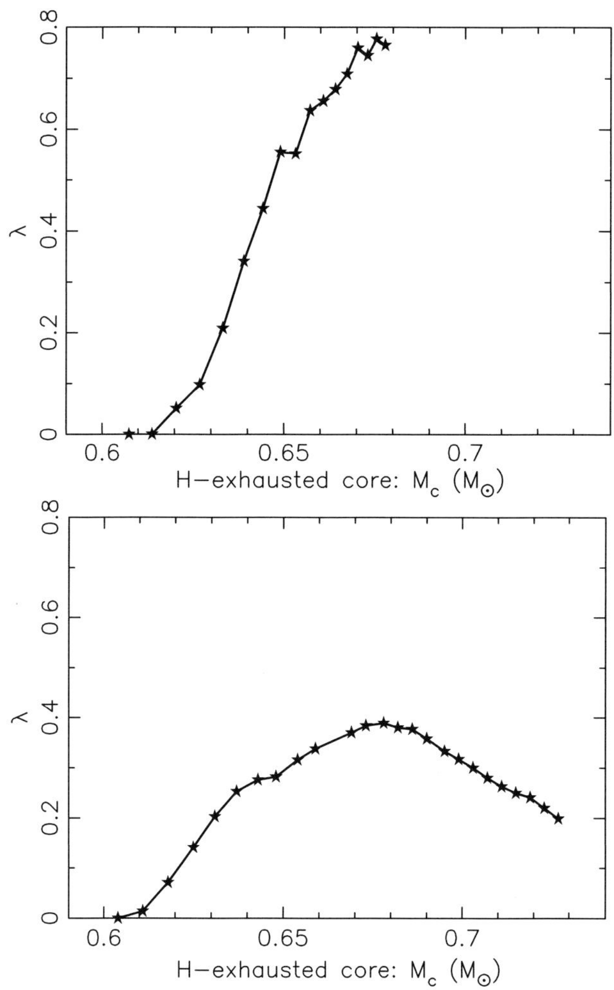

Fig. 15. (Top) λ as a function of the core mass for the $3M_\odot$, $Z = 0.02$ model. (Bottom) Same, but using data from [69]

AGB nucleosynthesis not only depends on the efficiency of the third dredge-up but also on the minimum core mass for thermal pulses, and on the minimum core mass for the onset of the third dredge-up. These three quantities were parameterised by Karakas et al. [36], and were found to vary as a function of the total mass, envelope mass, and metallicity in a complex

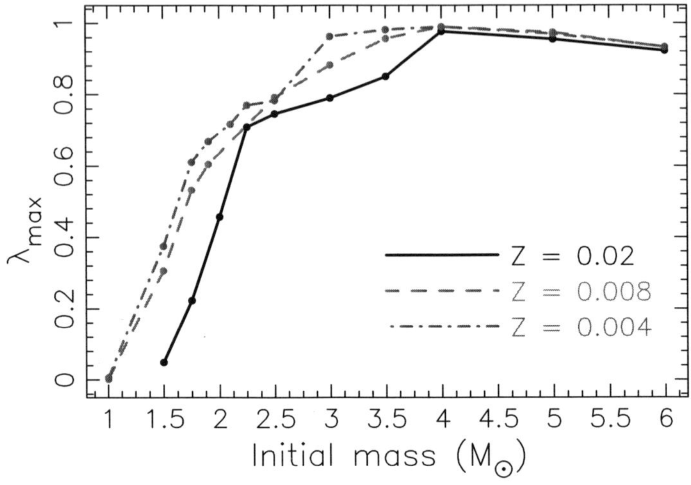

Fig. 16. The maximum value of λ calculated from the $Z = 0.02$, 0.008 and 0.004 models. From [36]

way. In Figure 16 we show the variation of the maximum λ calculated from grids of stellar models covering a range of mass and metallicity. The general trend is that λ increases with increasing stellar mass, at a given Z. λ also increases with decreasing metallicity, at a given mass (see also [70]). This means that it should be easier to make carbon stars in lower metallicity or higher mass models. These three quantities still suffer from large uncertainties, mostly because of our lack of understanding about how convection operates in stellar interiors. Different stellar codes predict different behaviour (Figure 15; [71, 72]).

It is important to know if the stellar models are providing an accurate description of the efficiency of mixing in AGB stars. For example, the models shown in Figure 16 do not predict any TDU for models less than $2M_{\odot}$ at $Z = 0.02$. This is in conflict with observations of carbon stars that suggest that the minimum mass for carbon stars in the Milky Way Galaxy is $\approx 1.5M_{\odot}$ [73]. Thus models such as those presented in Karakas et al. [36] do not predict an accurate picture of TDU mixing in low-mass AGB models. It is possible to constrain the efficiency of the third dredge-up using other observables? The two closest satellite galaxies of the Milky Way are the LMC and Small Magellanic Cloud (SMC) and they both have thousands of known carbon stars [74,75]. We know the distances to these two galaxies reasonably well, enabling us to construct carbon-star luminosity functions (CSLF).

The convenient fact that the stellar luminosity on the AGB is a nearly linear function of the H-exhausted core mass has stimulated the development of *synthetic* AGB evolution models, as a quick way of simulating populations of AGB stars. The main observational constraint which models must face is the CSLF for the Magellanic Clouds. Synthetic AGB evolution calculations performed by Groenewegen & de Jong [76] and Marigo, Bressan, & Chiosi [77], treat λ as a constant free parameter, calibrated by comparison with the CSLF. These studies suggest $\lambda \sim 0.6$ at a core mass of $0.58 M_{\odot}$ is required to fit the CSLF of the Magellanic Clouds (see Chapter 3 in [14] for further details). Other calculations take parameterizations of λ and the minimum core mass for TDU [78,36] and adjust as required to fit the CSLF [79]. Stancliffe, Izzard, & Tout [80] were able to reproduce the CSLF of the LMC by computing new AGB models using the Cambridge *STARS* code, which predicts deeper TDU at smaller core masses than Straniero et al. [69] or Karakas et al. [36].

4.1 Carbon stars

Thermal pulses and dredge-up may occur many times during the TP-AGB phase, as shown by Figures 11 and 13. Each TDU episode mixes ^{12}C from the He-intershell into the envelope and this has the effect of slowly increasing the C/O ratio of the surface. Repeated TDU episodes can explain the transition from M-type (C/O ≈ 0.5, similar to the Sun) to carbon star:

$$\mathrm{M} \rightarrow \mathrm{MS} \rightarrow \mathrm{S} \rightarrow \mathrm{SC} \rightarrow \mathrm{C}, \tag{3}$$

where SC-type stars have C/O equal to unity, and C-type stars have C/O > 1 by definition. Many carbon stars also have surface enrichments of heavy elements synthesized by the *s*-process (e.g., Ba, and Tc) [81, 82, 83, 84]. In particular, enrichments in the radioactive element technetium is an indication that the star is a real AGB star that has experienced "recent" nucleosynthesis and mixing [85,86]. This is because the half-life of ^{99}Tc (the isotope produced by the *s*-process) is 210,000 years, much shorter than the main sequence lifetime of low-mass stars.

There are other classes of carbon-rich stars including the cool and warm R-type stars, the ^{13}C-rich J-type stars, CH-stars, and dwarf carbon stars [73]. Barium stars also show enrichments in carbon and heavy elements, although they have C/O < 1 in general. Possibly $\sim 20\%$ of all very metal-poor with [Fe/H] < -2 are also carbon rich, with [C/Fe] ~ 2 in some cases [20]. Stars in these other classes are not on the AGB and are not responsible for producing their own carbon enrichments. Some of them, such as the barium and CH-type stars, are all known binaries [87,88] and thus presumably obtained their carbon from a former AGB companion. These stars are also devoid of Tc enrichments [89,90]. The warm R-type stars are all single stars [91], and may all result from some type of binary-star merger event (see Izzard, Jeffery & Lattanzio [92] and references therein). The rare J-type stars, with very low ^{12}C/^{13}C ratios, are still a mystery [93,94].

4.2 Luminosity variability

Luminosity variability is a common feature of AGB stars. The cause of the variability is mainly radial and non-radial pulsations of the envelope, although there is some evidence for variations caused by ejections of dust shells or orbiting dust clouds. While thermal pulses generally occur over a much longer timespan than the observed luminosity variability (10^4 years compared to a few hundred days), they can interact with the shorter-term pulsations. There are three groups of long-period variables that include AGB stars: Mira-type variables (large amplitude), and the semi-regular and irregular variables (smaller amplitude than the Miras). There are also dust-enshrouded variables that are too faint to be observed in the optical; e.g., OH/IR stars [95]. Stellar pulsation theory, when combined with pulsation periods and absolute luminosities, can be used to derive stellar masses [96]. The LMC and SMC provide large samples of variable AGB stars, where the entire mass range covered by AGB stars (from 0.8 to $8M_{\odot}$) is observed [4]. The radial pulsation modes of AGB stars are essentially confined to the convective envelope, with the nuclear-burning core acting as a point source of energy and gravity. Hence, energy transport is dominated by convection. We refer the reader to computations by Fox & Wood [97] and Ostlie & Cox [98] for details.

4.3 Mass loss

Mass loss is an essential component of any study involving AGB stars. This is because it is mass loss that terminates the AGB phase by removing the convective envelope, and this in turn determines the level of chemical enrichment by setting the number of thermal pulses and mixing events. Our knowledge of the dependence of AGB mass loss on stellar properties is rather poor, although advances are being made in the theory of AGB winds [99,100,101,102]. These winds are driven by the combination of stellar pulsation and radiation pressure acting on dust grains. In AGB evolutionary calculations (e.g., [103, 104, 105, 24]) the mass-loss laws [106, 107] are based on observational estimates and theoretical estimates, respectively. One important feature of these laws is that the mass-loss rate increases very rapidly with increasing stellar luminosity. This implies that most of the mass is lost right at the end of the AGB.

5 Nucleosynthesis during the AGB

It is on the TP-AGB that the richest nucleosynthesis occurs for low and intermediate-mass stars, even though stars spend such a short amount of time there compared to previous evolutionary phases. The nucleosynthesis is driven by thermal instabilities of the helium-burning shell, reviewed in Sect. 4. Of particular importance is the action of repeated third dredge-up events that mix the products of He-burning to the stellar surface. Material from the He-shell will become part of the next hydrogen shell, where they will experience proton captures during the next interpulse period. For this reason, not only do we need to consider nucleosynthesis in the thermal pulse itself but also the effect of subsequent nucleosynthesis of the dredged up material in the H-shell. Hot bottom burning occurs in the most massive AGB stars, which takes place when the base of the envelope becomes hot enough to sustain proton-capture nucleosynthesis. Overall, the TP-AGB gives rise to a combination of H and He-processed material that is expelled by the star as its envelope is lost through stellar winds.

5.1 Nucleosynthesis in the hydrogen-burning shell

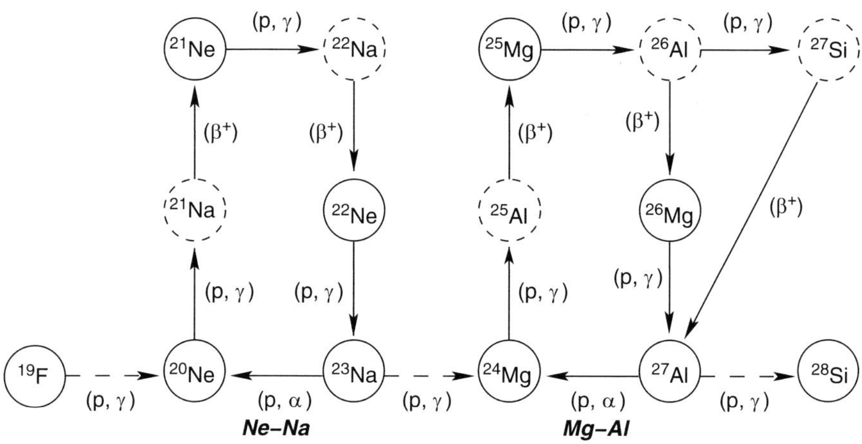

Fig. 17. Reactions of the Ne-Na and Mg-Al chains. Unstable isotopes are denoted by dashed circles

Hydrogen is burnt to ^{4}He via the CNO cycles, but the Ne-Na and Mg-Al chains also operate. The main result from the CNO cycle, besides the conversion of H to He, is an increase in the abundances of ^{13}C and ^{14}N from the destruction of other CNO species. The isotopes ^{12}C and ^{15}N are first destroyed by the CN cycle, which comes into equilibrium quickly. Later, the oxygen isotopes ^{16}O and ^{18}O are also destroyed to produce ^{14}N. The abundance of ^{17}O can be enhanced by the CNO cycle, depending on the uncertain rate of the ^{17}O + p branching reactions. The H-shell ashes are important because they affect the nucleosynthesis that occurs in the He-shell.

In the left part of Figure 17 we show the reactions of the Ne-Na chain [10, 44], where unstable isotopes are denoted by dashed circles. The main result of the Ne-Na chain is the production of ^{23}Na at the expense of the neon isotopes, primarily ^{22}Ne, which begins to be destroyed at about 20 million K. The production of sodium by the Ne-Na chain was examined in detail by [108], who predicted that AGB stars could play an integral role in the chemical evolution of sodium in the Galaxy. The rare neon isotope, ^{21}Ne is completely destroyed at temperatures over 40 million K. The dominant ^{20}Ne is not significantly altered by H-shell burning, but the destruction of ^{23}Na at temperatures over 80 million K can lead to a slight enhancement in the ^{20}Ne abundance. The rate by which ^{23}Na is destroyed is important for determining sodium yields (see [109, 110]). Whether there is leakage out of the Ne-Na chain into the Mg-Al chain depends on the relative rates of the uncertain ^{23}Na(p,α)^{20}Ne and ^{23}Na(p,γ)^{24}Mg reactions.

Magnesium and aluminium are altered in the H-burning shell via the activation of the Mg-Al chain, which begins operation at temperatures of about 30 million K [44]. This involves the radioactive nuclide ^{26}Al which has a ground state $^{26}Al_g$ that has to be considered a separate species from the short-lived ($\tau_{1/2} = 6.35\,s$) isomeric state $^{26}Al_m$, since they are out of thermal equilibrium at the relevant temperatures [44]. Hereafter, when we refer to ^{26}Al we are referring to the ground-state, $^{26}Al_g$. In the right part of Figure 17 we show the reactions involved in the Mg-Al chain [10, 44]. The first isotope in the Mg-Al chain to be affected is ^{25}Mg, which is burnt to ^{26}Al. The lifetime of β-decay relative to proton capture generally favours proton capture within the H-burning shell. This produces the unstable ^{27}Si which β-decays (with a lifetime on the order of a few seconds) to ^{27}Al. The rate of ^{26}Mg + p is slow until the temperature reaches about 60 million K, which leads to small reductions in ^{26}Mg in the H-shells of AGB models. The abundance of ^{26}Mg is enhanced by the β-decay of ^{26}Al in the H-shell ashes. Proton capture on ^{24}Mg requires higher temperatures than those required for the other reactions in the Mg-Al chain. For that reason AGB models show little change in the abundance of this isotope due to the slow rate of proton capture at temperatures below about 70 million K.

Even in the lowest mass AGB models the Ne-Na and Mg-Al chains are active in the H-shell during the TP-AGB, depleting ^{22}Ne and ^{25}Mg and producing ^{23}Na and ^{26}Al. The products of the H-shell are processed through He-shell burning before being dredged to the surface, therefore only those products that are not altered by α or neutron-captures will make it into the envelope (e.g., ^{17}O and ^{23}Na, depending on the temperature in the He-shell, whereas ^{14}N and ^{26}Al are destroyed).

5.2 Nucleosynthesis during thermal pulses

A He-shell flash occurs in the intershell region, which is composed mostly of the *ashes* of H-shell burning ($\sim$ 98% ^{4}He and 2% ^{14}N). The two main energy-generating reactions are:

1. **the triple-alpha process:** effectively 3 $^{4}\mathrm{He} \rightarrow {}^{12}\mathrm{C}$,
2. **the $^{12}C(\alpha,\gamma)^{16}O$ reaction**, which is relatively unimportant during thermal pulses owing to the short He-burning timescale.

The He-burning shell in AGB stars is a rich source of nucleosynthesis. The main result is that some of the ^{4}He in the shell is converted into ^{12}C by

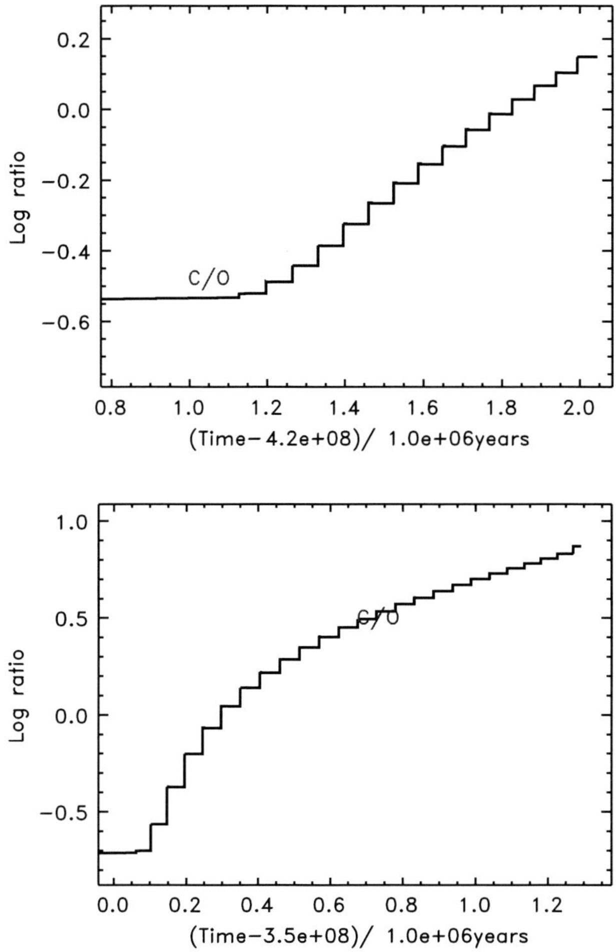

Fig. 18. Logarithm of the surface C/O ratio as a function of time during the TP-AGB for two $3M_{\odot}$ models of different metallicity. The top panel shows the C/O ratio from the $Z = 0.02$ model, the lower panel shows results for the lower metallicity $Z = 0.004$ model. Time has been scaled such that the start of the TP-AGB is $t = 0$

partial He burning, leaving the composition of the shell roughly 70–75% ^{4}He (by mass) and 20–25% ^{12}C. There is a few percent (by mass) each of ^{16}O and ^{22}Ne, and trace amounts of other species including ^{17}O, ^{23}Na, ^{25}Mg, ^{26}Mg, and ^{19}F. The exact composition of the He-intershell after a thermal pulse depends on the mass and composition of the He-shell before the pulse, the duration of the shell flash, as well as the peak temperature and density under which the burning takes place. These quantities in turn depend on the mass and metallicity of the star, and they evolve with time. This is because the core is contracting while the star evolves along the TP-AGB, causing the He-shell to become slightly more electron degenerate. Thus thermal pulses occur under somewhat more extreme conditions at the end of the TP-AGB (i.e., higher temperatures and densities in a thinner shell) than at the beginning.

AGB stars are one of the most important sources of ^{12}C in the Galaxy. An estimate of the contribution of ^{12}C from AGB stars suggests that they produce roughly one third of the Galaxy's inventory of ^{12}C, providing roughly the same amount as core-collapse supernovae and Wolf-Rayet stars [8]. These quantitative estimates are hindered by uncertainties in the depth and onset of the third dredge-up, as discussed in Sect. 4. In Figure 18 we show the log C/O ratio as a function of time during the TP-AGB for two $3M_\odot$ models, one of solar composition and the other with $Z = 0.004$ ([Fe/H] ~ -0.7). It is evident that the lower metallicity model becomes a carbon star in fewer thermal pulses, and that the final C/O ratio is higher. This occurs for two reasons. First, there is less oxygen in the lower metallicity model so it is easier to obtain the situation where C/O > 1. Second, dredge-up is more efficient.

During a thermal pulse, the ^{12}C$(\alpha,\gamma)^{16}$O reaction does not have time to produce much ^{16}O, resulting in a ^{16}O intershell composition that is $\approx$ 1%. This is the case for all standard model calculations that do not assume any convective overshoot of the flash-driven convective region [70]. Herwig [111] does include convective overshoot at the inner border of the flash driven convective zone and finds that some C and O from the C-O core is mixed into the intershell. This has the effect of increasing the C and O abundances to ~ 40 and ~ 20%, respectively. There is a wealth of other He-burning products including ^{22}Ne, ^{25}Mg, ^{26}Mg, and ^{27}Al [112, 113, 114], plus species produced through the combined operation of the He and H-burning shells including ^{19}F [115], and ^{23}Na [108].

The isotope ^{22}Ne is produced by the chain ^{14}N$(\alpha,\gamma)^{18}$F, where ^{18}F β-decays to ^{18}O allowing for the chain ^{18}O$(\alpha,\gamma)^{22}$Ne. The composition of ^{22}Ne in the He-intershell is fairly high, at ≈ 2%. This is because the abundant ^{14}N is completely converted into ^{22}Ne during the TP; see the top panel of Figure 19. From this figure we note that the He-intershell is also enriched in ^{23}Na. Sodium was not produced by He-burning, but was synthesized during H-shell burning during the previous interpulse (by the CNO cycle and Ne-Na chain, respectively). The next TDU episode will mix some fraction of the He-intershell into the envelope. The effect of many of these efficient TDU events will greatly affect the composition of the surface, shown in the bottom panel of Figure 19. The ^{22}Ne abundance is seen to increase by almost an order of magnitude (~ 1 dex). If the ^{22}Ne abundance exceeds or is equal to the ^{20}Ne abundance we should expect an enhancement in the elemental Ne composition and this is the case for the $4M_\odot$ model.

If the peak temperature of the TP reaches about 300×10^6 K, the neutron-rich Mg isotopes, ^{25}Mg and ^{26}Mg, can be synthesized by the ^{22}Ne$(\alpha,n)^{25}$Mg and ^{22}Ne$(\alpha,\gamma)^{26}$Mg reactions. These two ^{22}Ne$+\alpha$ reactions have similar although uncertain rates at He-shell burning temperatures (Figure 20; see also [117, 118]). Owing to the relatively high temperatures required for these two reactions, they are predicted to only occur efficiently in the most massive

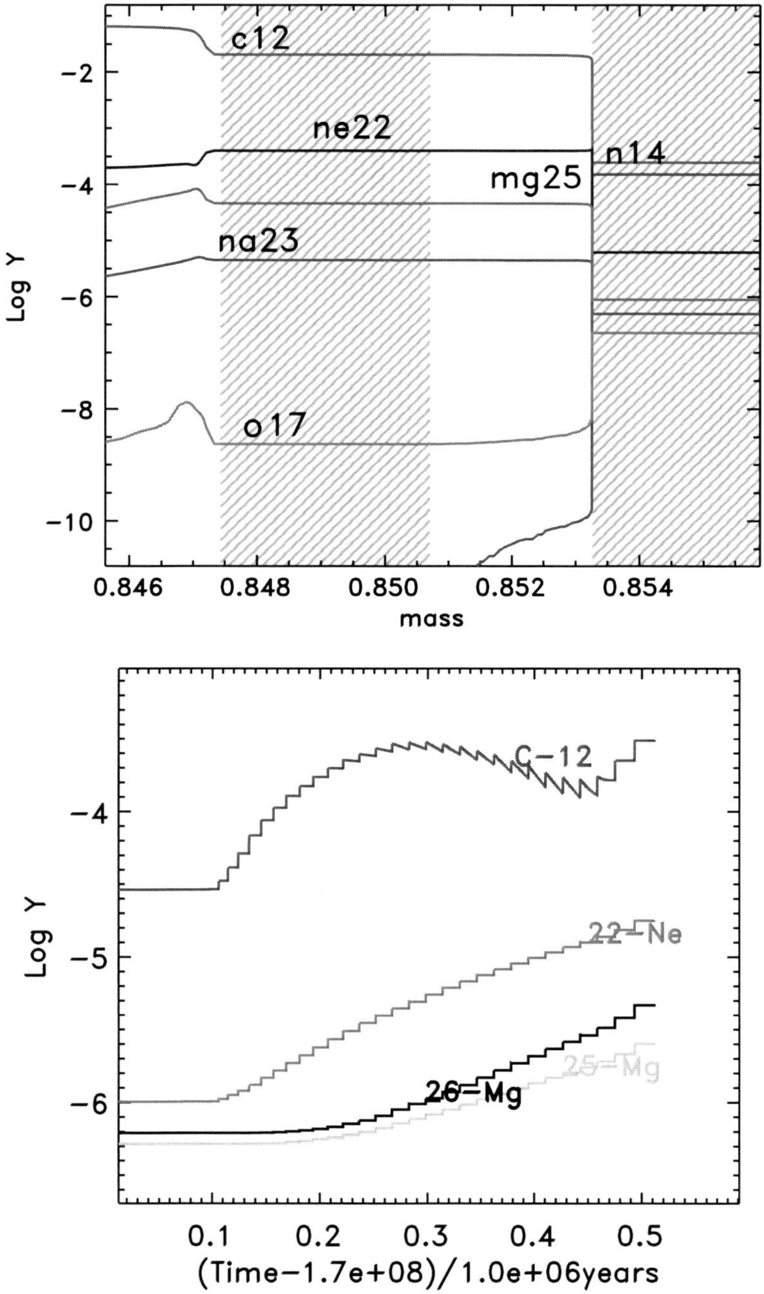

Fig. 19. (Top panel) Composition profile for the $4M_{\odot}$, $Z = 0.004$ model during the 20th thermal pulse. The peak temperature in the He-shell during this pulse is 360 million K. (Bottom panel) Evolution of the ^{12}C, ^{22}Ne, ^{25}Mg, and ^{26}Mg surface abundances of the $4M_{\odot}$, $Z = 0.004$ model during the TP-AGB. The effect of HBB can be seen on the surface abundance of ^{12}C

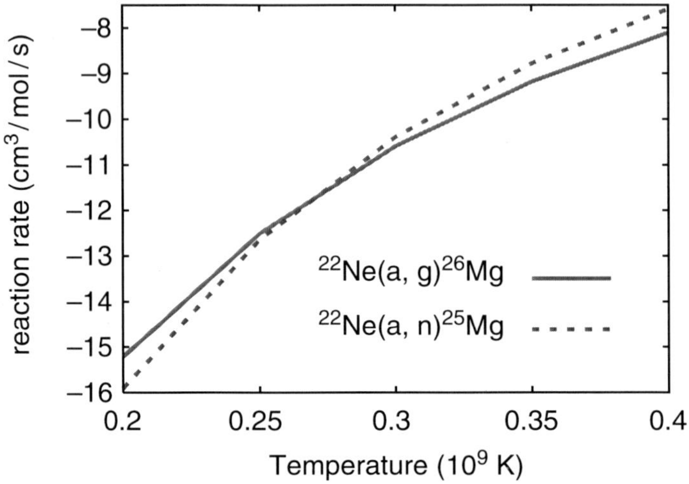

Fig. 20. The NACRE rates for the ^{22}Ne$(\alpha, n)^{25}$Mg and ^{22}Ne$(\alpha, \gamma)^{26}$Mg reactions at typical He-shell burning temperatures [116]. The reaction rate uncertainties are not shown but for the ^{22}Ne$(\alpha, n)^{25}$Mg reaction the uncertainties span more than an order of magnitude at 0.3 GK

AGB stars ($\geq 4M_{\odot}$). In Figure 21 we show the ^{22}Ne, ^{25}Mg, and ^{26}Mg intershell abundances during five thermal pulses for the $4M_{\odot}$, $Z = 0.004$ model. It is evident that some of the ^{22}Ne is being converted into the neutron-rich Mg isotopes. The initial decrease in the abundances of ^{22}Ne and the Mg isotopes is caused by the flash-driven convective zone moving into a region that was depleted in these species during the previous interpulse. Once the temperature of the TP increases, so do the abundances. The duration of the flash-driven convective zone for the $4M_{\odot}$ model is $\approx$ 26 years, and it covers a mass of approximately $6 \times 10^{-3} M_{\odot}$. The third dredge-up event that follows each of these pulses will enrich the surface, shown in Figure 19.

The He-intershell of lower mass AGB stars will only reach 300×10^6 K during the last few TPs, hence the ^{22}Ne$(\alpha, n)^{25}$Mg reaction is only marginally activated near the end of the AGB. This reaction is particularly important because it produces free neutrons that can be captured by iron-peak elements during the *s*-process. It is the dominant neutron-producing reaction in the He and C-burning shells of massive stars [119, 120, 121], and the dominant neutron source in massive AGB stars [122].

5.3 Comparison with observations: Intershell abundances

The PG1159 class of post-AGB stars are hot, helium-rich objects that show intershell matter of the preceding AGB phase on their surface. For this reason, their chemical analyses allow for a direct insight into nucleosynthesis processes during the AGB phase [123,124]. PG1159 stars are quite rare, with only about

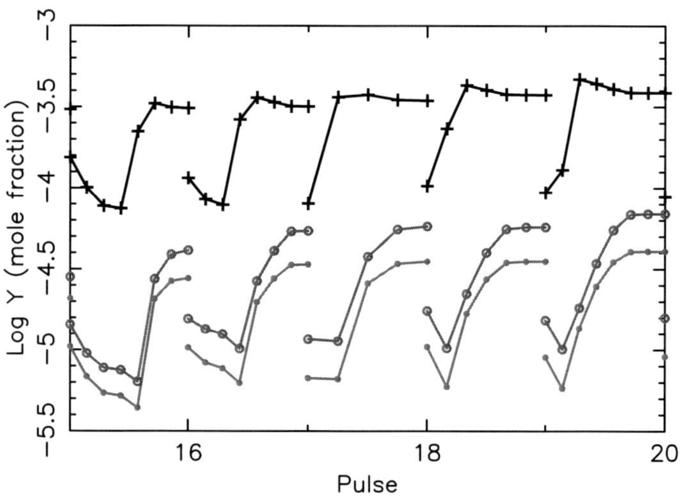

Fig. 21. The intershell abundances ^{22}Ne (plus signs), ^{25}Mg (open circles) and ^{26}Mg (closed circles) as a function of pulse number for the $4M_{\odot}$, $Z = 0.004$ model. In this diagram we show the intershell abundances during the 15$^{\rm th}$ to the 20$^{\rm th}$ pulse, but only during the time when the convective shell is present; the x-axis is the (scaled) duration of the convective pocket

two dozen known, and their atmospheres are mostly composed of the products of partial He-burning (helium, carbon and oxygen) [125, 124]. Spectroscopic observations of PG 1159 stars reveal oxygen mass fractions as high as 20% and carbon as high as 40% [125], at odds with standard stellar models that predict ≈ 1% and 25% O and C, respectively. Spectroscopic observations of neon reveal mass fractions of about 2 to 5% [125, 126], consistent with the models. Werner, Rauch, & Kruk [127] also find strong F overabundances (∼250 times solar) in a number of PG1159 stars, consistent with the F intershell abundances of Lugaro et al. [128]. Furthermore, the high F abundances in PG1159 stars confirm the conclusion from abundance determinations in red giants that F is synthesized in AGB stars.

To reproduce the large observed oxygen abundances, extra-mixing processes are required to bring ^{16}O from the C-O core into the intershell region. The diffusive convective overshoot models of Herwig [111] have intershell abundances that are consistent with the abundance patterns observed in PG 1159 central stars. The degenerate thermal pulses found by Frost, Lattanzio, & Wood [129] may also have a similar effect. It is unclear however, if this overshooting occurs for all AGB stars or only for AGB stars that produce PG 1159 stars as a consequence of late and very late TPs [2]. For example, the observed abundance analyses in intrinsic and extrinsic AGB stars, as well as in SiC grains, suggest that such an overshoot is uncommon [71].

5.4 Fluorine production in AGB stars

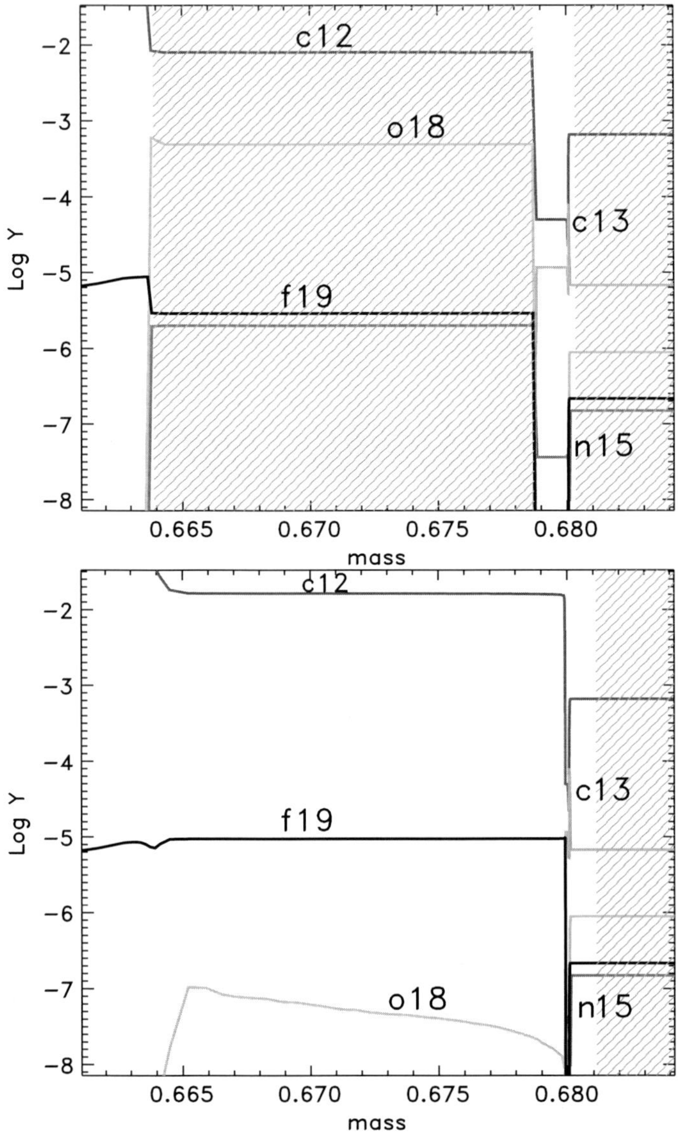

Fig. 22. The composition of the He-intershell for a $3M_{\odot}$, $Z = 0.02$ model. The upper panel shows the composition during a thermal pulse, where the flash-driven convective pocket is shown (at its full extent it reaches from $0.664M_{\odot}$ almost all the way to the H-shell at $0.679M_{\odot}$). The lower panel shows the composition of the He-intershell after the thermal pulse has died down, before the next third dredge-up takes place. The location of the base of the outer convective envelope is $\approx 0.681M_{\odot}$

The fragile element fluorine can be produced inside the He-intershell of AGB stars. Jorissen, Smith & Lambert [130] discovered that the [F/O] abundance correlates with the C/O ratio in AGB stars, and that some N-type carbon stars show surface enrichments a factor of 30 above the solar ratio. The increase in the C/O ratio is clearly a result of the third dredge-up and thermal pulses, hence it was concluded that the carbon and fluorine are produced in the same region in the star and mixed together to the surface. Jorissen et al. [130] examined the many pathways that fluorine, or more precisely the isotope ^{19}F, could be produced and concluded that the most likely chain is

$$^{14}\mathrm{N}(\alpha,\gamma)^{18}\mathrm{F}(\beta^{+}\nu)^{18}\mathrm{O}(p,\alpha)^{15}\mathrm{N}(\alpha,\gamma)^{19}\mathrm{F}. \tag{4}$$

Fluorine production takes place in the He-intershell, a region that is essentially devoid of protons, and has a very low abundance of ^{13}C and ^{15}N. Hence other reactions are required to produce the protons and the ^{15}N. These other reactions include $^{13}\mathrm{C}(\alpha,n)^{16}\mathrm{O}$ that is required to produce free neutrons; these are subsequently captured by $^{14}\mathrm{N}(n,p)^{14}\mathrm{C}$ to produce free protons. In Figure 22 we show the composition of the He-intershell during and after a thermal pulse. The increase in ^{15}N and ^{18}O during the pulse is evident; both of these species are subsequently destroyed and are absent in the lower panel. The resulting ^{19}F abundance in the He-intershell is more than an order of magnitude greater than in the envelope. Fluorine can also be destroyed via $^{19}\mathrm{F}(\alpha,p)^{22}\mathrm{Ne}$, which is more efficient at temperatures over $\approx 300 \times 10^{6}$ K [128,131]. In Figure 23 we show the surface abundance evolution of ^{19}F for the $3M_{\odot}$, $Z = 0.02$, noting the large increase over the TP-AGB through the repeated action of third dredge-up mixing events. The fluorine abundance increases by about a factor of four.

An example of the many reaction rates that affect the production of fluorine is the alternative proton production reaction $^{18}\mathrm{F}(\alpha,p)^{21}\mathrm{Ne}$ [132]. Including the $^{18}\mathrm{F}(\alpha,p)^{21}\mathrm{Ne}$ reaction reduces the abundance of ^{18}O because it competes with ^{18}O production via the $^{18}\mathrm{F}(\beta^{+}\nu)^{18}\mathrm{O}$ decay (the half life of ^{18}F is 109 minutes). However, the extra amount of protons from (α,p) enhances the $^{18}\mathrm{O}(p,\alpha)^{15}\mathrm{N}$ reaction rate, even though ^{18}O production has been deprived from the decay. This is shown in Figure 24 for the $3M_{\odot}$, $Z = 0.008$ model, where we also show an example of the effect of reaction rate uncertainties on the nucleosynthesis predictions. We refer the reader to [115], [133], [128], and [132] for more details on the complex production of fluorine in AGB stars.

The cosmic origin of fluorine is not yet completely understood, where Type II SN explosions [134] and stellar winds from Wolf-Rayet stars [135] play a significant role in producing this fragile element alongside AGB stars [136]. Certainly AGB stars and their progeny (e.g., post-AGB stars and planetary nebulae) are the only confirmed site of fluorine production thus far [130, 127, 137, 138], with no clear indication for enhanced F abundances resulting from the ν-process in a region shaped by past SNe [139]. The recent observations of a greatly enhanced fluorine abundance ([F/Fe] $= 2.90$) in a

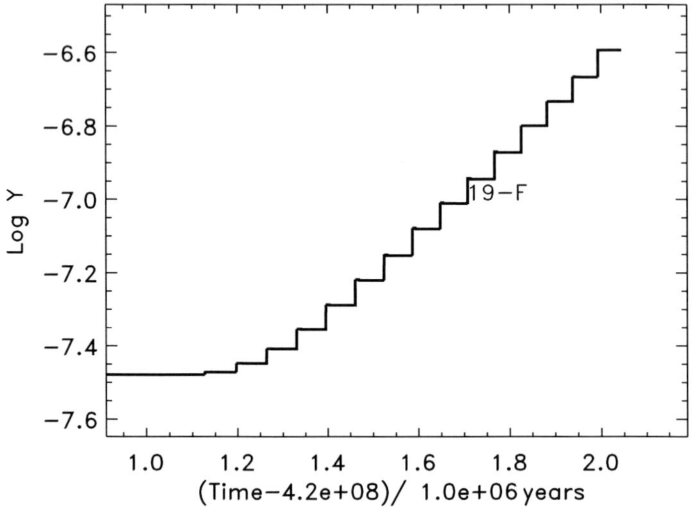

Fig. 23. The abundance evolution of ^{19}F at the surface for the $3M_\odot$, $Z = 0.02$ model during the TP-AGB. The ^{19}F abundance increases by about a factor of 4

carbon-enhanced metal-poor halo star [140] represents further strong motivation to better understand the details of fluorine production in AGB stars [141].

5.5 Extra-mixing process on the AGB

There is evidence that some type of *extra mixing* process operates between the H-shell and the base of the envelope in low-mass AGB stars. This mixing is sometimes referred to as *cool bottom processing*. Matter from the base of the envelope is slowly mixed down to a hot region near the H-shell, with the result that the surface abundances of the star are slowly altered. The main surface abundance changes are a decrease in the ^{12}C/^{13}C ratio and an increase in the nitrogen abundance. This is qualitatively similar to the extra mixing required on the first giant branch. The idea that extra-mixing processes are operating in low-mass AGB stars is supported by the lower than predicted ^{12}C/^{13}C ratios and enrichments in Li and N in some carbon-rich AGB stars [142]. The C, O, and Al isotopic ratios found in meteoritic silicon carbide and oxide grains from AGB stars also point toward some sort of extra-mixing process operating in the parent stars [46, 7, 143, 144, 145]. In these cases, the mixing should not be so efficient to prevent the formation of a carbon-rich atmosphere. We refer to Busso et al. [7] and Nollett, Busso, & Wasserburg [143] for more details.

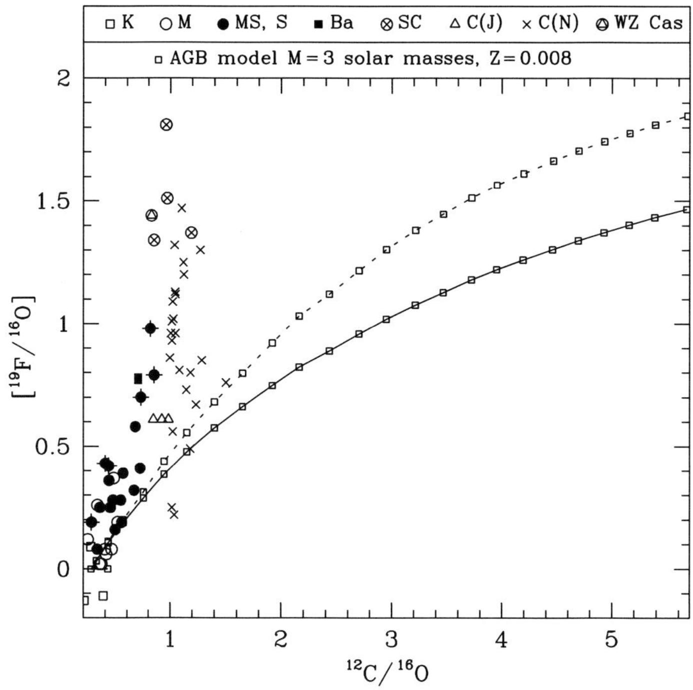

Fig. 24. Comparison of fluorine abundances observed by [130] and model predictions for the $3M_{\odot}$, $Z = 0.008$ model. The predictions are normalized such that the initial ^{19}F abundance corresponds to the average F abundance observed in K and M stars. Each symbol on the prediction lines represents a TDU episode. Solid lines represent calculations performed using no $^{18}\mathrm{F}(\alpha,p)^{21}\mathrm{Ne}$ reaction, which are equivalent to using the current lower limit, recommended value and Brussels library rate. Dotted lines are calculations performed using the current upper limit of the rate. Figure provided by Maria Lugaro

5.6 Hot bottom burning

In intermediate-mass AGB stars over about $4M_{\odot}$ (depending on Z), the base of the convective envelope can dip into the top of the H-burning shell, causing proton-capture nucleosynthesis to occur there (Figure 25). This phenomenon is known as hot bottom burning, and can change the surface composition because the entire envelope is exposed to the hot burning region a few thousand times per interpulse period. For a review of HBB nucleosynthesis see Lattanzio et al. [146]. If the base of the envelope is sufficiently hot, the Ne-Na and Mg-Al chains (Figure 17) may operate alongside the CNO cycle; ^{7}Li production is also possible via the Cameron-Fowler mechanism [63, 146]. HBB converts the ^{12}C dredged into the envelope to ^{14}N, which can prevent the C/O ratio exceeding unity, while keeping the ^{12}C/^{13}C ratio near the equilibrium value ~ 4. Frost et al. [147] noted that intermediate-mass AGB stars may become

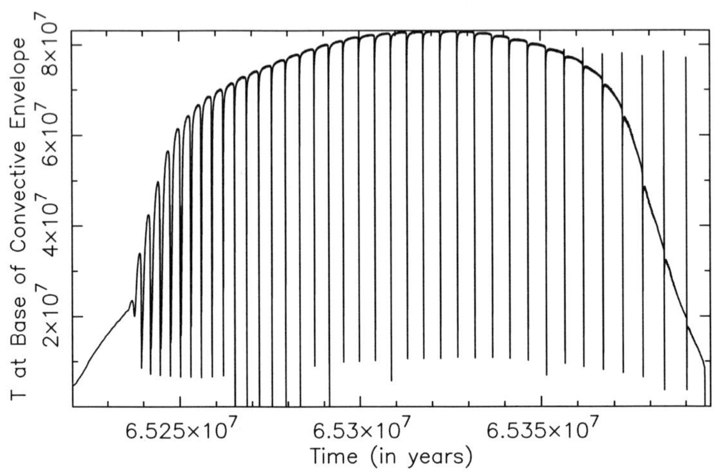

Fig. 25. The temperature at the base of the envelope during the TP-AGB phase for a $6M_\odot$, $Z = 0.02$ model

luminous, optically obscured carbon stars near the end of the TP-AGB, when mass loss has removed much of the envelope, extinguishing HBB but allowing dredge-up to continue. HBB stars may be important for the production of many elements including nitrogen, lithium, sodium, as well as the magnesium and aluminium isotopes.

5.7 The production of lithium by HBB

The observations of Wood, Bessell, & Fox [4] were the first to suggest that the oxygen-rich luminous AGB stars in the Magellanic Clouds are undergoing CNO cycling at the base of the convective envelope, converting the dredged up carbon to nitrogen. The discovery that these stars are also rich in Li [62,63,64] gave further credibility to the idea that HBB was actually occurring in massive AGB stars. The production of ^{7}Li is thought to occur via the Cameron-Fowler mechanism [148]: Some ^{3}He, created earlier in the evolution (during central H-burning), captures an α-particle to create ^{7}Be. The ^{7}Be can either 1) capture a proton to complete the PP III chain, or 2) capture an electron to produce ^{7}Li. Whether the ^{7}Be follows path 1) or path 2) depends critically on the temperature of the region. Owing to efficient mixing in the convective envelope (where the convective turnover time is ≈ 1 year), some of the ^{7}Be is mixed into a cooler region which prevents proton capture. The ^{7}Be will undergo electron-capture instead, producing ^{7}Li. The ^{7}Li is also subject to proton capture and is eventually mixed into the hot temperature region and subsequently destroyed. Once the envelope is depleted in ^{3}He, ^{7}Li production stops. The Li-rich regime lasts for $\sim$100,000 years for the $6M_\odot$, $Z = 0.02$ model shown in Figure 26. Time-dependent mixing is required to produce ^{7}Li in a HBB

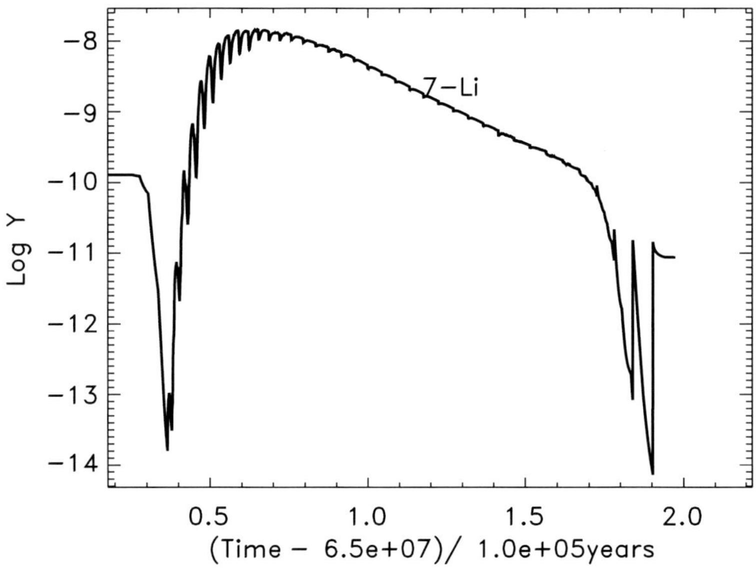

Fig. 26. The surface abundance of ^{7}Li during the TP-AGB phase for a $6M_\odot$, $Z = 0.02$ model

calculation because the nuclear timescale for the reactions involved in the Cameron-Fowler mechanism are similar to the convective turnover timescale (see Fig. 2 in [149]).

It is still an open question whether or not AGB stars contribute to the production of ^{7}Li in the Galaxy [150, 151]. There are many uncertainties involved in the production of ^{7}Li in AGB models, including the mass-loss rates used and the treatment of convective mixing [152]. Mass-loss rates for AGB stars, such as the formula given by Vassiliadis & Wood [106] and Blöcker [107], have a superwind phase which occurs during the final few thermal pulses. The superwind phase results in a period of rapid mass loss, and most of the convective envelope is lost during this time. Thus the composition of the envelope at the start of the superwind phase critically determines the contribution AGB stars make to the enrichment of the interstellar medium. In Figure 26 most of the ^{7}Li has been destroyed by the time the superwind phase starts.

5.8 HBB and the C, N, and O isotopes

Hot bottom burning first alters the envelope abundance of the CNO isotopes by the CN cycle and later, by the ON cycles. There is also a fourth cycle involving the destruction of ^{19}F to produce ^{16}O. The CN cycle burns ^{12}C first into ^{13}C and later into ^{14}N, which reduces the ^{12}C/^{13}C ratio from the pre-AGB value of ≈ 20 to close to $4 - 5$, see Figure 27. The ^{14}N abundance increases from CN cycling and the rare nitrogen isotope, ^{15}N, is destroyed by proton capture, resulting in large increases in the nitrogen isotope ratio. HBB quickly destroys the fragile ^{19}F in the envelope by proton captures via $^{19}\mathrm{F}(p,\alpha)^{16}\mathrm{O}$. The nitrogen enrichment expected from HBB has been observationally confirmed by McSaveney et al. [153] for stars in the LMC and SMC.

Intermediate-mass AGB stars with HBB are considered to be one of the major producers of ^{14}N in the Galaxy [154]. The ^{14}N produced by HBB has a primary and secondary component, depending on whether the carbon and oxygen used in the CNO reactions is produced by the He-burning shell (indicating primary production), or was initially present when the star first formed (indicating secondary production). Primary ^{14}N is produced when the TDU mixes primary ^{12}C from the He-shell into the envelope of an AGB star with HBB. Standard models of Type II supernovae cannot account for the primary nitrogen component in the Galaxy [155, 156, 157], which make low-metallicity intermediate mass AGB stars promising candidates. Low-metallicity, rapidly rotating massive stars may also supply a significant fraction of the nitrogen in the early Universe [158, 159]. If the principal source of primary nitrogen is intermediate-mass stars, then there should be a lag between the release of nitrogen into the interstellar medium and the release of oxygen, which is produced by Type II supernovae and released very soon after a period of star formation [160].

There are two burning cycles that involve the oxygen isotopes and these follow the same pattern as described for the H-shell. Both operate simultaneously but what effect each has on the envelope abundance depends on the branching ratio between the two $^{17}\mathrm{O}+p$ reaction rates. Overall, there is a net decrease in the ^{16}O and ^{18}O abundances, along with an increase in the ^{17}O abundance in the envelope. The decrease in the surface ^{16}O abundance is only significant in the most massive AGB models, depending on the convective model employed in the calculation [152, 104]. Hence the elemental oxygen abundances observed in AGB stars should be indicative of their initial abundances, except in the most massive objects.

When the mass of the envelope has been reduced below $\approx 1.5M_{\odot}$, HBB ceases but the continuation of dredge-up can turn the star into an obscured carbon star, depending on the initial mass and metallicity [147, 118]. Other helium burning products will also be mixed into the envelope, and in Figure 27 we see that while the ^{19}F abundance initially decreases, it increases again after the cessation of HBB. The final surface abundance of fluorine is similar to the initial. Hence, the composition of stars that are experiencing HBB depends crucially on the evolutionary state of the star, as well as on the mass. If near the beginning of the TP-AGB, the star is predicted to have a low $^{12}\mathrm{C}/^{13}\mathrm{C}$ ratio (≈ 5) and low ^{19}F abundance, and it would be enriched in nitrogen and lithium. Conversely, near the end of the TP-AGB that same star would be Li-poor and nitrogen rich, and it would have a $^{12}\mathrm{C}/^{13}\mathrm{C}$ ratio near ~ 10.

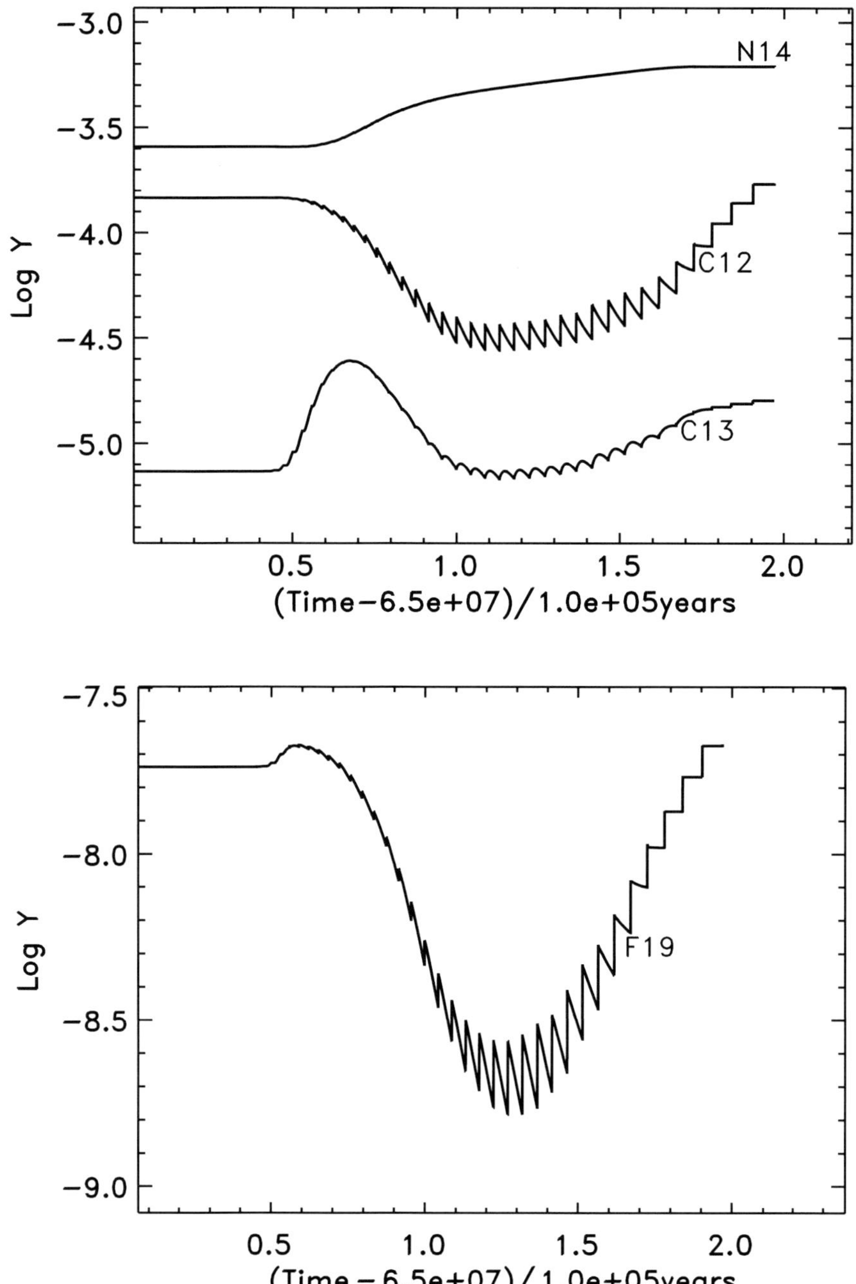

Fig. 27. The surface abundances evolution of C, N, and F during the TP-AGB phase for the $6M_{\odot}$, $Z = 0.02$ model. The final $^{12}C/^{13}C$ and C/O ratios are ~ 10 ~ 0.4, respectively

5.9 HBB and the Ne, Mg, and Al isotopes

When the temperature is high enough for the Ne-Na and Mg-Al chains to operate[1], both ^{23}Na and ^{26}Al are produced at the expense of ^{22}Ne and ^{25}Mg. HBB is responsible for the largest increase in the surface abundance of ^{23}Na, ^{25}Mg, and ^{26}Al, and is the only nucleosynthesis site in intermediate-mass stars capable of altering the ^{24}Mg abundance. The ^{26}Al production can be substantial with $^{26}\mathrm{Al}/^{27}\mathrm{Al}$ ratios of the order of a few $\times 10^{-1}$ predicted in the most massive models.

The Ne-Na and Mg-Al chains follow the same sequence as seen in the H-shell, except that temperatures of at least $\sim$90 million K are required before ^{24}Mg is substantially depleted. In Figure 28 we show the time variation of various surface abundances for the $6M_{\odot}$, $Z = 0.02$ and $Z = 0.004$ models. Significant enrichments of ^{22}Ne, ^{25}Mg, and ^{26}Mg are seen in both models, although the behaviour exhibited by the lower metallicity $Z = 0.004$ model is more extreme. This is because the temperature reached at the base of the envelope is 98×10^6 K (compared to 82 million for the $Z = 0.02$ model), and the lower metallicity model experiences 100 thermal pulses with efficient mixing. The final $^{26}\mathrm{Al}/^{27}\mathrm{Al}$ ratios are 0.015 and 0.6, respectively.

The model calculations shown in Figure 28 are subject to severe reaction rate uncertainties (along with uncertainties that affect the structure of the model, such as convection and mass loss). In Figure 29 we show results of calculations using the NACRE [116] rates for the Ne-Na and Mg-Al chains compared to the rates from [109] used in Figure 28. The main difference is that the models using the NACRE rates produce significant amounts of sodium at the expense of ^{22}Ne. Izzard et al. [110] examined the effect of reaction rate uncertainties on the operation of the Ne-Na and Mg-Al chains during HBB in intermediate-mass AGB stars of various compositions. The main conclusion is that more accurate determinations of the following reaction rates: $^{22}\mathrm{Ne}(p,\gamma)^{23}\mathrm{Na}$, $^{23}\mathrm{Na}(p,\gamma)^{24}\mathrm{Mg}$, $^{25}\mathrm{Mg}(p,\gamma)^{26}\mathrm{Al}$, and $^{26}\mathrm{Mg}(p,\gamma)^{27}\mathrm{Si}$, are required to obtain reliable estimates of the Ne, Na, Mg, and Al yields from AGB stars.

5.10 Yields from AGB stars

To study the chemical evolution of galaxies and stellar systems, it is essential to have an estimate of what elements are produced (or destroyed) by stars of different mass ranges. *Stellar yields* provide this information, by quantifying the amount (in mass) of species i that is expelled into the interstellar medium over the course of a star's life. The definition of the yield that we use is given by the following expression

$$M_i = \int_0^{\tau} [X(i) - X_0(i)] \frac{dM}{dt} dt, \tag{5}$$

[1] higher for HBB than in the H-shell, because the density is lower at the base of the convective envelope

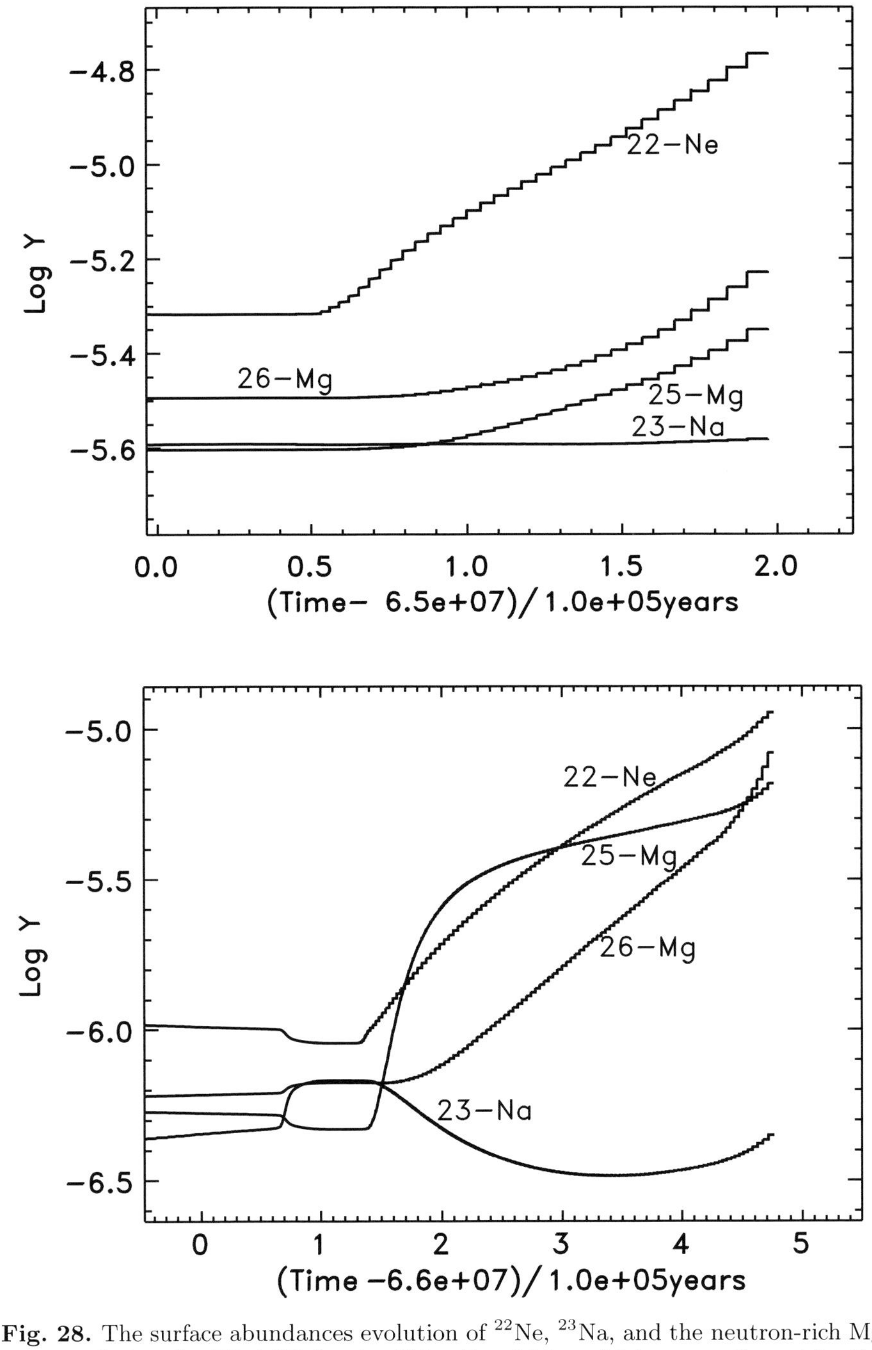

Fig. 28. The surface abundances evolution of ^{22}Ne, ^{23}Na, and the neutron-rich Mg isotopes during the TP-AGB for the $6M_{\odot}$, $Z = 0.02$ model (top panel), and for the $6M_{\odot}$, $Z = 0.004$ model (bottom panel)

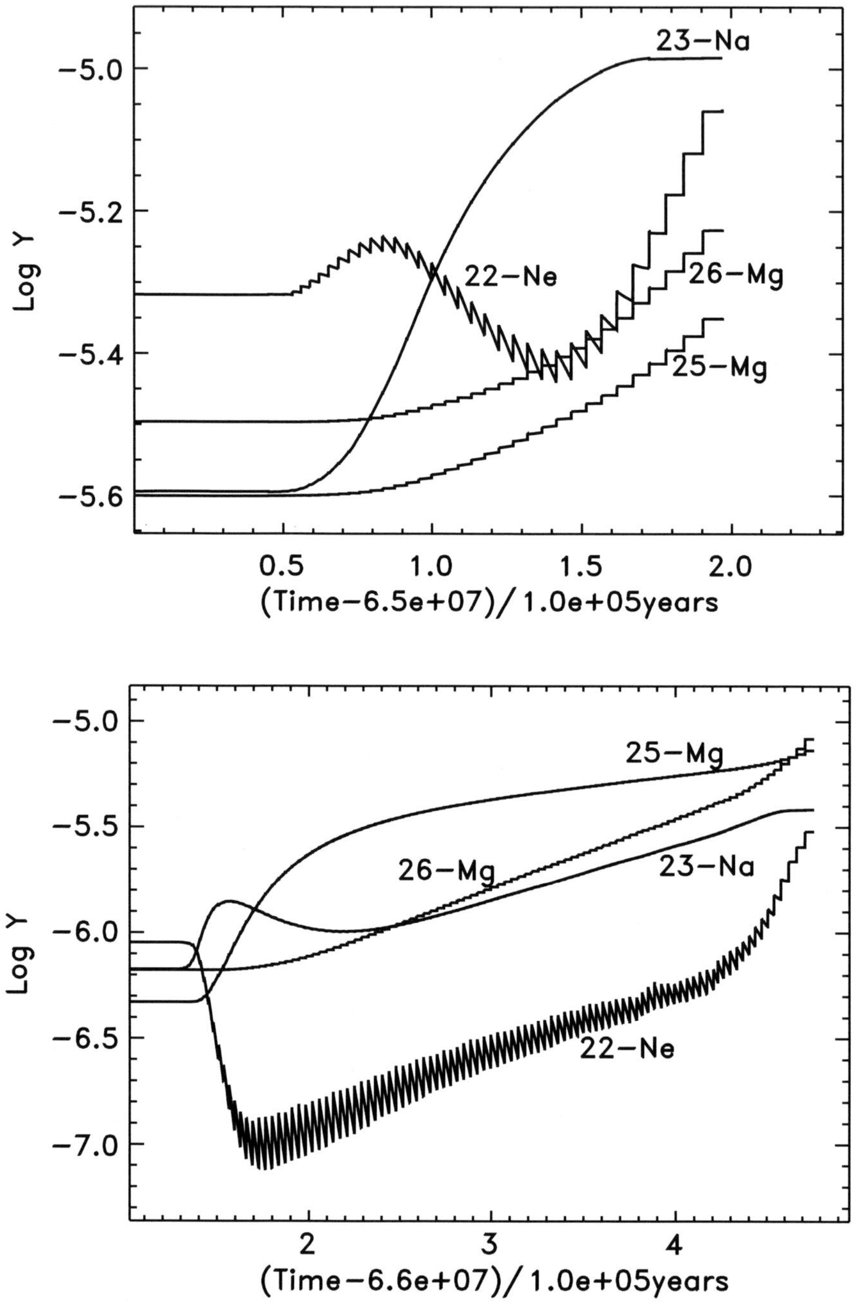

Fig. 29. Same as Figure 28, but using the NACRE rates for the Ne-Na and Mg-Al chains

where M_i is the yield of species i (in solar masses), dM/dt is the current mass-loss rate, $X(i)$ and $X_0(i)$ refer to the current and initial mass fraction of species i, and τ is the total lifetime of the stellar model. The yield can be negative, in the case where the element is destroyed, and positive if it is produced.

Equation (5) can be re-written as

$$M_i = \int_0^\tau X(i)\frac{dM}{dt}dt - \int_0^\tau X_0(i)\frac{dM}{dt}dt. \tag{6}$$

The first integral gives the total amount of species i expelled into the interstellar medium:

$$M_i^{\text{tot}} = \int_0^\tau X(i)\frac{dM}{dt}dt, \tag{7}$$

where M_i^{tot} is in solar masses, and is always positive.

Yields from AGB stars are provided by a number of authors. Many AGB yields available in the literature are from *synthetic* AGB computations that use fitting formula to estimate the evolution during the TP-AGB. Synthetic AGB models have successfully been used to model AGB populations [76], and compute stellar yields [161, 162, 163, 79]. The yields published by Forestini & Charbonnel [112] employed a combination of full, or detailed, AGB models and synthetic. The AGB yields of [164], [103], [105], and [24] were computed from detailed AGB model computations. An example of the yields from Karakas & Lattanzio [24] are shown in Figure 30. The yields have been weighted by the initial mass function of Kroupa, Tout, & Gilmore [165], and we show results from the $Z = 0.02$ (solar), 0.008 (LMC) and 0.004 (SMC) metallicity models. For comparison we also show the yields from a number of different synthetic AGB calculations, and from [164]; see the figure caption for details.

Many of the parameterizations used in synthetic evolution studies are derived from detailed stellar models, such as the growth of the H-exhausted core with time, and as such are only accurate over the range in mass and metallicity of the stellar models they are based upon. An example is provided by van den Hoek & Groenewegen [162] who computed AGB yields for initial masses between 0.9 and $8M_\odot$ whereas the interpulse-period-core mass relation [70] they use was only derived for stars with initial masses between 1 and $3M_\odot$. What affect this has on the yields is unclear since this relation will affect the number of TPs during the TP-AGB phase and hence the level of chemical enrichment. Recent improvements in computer power mean that grids of detailed AGB models can be produced in a reasonable time [114, 118, 103]; however, producing yields from more than ~ 20 AGB stars for any given metallicity range is still challenging. For this reason synthetic models are still preferred for some applications e.g., [110].

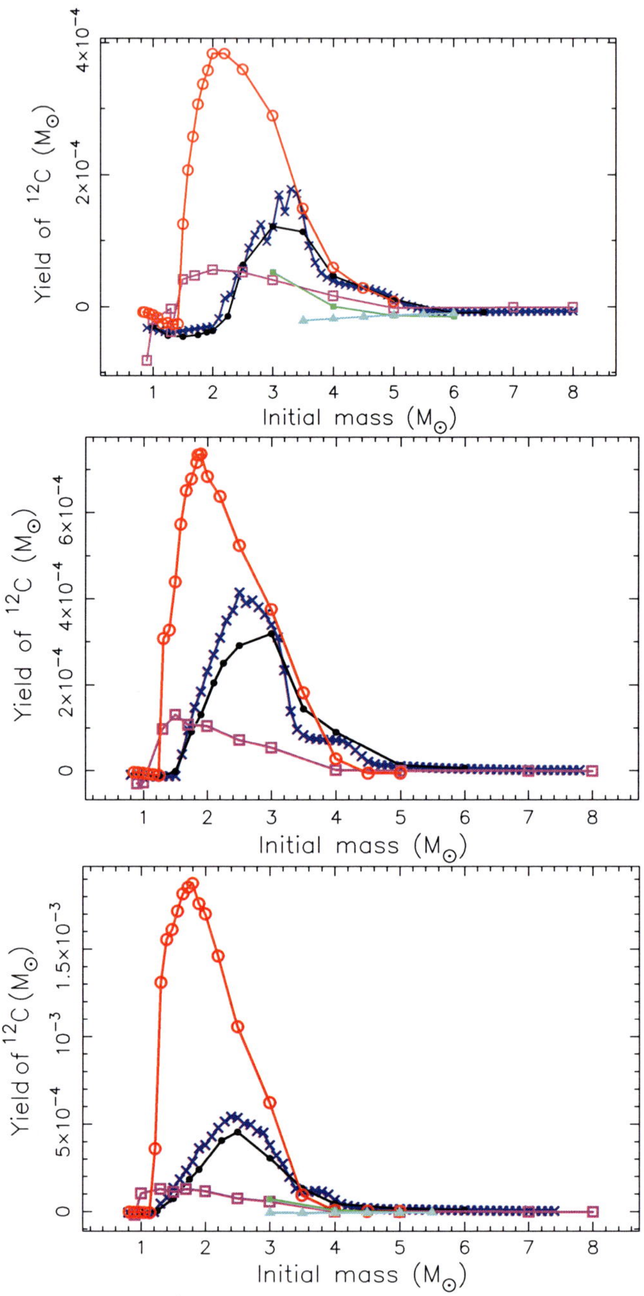

Fig. 30. Weighted yield of ^{12}C as a function of the initial mass for the $Z = 0.02$ (top), the $Z = 0.008$ (middle) and the $Z = 0.004$ models (bottom). We show results from [24](black solid points), [162] (open magenta squares), [112] (solid green squares), [163] (open red circles), [164] (solid aqua triangles), and [79] (blue crosses)

6 The *slow* neutron-capture process

For heavy elements with atomic masses greater than ≈ 56 their proton numbers are so large that electrostatic repulsion inhibits charged particle reactions (e.g., proton and α captures) except under very specific conditions. Most heavy nuclei are instead formed by neutron addition onto abundant Fe-peak elements. For this reason, neutron-capture processes are *secondary*. That is, they require that some Fe-peak nuclei (e.g., ^{56}Fe) is already present in the star. The solar abundance distribution is characterized by peaks that can be explained by

1. **the slow-neutron-capture process**, the s-process,
2. **the rapid-neutron-capture process**, the r-process,
3. **proton-rich nuclei** are much less abundant in the solar system than nuclei synthesized by the s- and r-processes.

The seminal paper by Burbidge, Burbidge, Fowler, & Hoyle [166] (hereafter B^{2}FH) laid down the foundations for these processes, and Wallerstein et al. [167] provides an updated review on B^{2}FH. An historical overview and details of how the s-process operates in AGB stars has been reviewed by Meyer [168], Busso et al. [7], and Herwig [1]. The textbooks by Clayton [9] and Lugaro [15] provide excellent discussions of the s-process at a level aimed at graduate students.

We will concern ourselves with the s-process, which occurs under conditions of relatively low neutron densities ($N_{\rm n} \sim 10^7$ neutrons/cm^3). In this case the timescale for neutron-capture is much slower, in general, than the β-decay rate of unstable isotopes. Hence, the s-process will produce isotopes along the *valley of β-stability*. Examples of s-process elements include Sr, Zr, Nb, Ba, and La, where the dominant isotopes are mainly produced by the s-process. There are also some isotopes that are only produced by the s-process and examples include ^{86}Sr, ^{96}Mo, ^{104}Pd, and ^{116}Sn. During the r-process, neutron densities as high as $N_{\rm n} \sim 10^{25}$ neutrons/cm^3 ensures that the timescale for neutron-capture is much faster than the β-decay rates. The r-process will produce isotopes right up to the neutron drip line, which decay to stable, neutron-rich isotopes once the neutron flux is gone. Given the extreme conditions required for the r-process, it has been hypothesized to occur during supernovae explosions [169, 170] but other sites have also been proposed including colliding neutron stars [171], and black hole/neutron star mergers [172]. We refer to the reviews by Meyer [168] and Arnould, Goriely, & Takahashi [173] for further details.

The solar abundance distribution shows peaks that are constrained by nuclei with a *magic number* of neutrons ($n = 50, 80, 82, 126$). Note that for lighter elements there are also peaks at $n = 2, 8, 20$, and 28. A nucleus composed of a magic number of protons and neutrons is very stable and considered to be "doubly magic". Examples include ^{16}O (with 8 protons and neutrons), and ^{208}Pb, with 82 protons and 126 neutrons. The stability of nuclei with a magic number of neutrons can be explained when nuclei are described as systems ruled by quantum mechanics, as done for atomic systems. In practice, nuclei with a magic number of neutrons are very stable against neutron-capture and have low *neutron-capture cross sections.* Hence these nuclei act as bottlenecks and are consequently seen as *s*-process peaks. Note that ^{56}Ni, made in abundance by supernovae, is doubly magic, with 28 protons and 28 neutrons.

The neutron exposure, τ, is defined by

$$\tau = \int N_n v_T dt, \tag{8}$$

where N_n is the neutron density and v_T the thermal velocity of the neutrons. The neutron exposure is a time-averaged neutron flux with units of mbarn^{-1} (where 1 barn is equal to 10^{-24} cm^2), see [7] for further details. There are three distinct components of the *s*-process that require distinct neutron exposures:

1. *The weak component:* This produces most of the *s*-isotopes with atomic mass $\leq$ 90, from Fe to Sr, and can be described by a mean neutron exposure of $\tau \approx 0.06$ mbarn^{-1}.
2. *The main component:* This is responsible for the *s*-process isotopes from $90 \leq A \leq 204$, from Sr to Pb, where the mean $\tau \approx 0.3$ mbarn^{-1}.
3. *The strong component:* Required to reproduce the solar system Pb abundance, requiring $\tau \approx 7.0$ mbarn^{-1}.

The weak component is thought to originate in the He and C-burning shells of massive stars [121], the main component in low-mass AGB stars, and the strong component, that produces Pb, has been hypothesized to occur in very low metallicity, low-mass AGB stars [168, 7].

6.1 Neutron sources operating in AGB stars

Neutron-capture processes require a source of free neutrons, given that neutrons are unstable and decay in $\approx$ 10 minutes. There are two important neutron sources available during He-shell burning in AGB stars:

1. ^{14}N$(\alpha,\gamma)^{18}$F$(\beta^+\nu)$ ^{18}O$(\alpha,\gamma)^{22}$Ne$(\alpha,n)^{25}$Mg.
2. ^{12}C$(p,\gamma)^{13}$N$(\beta^+\nu)^{13}$C$(\alpha,n)^{16}$O.

The ^{22}Ne$(\alpha,n)^{25}$Mg was first identified as a neutron source for AGB stars by [174]. Note that B^{2}FH suggested that the ^{21}Ne$(\alpha,n)^{24}$Mg reaction might be an important neutron source, but the scarcity of ^{21}Ne in the He-intershell meant that the reaction is not efficient at providing neutrons. The intershell region is rich in ^{14}N from CNO cycling, and during a thermal pulse, ^{14}N can suffer successive α captures to produce ^{22}Ne. If the temperature exceeds about 300×10^6 K, ^{22}Ne can capture an α particle to produce ^{25}Mg and ^{26}Mg, in almost equal proportions (see Figure 20). Neutrons are released by the ^{22}Ne$(\alpha,n)^{25}$Mg reaction. Given the high temperatures required for the ^{22}Ne$(\alpha,n)^{25}$Mg to operate efficiently, it is theoretically predicted to be efficient in the most massive AGB stars, with initial masses $\geq 4M_{\odot}$. The temperatures required for the ^{22}Ne$(\alpha, n)^{25}$Mg may also be reached during the last few thermal pulses of lower-mass AGB stars, where it has a marginal influence on the final s-process abundance distribution prior to the subsequent third dredge-up [175].

The ^{13}C$(\alpha, n)^{16}$O reaction requires the operation of both proton and α-capture reactions to occur in the He-shell, a region normally devoid of protons. There is some ^{13}C left over in the He-intershell after the end of CNO cycling, but the study by Gallino et al. [176] has shown that this is not enough to account for the s-process enrichments of AGB stars. Hence, some mixing of protons from the convective envelope into the top layers of the He-intershell is required. Straniero et al. [177] discovered that the ^{13}C burns under radiative conditions because the ^{13}C$(\alpha, n)^{16}$O reaction occurs at relatively low temperatures ($T \geq 90 \times 10^6$ K). The neutrons are released in the ^{13}C pocket, and the s-processing occurs between pulses in the same layers where the ^{13}C was produced. When the next convective thermal pulse occurs, it ingests this s-element rich layer, mixing it over most of the intershell. Observational [81, 178, 84] and theoretical evidence [176, 179, 175] suggests that the ^{13}C$(\alpha, n)^{16}$O reaction is the dominant neutron source in low-mass AGB stars, with initial masses $\leq 3M_{\odot}$. The timescales for neutron production during the interpulse are much longer ($\geq 10^3$ years) than during the convective pulse (~ 10 years), resulting in much lower neutron densities ($\sim 10^8$ neutrons/cm^3) than the ^{22}Ne source ($\sim 10^{13}$ neutrons/cm^3). Together, the timescale for neutron production and the neutron source determine the resulting s-process element distribution.

6.2 Partial mixing and the formation of ^{13}C pockets

For the ^{13}C$(\alpha, n)^{16}$O reaction to occur efficiently, some partial mixing is required at the border between the H-rich envelope and the C-rich intershell. Protons are mixed down into the top 1/10th of the ^{4}He and ^{12}C-rich region, where the CN cycle is able to partially operate to produce a ^{13}C pocket (along with a ^{14}N pocket). In Figure 31 we show an example of calculations performed with the inclusion of a partially-mixed zone, where the resulting ^{13}C pocket covers $\approx 10\%$ of the He-intershell. Note the large increase in Ge isotopes in the model with a partially mixed zone, and the lack of any s-process enrichments in the model with no pocket. It is important that there not be too many protons, because in this case the CN cycle goes to completion and ^{14}N is mostly synthesized. Both ^{14}N and ^{26}Al are *neutron poisons*, which means that they are efficient neutron absorbers and will change the resulting abundance distribution. The details of how the ^{13}C pocket forms and its extent in mass in the He-intershell are still unknown, although various mechanisms have been proposed including convective overshoot, rotation, and gravity waves (see [1] for a discussion of the relative merits of each mechanism). Models that include artificial ^{13}C pockets produce s-process abundance distributions that fit the

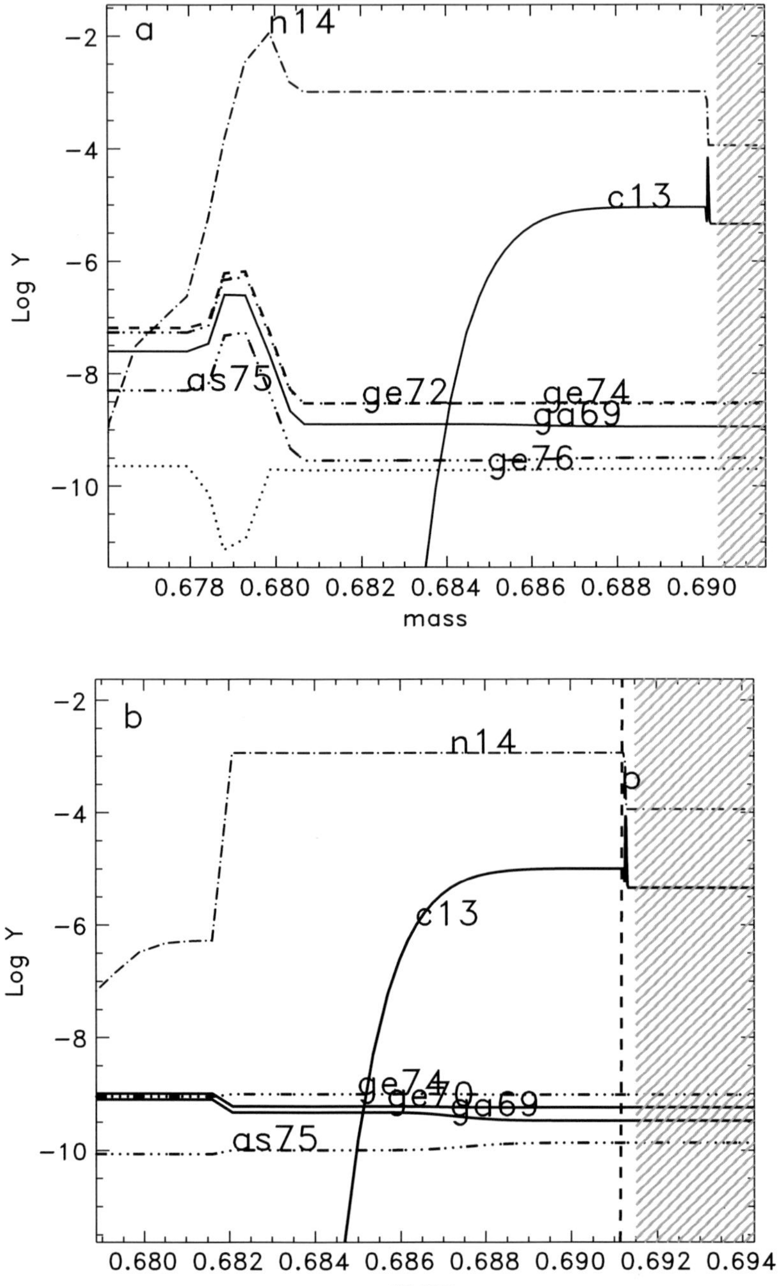

Fig. 31. Composition profile showing the intershell abundances just before the last computed thermal pulse. The shaded region is the inner edge of the convective envelope. We show abundances from the $3M_{\odot}$, $Z = 0.012$ model with a partially mixed zone (a), and without (b). Using models from [180]

observational data reasonably well, with a small spread in the pocket sizes required [181, 182].

6.3 The *s*-process in massive AGB stars

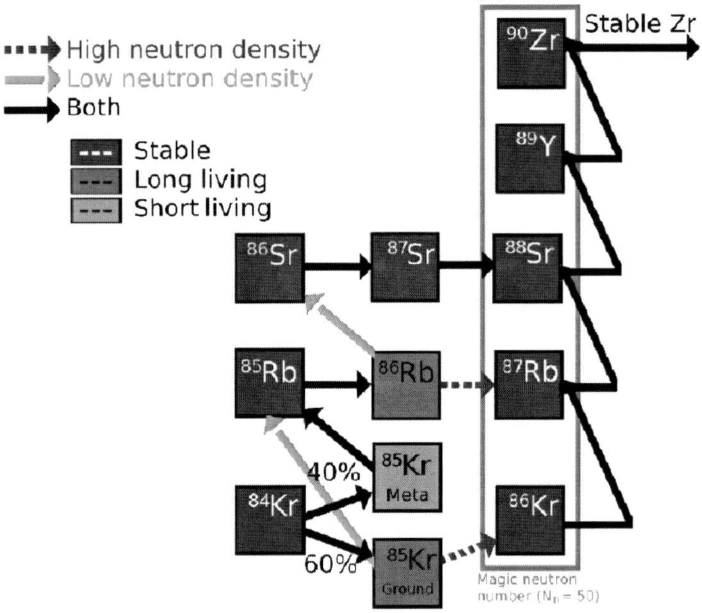

Fig. 32. Section of the chart of the nuclides from Kr to Zr. Stable isotopes are depicted with dark backgrounds and white labels. Unstable isotopes are depicted on a lighter background with black labels. The light arrow shows the neutron flow for low-neutron densities, whereas the dark, dashed arrows show the flow for high neutron densities ($N_{\rm n} \geq 10^8$ n/cm^{-3}). The black solid arrows shows the direction of the flow for all densities. Isotope with magic neutron number = 50 are surrounded by the open rectangle. Figure provided by Mark van Raai

Luminous, O-rich AGB stars in the LMC and SMC exhibit strong molecular bands of ZrO, indicating that the atmospheres of these stars are enriched in the *s*-process element Zr [4]. Recent observations by García-Hernández et al. [122] showed that luminous OH/IR stars in the Galaxy have enhanced rubidium abundances. This evidence suggest that massive AGB stars do produce *s*-process elements and that the neutron source operating in these stars is the ^{22}Ne$(\alpha, n)^{25}$Mg reaction. The contribution of these stars to the total Galactic inventory of *s*-process elements is still expected to be small [183]. Travaglio et al. [183] estimate that intermediate-mass stars contribute $\approx$ 8, 6, 6, 1, and 5% toward the solar-system composition of Sr, Y, Zr, Nb, and Mo, respectively.

Rubidium production is an indicator of high neutron densities and the operation of the $^{22}\mathrm{Ne}(\alpha, n)^{25}\mathrm{Mg}$ reaction. This is because at high neutron densities two branching points open that allow Rb to be synthesized (Figure 32). At $N_{\mathrm{n}} = 5 \times 10^8$ n/cm^3, $\sim$ 86% of the neutron flux goes through ^{85}Kr allowing for $^{85}\mathrm{Kr}(n, \gamma)^{86}\mathrm{Kr}(n, \gamma)^{87}\mathrm{Kr}$ that decays to ^{87}Rb. Note that ^{87}Rb has a magic number of neutrons and is fairly stable against neutron-capture. Also, high neutron densities allow neutrons to bypass the branching point at ^{86}Rb (in low-mass stars this isotope decays to ^{86}Sr), allowing for the chain $^{86}\mathrm{Rb}(n, \gamma)^{87}\mathrm{Rb}$. In low-mass stars, the magic ^{88}Sr is produced instead. For this reason, the elemental ratios of Rb/Sr and Rb/Zr are indicators of the neutron densities, and was used as evidence that the $^{13}\mathrm{C}(\alpha, n)^{16}\mathrm{O}$ reaction is the major neutron source in low-mass AGB stars [178].

7 Concluding remarks

The AGB phase is the last nuclear burning phase for stars with initial masses between about $0.8M_{\odot}$ to $8M_{\odot}$. The AGB phase of evolution is very short, comprising less than 1 per cent of the main-sequence lifetime, nevertheless, it is on the AGB that the richest nucleosynthesis occurs for this mass range. The nucleosynthesis is driven by thermal instabilities of the helium-burning shell, the products of which are dredged to the stellar surface by recurrent mixing episodes. Hot bottom burning occurs in the most massive AGB stars, and this also alters the surface composition. AGB stars are important factories for producing many elements including carbon, nitrogen, fluorine, and heavy elements synthesized by the *s*-process. It is now estimated that perhaps up to half of all elements heavier than iron are made by the *s*-process in low-mass AGB stars [7]. While much is known about the inner workings of AGB stars, there are many unknowns that render predictions uncertain. In particular, our lack of knowledge about convective mixing processes in stars is perhaps the greatest uncertainty, since it overshadows so much of AGB evolution. Other modelling uncertainties such as mass-loss rates, opacities, and reaction rate uncertainties are also important.

In these lecture notes, we have reviewed the structure and evolution of low and intermediate-mass stars prior to the AGB phase, along with the surface abundance changes predicted by models for the first and second dredge-up mixing events. We reviewed the evolution and nucleosynthesis during the AGB, including a brief discussion of the *slow* neutron-capture process. There are many difficult problems left unsolved including the formation of ^{13}C pockets, and the extent of extra-mixing processes during the first and asymptotic giant branch phases. The process driving mass loss in AGB stars also poses a severe challenge for the future. It will be important to find techniques to accurately measure many of the reaction rates of interest for AGB nucleosynthesis. It is also essential that we understand nucleosynthesis in low-metallicity AGB stars if we are to unravel the puzzle surrounding the abundances of many carbon-rich metal-poor stars in the Galaxy. The mass-transfer processes that occur in binary star systems are poorly understood but these need to be untangled. Finally, AGB stars are a fascinating, challenging phase of evolution to study with much progress yet to be made.

Acknowledgments

AIK warmly thanks the organisers of the Kodai School on *Synthesis of the Elements in Stars* for the hospitality and the opportunity to attend and give lectures at the school. AIK also thanks John Lattanzio, Maria Lugaro, Simon Campbell, Peter Wood, Pilar Gil-Pons, Lionel Siess, and Mark van Raai for discussions and/or material that helped in the preparation of the lecture notes given at the School. She thanks Robin Humble, Ross Church, and Maria Lugaro for help with proof-reading her lecture notes.

References

1. F. Herwig, ARA&A **43**, 435 (2005)
2. T. Blöcker, Ap&SS **275**, 1 (2001)
3. H. van Winckel, ARA&A **41** 391 (2003)
4. P. R. Wood, M. S. Bessell, M. W. Fox, ApJ **272**, 99 (1983)
5. D. A. VandenBerg, F. J. Swenson, F. J. Rogers, C. A. Iglesias, D. R. Alexander, ApJ **532**, 430 (2000)
6. R. Buonanno, C. E. Corsi, A. Buzzoni, C. Cacciari, F. R. Ferraro, F. Fusi Pecci, A&A **290**, 69 (1994)
7. M. Busso, R. Gallino, G. J. Wasserburg, ARA&A **37**, 239 (1999)
8. L. M. Dray, C. A. Tout, A. I. Karakas, J. C. Lattanzio, MNRAS **338**, 973 (2003)
9. D. D. Clayton in *Principles of stellar evolution and nucleosynthesis*, University of Chicago Press (1983)
10. C. E. Rolfs, W. S. Rodney in *Cauldrons in the cosmos: Nuclear astrophysics*, University of Chicago Press, (1988)
11. C. Iliadis in *Nuclear Physics of Stars*, Wiley-VCH Verlag (2007)
12. C. J. Hansen, S. D. Kawaler, V. Trimble in *Stellar interiors : physical principles, structure, and evolution*, Springer-Verlag, (2004).
13. B. E. J. Pagel in *Nucleosynthesis and Chemical Evolution of Galaxies*, Cambridge University Press (1997)
14. H. J. Habing, H. Olofsson in *Asymptotic Giant Branch Stars*, Springer (2004)
15. M. Lugaro in *Stardust from meteorites. An introduction to presolar grains*, World Scientific (2005)
16. S. E. Woosley, A. Heger, T. A. Weaver, Rev. Mod. Phys. **74**, 1015 (2002)
17. E. Anders, N. Grevesse, Geochim. Cosmochim. Acta **53**, 197 (1989)
18. M. Asplund, N. Grevesse, A. J. Sauval, ASPS **336**, 25 (2005)

19. C. Allende Prieto, D. L. Lambert, M. Asplund, ApJ **556**, L63 (2001)
20. T. C. Beers, N. Christlieb, ARA&A **43**, 531 (2005)
21. A. Frebel, W. Aoki, N. Christlieb, H. Ando, M. Asplund, P. S. Barklem et al., Nature **434**, 871 (2005)
22. E. E. Salpeter, ApJ, **121**, (1955)
23. P. Kroupa, Science **295**, 82 (2002)
24. A. I. Karakas, J. C. Lattanzio, PASA **24**, 103 (2007)
25. G. C. Sloan, K. E. Kraemer, P. R. Wood, A. A. Zijlstra, J. Bernard-Salas, D. Devost et al., ApJ **686**, 1056 (2008)
26. G. Marconi, R. Buonanno, E. Carretta, F. R. Ferraro, P. Montegriffo, F. Fusi Pecci et al., MNRAS **293**, 479 (1998)
27. A. I. Karakas in *Asymptotic Giant Branch Stars: their influence on binary systems and the interstellar medium*, PhD thesis, Monash University (2003)
28. F. Fagotto, A. Bressan, G. Bertelli, C. Chiosi, A&AS **104**, 365 (1994)
29. C. Frost, J. Lattanzio, ArXiv Astrophysics e-prints, astro-ph/9601017 (1996)
30. J. C. Lattanzio A. I. Boothroyd, American Institute of Physics Conference Series **402**, ed. by T. J. Bernatowicz, E. Zinner, 85 (1997)
31. C. Charbonnel, A&A **282**, 811 (1994)
32. A. I. Boothroyd, I.-J. Sackmann, ApJ **510**, 232 (1999)
33. H. Schlattl, S. Cassisi, M. Salaris, A. Weiss, ApJ **559**, 1082 (2001)
34. I. Picardi, A. Chieffi, M. Limongi, O. Pisanti, G. Miele, G. Mangano, G. Imbriani, ApJ **609**, 1035 (2004)
35. T. Suda, M. Y. Fujimoto, N. Itoh, ApJ **667**, 1206 (2007)
36. A. I. Karakas, J. C. Lattanzio, O. R. Pols, PASA **19**, 515 (2002)
37. G. Bertelli, A. Bressan, C. Chiosi, K. Angerer, A&AS **66**, 191 (1986)
38. V. Castellani, P. Giannone, A. Renzini, A&AS **10**, 355 (1971)
39. V. Trimble, PASP **104**, 1 (1992)
40. C.-W. Chin, R. B. Stothers, ApJ **77**, 299 (1991)
41. V. Castellani, S. Degl'Innocenti, M. Marconi, P. G. Prada Moroni, P. Sestito, A&A **404**, 645 (2003)
42. P. Ventura, M. Castellani, C. W. Straka, A&A **440**, 623 (2005)
43. G. Imbriani, M. Limongi, L. Gialanella, F. Terrasi, O. Straniero, A. Chieffi, ApJ **558**, 903 (2001)
44. M. Arnould, S. Goriely, A. Jorissen, A&A **347**, 572 (1999)
45. M. F. El Eid, A&A **285**, 915 (1994)
46. G. J. Wasserburg, A. I. Boothroyd, I.-J. Sackmann, ApJ **447**, L37 (1995)
47. K. K. Gilroy, ApJ **347**, 835 (1989)
48. M. D. Shetrone, AJ **112**, 2639 (1996)
49. C. A. Pilachowski, C. Sneden, R. P. Kraft, G. E. Langer, AJ **112**, 545 (1996)
50. R. Gratton, C. Sneden, E. Carretta, ARA&A **42**, 385 (2004)
51. J. G. Cohen, M. M. Briley, P. B. Stetson, AJ **130**, 1177 (2005)
52. A. V. Sweigart, J. G. Mengel, ApJ **229**, 624 (1979)
53. P. P. Eggleton, D. S. P. Dearborn, J. C. Lattanzio, Science **314**, 1580 (2006)
54. P. P. Eggleton, D. S. P. Dearborn, J. C. Lattanzio, ApJ **677**, 581 (2008)
55. C. Charbonnel, J.-P. Zahn, A&A **467**, L15 (2007)

56. G. H. Smith, C. A. Tout, MNRAS **256**, 449 (1992)
57. C. Charbonnel, ApJ **453**, L41 (1995)
58. P. A. Denissenkov, C. A. Tout, MNRAS, **316**, 395 (2000)
59. I. Iben, Jr. In *IAU Symp. 145: Evolution of Stars: the Photospheric Abundance Connection* ed. by G. Michaud, A. V. Tutukov, 257 (1991)
60. P. R. Wood. In *Planetary Nebulae*, IAU Symposium **vol. 180**, ed. by H. J. Habing, H. J. G. L. M. Lamers, 297 (1997)
61. J. M. Scalo, K. H. Despain, R. K. Ulrich, ApJ **196**, 805 (1975)
62. V. V. Smith, D. L. Lambert, ApJ **345**, L75 (1989)
63. V. V. Smith, D. L. Lambert, ApJ **361**, L69 (1990)
64. B. Plez, V. V. Smith, D. L. Lambert, ApJ, **418**, 812 (1993)
65. T. Bloecker, D. Schoenberner, A&A **244**, L43 (1991)
66. J. C. Lattanzio, PASA **10**, 120 (1992)
67. B. Paczyński, Acta Astronomica **20**, 47 (1970)
68. A. I. Boothroyd, I.-J. Sackmann, S. C. Ahern, ApJ **416**, 762 (1993)
69. O. Straniero, A. Chieffi, M. Limongi, M. Busso, R. Gallino, C. Arlandini, ApJ **478**, 332 (1997)
70. A. I. Boothroyd, I.-J. Sackmann, ApJ **328**, 653 (1988)
71. M. Lugaro, F. Herwig, J. C. Lattanzio, R. Gallino, O. Straniero, ApJ **586**, 1305 (2003)
72. R. J. Stancliffe, C. A. Tout, O. R. Pols, MNRAS **352**, 984 (2004)
73. G. Wallerstein, G. R. Knapp, ARA&A **36**, 369 (1998)
74. J. A. Frogel, J. Mould, V. M. Blanco, ApJ **352**, 96 (1990)
75. M. A. T. Groenewegen, A&A **425**, 595 (2004)
76. M. A. T. Groenewegen, T. de Jong, A&A **267**, 410 (1993)
77. P. Marigo, A. Bressan, C. Chiosi, A&A **313**, 545 (1996)
78. N. Mowlavi, A&A **344**, 617 (1999)
79. R. G. Izzard, C. A. Tout, A. I. Karakas, O. R. Pols, MNRAS **350**, 407 (2004)
80. R. J. Stancliffe, R. G. Izzard, C. A. Tout, MNRAS **356**, L1 (2005)
81. V. V. Smith, D. L. Lambert, ApJ **311**, 843 (1986)
82. V. V. Smith, D. L. Lambert, A. McWilliam, ApJ **320**, 862 (1987)
83. V. V. Smith, D. L. Lambert, ApJS **72**, 387 (1990)
84. C. Abia, I. Domínguez, R. Gallino, M. Busso, S. Masera, O. Straniero et al., ApJ, **579**, 817 (2002)
85. I. R. Little-Marenin, S. J. Little, AJ **84**, 1374 (1979)
86. S. Uttenthaler, J. Hron, T. Lebzelter, M. Busso, M. Schultheis, H. U. Käufl, A&A **463**, 251 (2007)
87. R. D. McClure, A. W. Woodsworth, ApJ **352**, 709 (1990)
88. R. D. McClure, PASP **109**, 536 (1997)
89. V. V. Smith, D. L. Lambert, ApJ **333**, 219 (1988)
90. J. A. Brown, V. V. Smith, D. L. Lambert, E. J. Dutchover, K. H. Hinkle, H. R. Johnson, AJ **99**, 1930 (1990)
91. R. D. McClure, PASP **109**, 256 (1997)
92. R. G. Izzard, C. S. Jeffery, J. Lattanzio, A&A **470**, 661 (2007)
93. T. Lloyd Evans, MNRAS **220**, 723 (1986)
94. C. Abia, J. Isern, ApJ **536**, 438 (2000)
95. J. Herman, H. J. Habing, A&AS **59**, 523 (1985)
96. T. Lebzelter, P. R. Wood, A&A **475**, 643 (2007)

97. M. W. Fox, P. R. Wood, ApJ **259**, 198 (1982)
98. D. A. Ostlie, A. N. Cox, ApJ **311**, 864 (1986)
99. G. H. Bowen, ApJ **329**, 299 (1988)
100. J. M. Winters, T. Le Bertre, K. S. Jeong, L.-Å. Nyman, N. Epchtein, A&A, **409**, 715 (2003)
101. P. Woitke, A&A, **452**, 537 (2006)
102. B. Freytag, S. Höfner, A&A **483**, 571 (2008)
103. F. Herwig, ApJS **155**, 651 (2004)
104. P. Ventura, F. D'Antona, A&A **439**, 1075 (2005)
105. R. J. Stancliffe, C. S. Jeffery, MNRAS **375**, 1280 (2007)
106. E. Vassiliadis, P. R. Wood, ApJ **413**, 641 (1993)
107. T. Blöcker, A&A **297**, 727 (1995)
108. N. Mowlavi, A&A **350**, 73 (1999)
109. C. Iliadis, J. M. D'Auria, S. Starrfield, W. J. Thompson, M. Wiescher, ApJS **134**, 151 (2001)
110. R. G. Izzard, M. Lugaro, A. I. Karakas, C. Iliadis, M. van Raai, A&A **466**, 641 (2007)
111. F. Herwig, A&A **360**, 952 (2000)
112. M. Forestini, C. Charbonnel, A&AS **123**, 241 (1997)
113. A. I. Karakas J. C. Lattanzio, PASA **20**, 393 (2003)
114. A. I. Karakas, J. C. Lattanzio, PASA **20**, 279 (2003)
115. M. Forestini, S. Goriely, A. Jorissen, M. Arnould, A&A **261**, 157 (1992)
116. C. Angulo, M. Arnould, M. Rayet, P. Descouvemont, D. Baye, C. Leclercq-Willain et al., Nucl. Phys. A **656**, 3 (1999)
117. F. Kaeppeler, M. Wiescher, U. Giesen, J. Goerres, I. Baraffe, M. El Eid et al., A. Chieffi, ApJ **437**, 396 (1994)
118. A. I. Karakas, M. Lugaro, M. Wiescher, J. Goerres, C. Ugalde, ApJ **643**, 471 (2006)
119. L.-S. The, M. F. El Eid, B. S. Meyer, ApJ **533**, 998 (2000)
120. L.-S. The, M. F. El Eid, B. S. Meyer, ApJ **655**, 1058 (2007)
121. M. Heil, F. Käppeler, E. Uberseder, R. Gallino, M. Pignatari, Phys. Rev. **C 77(1)**, 015808 (2008)
122. D. A. García-Hernández, P. García-Lario, B. Plez, F. D'Antona, A. Manchado, J. M. Trigo-Rodríguez, Science **314**, 1751 (2006)
123. R. Napiwotzki, Rev. Mod. Astron. **11**, 3 (1998)
124. K. Werner, Ap&SS **275**, 27 (2001)
125. K. Werner, T. Rauch, A&A **284**, L5 (1994)
126. K. Werner, B. Wolff, A&A **347**, L9 (1999)
127. K. Werner, T. Rauch, J. W. Kruk, A&A **433**, 641 (2005)
128. M. Lugaro, C. Ugalde, A. I. Karakas, J. Görres, M. Wiescher, J. C. Lattanzio, R. C. Cannon, ApJ **615**, 934 (2004)
129. C. A. Frost, J. C. Lattanzio, P. R. Wood, ApJ **500**, 355 (1998)
130. A. Jorissen, V. V. Smith, D. L. Lambert, A&A, **261**, 164 (1992)
131. C. Ugalde, R. E. Azuma, A. Couture, J. Görres, H. Y. Lee, E. Stech et al., Phys. Rev. **C 77(3)**, 035801 (2008)
132. A. I. Karakas, H. Y. Lee, M. Lugaro, J. Görres, M. Wiescher, ApJ **676**, 1254 (2008)
133. N. Mowlavi, A. Jorissen, M. Arnould, A&A **311**, 803 (1996)
134. S. E. Woosley, T. A. Weaver, ApJS **101**, 181 (1995)

135. G. Meynet, M. Arnould, A&A **355**, 176 (2000)
136. A. Renda, Y. Fenner, B. K. Gibson, A. I. Karakas, J. C. Lattanzio, S. Campbell, A. Chieffi et al., MNRAS **354**, 575 (2004)
137. Y. Zhang, X.-W. Liu, ApJ **631**, L61 (2005)
138. G. Pandey, ApJ **648**, L143 (2006)
139. S. R. Federman, Y. Sheffer, D. L. Lambert, and V. V. Smith, ApJ **619**, 884 (2005)
140. S. C. Schuler, K. Cunha, V. V. Smith, T. Sivarani, T. C. Beers, Y. S. Lee, ApJ **667**, L81 (2007)
141. M. Lugaro, S. E. de Mink, R. G. Izzard, S. W. Campbell, A. I. Karakas, S. Cristallo et al., A&A **484**, L27 (2008)
142. C. Abia, J. Isern, MNRAS **289**, L11 (1997)
143. K. M. Nollett, M. Busso, G. J. Wasserburg, ApJ **582**, 1036 (2003)
144. E. Zinner, L. R. Nittler, R. Gallino, A. I. Karakas, M. Lugaro, O. Straniero, J. C. Lattanzio, ApJ **650**, 350 (2006)
145. E. Zinner, PASA **25**, 7 (2008)
146. J. Lattanzio, C. Frost, R. Cannon, P. R. Wood, Mem. Soc. Astron. Italiana **67**, 729 (1996)
147. C. A. Frost, R. C. Cannon, J. C. Lattanzio, P. R. Wood, M. Forestini, A&A **332**, L17 (1998)
148. A. G. W. Cameron, W. A. Fowler, ApJ **164**, 111 (1971)
149. A. I. Boothroyd, I.-J. Sackmann, ApJ **393**, L21 (1992)
150. D. Romano, F. Matteucci, P. Ventura, F. D'Antona, A&A **374**, 646 (2001)
151. C. Travaglio, S. Randich, D. Galli, J. Lattanzio, L. M. Elliott, M. Forestini, F. Ferrini, ApJ **559**, 909 (2001)
152. P. Ventura, F. D'Antona, A&A **431**, 279 (2005)
153. J. A. McSaveney, P. R. Wood, M. Scholz, J. C. Lattanzio, K. H. Hinkle, MNRAS **378**, 1089 (2007)
154. R. B. C. Henry, M. G. Edmunds, J. Köppen, ApJ **541**, 660 (2000)
155. F. X. Timmes, S. E. Woosley, T. A. Weaver, ApJS **98**, 617 (1995)
156. M. Samland, ApJ **496**, 155 (1998)
157. A. Goswami & N. Prantzos, A&A **359**, 191 (2000)
158. C. Chiappini, R. Hirschi, G. Meynet, S. Ekström, A. Maeder, F. Matteucci, A&A **449**, L27 (2006)
159. R. Hirschi, A&A **461**, 571 (2007)
160. M. Pettini, S. L. Ellison, J. Bergeron, P. Petitjean, A&A **391**, 21 (2002)
161. A. Renzini, M. Voli, A&A **94**, 175 (1981)
162. L. B. van den Hoek, M. A. T. Groenewegen, A&AS **123**, 305 (1997)
163. P. Marigo, A&A **370**, 194 (2001)
164. P. Ventura, F. D'Antona, I. Mazzitelli, A&A **393**, 215 (2002)
165. P. Kroupa, C. A. Tout, G. Gilmore, MNRAS **262**, 545 (1993)
166. E. M. Burbidge, G. R. Burbidge, W. A. Fowler, F. Hoyle, Rev. Mod. Phys. **29**, 547 (1957)
167. G. Wallerstein, I. J. Iben, P. Parker, A. M. Boesgaard, G. M. Hale, A. E. Champagne et al., Rev. Mod. Phy. **69**, 995 (1997)
168. B. S. Meyer, ARA&A **32**, 153 (1994)
169. C. L. Fryer, F. Herwig, A. Hungerford, F. X. Timmes, ApJ **646**, L131 (2006)

170. S. Wanajo, ApJ **650**, L79 (2006)
171. D. Argast, M. Samland, F.-K. Thielemann, Y.-Z. Qian, A&A **416**, 997 (2004)
172. R. Surman, G. C. McLaughlin, M. Ruffert, H.-T. Janka, W. R. Hix, ApJ **679**, L117 (2008)
173. M. Arnould, S. Goriely, K. Takahashi, Phys. Rep. **450**, 97 (2007)
174. A. G. W. Cameron, AJ **65**, 485 (1960)
175. M. Busso, R. Gallino, D. L. Lambert, C. Travaglio, V. V. Smith, ApJ **557**, 802 (2001)
176. R. Gallino, C. Arlandini, M. Busso, M. Lugaro, C. Travaglio, O. Straniero, A. Chieffi, M. Limongi, ApJ **497**, 388 (1998)
177. O. Straniero, R. Gallino, M. Busso, A. Chiefei, C. M. Raiteri, M. Limongi, M. Salaris, ApJ **440**, L85, (1995)
178. C. Abia, M. Busso, R. Gallino, I. Domínguez, O. Straniero, J. Isern, ApJ **559**, 1117 (2001)
179. S. Goriely, N. Mowlavi, A&A **362**, 599 (2000)
180. A. I. Karakas, M. Lugaro, and R. Gallino, ApJ **656**, L73 (2007)
181. S. Goriely, L. Siesu, A&A **421**, L25 (2004)
182. A. Bonačić Marinović, M. Lugaro, M. Reyniers, H. van Winckel, A&A **472**, L1, (2007)
183. C. Travaglio, R. Gallino, E. Arnone, J. Cowan, F. Jordan, C. Sneden, ApJ **601**, 864 (2004)

Spectral Classification: Old and Contemporary

Sunetra Giridhar

Indian Institute of Astrophysics, Bangalore 560034, India

Summary. Beginning with a historical account of the spectral classification, its refinement through additional criteria is presented. The line strengths and ratios used in two dimensional classifications of each spectral class are described. A parallel classification scheme for metal-poor stars and the standards used for classification are presented. The extension of spectral classification beyond M to L and T and spectroscopic classification criteria relevant to these classes are described. Contemporary methods of classifications based upon different automated approaches are introduced.

Keywords: Spectral classification, luminosity classes, metal-poor stars, M-L-T spectral classes

1 Historical Account of Spectral Classification

In 1866, Fr Angelo Secchi a Jesuit astronomer working in Italy observed prismatic spectra of about 4000 stars visually and divided stars in four broad spectral classes using common absorption features of hydrogen. During 1886-97, Henry Draper Memorial Survey at Harvard carried out a systematic photographic spectroscopy of stars brighter than 9th magnitude covering entire sky using telescopes at Harvard and Arequipa, Peru under the leadership of E.C. Pickering. The Henry Draper Catalog was published in 9 volumes of the Annals of Harvard College Observatory between 1918 and 1924. It contains rough positions, magnitudes, spectral classifications for 225,300 stars. Earlier work by W. Fleming essentially subdivided the previously used Secchi classes (I to IV) into more specific classes, by giving letters from A to N. The strength of hydrogen lines being the main classifier, the spectral type A was assigned to stars with strongest hydrogen lines followed by B, C with weaker hydrogen lines. This system was found to be unsatisfactory since the line strengths of other lines varied irregularly and so did the B−V color. This system was improvised by A. Maury, A.J. Cannon and E. Pickering who re-arranged the

A. Goswami and B.E. Reddy (eds.), *Principles and Perspectives in Cosmochemistry*, Astrophysics and Space Science Proceedings,
DOI 10.1007/978-3-642-10352-0_3,

spectral sequence taking into consideration the changes in other lines and this new spectral sequence was also a sequence according to the color of the stars. But well-known stars had been already assigned the older spectral classes for long hence it was not possible to change them. We, therefore, have a spectral sequence essentially temperature dependent but goes like OBAFGKM. At the cool end the classification becomes more complex with parallel branches of R, N, S stars. While the M stars have TiO bands, S stars display ZrO bands, while R and N are carbon stars showing strong bands of various molecules with carbon. These have more recently been merged into a unified carbon classifier C scheme of C1, C2 etc with the old N0 starting at roughly C6. Another subset of cool carbon stars are the J-type stars, which are characterized by the strong ^{13}CN molecules in addition to those of ^{12}CN.

Each of the above mentioned spectral classes OBAFGKM have been subdivided into ten subclasses e.g. A0, A1 ... A9.

1.1 Luminosity Effects in Stellar Spectra

E. Hertzsprung suggested in 1905 that spectral line widths were related to the luminosity of the stars. He pointed out that at a given apparent magnitude, the low proper motion stars would be at larger distance from us than the high proper motion stars of the same apparent magnitude and hence of higher intrinsic luminosity. These low proper motion stars were found to exhibit narrower spectral lines, so, Hertzsprung concluded that these narrow line stars have larger intrinsic luminosity than the broad line stars.

In 1943, W. Morgan, P.C. Keenan and E. Kellman introduced luminosity as second classification parameter. Morgan noticed the near constancy of the gravity along the main sequence in HR diagram and luminosity class parameter was an attempt to identify stars of different gravities and hence radii at nearly constant temperature.

The above mentioned system also known as Yerkes Spectral Classification. Within the system, six luminosity classes are defined on the basis of standard stars over the observed luminosity range.

The Six classes are:

Ia: most luminous supergiants
Ib: less luminous supergiants
II: luminous giants
III: normal giants
IV: subgiants
V: main sequence stars

The main sequence class (dwarfs) are the stars at the main sequence, sustaining themselves through the conversion of hydrogen to helium by nuclear fusion in the stellar core, giant is a post main sequence star which is no longer burning hydrogen at the core but has H-burning shell outside the core. Giants as well as supergiants have comparable mass to dwarfs but have expanded to a much larger radii resulting in a decrease in their surface gravities. The spectral lines appear broad in the dense atmospheres of dwarfs primarily due to pressure broadening and Stark broadening while the same line would appear narrow in the low gravity atmospheres of supergiants.

The luminosity effects are not restricted to the narrowing of strong lines. The line strengths and ratios of line strengths of neutral and ionized species also show remarkable variations over spectral classes and luminosity types and have been used for defining the subclasses and luminosity types. In addition, there are well-known luminosity indicators such as the emission components in the lines of Ca II H and K in late type stars which are related to the luminosity (absolute magnitude) of the stars and the calibration of this relationship has been carried out by Wilson and Bappu [1]. The strength of near IR OI triplet at 7771-75 Å has been used by Osmer, Arellano and collaborators [2, 3, 4] for absolute magnitude calibration of A–G stars. In the next section, we will describe the line strengths and their ratios which are used to define the spectral classes and luminosity types.

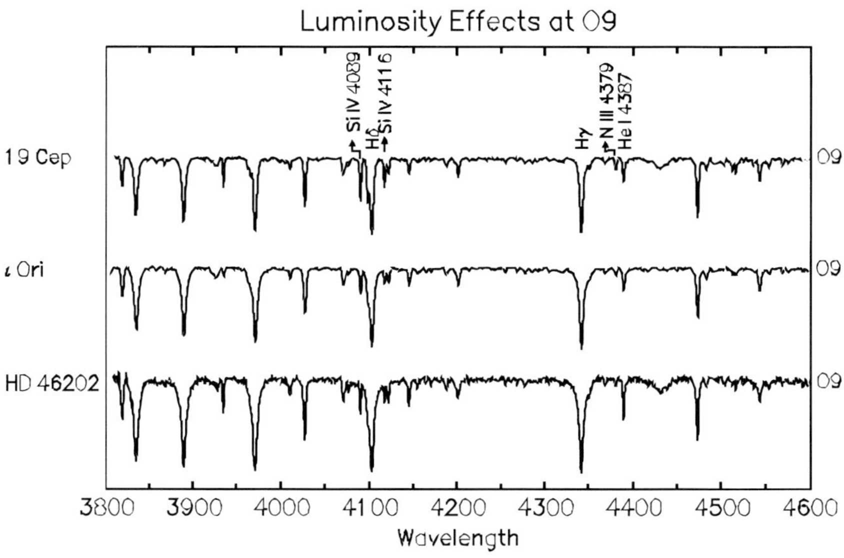

Fig. 1. The luminosity effect in late O stars from [5] with author's permission

2 Classification Criteria for various spectral types

Beginning from O stars which are hottest, we will briefly describe the spectral characteristic for each spectral class and also list the spectral type and luminosity class indicators. Most of the information for this chapter is taken from Jaschek and Jaschek [6]. The figures used in this section illustrating the different spectral types and also the luminosity effects at various spectral types are taken from 'A Digital Spectral Classification Atlas v1.02' by R.O. Gray with his permission. More figures can be found on the website http://nedwww.ipac.caltech.edu/level5/Gray/frames.html.

O-type Stars

These are hot stars with temperature range of 28,000 K to 50,000 K. These stars exhibit strong lines of neutral and ionized helium. The strength of He II lines decrease and He I as well as H I lines increase in strength towards later O-types. Over the spectral range O4 to B0 the line strength of He II line at 4541 Å decreases from 800 mÅ to 200 mÅ, while He I line at 4471 Å increases from 100 mÅ to 1000 mÅ and that of H I line at 4341 Å increases from 1.5 Å to 2.5 Å. The O-type spectra also exhibit the features of doubly and triply ionized carbon, nitrogen, oxygen and silicon. The line ratios such as $\frac{C\ III\ \lambda4649}{He\ II\ \lambda4686}$ are used for luminosity classification. Similarly, the ratio $\frac{Si\ IV\lambda4089}{He\ I\ \lambda4143}$ serves as good luminosity indicator in late O type stars. Wolf-Royet stars are a special family of O-type stars that are characterized by broad emission lines of ionized helium, carbon (WC sequence) or nitrogen (WN sequence). The WC stars exhibit emission lines of He II such as He II λ4686, ionised carbon such as C II λ4267, C III $\lambda\lambda$3609, 4187, 4325, 4650 etc. C IV $\lambda\lambda$4441, 4658, 4758 etc. and lines of O II, O III, O IV and O V. These stars are subdivided into WC classes from WC2−WC10 based upon the line ratios of $\frac{C\ III\ \lambda5696}{O\ V\ \lambda5592}$, $\frac{C\ III\ \lambda5696}{C\ IV\ \lambda5805}$ etc.

The WN stars exhibit emission lines of He II, N III $\lambda\lambda$4097, 4640, 5314, N IV $\lambda\lambda$3483, 4057 and N V $\lambda\lambda$4605, 4622. These stars are also subdivided into WN subclasses using the line ratios such as $\frac{N\ III\ \lambda4640}{He\ II\ \lambda4686}$.

B-type Stars

The B type spectra contains lines of He I, H I, C II, C III, N II, N III, O II, Si II, Si III, Si IV, Mg II. The lines of higher ionized states of C, N, O are present in early B stars. The maximum strength of He I line reaches near B2. Many B stars are fast rotators and emission lines are present in some of them.

The line ratios $\frac{Si\ III}{Si\ IV}$, $\frac{SiII\ \lambda4128-30}{He\ I\ \lambda4121}$ and $\frac{Si\ II\ \lambda4128-30}{He\ I\lambda4144}$ are used for spectral class determination. The luminosity criteria used include the ratios of features at $\frac{\lambda4119\ (Si\ IV+He\ II)}{\lambda4144\ (He\ I)}$, $\frac{\lambda4481\ (Mg\ II)}{\lambda4471\ (He\ I)}$ which increase with luminosity. Profiles of Balmer lines become narrower with luminosity. These are blue white stars with temperature range of 10,000 K to 28,000 K.

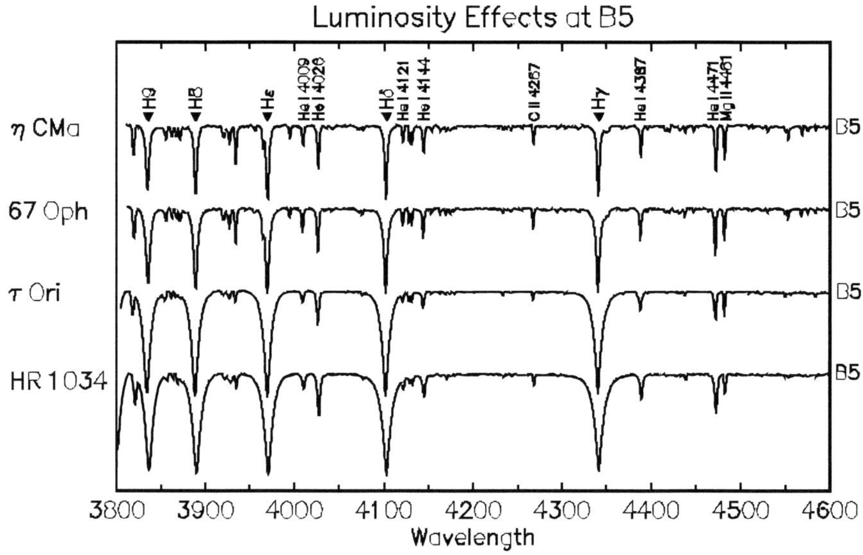

Fig. 2. The luminosity effect in B stars is illustrated. The figure is taken from [5] with author's permission

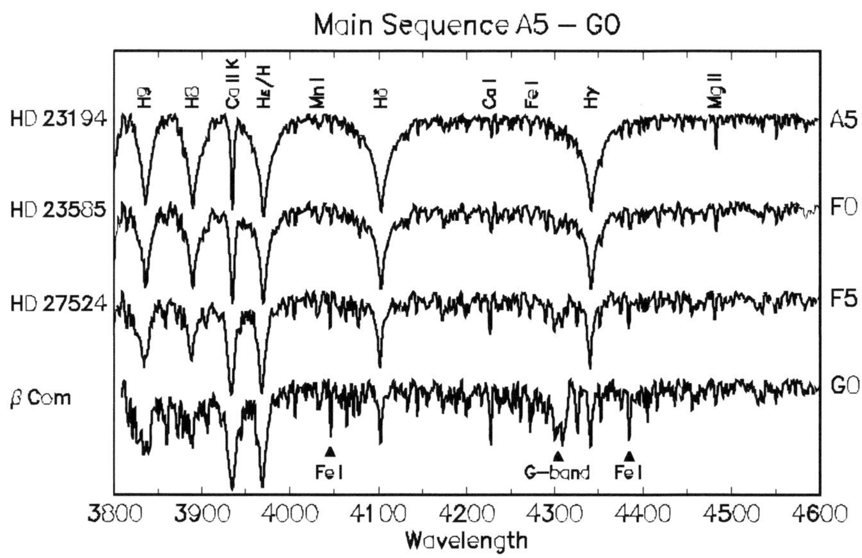

Fig. 3. The luminosity effect in A5 to G0 stars are illustrated. The figure is taken from [5] with author's permission

A-type Stars

These are white stars with temperature range of 7,500 K to 10,000 K. The A type stars exhibit strong hydrogen lines of Balmer series. The hydrogen lines are strongest at spectral type A2. The H_γ has a strength of 13.6 Å at A0, 17 Å at A2 and decreases to 13 Å at A7 and to 8Å at F0. The similar pattern is followed by other lines of Balmer series. The metallic lines gradually increase in strength from A0 to A9. The helium lines are absent. To assign subclasses the line ratios $\frac{Ca\ I\lambda 4227}{Mg II\lambda 4481}$, $\frac{Fe\ I\lambda 4045}{Fe\ II\ \lambda 4173}$, $\frac{Mg\ II\lambda 4481}{Fe\ I\lambda 4485}$, $\frac{Mg\ II\lambda 4481}{Fe\ II\lambda 4416}$ are useful. However other line ratios are also used. The luminosity criteria used are blend ratios such as $\frac{Fe\ I\lambda\lambda 4383-85}{Mg\ II\lambda 4481}$, $\frac{Fe\ II\lambda 4417}{Mg\ II\lambda 4481}$ which become stronger towards higher luminosity. The hydrogen lines also become narrower towards higher luminosity. The additional luminosity indicators are $\frac{Sr\ II\lambda 4215}{Ca\ I\lambda 4226}$, $\frac{Fe\ II\lambda 4351}{Mg\ II\lambda 4481}$ which increase at higher luminosity. The near infrared O I triplet at λλ 7771-75 is a very good indicator of luminosity for A-F stars.

Additional sub-classification of A type stars such as Am and Ap is done based on their chemical peculiarities, magnetic fields, rotation and the presence of emission lines in their spectra.

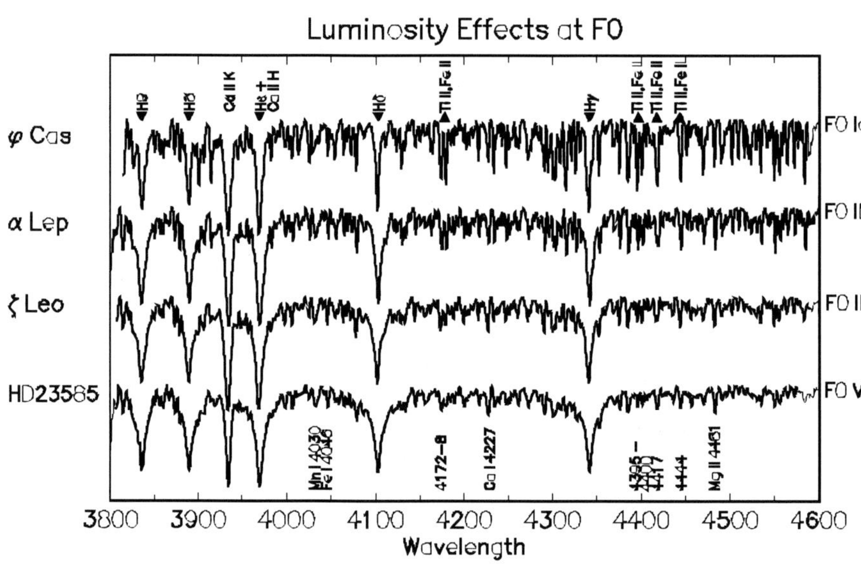

Fig. 4. The luminosity effect in F stars is illustrated. The figure is taken from [5] with author's permission

F-type Stars

These are white-yellow stars with temperature range of 6,000 K to 7,500 K. The F type spectra have large number of metallic lines, Ca II H and K lines are very strong which become stronger than hydrogen lines of Balmer series. The hydrogen lines although present, are on decline from F0–F9. The G band of CH molecule makes its appearance around F3. The Ca II at $\lambda3933$ increases from 6.5 Å at F0 to 17.0 Å at G0, Ca I at $\lambda4226$ increases from 0.25 Å at F0 to 1.1 Å at G0 while H_γ decreases from 8 Å at F0 to 3 Å at G0. Due to large number of lines present, at the resolution used for classification, most feature are blended. For the spectral class the criteria used are $\frac{Fe\ I\ \lambda4045}{H_\delta}$, $\frac{Ca\ I\ \lambda4226}{H_\gamma}$, $\frac{Mn\ I\ \lambda\lambda4030-34}{Si\ II\ \lambda\lambda4128-32}$ etc.

The Ca II lines show positive luminosity effect, ratios $\frac{Ti\ II\ \lambda4444}{Mg\ II\ \lambda4481}$, $\frac{Sr\ II\ \lambda4077}{Fe\ I\ \lambda4045}$, $\frac{Sr\ II\ \lambda4077}{H_\delta}$ are used for luminosity classification.

G-type Stars

These are yellow stars with temperature range of 4,900 K to 6,000 K. In G type stars the hydrogen lines are further weakened and become comparable to the strength of metal lines. Metal lines are stronger and more numerous towards later G, molecular bands of CH and CN become visible. The spectral types are obtained by taking the ratio of metal lines with those of hydrogen lines e.g. using the ratios such as $\frac{Fe\ I\ \lambda4384}{H_\gamma}$, $\frac{Fe\ I\ \lambda4143}{H_\delta}$, $\frac{Fe\ I\ \lambda4045}{H_\delta}$, $\frac{Ca\ I\ \lambda4226}{H_\delta}$ etc. For spectral types later than G5 the $\frac{Ca\ I\ \lambda4226}{H_\delta}$ can be used. The line ratios such as $\frac{Cr\ I\lambda4254}{Fe\ I\lambda4250}$ or $\frac{Cr\ I\ \lambda4742}{Fe\ I\ \lambda4271}$ are recommended for stars showing compositional anomalies such as weak metal line stars or weak G band stars.

Luminosity effects at low dispersion can be seen through CN bands. The ratio of the Sr I+Fe I blend at $\frac{\lambda4216}{Ca\ I\ \lambda4226}$ is known to be luminosity sensitive. Ratios of Y II+Fe I at $\frac{\lambda4376}{Fe\ I\lambda4383}$, $\frac{Sr\ II\ \lambda4077}{H_\delta}$ is also known to be luminosity sensitive but will not be suitable for stars with anomalous s–process abundances. Mg I triplet at $\lambda\lambda5167-83$ are luminosity sensitive for spectral type range G8–K5.

K-type Stars

These are orange stars with temperature range of 3,500 K to 4,900 K. In these stars the hydrogen lines are very weak but strong numerous metal lines are seen. The Ca II lines are very strong and CH molecular band like G band becomes very strong. In late K stars TiO and VO bands are also seen.

The line ratios for spectral type used are $\frac{Cr\ I\ \lambda4254}{Fe\ I\ \lambda4250}$, $\frac{Cr\ I\ \lambda4254}{Fe\ I\ \lambda4260}$, $\frac{Cr\ I\ \lambda4254}{Fe\ I\ \lambda4271}$. Additional qualifiers are $\frac{Ti\ I\ \lambda3999}{Fe\ I\ \lambda4005}$, $\frac{Fe\ I\ \lambda4144}{H_\delta}$, $\frac{Ca\ I\ \lambda4226}{Fe\ I\ \lambda4250}$. The TiO band becomes visible at K7 and MgH at K5.

The CN band increases with luminosity, so does $\frac{H_\delta\ 4101}{Fe\ I\ \lambda4071}$, $\frac{Sr\ II\ \lambda4077}{Fe\ I\ \lambda4063}$, $\frac{Ti\ II\ \lambda4400{,}08}{Fe\ I\ \lambda4405}$ also increase with luminosity.

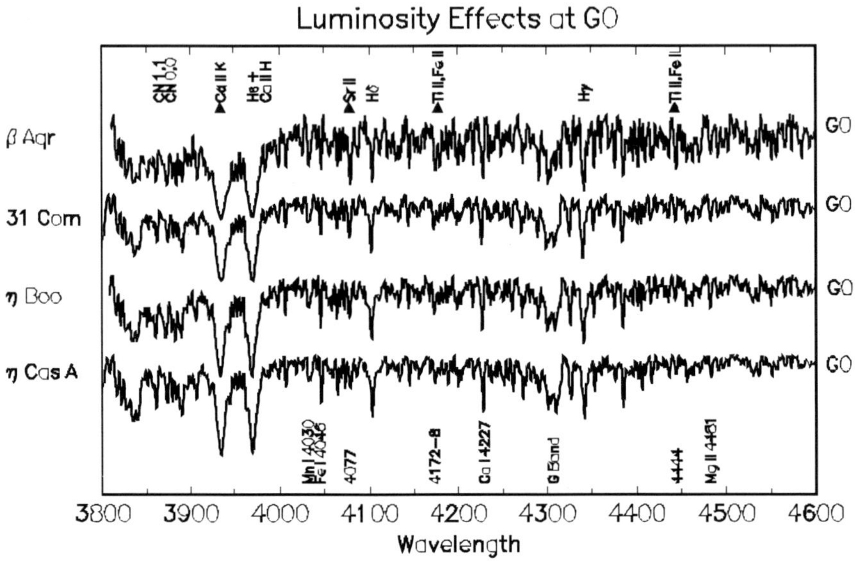

Fig. 5. The luminosity effect in G stars is illustrated. The figure is taken from [5] with author's permission

Carbon Stars

A carbon star is a late type giant with strong bands of carbon compound but no metallic oxide band. In their spectra, very intense bands of C_2, CN and CH are present but no bands of TiO, VO are seen. The carbon stars as a group have been studied by Secchi [7]. Although in older classification they were given two different types R and N, the R type stars were similar to late G and early K but exhibited very strong Swan C_2 bands around $\lambda4700$ and $\lambda4395$ which are as strong as G-band of CH. In N stars the Swan band is so strong that the spectrum appears chopped up into sections of different intensities. Although based upon CN band (at $\lambda4216$ and $\lambda3833$) strengths a sequence R0, R1 to R10=N0 and subsequently upto N7 can be defined, the same pattern is not shown by Swan C_2 bands, which become strongest at R5 and weakest at N0 and further strengthening towards later N types. A more detailed study by Shane [8] revealed that temperature variations over the subclasses are not large and branching of these stars R and N stars is caused by abundance difference in C and O. If oxygen is more abundant than carbon then the spectrum is dominated by oxides like TiO (M stars). Keenan and Morgan [9] used the term 'carbon stars' instead of R and N wherein the overabundance of carbon varied from star to star. They established the sequence of carbon stars by means of temperature index of $\frac{Cr\ I\lambda4254}{Fe\ I\ \lambda4250}$ and the strength of resonance lines of Na I at $\lambda\lambda5890-96$ which are good temperature indicators for the late spectral type stars. The temperature sequence C0 to C7 covers full range of R0 to N7 which in temperature is very similar to the sequence of G4–M4. J stars are another group of carbon stars characterized by unusually strong isotopic bands of carbon which implies very low $^{12}C/^{13}C$ ratio.

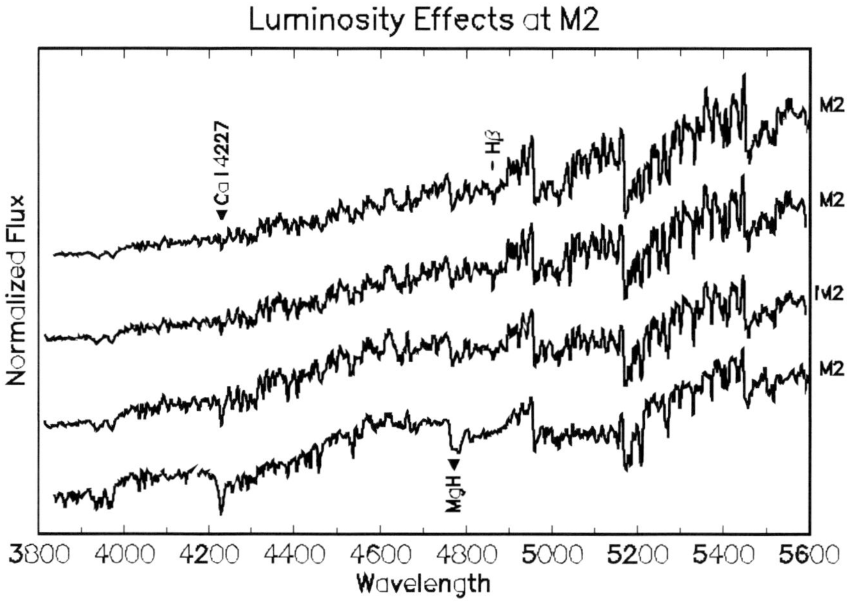

Fig. 6. The luminosity effect in M2 stars is illustrated. The figure is taken from [5] with author's permission

M-type Stars

These are red stars with temperature range of 2,000 K to 3,500 K. The spectra of M stars are dominated by strong bands of TiO, VO, LaO. The bands used for classification upto M2 are those of TiO at $\lambda\lambda$4584, 4761, 4954, 5448. After M3 these bands saturate, those at $\lambda\lambda$5759, 5810 saturate after M5. The VO bands at $\lambda\lambda$5737, 7373, 7865, 7896 become conspicuous after M7.

The luminosity effect in early M type stars can be seen through decrease of Ca I λ4226 at higher luminosity; a similar negative luminosity dependence is exhibited by Cr I feature at $\lambda\lambda$4254–74–90. Ratios $\frac{Sr\ II\ \lambda 4077}{Fe\ I\ \lambda 4263}$ and $\frac{(Y\ II+Fe\ I)\lambda 4376}{Fe\ I\ \lambda 4386}$ increase with luminosity. The K I λ7699, Na I λ8183 and λ8185 decrease in intensity from dwarfs to giants. The Ca II triplet at $\lambda\lambda$8498, 8542 and 8662 is a very important luminosity indicator in these stars. It is very weak in dwarfs but becomes very strong in giants and supergiants.

S-type Stars

These stars are similar in temperature to K5 to M stars but exhibit strong bands of ZrO with band heads at $\lambda\lambda$4640, 5551 and 6474.

3 New Spectral types L and T

These are the coolest stars with temperatures less than 2500 K. Since they are very cool, they emit mostly in infrared wavelengths. The original classification scheme of L dwarfs based on red to far-red spectral region has been described in [10, 11]. The L stars have a temperature range of 2500 K to 1300 K and exhibit some overlapping features with M stars such as TiO, VO although these features do not remain as strong. These features are very weak in L3 and almost disappear at L6 leaving only metal hydride bands of FeH, CaH, CrH. These bands have maximum strength at mid L but become weak in late L types where H_2O presents a strong feature at 9300 Å. The atomic features such as resonance lines of Rb I at 7800 Å, 7948 Å and Cs I at 8521 Å, 8945 Å can be seen in early L and grow in strength throughout the sequence. The resonance line of Li I at 6708 Å is present in L dwarfs. The K I resonance doublet shows very remarkable change. Being somewhat narrower and weak at M9, it broadens and becomes strong through L0 onwards such that at L5 two separate cores are indistinguishable and doublet looking like a broad trough $\sim$ 600 Å wide; at L8 it is $\sim$ 1000 Å wide. The Na I resonance doublet also follow the same pattern by becoming broad and shallow at L5 .

These spectral variations like disappearance of TiO, VO and other molecular features at later L subclasses are believed to be caused by formation of various types of condensates at the temperatures lower than 2600 K. Lodders [12] have made extensive chemical equilibrium calculations and the atmospheric composition changes as the material is removed from gas to solid phase.

3.1 The T dwarfs

As the brown dwarfs cool to temperatures below 1400 K, a drastic change occurs between 1400 K to 1200 K when CH_4 bands become strong while CO band weaken. The CH_4 bands are extremely broad in near infrared and they even modify the near infrared broad band colors. Strong methane absorption at 1.6 and 2.2 μ reduces more than half flux from H and K pass-band giving them unrealistic colors.

A set of flux ratios measuring the strengths of methane and H_2O have been attempted to classify the T dwarfs by Burgasser and co-workers [14] and others. Considerable progress will be made through deeper IR surveys.

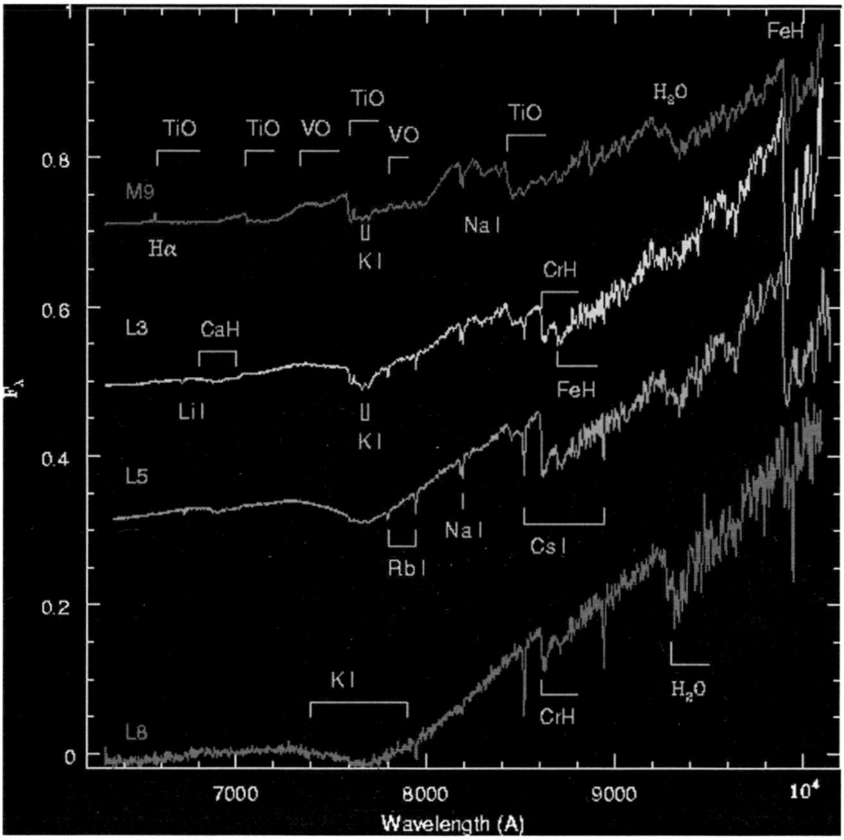

Fig. 7. The spectral variations seen from late M to L stars are illustrated. The figure is taken from the website www.stsci.edu/inr/ldwarfs originally published in [10]

4 Modification of MK system

The MK system has gone through extensive research and refinement. The important work in refining and extending the MK system in dimensions beyond the traditional two dimensional spectral type luminosity class grid has been carried out by Keenan and co-workers and the use of third 'metallicity parameter' is recommended. Grey [16] has carried out the extension of MK system to metal-weak F and G stars. A grid of metal-poor standard stars has been prepared with the objective of providing a mean of classifying metal-weak stars using a standard of similar metallicity rather than the old method of comparing the metallic line strengths of the program stars with that of a considerable earlier solar composition standard.

The earlier attempt of deriving spectral types for metal-poor stars relied upon hydrogen line strengths or G-band strength due to their strong

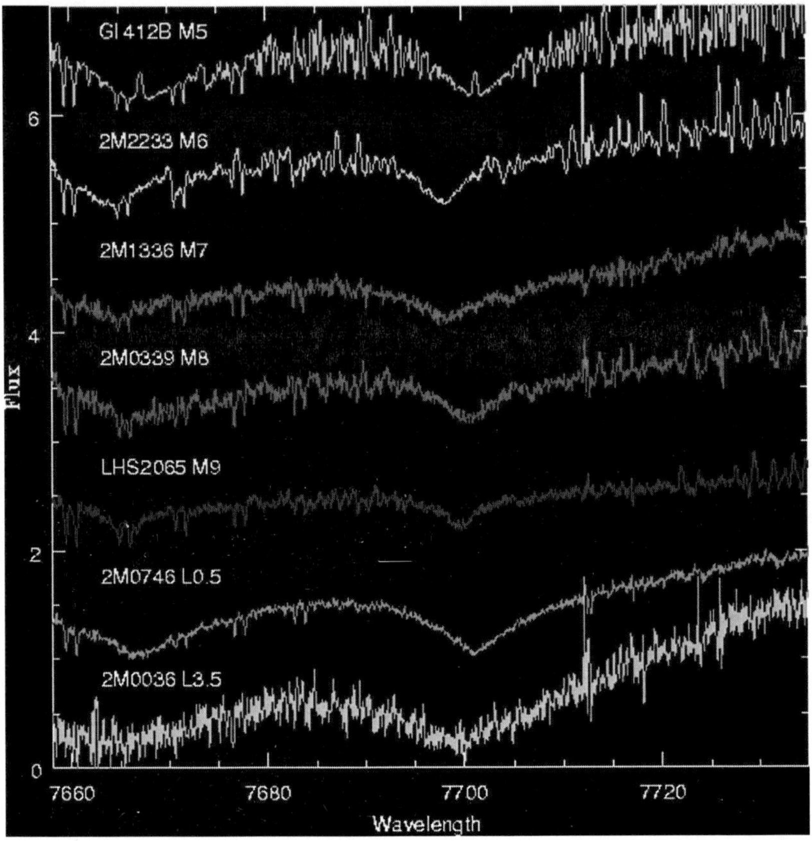

Fig. 8. The drastic variations in KI resonance doublet is illustrated above. The figure was originally published in [13]

temperature dependence. However these classifiers are not adequate; the hydrogen line strengths do depend upon other stellar parameters and are not suitable at the later spectral types. Corbally [17] has pointed out that the ratio of Cr I resonance triplet at 4254, 4274, 4290 Å to the neighboring lines of Fe I at 4251, 4272, 4362 Å arising from excited level is nearly independent of metallicity and hence can serve as temperature indicator. For luminosity classification the ratios of features like $\frac{Ti\ II,\ Fe\ II\lambda\lambda4172-9}{Fe\ I\ \lambda4203,\ \lambda4271}$ are useful for F stars while for G stars $\frac{Sr\ II\ \lambda4077}{Fe\ I\ \lambda4063,\ \lambda4046}$ and $\frac{Fe\ I\lambda4383}{Y\ II\ \lambda4375}$ features may be useful since these lines (blends) are very strong and remain discernible even at low metallicity. A list of standard stars for metallicity classification can be found in [16].

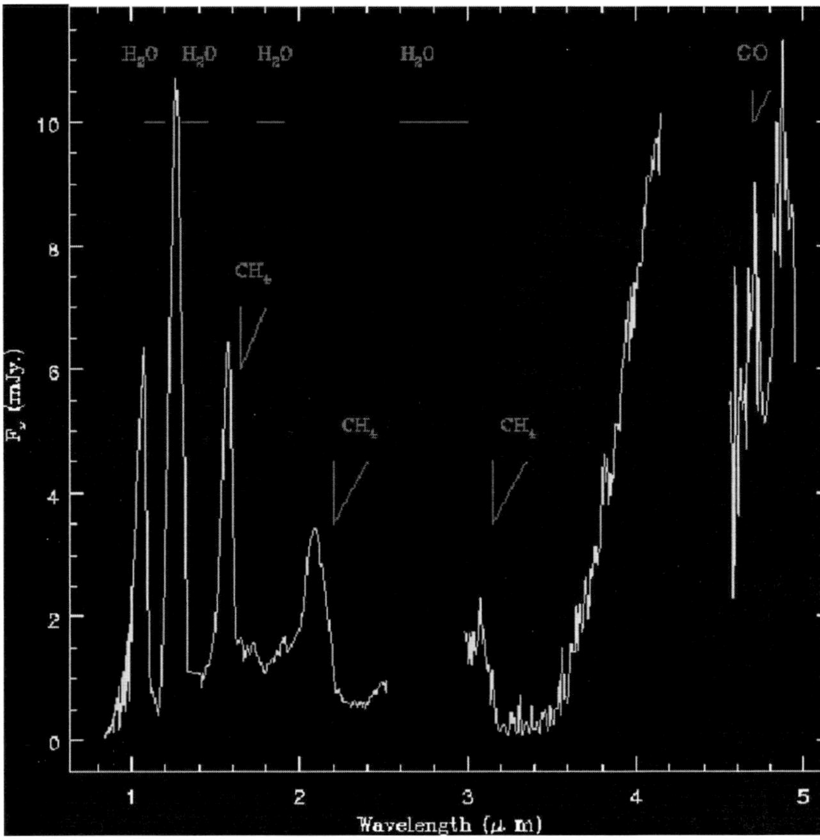

Fig. 9. The strengthening of CH_4 bands in T stars is illustrated above. The figure was originally published in [15]

5 Contemporary methods of spectral classification

Stellar classification has been an important tool in stellar and Galactic astronomy since it provides empirical measure of the fundamental stellar parameters such as temperature, luminosity and metallicity.

The massive surveys both ground based as well as from space missions provide large number of stellar spectra covering distant components of Galaxy. To understand the complex evolutionary history of our Galaxy, rapid and accurate methods of stellar classification are necessary. A short review of the automated procedures are presented here. The most commonly used automated spectral classification methods are based on (a) Minimum Distance Method (MDM) (b) Gaussian Probability Method (GPM) (c) Principal Component Analysis (PCA) and (d) Artificial Neural Network (ANN). We chose to describe only two of them to introduce the automated approach of classification.

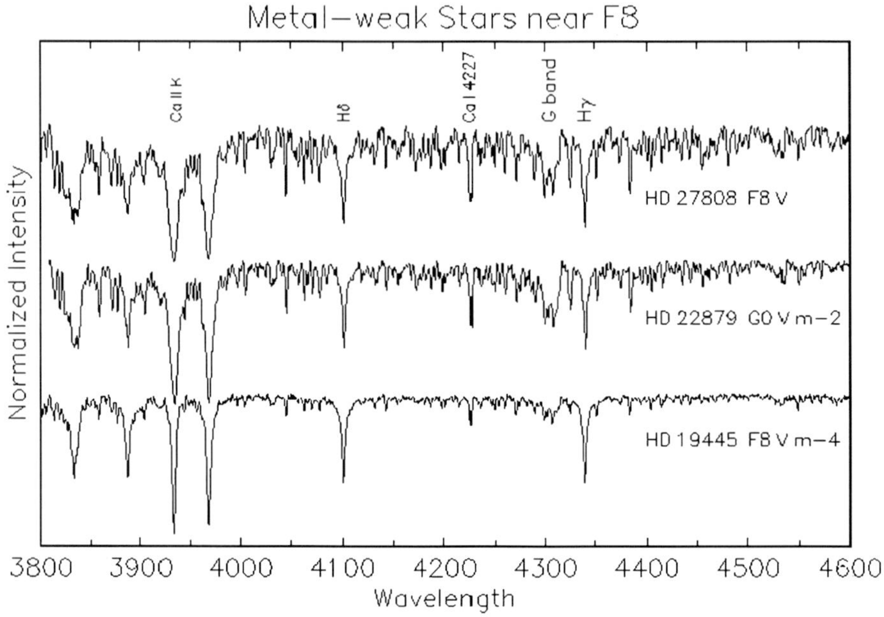

Fig. 10. The spectra of a few metal-poor stars are shown above

In MDM, the classification is done by minimizing the distance metric between the object to be classified and each member of a set of templates. The object is assigned the class of the template, which gives the smallest distance. In this approach the number of templates used to define subclasses limit the accuracy of classification. Interpolation can be made to make intra-class assignment. Katz [18] used this method with χ^2 minimizing on high resolution Elodie spectra using a large number of reference stars of known T_{eff}, log g and [M/H] to derive atmospheric parameters of target stars. These authors achieved accuracy of 86 K in T_{eff}, 0.28 in log g and 0.35 in [M/H].

The Neural Network methods have become very popular due to their speed and objectivity as explained in the papers [19, 20, 21, 22] and more recently in [23]. These are computational methods which can provide non-linear parametrized mapping between an input vector (a spectrum for example) and one or more outputs like SpT, LC or T_{eff}, log g and [M/H]. For the network to give required input-output mapping, it must be trained with the help of representative data patterns. These could be stellar spectra (or a set of line strengths measured from a spectrum) for which classification or stellar parameters are well determined. The training procedure is a numerical least square error minimization method. The training proceeds by optimizing the network parameters (weights) to give minimum classification error. Once the network is trained the weights are fixed, the network can be used to produce output SpT, LC or T_{eff}, log g and [M/H] for an unclassified spectrum.

The ANN has been used in very large number of stellar applications. Vieira and Ponz [24] have used ANN on low-resolution IUE spectra and have determined SpT with an accuracy of 1.1 subclass. Bailer-Jones and Irwin [19] used ANN to classify spectra from Michigan Spectral Survey with an accuracy of 1.09 SpT. Prieto and co-workers [25] used ANN in their search of metal-poor stars. Snider and co-workers [26] used ANN for the three dimensional classification of metal-poor stars.

We have made a modest effort to use ANN for parametrization of a sample of stars in the temperature range 4500 to 8000 K. We have used a medium resolution Cassegrain spectrograph with the 2.3 m Vainu Bappu Telescope at VBO, Kavalur, India to get spectra at resolution (R) of 2000. Using 90 spectra for stars of known parameters, and ANN of 680:11:3 architecture we could attain an accuracy of ± 200 K in temperature and ± 0.3 dex in metallicity.

It is very important to envisage an approach that would give quick, reliable spectral classifications (or stellar parameters) for stars falling in all regions of HR diagram. The pipeline procedures are being developed for the future ambitious missions such as GAIA and PAN-STARS.

References

1. O. C. Wilson, M. K. V. Bappu, ApJ **125**, 661 (1957)
2. P. S. Osmer, ApJS **24**, 247 (1972)
3. A. Arellano Ferro, S. Giridhar, A. Goswami, MNRAS **250**, 1, (1991)
4. A. Arellano Ferro, S. Giridhar, E.Rojo Arellano, RMxAA **39**, 3, (2003)
5. R. O. Gray, *A Digital Spectral Classifcation Atlas V1.02*, (2003)
6. C. Jaschek, M. Jaschek, *The classification of stars*, Cambridge University Press, Cambridge (1987)
7. A. Secchi, Mem. Soc. Ital. Sciencze (2) **3**, 73 (1868)
8. C. D. Shane, Lick Bull. **13**, 123 (1928)
9. C. P. Keenan, W. W. Morgan, ApJ **94**, 501 (1941)
10. J. D. Kirkpatrick, I. N. Reid, J. Liebert et al., ApJ **519**, 802 (1999)
11. J. D. Kirkpatrick, I. N. Reid, J. Liebert et al., AJ **120**, 447 (2000)
12. K. Lodders, ApJ **519**, 793 (1999)
13. I. N. Reid, J. D. Kirkpatrick et al., AJ **119**, 369 (1999)
14. A. J. Burgasser et al. ApJ **524**, 421 (2002)
15. B. R. Oppenheimer, S. R. Kulkarni et al., ApJ **502**, 932 (1998)
16. R. O. Gray, AJ **98**, 1049 (1989)
17. C. J. Corbally, AJ **94**, 161 (1987)
18. D. Katz, C. Soubiran et al., A&A **338**, 151 (1998)
19. C. A. L. Bailer-Jones, M. Irwin, T. von Hippel, MNRAS **298**, 361, (1998)
20. C. A. L. Bailer-Jones, An Introduction to Artificial Neural Networks in *Automated Data Analysis in Astronomy* ed. by R. Gupta, H. P. Singh et al., (New Delhi, London : Narosa Pub. House), p83 (2002)
21. T. von Hippel, L. J. Storrie-Lombardi et al., MNRAS **269**, 97 (1994)

22. H. P. Singh, C. A. L. Bailer-Jones, R. Gupta, Principal Component Analysis and its Application to Stellar Spectra in *Automated Data Analysis in Astronomy* ed. by R. Gupta, H. P.Singh et al., (New Delhi, London : Narosa Pub. House), p69 (2002)
23. S. Giridhar, S. Muneer & A. Goswami, MmSAI **77**, 1130, (2006)
24. E. F. Vieira, J. D. Ponz, A&AS **111**, 393 (1995)
25. C. Allende Prieto, R. Rebolo, et al., AJ **120**, 1516 (2000)
26. S. Snider, C. Allende Prieto, T. von Hippel et al., ApJ **562**, 528 (2001)

Part II

Massive Stars, Core Collapse, Explosive Nucleosynthesis

Weak Interaction Rates for Stellar Evolution, Supernovae and r-Process Nucleosynthesis

Kamales Kar

Saha Institute of Nuclear Physics, Bidhannagar, Kolkata - 700064, INDIA

Summary. The role of weak interaction processes in the late stage of evolution of massive stars and during supernova explosions is described. Models for calculating the β-decay, electron capture and neutrino capture rates on nuclei are reviewed. The neutrinos produced in the sun in the weak interaction reaction processes when detected on earth gave rise to the solar neutrino deficit problem. The resolution of this problem is also mentioned. Roughly half the elements heavier than iron are produced through r-process nucleosynthesis. The methods of calculating the β-decay rates of very neutron-rich nuclei taking part in the r-process are also presented.

Keywords: Weak interaction rate, β-decay, Electron capture, Stellar collapse, Supernovae, r-process nucleosynthesis

1 Introduction

Different weak interaction processes like β-decay and electron capture are very crucial at the late stage of stellar burning and gravitational collapse for the type II supernovae. The energy of the shock wave that eventually causes the explosion and the outward propagation of the shock depends on the rates of these processes. This article describes this aspect and also presents the nuclear structure models used to calculate the rates of these processes. The fusion reactions in the sun also involve some of the weak reactions that produce neutrinos. The detection of the solar neutrinos led to the solar neutrino puzzle. That is also covered in this article along with the β-decay rates needed for r-process nucleosynthesis. In Sect. 2 we mention some basic facts of nuclear physics that will be repeatedly used. Section 3 gives an overview of core collapse supernovae. Section 4 points out the stages of the supernova where electron capture, β-decay and neutrino capture become important. Section 5 describes the different structure models used to calculate the rates and Sect. 6 gives the results of incorporating some improved electron and β-decay rates

A. Goswami and B.E. Reddy (eds.), *Principles and Perspectives in Cosmochemistry*, Astrophysics and Space Science Proceedings,
DOI 10.1007/978-3-642-10352-0_4,

for the supernova simulations. Section 7 is concerned with the weak interaction reactions that produce neutrinos in the pp-chain of stellar burning and how one explains the observed deficit of neutrinos in terrestrial detectors. Finally in Sect. 8 we discuss the r-process nucleosynthesis through which highly neutron-rich heavy nuclei are produced. The possible sites are discussed and the models for calculating the β-decay rates are presented.

2 Some Nuclear Physics Basics

A nucleus X with atomic mass A and atomic number Z will be denoted as ${}^A_Z X$. Sometimes we also write it as (N, Z) where the number of neutrons in the nucleus is given by $N = A - Z$. The binding energy per nucleon for the nuclei have a range from a fraction of MeV, for the nuclei very close to drip line to 8-9 Mev, for the iron type of nuclei. The nuclear energy eigenstates have angular momentum, isospin and parity as good quantum numbers. The total angular momentum is the sum of the angular momenta of all the nucleons. The ground state angular momenta of even(Z)-even(N) nuclei are always zero. The ground state angular momentum for the odd A nuclei or the odd-odd nuclei are predicted by simple rules based on the shell structure of the nucleons.

Isospin

The neutron and the proton are considered to form the isospin doublet with $T = 1/2$ and with M_T 1/2 and -1/2 respectively. Denoting $|n >= |1/2, 1/2 >$ and $|p >= |1/2, -1/2 >$ one gets $t_+|p >= |n >$ and $t_-|n >= |p >$ where $t_+ = t_x + it_y$ and $t_- = t_x - it_y$. A nucleus (N,Z) has isospin $|T, M_T >$ with the eigenvalue $M_T = (N - Z)/2$.

2.1 Shell Model

The nucleons inside a nucleus, due to the dominant mean field, have a shell structure with the quantum numbers (nlj) (standing for the radial, orbital angular momentum and total angular momentum quantum numbers respectively) defining each orbit, similar to atomic spectroscopy. In the limit of non-interacting fermions the nucleons occupy the orbits of lowest possible energies filling up the shells $(1s_{1/2})$, $(1p_{3/2}, 1p_{1/2})$, $(1d_{5/2}, 2s_{1/2}, 1d_{3/2})$, ...and the ground state energy is the sum of the energy of the nucleons. Each shell differs in energy from the lower by $1\hbar\omega$. The 1-body Hamiltonian is given by

$$H_0 = \Sigma_\lambda \epsilon_\lambda a_\lambda^\dagger a_\lambda \qquad (1)$$

where the summation is over the orbits λ. For example ^{16}O has the first two shells completely filled and is one of the magic nuclei. ^{20}Ne has ^{16}O as a closed core and has 4 nucleons outside ($M_T = 0$ for the ground state) occupying the

$d_{5/2}$ orbit. The nuclear ground state wavefunction is the antisymmetrised product of the single particle states.

But the nuclei indeed have residual interaction among the nucleons but that does not destroy the shell structure. The additional Hamiltonian is given by

$$H^{'} = \Sigma < kl|H|ij > a_l^{\dagger} a_k^{\dagger} a_i a_j \tag{2}$$

giving rise to non-diagonal matrix elements in the two-particle matrix representation. Now the numerical procedure of calculating the energy eigenvalues and wavefunctions for a nucleus given a model space with specified active single particle orbits and two body nuclear matrix elements is known as the 'shell model' in nuclear structure [1]. The model also can calculate the transition matrix elements of different excitation operators. As nuclear states have (J, T) and parity as conserved quantities the shell model consists of the following steps: a) construct all the m-particle basis states with fixed (J, T) for a nucleus with m valence nucleons by distributing them over the active orbits, b) calculate the matrix elements of the total Hamiltonian in this m-particle basis and c) finally diagonalising the matrix to obtain the eigenvalues and eigenfunctions. Each state in the original space is called a configuration and this shell model diagonalisation gives rise to *configuration mixing*. These spaces often have large dimensions. For example, for ^{28}Si for states with (J, T) as $(2, 0)$ and $(2, 1)$, the matrix dimensions are 3276 and 5768 respectively, staying within the sd-shell. For highly excited states in shell model studies one may need to include single particle orbits from the higher shell in the process enlarging the space and its dimension.

2.2 β-decay

The β^- and the β^+ are given as :

$$(N, Z) \rightarrow (N - 1, Z + 1) + e^- + \bar{\nu}_e \tag{3}$$

$$(N, Z) \rightarrow (N + 1, Z - 1) + e^+ + \nu_e \tag{4}$$

with the release of energy $Q_{\beta^-} = [M(N, Z) - M(N - 1, Z + 1)]c^2$ and $Q_{\beta^+} = [M(N, Z) - M(N + 1, Z - 1) - 2m_e]c^2$ respectively. The process of electron capture [EC] has the reaction

$$(N, Z) + e^- \rightarrow (N + 1, Z - 1) + \nu_e \tag{5}$$

Here the release of energy is $Q_{EC} = [M(N, Z] - M[N + 1, Z - 1)]c^2 + E_e$, where E_e is the energy of the electron absorbed. Finally the neutrino capture reaction is

$$(N, Z) + \nu_e \rightarrow (N - 1, Z + 1) + e^- \tag{6}$$

with the corresponding energy release of $Q_\nu = [M(N, Z) - M(N - 1, Z + 1)]c^2 + E_\nu$. The half-life $t_{1/2}$ for a β-decay from the ground state $|i>$ is given by

$$t_{1/2} = ln\ 2/(\Sigma_f w_{fi}) \tag{7}$$

where the transition probability w_{fi} is summed over all final states to which the transitions are possible energetically and given by the expression

$$w_{fi} = (mc^2/\hbar)\ \ (\Gamma^2/\ \pi^2)\ \ |M_{fi}|^2 f(Z,\epsilon) \tag{8}$$

with $\epsilon_0 = E_0/(m_e c^2)$, $\epsilon = E/(m_e c^2)$ and Γ is related to the weak interaction constant. The phase factor $f(Z,\epsilon)$ has the expression [2]

$$f(Z,\epsilon) = \int_0^{\epsilon_0} d\epsilon F(Z,\epsilon)(\epsilon_0 -\ \epsilon)^2 \epsilon(\epsilon^2 - 1)^{1/2} \tag{9}$$

where $F(Z,\epsilon)$ is the Coulomb distortion factor. The squared nuclear matrix is

$$|M_{fi}|^2 = |C_F|^2 |M_F(fi)|^2 + |C_{GT}|^2 |M_{GT}(fi)|^2 \tag{10}$$

This has contribution from the allowed transitions by Fermi and Gamow-Teller (GT) operators. They have the form

$$M_F(fi)|^2 = | < f|\Sigma_k t_\pm(k)|i > |^2/(2J_i + 1) for \beta^\pm decay \tag{11}$$

$$|M_{GT}(fi)|^2 = | < f|\Sigma_k \boldsymbol{\sigma}(k) t_\pm(k)|i > |^2/(2J_i + 1) for \beta^\pm decay \tag{12}$$

The transition probability for the process of electron capture in stellar temperatures T is given by

$$\lambda_{EC} = (ln2/K)\Sigma_i((2J_i + 1)exp(-E_i/kT)/G(Z,A,T)) \times \Sigma_j B_{ij}(GT)\Phi_{ij} \tag{13}$$

where Φ_{ij} is the phase space factor for EC and K is a constant.

The electron capture transitions change the initial isospin T to $T' = T+1$.

3 Overview of Core Collapse Supernovae

Supernova (SN) explosions are one of the most violent explosions in the universe where energies of a few times 10^{53} ergs are released in a few tens of seconds. Though long ago Baade and Zwicky associated this with the death of large stars but the exact mechanism of the explosions was one of the outstanding problems of the last century which is still not understood completely. Here we shall be concerned with the explosions that start with the collapse of the core. Though from the outer regions of the star different forms of electromagnetic radiation give information about the physical processes in the envelope but nothing except the neutrinos can come from the much denser core region. The production of neutrinos in SN was spectacularly confirmed by the 11 and 8 neutrinos detected in two different terrestrial neutrino detectors during the explosion of Supernova 1987A. The theory of core collapse

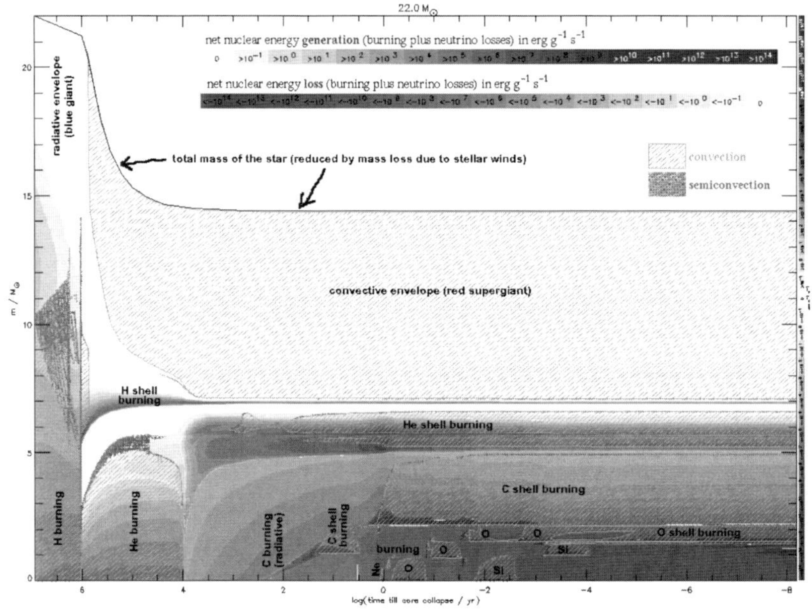

Fig. 1. The stellar evolution leading to collapse for stars of different masses (taken from [6])

supernova involves elements of astrophysics, particle physics, nuclear physics and statistical mechanics [3, 4, 5].

The stellar burning stages leading to the late silicon burning in the pre-supernova stage is depicted in Figure 1. Massive stars (with mass $> 8M_{\odot}$) collapse once the nuclear burning in the centre of the core stops and the core mass exceeds the Chandrasekhar mass. A combination of electron capture and photodissociation of the iron type nuclei initiates the process. The fraction of neutrons are small at this stage and the fraction of protons much smaller. The radius of the core is a few thousand kilometres and the temperature less than 1 MeV. The density of matter at the start of the collapse is around 10^9 g/cm^3. At this stage electron capture on the nuclei and on the very small fraction of free protons reduce the electron density and hence the pressure. The electron fraction is expressed by Y_e, the ratio of the number of electrons to the number of nucleons.

When the density reaches the value around 10^{11} g/cm^3, the neutrino mean free path becomes close to the core radius and so the neutrinos that were escaping the star earlier get trapped inside. So beyond this density the lepton fraction Y_l becomes constant. The entropy change during this period is not enough to break up the nuclei and hence the nuclei survive through this stage. Beyond the nuclear matter density the nucleons touch each other and pressure waves start travelling from the centre, the region of highest density. Matter becomes stiffer, the pressure waves stop going outward at the 'sonic point' and eventually a shock wave develops there. The timescale of this whole process is tens of milliseconds. Eventually the shock gathers enough energy to start moving outward. In the early eighties it was thought that this leads to the explosion. But the simulations done showed otherwise. The shock in almost all simulations lost energy fast by dissociating the nuclei in its path and soon became a static shock after moving a few hundred kilometres. The typical dissociation reaction on a iron nucleus is:

$$^{56}Fe \rightarrow 13\alpha + 4n \tag{14}$$

$$\alpha \rightarrow 2n + 2p \tag{15}$$

Cooling through neutrino emission also worked against the shock motion. For every 0.1 $M_{\odot}$, dissociation energy $\sim 1.7 \times 10^{51}$ ergs is lost and that clearly indicates why the shock stops moving out. Except for a few special cases known as 'prompt' explosions for smaller star masses with lower entropies and softer equations of state, all simulations gave no explosion. Now one believes that the energising of the stalled shock takes place through a process known as 'delayed neutrino heating' where the neutrinos and the weak reactions involved are crucial.

From the central region of the core a huge number of neutrinos of all three flavors radiate out. They originate from the high temperature region deep inside where $\nu\bar{\nu}$ pairs are produced thermally. These neutrinos through charge current and neutral current interactions heat up the matter at the stalled region over a timescale of a few seconds. Though a very tiny fraction of the neutrinos interacts but their number is still large enough to revitalise the shock over a long timescale. Unfortunately as yet very few simulations lead to the shock reaching the edge of the core. In other words, even today with a lot of improvements in the simulations taking into account large-scale convection, relativistic treatment of the matter etc. very few simulations give rise to an explosion. One of the few cases where explosion is seen is given in Figure 2 [7].

One should mention here that some stars with mass in the range of 9-10$M_{\odot}$ end up with a $O-Ne$ core instead of the normal iron core. These stars even in 1 dimensional simulations, are seen to have the stalled shock moving out in 150 ms after bounce and mass shells start 'being ablated' from the protoneutron star about 50 ms later giving rise to a successful explosion [8].

The research work today concentrates on better treatment of convection by combing 2D/3D hydrodynamics for the matter coupled with the neutrino radiative transport. Improved pre-supernova and collapse conditions and more detailed treatment of the neutrino-matter interaction are the other areas of activity. One also envisages that an instability of accretion shock to non-radial deformation (particularly $l = 1, 2$) known as Standing Accretion Shock Instability (SASI) can lead to a breakthrough in the understanding of the energy transport [9].

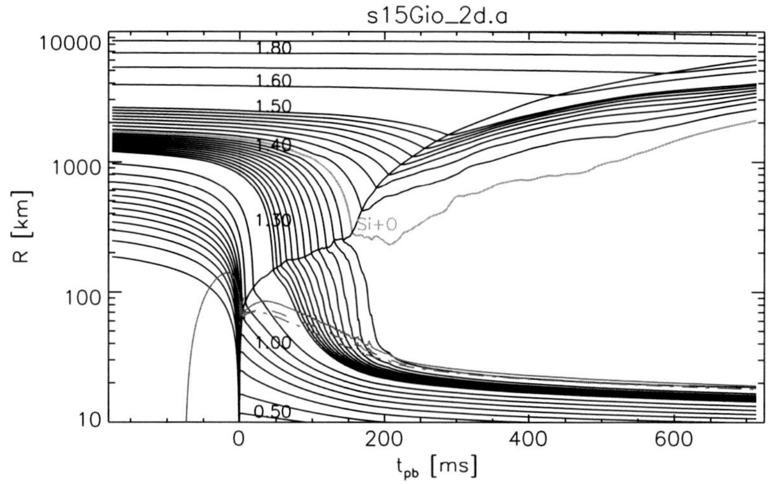

Fig. 2. The results of a hydrodynamic simulation showing an explosion (taken from [7])

4 Weak Interaction Processes in Supernova Evolution

We discuss the different stages of stellar evolution and core collapse supernovae where the β-decay and its inverse processes play important roles:

4.1 At the Pre-SN Stage

As already described electron capture on nuclei start at the late stages of stellar burning. It is also realised that the β-decay of the unstable nuclei increase Y_e and compete with the electron capture. We have seen in equations (5) and (3) that in the case of EC only GT transitions participate ($T \rightarrow T+1$ is not possible by the Fermi operator) where for β-decay, Fermi transitions are, in principle, allowed. Figure 3 gives a sketch of the contributions from Fermi and GT for β-decay and EC. The timescale of silicon burning is a few days to a month. During the late phase of Si burning successive addition of alpha particles to silicon and other nuclei around that mass take place at temperatures $T = 4 \times 10^9$ (i.e. 350 keV) with the end product being the 'iron group' nuclei (like Mn, Fe, Co, Ni,...). Near the end of this burning stage Nuclear Statistical Equilibrium (NSE) is reached and the system is then characterised by ρ, T and Y_e.

4.2 At the Collapse Stage

As the collapse proceeds the electron captures along with the β-decays of the neutron-rich nuclei control the change in Y_e. But as the density increases so does the electron chemical potential and when it exceeds the Q-value of the

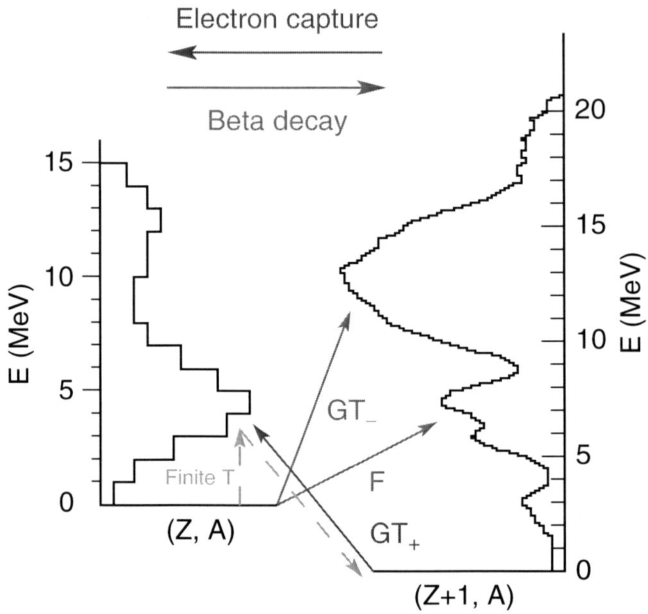

Fig. 3. The Fermi and GT resonances observed through β-decay and electron capture between nuclei (N,Z) and (N-1,Z+1) (taken from [6])

β-decay, the decay essentially stops. Also the equation of state of the matter makes the average nucleus heavier i.e. the mass number slowly increases from the initial value of 56. When the neutron number crosses the value of 40, the allowed electron capture starts getting blocked. This nuclear structure effect, first pointed out by Fuller [10], reduces the rate of decrease of Y_e substantially. We can understand this effect with the example of the nucleus ^{74}Ge given in Figure 4. ^{74}Ge has 32 protons and 42 neutrons. In the extreme single particle model i.e. with 1-body H only, the neutrons fill the fp-shell with the 4 orbits $1f_{7/2}$, $2p_{3/2}$, $2p_{1/2}$ and $1f_{5/2}$ completely. Two extra neutrons occupy the $1g_{11/2}$ orbits. The protons on the other hand outside the closed Ca core occupy the $1f_{7/2}$ and $2p_{3/2}$ orbits. On these valence protons no allowed electron capture takes place as the corresponding 4 neutron orbits are full. Of course the contribution from the closed Ca core is zero too. Hence one sees that for these nuclei with $N > Z$ once N crosses 40 allowed electron captures stop. The rate of 'forbidden' electron capture is 2 to 3 orders lower. This effect is known as the 'blocking' of the EC rate.

4.3 Mechanisms of Unblocking

But actually complete blocking of the rates does not start suddenly. Below we describe the mechanisms for unblocking which operate in reality. Firstly, there is always an ensemble of nuclei and not just one. So while ^{74}Ge may

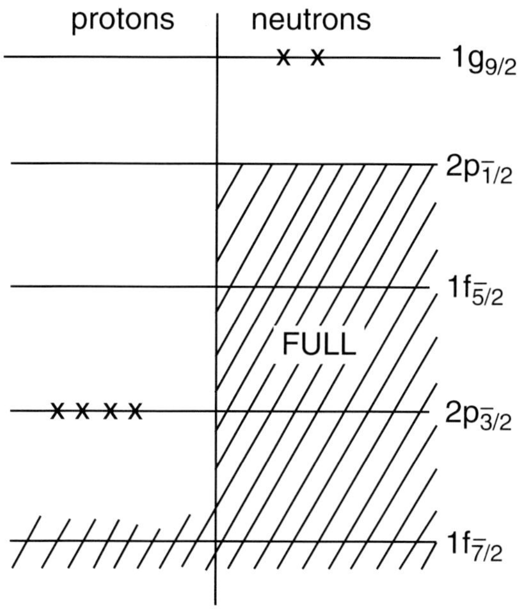

Fig. 4. The occupation of the different orbits for the nucleus ^{74}Ge

have its EC rate blocked lighter nuclei in the ensemble with N $<$ 40 with smaller abundances will have rates with allowed transitions. Secondly the extreme single particle model does not take into account the residual interaction and the configuration mixing. For the ^{74}Ge case, a small contribution in the n-sector of $(2p_{1/2})^1(2g_{9/2})^3$ to the dominant $(2p_{1/2})^2(1g_{9/2})^2$ will make a $2p_{3/2} \rightarrow 2p_{1/2}$ transition possible. Thirdly one can also thermally excite a neutron from the $2p_{1/2}$ to the $1g_{9/2}$ orbit causing the allowed transition. Finally a small deformation of the nucleus from spherical shape will lead to the Nilsson model being operative instead of the spherical shell model. There each orbit no longer has the $(2j + 1)$ degeneracy but each $|m|$ component has different energy depending on the deformation where the degeneracy for each one is 2 coming from the states with quantum numbers 'm' and '-m' having the same energy. This also leads to non-zero allowed transition rates.

4.4 At the late time neutrino heating stage

The heating of the matter through neutrinos take place through the charged current electron neutrino capture reactions given by Equation (6). As matter behind the shock is heated by neutrinos coming from deep inside the core, its temperature rises and that leads to radiation of neutrinos. Bethe and Wilson [11] first pointed out that the electron neutrino absorption on free nucleons and neutrino-electron elastic scattering (both charged and neutral current contributions) is able to make the shock move out again. A self-consistent treatment of the heating and the dissociation of the nuclei was done [12]. However Haxton [13] stressed the importance of neutral current contribution in elastic scattering off nuclei given by

$$(N, Z) + \nu_x = (N, Z) + \nu_x \tag{16}$$

where 'x' stands for all three flavors, particularly the part that arises due to the excited states of the nucleus. The net rate of gain of energy by a gram of matter is given by

$$\dot{E} = \dot{E}_1 - \dot{E}_2 + \dot{E}_{scatt} \tag{17}$$

where $\dot{E}_1$ is the heating rate and $\dot{E}_2$ is the rate of energy loss due to emission of neutrinos. $\dot{E}_{scatt}$ is the heating through the scattering processes mentioned above. The nuclei involved are the 'iron group' nuclei near the shock front, in particular. The amount of heating through these reactions are carefully calculated. Also processes that can help in getting extra heating are also investigated. One such idea was the flavor mixing through matter enhanced oscillation increasing the heating. But one finds that with the realistic masses the mixing cannot take place within a few hundred kilometres as needed in the core collapse scenario.

5 Nuclear Models for Calculation of the Weak Interaction Rates

5.1 Systematics with simple shell structure

The first realistic estimation of the electron capture and β-decay rates including the contribution from the giant resonance was by Fuller, Fowler and Newman (FFN) [14]. While for the *sd*-shell nuclei shell model results were used, for the *fp*-shell nuclei low-lying experimental values of $log ft$ were supplemented by contributions from a collective strength whose energy position and magnitude were estimated using systematics and single particle shell structure for nuclei with $A < 60$.

Working on improved FFN ideas Aufderheide et al. [15] calculated the EC and β-decay rates for nuclei beyond $A = 60$ and for different density-temperature grid points for the presupernova evolution and collapse. They also tabulated the important nuclei that need to be considered in the ensemble.

5.2 Statistical models for β strength

One of the earliest models that treat the β-decay strength in a statistical way is the 'Gross Theory'. Here assuming enough final states, the strength function is given as an integral of a one-particle strength function multiplied by a weight function to take into account the Pauli blocking and the density of single particle states [16, 17]. The strength function took into account the allowed and forbidden transitions and also the pairing and sum rules in a simple approach. It was almost independent of the nucleus considered except for considering its Q-value. Later the theory was improved [18,19] by taking into account the shell effects and modifying the strength function. This is called the 'Semi-gross Theory' or the second generation gross theory.

The other approach involved constructing the giant resonance using elements of spectral distribution theory [20,21,22] and experimental information on the resonance. Here one first recognises the sum rule quantities corresponding to the transition strength due to the excitation operator O. The sum rule strength S_O is given by

$$S_O = \Sigma_f | < f|O|i > |^2 = \Sigma_f < i|O^\dagger|f >< f|O|i >=< i|O^\dagger O|i > \qquad (18)$$

This is the total strength of the giant resonance plus the strength of the discrete states in the low-lying excitation region. For the Fermi and GT β-decay transitions these sum rules are exactly given by the Ikeda sum rule expressions [23]:

$$S_{\beta^-}(GT) - S_{\beta^+}(GT) = 3(N - Z) \qquad (19)$$

$$S_{\beta^-}(F) - S_{\beta^+}(F) = (N - Z) \qquad (20)$$

However these sum rules give the difference of the total electron β-decay and the total positron β-decay strengths. For finding the β strength sums individually, Kar, Ray and Sarkar [24] expressed the strength sum in terms of the average neutron and average proton occupancies. These occupancies averaged over the configurations were evaluated using the spectral distribution theory. One finds that the assumption that the GT strength distribution in final energy is close to a Gaussian is not a bad approximation to start with and fixes the centroid by the best fit to observations in (p, n) reactions [25]

$$\epsilon_{GT} = \epsilon_{IAS} + 26A^{-1/3} - 18.5(N - Z)/A \qquad (21)$$

where ϵ_{GT} is the GT centroid energy and ϵ_{IAS} is the energy of the Isobaric Analog State (IAS). Deviation from the Gaussian can also be handled here by using a skewness. The width of the GT resonance is left as a free parameter and fixed by the best fit to observed half-lives of nuclei in this region. This form of the β strength function is then used to calculate the rates for the presupernova β-decay. There are also some special excited states of the mother nucleus which have large overlaps with states in the ground state region of the daughter and their contributions are also taken into account. These so-called 'back resonances', first pointed out by Fuller, can become important for the higher temperatures.

The statistical method using the spectral distribution theory including orbits higher than the fp-shell was recently used to evaluate the β-decay rates for nuclei involved in the collapse stage but with $A > 65$. The half-lives using this method are compared to the experimental values in Table 1 [26]. The rates are given in Table 2.

Table 1. Half-lives of nuclei with $A > 65$ using statistical model [26]

Sr No.	Nucleus	Z	Q_{val} (MeV)	$T_{1/2}^{expt.}$	$T_{1/2}^{calc}$
1	^{66}Co	27	10.0	0.23	0.32
2	^{67}Ni	28	3.385	21.0	26.91
3	^{68}Co	27	9.30	0.18	0.20
4	^{68}Ni	28	2.06	19.0	18.50
5	^{68}Cu	29	4.46	31.1	31.23
6	^{69}Co	27	9.30	0.27	0.36
7	^{70}Ni	28	5.36	11.4	12.04
8	^{71}Ni	28	6.90	1.86	1.62
9	^{72}Cu	29	8.22	6.60	6.13
10	^{74}Cu	29	9.99	1.59	0.40

Table 2. Rates for β^- decay for typical supernova densities and temperatures [26]

Nucleus	ρ(gms/cc)	Y_e	Temperature in ^{o}K			
			2×10^9	3×10^9	4×10^9	5×10^9
			Rates (s^{-1})			
^{69}Co	10^9	0.47	7.25×10^{-3}	1.18×10^{-1}	2.96×10^{-1}	3.63×10^{-1}
	10^8	0.47	1.64×10^{-2}	2.62×10^{-1}	6.37×10^{-1}	7.53×10^{-1}
	10^7	0.47	1.94×10^{-2}	3.05×10^{-1}	7.32×10^{-1}	8.55×10^{-1}
	10^9	0.45	7.47×10^{-3}	1.22×10^{-1}	3.05×10^{-1}	3.73×10^{-1}
	10^8	0.45	1.66×10^{-2}	2.64×10^{-1}	6.41×10^{-1}	7.57×10^{-1}
	10^7	0.45	1.94×10^{-2}	3.06×10^{-1}	7.33×10^{-1}	8.56×10^{-1}
^{68}Ni	10^9	0.47	8.38×10^{-6}	7.81×10^{-4}	6.00×10^{-3}	1.59×10^{-2}
	10^8	0.47	2.18×10^{-3}	4.38×10^{-2}	1.34×10^{-1}	1.93×10^{-1}
	10^7	0.47	4.55×10^{-3}	8.08×10^{-2}	2.20×10^{-1}	2.90×10^{-1}
	10^9	0.45	1.08×10^{-5}	9.18×10^{-4}	6.76×10^{-3}	1.75×10^{-2}
	10^8	0.45	2.25×10^{-3}	4.50×10^{-2}	1.36×10^{-1}	1.97×10^{-1}
	10^7	0.45	4.57×10^{-3}	8.11×10^{-2}	2.20×10^{-1}	2.91×10^{-1}

5.3 Microscopic Models

Among the microscopic models, the most accurate one for the presupernova and supernova calculations is the shell model. Over the years very large space shell model calculations for nuclei in the fp-shell were found to very successful and weak interaction rates have been calculated extensively using the shell model results [27, 6].

Shell model calculations in the fp-shell using the modified Kuo-Brown (KB3) interaction with the valence orbits $f_{7/2}$, $p_{3/2}$, $p_{1/2}$ and $f_{5/2}$ have been carried out using the Strasbourg-Madrid [28] code. They reproduce the experimental spectra well. Information on the GT matrix elements come from the (p, n) and (n, p) reactions for the β^- and β^+ decays. Calculations for GT transition matrix elements using wavefunctions with the KB3 interaction using the same code again showed good agreement. This is seen in Figure 5. The figure compares the GT strength for β^+ transition as a function of en-

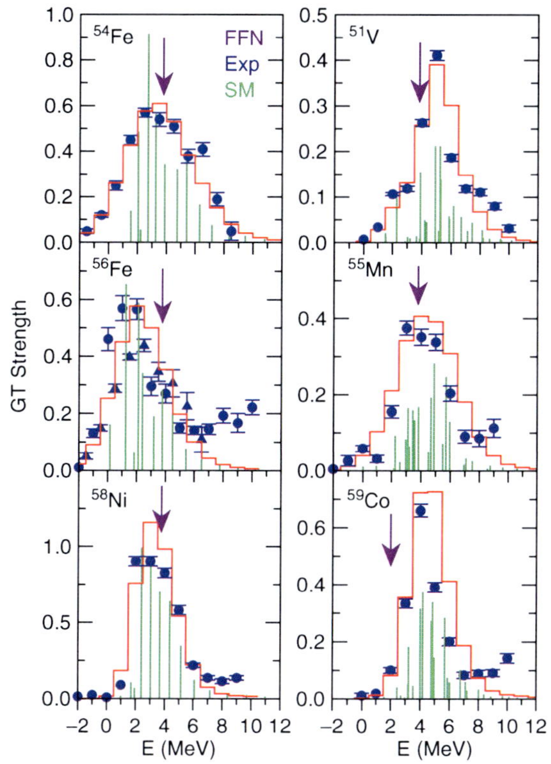

Fig. 5. Comparison of GT matrix elements calculated by shell model and observed in (n, p) experiments (taken from [6])

ergy for the shell model with the same obtained through (n,p) reactions for the 6 nuclei ^{54}Fe, ^{56}Fe, ^{58}Ni, ^{51}V, ^{55}Mn and ^{59}Co. The figure also shows the centroid of the GT distribution used in FFN. Similar comparisons were done for β^- transitions as well. The success of the shell model results motivated the authors in calculating β^-, β^+ and EC rates for a large of nuclei in the mass range of $40 < A < 65$ for a wide range of density and temperature. For the nuclei in the range $65 < A < 80$ Pruet and Fuller [29] did the calculations extending the FFN model alongwith taking into account the GT resonance.

5.4 Calculations with Improved Rates

The detailed shell model results in agreement with experiments led to better EC and β-decay rates. The EC rates were often reduced compared to the FFN earlier values. For example the EC rate on ^{60}Co was two orders of magnitude lower for typical presupernova conditions! Presupernova evolution of stars in the mass range of 11M$_\odot$ to 40M$_\odot$ were carried out using the full FFN and the rates based on the shell model results (LMP) by redoing the Woosley and Weaver calculations [31] only changing the weak rates from FFN to LMP. These results are displayed in Figures 6 to 9 [30, 6].

Figure 6 plots the change of Y_e for 15M$_\odot$ and 25M$_\odot$ stars as a function of time till core bounce. The different nuclei that are most important at a particular time are also given both for EC and β-decay.

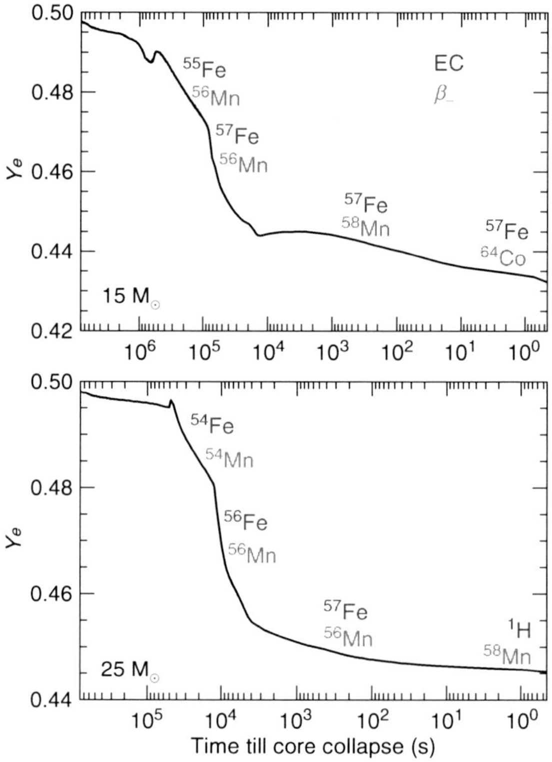

Fig. 6. The change of Y_e with time using the improved weak interaction rates (taken from [6])

Figure 7 displays the comparison of the central value of Y_e at the onset of collapse with the two rates FFN and LMP for the range of star masses from 11M$_\odot$ to 40M$_\odot$. This clearly shows that for LMP the central Y_e is higher for all stars considered. The next two Figures 8 and 9 shows the iron core size and the central entropy for the stars again for the two different weak rates. Now let us remind us that the expression for the Chandrasekhar mass generalised for finite temperature is [30]

$$M_{Ch} = 5.83Y_e^2[1 + (S_e)^2/(\pi Y_e)^2] \tag{22}$$

where S_e is the electron entropy at the non-zero temperature.

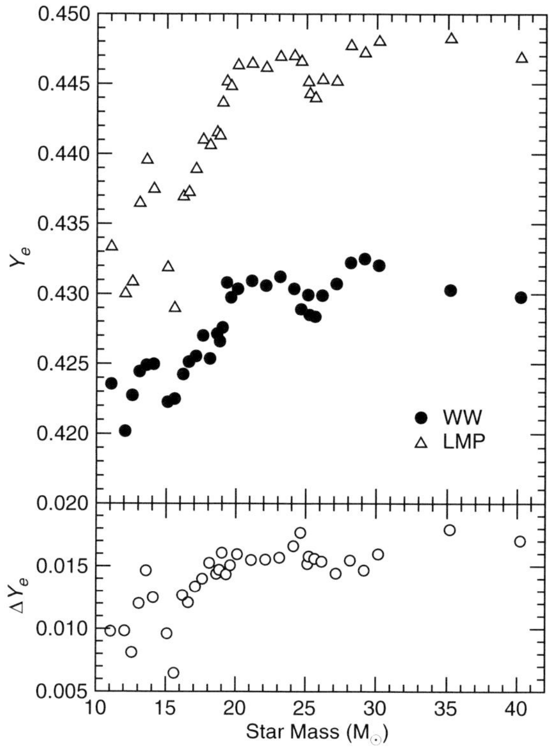

Fig. 7. Comparison of the central value of Y_e using the FFN and LMP rates; the lower panel shows the change in the values. The most important nuclei at each stage is also given (taken from [6])

In Figure 9 it is seen that for star masses upto $20M_\odot$ the central entropy is lower for the LMP rates. So in this mass range the increase in the Chandrasekhar mass due to higher Y_e gets neutralised by the lower S_e and no strong effect is seen upto $20M_\odot$. On the other hand the mass of the homologous core which depends on Y_e^2 at the time of neutrino trapping is expected to increase for these lower mass stars with LMP rates. The shock energy also increases with Y_e and with less mass to travel through, the new rates are supposed to help the shock propagation for this mass range. But this effect may not be enough to result in 'prompt' explosions. But more studies on numerical stellar evolution both at the preSN and the SN stages with realistic weak interaction rates are needed.

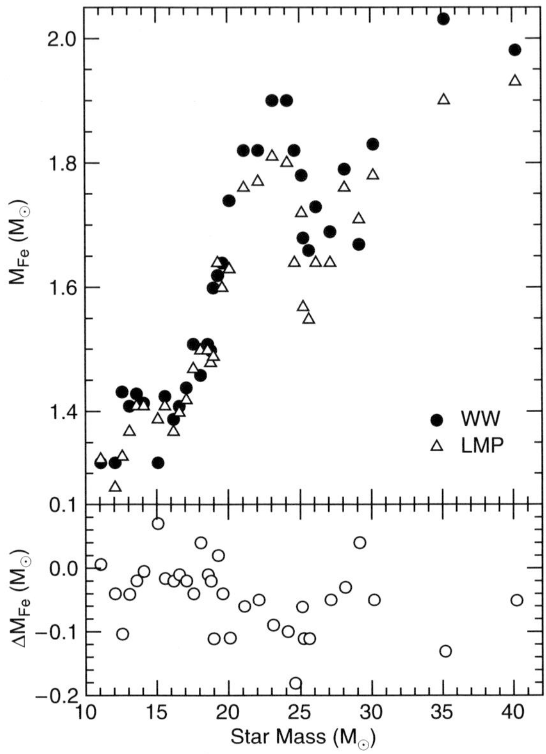

Fig. 8. Comparison of the iron core sizes using the FFN and LMP rates; the lower panel shows the change in the values (taken from [6])

6 Weak Interaction Processes During pp-Chain and Solar Neutrino Problem

In the pp-chain i.e. the first stage of nuclear burning there are a number of reactions that produce neutrinos. We list them below:

Neutrinos produced in the fusion reactions in the sun (pp chain):

1. $p + p \rightarrow^2 H + e^+ + \nu_e$ (pp) $[E_\nu < 0.42$ MeV$]$
2. $p + e^- + p \rightarrow^2 H + \nu_e$ (pep) $[E_\nu = 1.442$ MeV$]$
3. ${}^2H + p \rightarrow^3 H + \gamma$
4. ${}^3He +^3 He \rightarrow \alpha + 2p$
5. ${}^3He +^4 He \rightarrow^7 Be + \gamma$

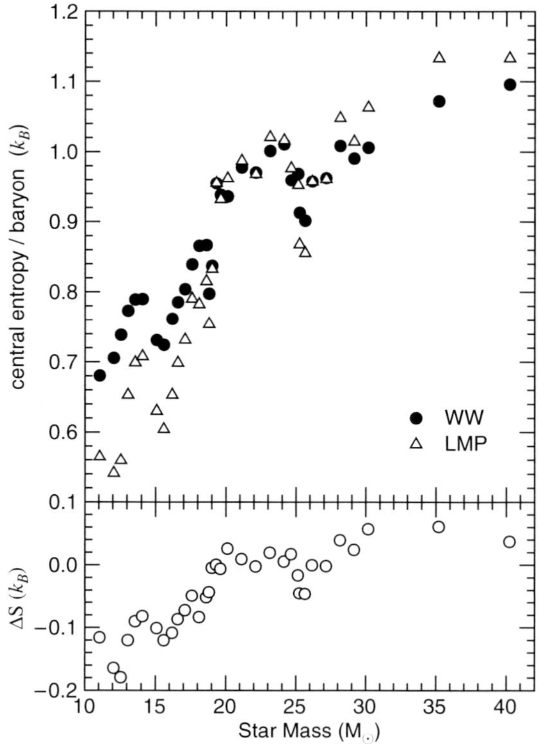

Fig. 9. The central entropy for 11-40$M_\odot$ star mass using the LMP rates compared to the same using the old FFN rates; the lower panel shows the change in the values (taken from [6])

6. $^3He + p \rightarrow \alpha + e^+ + \nu_e$ (hep) $[E_\nu < 18.77$ MeV$]$

7. $^7Be + e^- \rightarrow ^7 Li + \nu_e$ $[E_\nu = 0.861$ MeV(90%) or 0.331 MeV(10%)$]$

8. $^7Be + p \rightarrow ^8 B + \gamma$

9. $^8B \rightarrow ^8 Be^* + e^+ + \nu_e$ $[E_\nu = 15$ MeV$]$

10. $^8Be^* \rightarrow 2\alpha$

The reactions all put together give rise to the following fusion reaction:

$$4p \rightarrow \alpha + 2e^+ + 2\nu_e + 25MeV \tag{23}$$

The spectra of the neutrinos produced in these reactions are given in Figure 10.

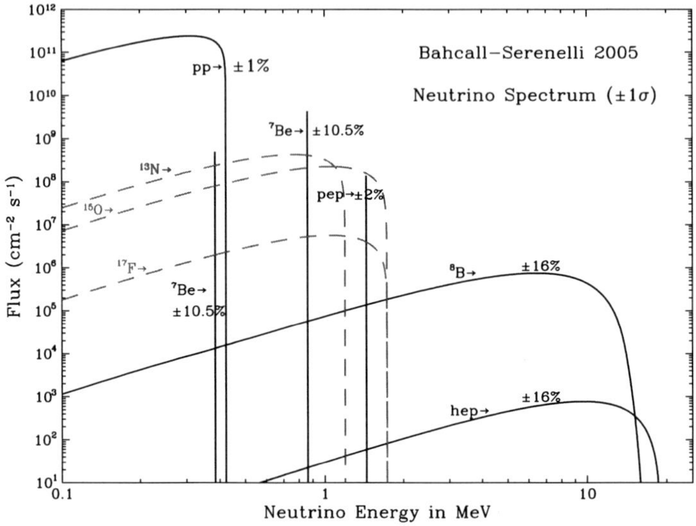

Fig. 10. The spectrum of the neutrinos produced by different reactions in the pp-chain in the sun (taken from [33])

The energy threshold of the detectors with chlorine (Homestake), water (Kamiokande/SuperK) and gallium (GALLEX/SAGE) are 0.814 MeV, 7.5 MeV and 0.233 MeV respectively. So the neutrinos produced by the first reaction, known as the 'pp' reaction, with a maximum energy of 0.42 MeV were not seen with chlorine or water and first detected by the gallium detectors. Now for all these three reactions the number of neutrinos detected were substantially lower, by a factor of 2-3 than the predictions of theory based on the Standard Solar Model (SSM) [32]. This is seen clearly in Figure 11. On the other hand for the Sudbury Neutrino Observatory (SNO) though the charged current reaction (CC) $(d + \nu_e \rightarrow p + p + e^-)$ with a threshold of 5 MeV sees only 35% of the predicted events, but the neutral current events (NC) $(d + \nu_x \rightarrow p + n + \nu_x)$ with a threshold of 2.2 MeV is able to register all the neutrinos. This proved once for all that the solar neutrinos in their transit get converted from the electron flavor to other flavors but their total flux remains unchanged. As beyond 5 MeV the neutrinos come almost fully from the ^{8}B decay one gets from the SNO NC results, the neutrino flux $\Phi_{NC} = [5.09 \pm 0.44(stat)^{+0.46}_{-0.43}(sys)] \times 10^6 cm^{-2} s^{-1}$.

6.1 Neutrino Oscillation

The mechanism of conversion of neutrinos from one flavor to another is seen to be the neutrino flavor oscillation. We briefly describe below how in a two

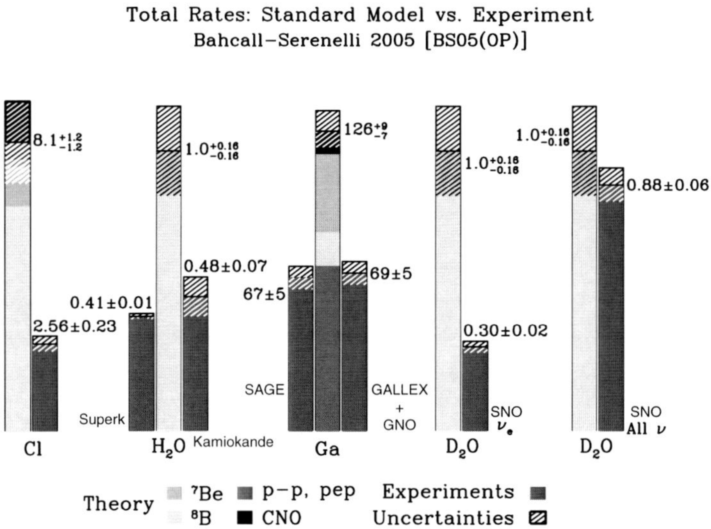

Fig. 11. The total rates as predicted by theory for the different detectors compared with the observed rates (taken from [33])

flavor scenario in vacuum the neutrino oscillation occurs. Let the two flavor eigenstates be $|\nu_e>$ and $|\nu_\mu>$ whereas the mass eigenstates are $|\nu_1>$ and $|\nu_2>$. If the oscillation angle is θ then we have

$$\begin{pmatrix} \nu_e \\ \nu_\mu \end{pmatrix} = \begin{pmatrix} cos\theta & sin\theta \\ -sin\theta & cos\theta \end{pmatrix} \begin{pmatrix} \nu_1 \\ \nu_2 \end{pmatrix} \tag{24}$$

As the mass eigenstates evolve in time as e^{-iEt} one gets

$$|\nu_e(t)>= cos\theta \ e^{-iE_1t}|\nu_1> +sin\theta \ e^{-iE_2t}|\nu_2> (E_i^2 = m_i^2 + p^2) \tag{25}$$

leading to

$$P_{\nu_\mu\nu_e} = sin^2(2\theta) \ \ sin^2(\Delta m^2 L/(4E)) \tag{26}$$

This shows that the probability of a electron type neutrino getting converted into a mu type neutrino has an oscillatory dependence on (L/E) where L is the length travelled and E is the energy. The two mass eigenstates have the same momentum here so that the energies are given by $E_i^2 = m_i^2 + p^2$. The mass square difference is $\Delta m^2 = m_2^2 - m_1^2$. With three neutrino flavors instead of two, one has three oscillation angles and two mass squared differences but the probabilities can be similarly be expressed in terms of the oscillation angles and the mass squared differences.

When there is matter in the travel path of the neutrinos the ν_es interact with the electrons in the matter and give rise to an extra potential (proportional to n_e) in the mass matrix. This results in matter-enhanced neutrino oscillation (MSW effect). The observed depletion of the solar electron type neutrino by different detectors has been explained using certain mass differences and oscillation angles [34]. The allowed region in the mass squared difference and oscillation angle with a global fit to the solar and reactor experiments all put together is shown in Figure 12. This shows how the uncertainties got reduced with more precise experimental input in 2004.

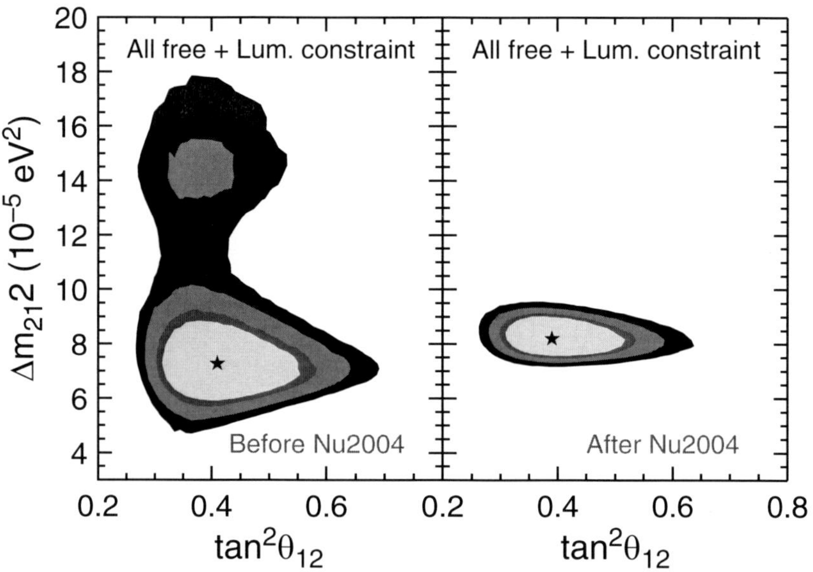

Fig. 12. More accurate determination of the oscillation parameters with time (taken from [33])

7 β-decay Rates for r-Process Nucleosynthesis

One of the most basic questions in nuclear astrophysics is: How do the nuclei heavier than iron get produced? This question was first answered by Burbidge, Burbidge, Fowler and Hoyle in 1957 [35]. They proposed that these elements are produced through the slow (s) and rapid (r) neutron-capture processes. The words rapid/slow refer to the rate of neutron capture compared to the rate of β-decay in the astrophysical conditions. Figure 13 shows the nuclei involved in the r- and s-processes. The s-process path stays close to the valley of stability whereas the r-process path moves staying close to the drip line. The figure also shows the nuclei involved in the rp-process; these are proton rich nuclei where capture of protons are involved and that the rate is compared to the β^+ rates.

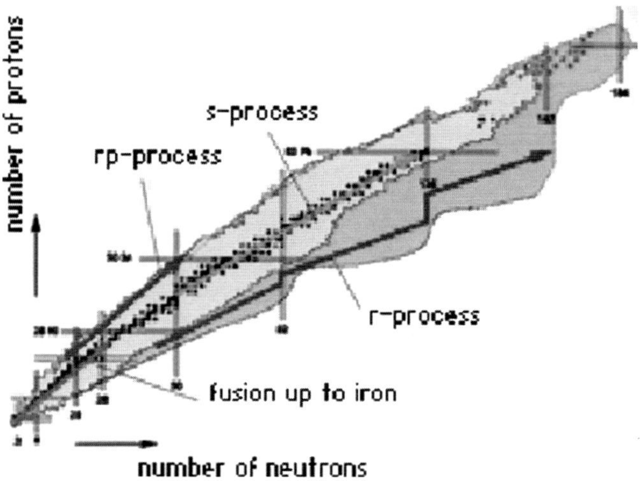

Fig. 13. The s- and r-process path and the nuclear chart

7.1 s-process and r-process

For the s-process nuclei the neutrons get captured by one stable nucleus going to another stable nucleus and so on. Once an unstable isotope is reached, there is a β-decay. So the yield $Y_s(A)$ of these nuclei can be expressed as

$$dY_s(A)/dt = \phi_n[\sigma_{n,\gamma}(A-1)Y_s(A-1) - \sigma_{n,\gamma}(A)Y_s(A)] \tag{27}$$

At equilibrium then one gets $Y_s(A)/Y_s(A-1) = \sigma_{n,\gamma}(A-1)/\sigma_{n,\gamma}(A)$. As the n-capture rates are very small for magic neutron number nuclei like ^{88}Sr (N=50), ^{138}Ba (N=82), ^{208}Pb (N=82) that explains the observed large abundances for shell closure nuclei.

For the rapid process, if one takes into account the different competing processes the general equation for the evolution of the yield is given by [36]

$$\begin{aligned} dY(Z,A)/dt = n_n < v\sigma_{n,\gamma}(Z,A-1) > Y(Z,A-1) + \lambda_{\gamma,n}(Z,A+1)Y(Z,A+1) \\ +\lambda_{\beta 0}(Z-1,A)Y(Z-1,A) + \lambda_{\beta 1}(Z-1,A+1)Y(Z-1,A+1) \\ +\lambda_{\beta 2}(Z-1,A+2)Y(Z-1,A+2) + \lambda_{\beta 3}(Z-1,A+3)Y(Z-1,A+3) \\ -n_n < v\sigma_{n,\gamma}(Z,A) > Y(Z,A) - \lambda_{\gamma,n}(Z,A)Y(Z,A) \\ -[\lambda_{\beta 0}(Z,A) + \lambda_{\beta 1}(Z,A) + \lambda_{\beta 2}(Z,A) + \lambda_{\beta 3}(Z,A)]Y(Z,A) \end{aligned} \tag{28}$$

where n_n is the neutron number density, $n_n < v\sigma_{n,\gamma}(Z,A) >$ is the thermally averaged neutron capture rate and $\lambda_{\beta n}(Z,A)$ is the rate for β-decay followed by the emission of 'n' neutrons. Here the charged particle reactions are considered unimportant. When the $n-$capture and photodisintegration occur much faster than the β-decay, one gets for equilibrium

$$Y(Z, A+1)/Y(Z, A) = n_n < v\sigma_{n,\gamma}(Z, A) > /\lambda_{\gamma,n}(Z, A+1) \qquad (29)$$

Then using the expressions for the averaged (n, γ) and (γ, n) processes one sees that the most abundant isotopes in different isotopic chains have approximately the same separation energy of about 2-3 MeV for r-process conditions. Later network calculations get the equilibrium at neutron density greater than $10^{20} cm^{-3}$.

It was first seen that $(n, \gamma) \leftrightarrow (\gamma, n)$ equilibrium is obtained at T$> 2 \times 10^9$ K and $N_n > 10^{20}$ cm^{-3} [37]. Later network calculations get the equilibrium at neutron density of 10^{20} cm^{-3} for temperatures 2×10^9 K and at a density of 10^{28} cm^{-3} for half the temperature. One also notes that at temperatures greater than 6×10^9 K Nuclear Statistical Equilibrium (NSE) is achieved where the forward strong and electromagnetic reactions are balanced by their reverse reactions.

In this description Fission Cycling has not been taken into account. But if the r-process involves nuclei with Z $>$ 80, then fission should be included. In that case the heaviest nucleus fissions and a cyclic flow occurs between this nucleus and its fission fragments in the presence of large neutron abundance.

The solar r-process abundances are normally obtained by subtracting the calculated s-process abundances from the observed solar abundances. A typical example is given in Figure 14.

7.2 Possible r-process site

Curiously enough no r-process site is clearly identified yet. But, of course, there are good candidates. A neutron rich environment is the primary necessity for the r-process. The delayed neutrino driven core collapse is considered to be one of more promising ones. The neutrino driven wind starting from the protoneutron star surface is hot (temperatures more than 10^{10} K) and has low density and high entropy. An entropy reaching a value of 400 k_B per baryon in about ten seconds showed the onset of r-process [38] in one dimensional simulation. But another calculation got entropy much lower and not showing the development of the r-process [40]. A better understanding of the neutrino driven wind is required to establish that as a r-process site. Several analytical models to treat this wind exist in literature. Among the spherically symmetric steady-state models both Newtonian adiabatic and general relativistic versions exist. There are also numerical models available [39]. On the other hand there are claims that the heavy r-nuclei with mass number $A > 130$ cannot be produced in the neutrino-driven winds [41].

The possibility of r-process in accretion induced delayed core collapse SN has also been studied [42]. The other possibility lies in the neutron star merger

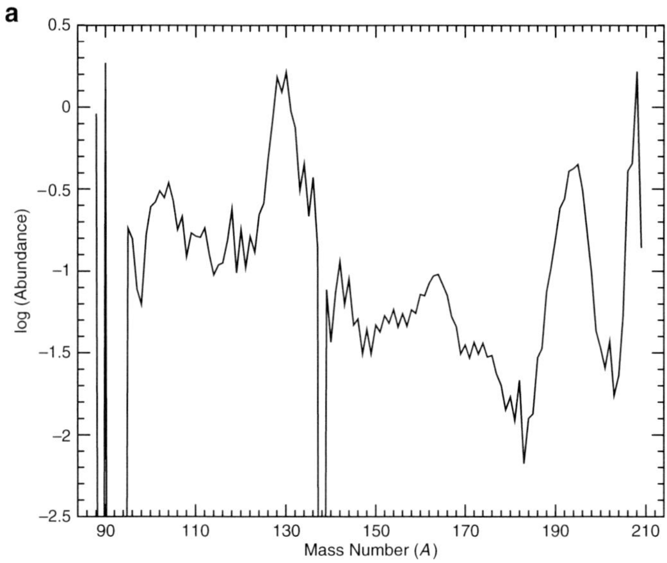

Fig. 14. The r-processes abundances by subtracting the s-process abundances from the observed values (taken from [36])

models for the r-process. During the merger of two neutron stars or a neutron star and a black hole decompression of the cold neutron star matter takes place which can be considered as a site for high density r-process. But this has the drawback that the rates of the mergers are much less compared to core collapse SN rates and it has difficulty in reproducing the abundances of nuclei in the mass region 90 - 110.

7.3 Models for calculation of β-decay rates for r-process nuclei

For the r-process the models for calculating β-decay rates can again be divided into microscopic and statistical categories. Among the microscopic ones shell model is of limited use as this involves very neutron-rich nuclei all over the periodic table. Beyond the fp-shell nuclei shell model has been applied to nuclei with either a few valence particles or with more particles but with not too many valence orbits. The microscopic theory that has been widely used is the 'Random Phase Approximation' (RPA) and its different improved version. We refer here to the review by Arnould, Goriely and Takahashi [39] for a detailed description and references. The effective nucleon-nucleon interaction is often taken to be of the spin-isospin type ($\boldsymbol{\sigma\tau}.\boldsymbol{\sigma\tau}$) and one builds the particle-hole excitations of the charge exchange mode. The Quasi-particle RPA to take into account the pairing along with the models for masses like the extension of the droplet model, FRDM [43] has also been used extensively.

Among the statistical models one uses the 'Gross Theory' or the 'Semi-gross Theory' described before. Here we shall briefly mention the beginning of another statistical approach following the spectral distribution methods applied earlier for the presupernova problem. In this treatment [44, 45], the GT sum rule strength has contributions from the low-lying states and the giant resonance. Nuclei involved are divided into two categories- one set for which some experimentally measured values of log ft are available in the ground state region and the other set, normally the nuclei with larger neutron fractions, where no experimental values of log ft exists. The sharp Fermi Resonance above the IAS has very little contribution but is taken into account. The GT resonance is again taken as a Gaussian with the strength centroid given by the Bertsch and Esbensen expression [46]

$$E_{GT^-} = E_{IAS} + \Delta E_{s.o.} + 2[[k_{\sigma\tau} S_{GT^-}]/3 - (N-Z)k_\tau] \tag{30}$$

Here the $\Delta E_{s.o.}$ is the contribution coming from the spin-orbit interaction and one takes $k_\tau = 28.5/\mathrm{A}$ MeV and $k_{\sigma\tau} = 23/\mathrm{A}$ MeV [29]. The method has been used to calculate the β-decay rates of nuclei in the range $115 < A < 140$. The width of the GT giant resonance is set to give a best fit to the observed half-lives. The two sets of nuclei with known values of log ft and no known ones are treated separately and the best fit value for the width for the second case naturally comes out larger. The calculated and the experimental half-lives are given for all the nuclei in Figure 14. The figure also give a straight line fit of half-life as a function of Q−value given by

$$\lambda_{\beta^-} = ln2/t_{1/2} = 10^{-4} \times (Q_{\beta^-}/MeV)^5 \tag{31}$$

where $t_{1/2}$ is the half-life and the Q value is expressed in MeV. This line actually represents the case where the GT matrix element has a fixed value independent of energy [39]. The rates are then calculated for typical r-process densities and temperatures and the method is used to predict half-lives of nuclei very close to the drip line where no experimental information available.

8 Concluding Remarks

This article discusses the importance of the weak interaction rates in different astrophysical processes. Though a lot of effort has gone in to calculate these rates accurately, more work is still needed, particularly for the highly neutron-rich nuclei. Radioactive ion beam experiments are also expected to give valuable information for these nuclei. So one hopes that these will help in a more accurate description of these processes in the near future.

The author thanks Palash B. Pal for help in preparing the manuscript.

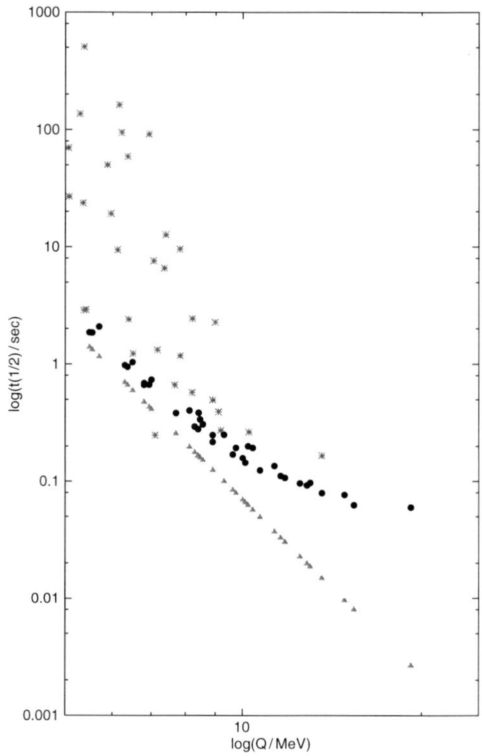

Fig. 15. The half-lives of nuclei with $115 < A < 140$ calculated by the statistical method [45] compared with a form with only the Q^5 dependence

References

1. A. deShalit, H. Feshbach, Theoretical Nuclear Physics Vol **I**, John Wiley and Sons (New York) (1974)
2. C. S. Wu, S. Mozkowski, Beta Decay, Inter Science Publishers, New York (1966)
3. H. A. Bethe, G. E. Brown, Sc. American **25**, 60 (1985)
4. H. A. Bethe, Rev. Mod. Phys. **60**, 801 (1990)
5. H-Th. Janka, K. Langanke, A. Marek, G. Martinez-Pinedo, B. Muller, Phys. Rep. **442**, 38 (2007)
6. K. Langanke, G. Martinez-Pinedo, Rev. Mod. Phys. **75**, 819 (2003)
7. G. G. Raffelt, M. Th. Keil, R. Buras, H.-Th. Janka, M. Rampp, astro-ph/0303226; *Proc. 4th Workshop on Neutrino Oscillations and their Origin NOON* (2003)
8. F. S. Kitaura, H.-Th. Janka, W. Hillebrandt, A&A **450**, 345 (2006)
9. J. M. Blondin, A. Mezzacappa, C. Demarino, ApJ **584**, 971 (2004)
10. G. M. Fuller, ApJ **252**, 741 (1982)
11. H. A. Bethe, J. R. Wilson, ApJ **295**, 14 (1985)
12. A. Ray, K. Kar, ApJ **319**, 143 (1987)
13. W. C. Haxton, Phys. Rev. Lett. **60**, 1999 (1988)

14. G. M. Fuller, W. A. Fowler, M. J. Newman, ApJS **42**, 447 (1982); **48**, 279 (1975); ApJ **252**, 715 (1982); **293** 1 (1985)
15. M. B. Aufderheide, G. E. Brown, T. T. S. Kuo, D. B. Stout, P. Vogel, ApJ **362**, 241 (1990)
16. K. Takahashi, M. Yamada, Prog. Theor. Phys. **41**, 1470 (1969)
17. K. Takahashi, M. Yamada, T. Kondoh, At. Data Nucl. Tables **12**, 101 (1973)
18. T. Tachibana, M. Yamada, Y. Yoshida, Prog. Theor. Phys. **84**, 641 (1990)
19. H. Nakata, T. Tachibana, M. Yamada, Nucl. Phys. **A 625**, 521 (1997)
20. J. B. French, V. K. B. Kota, Ann. Rev. Nuc. Part. Sci. **32**, 35 (1982)
21. V. K. B. Kota, K. Kar, Prmana- J. Phys. **32**, 647 (1989)
22. K. Kar, Nucl. Phys. **A 368**, 285 (1981)
23. K. Ikeda, S. Fujii, J.I. Fujita, Phys. Lett. **3**, 271 (1963)
24. K. Kar, A. Ray, S. Sarkar, ApJ **434**, 662 (1994)
25. K. Nakayama, A. Pio Galeao, F. Krmpotic, Phys. Lett. **B 114**, 217 (1982)
26. D. Majumdar, K. Kar, Pramana- Jour. Phys. **68**, 423 (2007)
27. K. Langanke, G. Martinez-Pinedo, At. Nucl. Data Tables **79**, 1 (2001)
28. E. Caurrier, G. Martinez-Pinedo, F. Nowacki, A. Poves, J. Retamosa, A. P. Zuker, Phys. Rev. **C 59**, 2033 (1999)
29. J. Pruet, G. M. Fuller, ApJS **149**, 189 (2003)
30. A. Heger, K. Langanke, G. Martinez-Pinedo, S. E. Woosley, ApJ **560**, 307 (2001)
31. S. E. Woosley, T. A. Weaver, ApJS **101**, 181 (1995)
32. J. N. Bahcall, A. Serenelli, S. Basu, ApJ **621**, l85 (2005)
33. Homepage of John Bahcall: http://www.sns.ias.edu/ jnb/
34. A. Bandyopadhyay, S. Choubey, S. Goswami, K. Kar, Phys. Lett. **B 519**, 83 (2001)
35. E. M. Burbidge, G. M. Burbidge, W. A. Fowler, F. Hoyle, Rev. Mod. Phys. **29**, 547 (1957)
36. Y.Z. Qian, Prog. Part. Nucl. Phys. **50**, 153 (2003); astro-ph/0301422
37. A. G. Cameron, J. J. Cowan and J. W. Truran, Ap&SS **91**, 235 (1983)
38. S. E. Woosley, J. R. Wilson, G. J. Mathews et al., ApJ **433**, 229 (1994)
39. M. Arnould, S. Goriely, K. Takahashi, Phys. Rep. **450**, 97 (2007)
40. K. Takahashi, J. Witti, H.-Th. Janka, A&A **286**, 857 (1994)
41. Y. Z. Qian, G. J. Wasserburg, Phys. Rep. **442**, 237 (2007)
42. L. Dessart, A. Burrows, C. D. Ott et al., ApJ **644**, 1063 (2006)
43. P. Moller, J. Nix, W. D. Myers, W. J. Swiatecki, At. Data Nucl. Data Tables **59**, 185 (1995)
44. K. Kar, S. Chakravarti, V. R. Manfredi, Pramana -J. Phys. **48**, 123 (2007)
45. K. Kar, S. Chakravarti (unpublished)
46. G. F. Bertsch, H. Esbensen, Phys. Prog. Phys. **50**, 607 (1987)

Massive stars as thermonuclear reactors and their explosions following core collapse

Alak Ray

Tata Institute of Fundamental Research, Mumbai 400 005, India akr@tifr.res.in

Summary. Nuclear reactions transform atomic nuclei inside stars. This is the process of stellar nucleosynthesis. The basic concepts of determining nuclear reaction rates inside stars are reviewed. How stars manage to burn their fuel so slowly most of the time are also considered. Stellar thermonuclear reactions involving protons in hydrostatic burning are discussed first. Then I discuss triple alpha reactions in the helium burning stage. Carbon and oxygen survive in red giant stars because of the nuclear structure of oxygen and neon. Further nuclear burning of carbon, neon, oxygen and silicon in quiescent conditions are discussed next. In the subsequent core-collapse phase, neutronization due to electron capture from the top of the Fermi sea in a degenerate core takes place. The expected signal of neutrinos from a nearby supernova is calculated. The supernova often explodes inside a dense circumstellar medium, which is established due to the progenitor star losing its outermost envelope in a stellar wind or mass transfer in a binary system. The nature of the circumstellar medium and the ejecta of the supernova and their dynamics are revealed by observations in the optical, IR, radio, and X-ray bands, and I discuss some of these observations and their interpretations.

Keywords: nuclear reactions, nucleosynthesis, abundances; stars: interiors; supernovae: general; neutrinos; circumstellar matter; X-rays: stars

1 Introduction

The sun is not commonly considered a star and few would think of stars as nuclear reactors. Yet, that *is* the way it is, and even our own world is made out of the "fall-out" from stars that blew up and spewed radioactive debris into the nascent solar system.

Nuclear Astrophysics is the field concerning "the synthesis and Evolution of atomic nuclei, by thermonuclear reactions, from the Big Bang to the present. What is the origin of the matter of which we are made?" [1]. Our high entropy universe resulting from the Big Bang, contains many more photons per particle

A. Goswami and B.E. Reddy (eds.), *Principles and Perspectives in Cosmochemistry*, Astrophysics and Space Science Proceedings,
DOI 10.1007/978-3-642-10352-0_5,

of matter with mass, e.g. protons and neutrons. Because of the high entropy as the universe expanded, there was time to manufacture elements only upto helium and the major products of cosmic nucleosynthesis remained the light elements hydrogen and helium[1]. Stars formed from this primordial matter. They used these elements, hydrogen and helium as fuel to generate energy like a giant nuclear reactor[2]. In the process, the stars could shine and manufacture higher atomic number elements like carbon, oxygen, calcium and iron which we and our world are made of. The heavy elements are either dredged up from the core of the star to the surface of the star from which they are dispersed by stellar wind or directly ejected into the interstellar medium when a (massive) star explodes. The stardust is the source of heavy elements for new generation of stars and sun-like systems.

Our sun is *not* a massive star. It burns hydrogen in a set of nuclear reactions called the pp-chain, whereas the more massive stars *presently* burn their hydrogen by the so-called CNO-cycle[3]. Nevertheless, to put nuclear reactions in stars in perspective, we shall start with a discussion of how these reactions proceed within the sun. There is a correspondence between the evolutionary state of a star, its external appearance[4] and internal core conditions and the nuclear fuel it burns, – a sort of a mapping between the astronomers Hertzsprung-Russel diagram and the nuclear physicist's chart of the nuclide [4], until nuclear burning takes place on too rapid a time scale.

The problem of evolution of stars to their explosion and subsequent interaction with the circumstellar medium has many time scales (ranging from tens of milliseconds to tens of thousands of years) and macroscopic length scales (from dimensions effectively that of a white dwarf to that of a supernova remnant, i.e from few thousand kilometers to many tens of light years). The physics of supernova explosions is complex and diverse and in many parts, the explosion mechanism is still an unsolved problem. Even the constraining parameters and ingredients which makes the SN explode are still controversial (see e.g. the discussion in [96]). It is possible that the identification of the key

[1] Note however suggestions [5,6] that early generation of stars called Pop III objects can also contribute to the abundance of ^4He seen in the universe today and the entire helium may not be a product of big bang nucleosynthesis alone.

[2] Our sun is slowly burning hydrogen into helium and is presently not exactly the same when it just started burning hydrogen. It will appear different once it exhausts all hydrogen it *can* burn in its core. In other words, nuclear reactions in stellar interiors determine the life-cycle of stars, apart from providing them with internal power for heat and light and manufacturing all the heavier elements that the early universe could not.

[3] Note however that CN cycle may have driven an early stage convection in the young Sun.

[4] Astronomers classify stars according to their colors and (absorption) line spectra and luminosities. Meghnad Saha showed [2,3] the link between the classification scheme and temperature (and thermal ionization) of stellar atmosphere.

aspects in the explosion may require seminal observations about the conditions in the supernova core other than the indirect evidence such as explosion asymmetries, pulsar kicks or nucleosynthetic yields. Such observations may involve the future detection of strong neutrino signals and gravitational waves from a galactic supernova in future. Detectable neutrino signals from a supernova was seen in the case of SN 1987A, but since that target was in a neighboring satellite galaxy (the Large Magellanic Cloud), the number of neutrinos were too small to distinguish the fine points of theoretical issues.

Since nuclear astrophysics is not usually taught at the master's level nuclear physics specialization in our universities, these lecture notes are meant to be an introduction to the subject and a pointer to literature and Internet resources[5]. The emphasis in the first part of these lecture notes is on the nuclear reactions in the stars and how these are calculated, rather than how stars evolve. The latter usually forms a core area of stellar astrophysics.

This article is organized essentially in the same sequence that a massive star burns successively higher atomic number elements in its core, until it collapses and explodes in a supernova. The introductory part discusses how the rates of thermonuclear reactions in (massive) stars are calculated, what the different classes of reactions are and how the stars (usually) manage to burn their fuels so slowly[6]. The middle part describes the nuclear physics during the collapse phase of the massive star. The last part describes a few typical examples of what can be learned by optical, IR and X-ray studies about nucleosynthesis and dynamics of explosion in supernovae and supernova remnants such as Cassiopeia A, SN 1987A etc. Only core-collapse supernovae are discussed in these lectures, those that arise from massive stars (e.g. stars more massive than $8M_{\odot}$ with typical solar metallicity at the time they start

[5] See for example, [7] for a course of nuclear astrophysics, and the International Conference Proceedings under the title: "Nuclei in the Cosmos" for periodic research conferences in the field. Valuable nuclear astrophysics datasets in machine readable formats can be found at sites: [8], [9]. A new and updated version of the nuclear reactions rate library REACLIB for astrophysics is now being maintained as a public, web-based version at [10]. A complementary effort to develop software tools to streamline the process of getting the latest and best information into this new library is available at [11] (see [12]). Much of the material discussed in the first part of these notes can be found in textbooks in the subject, see e.g. [13], [14], [4], [15], [1] etc. There is also a recent book on the subject by Richard Boyd [16] that among other topics describes terrestrial and space born instruments operating in service to nuclear astrophysics. A Workshop on Solar fusion cross sections for the pp chain and CNO cycle held in 2009 by the Institute of Nuclear Theory is expected to result in a Reviews of Modern Physics article. Supernovae of various types are the sites where nuclear reactions in stars or explosions are of prime importance. For Internet resources to two recent Schools on these topics, see http://icts.tifr.res.in/sites/Sgrb/Programme and http://www.tifr.res.in/~sn2004.

[6] These issues were discussed in an earlier SERC School [17].

burning hydrogen in their cores, i.e. at the "Zero Age Main Sequence", before any mass was lost from their surface). We shall not discuss the type Ia SNe[7] in these lectures. Abundance of elements in our galaxy Milky Way give important information about how stars affect the element and isotopic evolution in various parts of the galaxy. For a study of the evolution of elements from C to Zn in the galactic halo and disk, with yields of massive stars and type Ia SNe, see [20] and elsewhere in these Proceedings.

2 Stars and their thermonuclear reactions

While referring to Sir Ernest Rutherford "breaking down the atoms of oxygen and nitrogen", Arthur Eddington remarked: "what is possible in the Cavendish Laboratory may not be too difficult in the sun" [21]. Indeed this is the case, but for the fact that a star does this by fusion reactions, rather than by transfer reactions, – in the process giving out heat and light and manufacturing fresh elements. Of all the light elements, hydrogen is the most important one in this regard, because: a) it has a large universal abundance, b) considerable energy evolution is possible with it because of the large binding energies of nuclei that can be generated from its burning and c) its small charge and mass allows it to penetrate easily through the potential barriers of other nuclei. A star burns its fuel in thermonuclear reactions in the core where the confinement of the fuel is achieved in the star's own gravitational field. These reactions remain "controlled", or self-regulated[8], as long as the stellar material remains non-degenerate.

[7] The supernovae are classified by astronomers on the basis of their optical spectra at the time of their maximum light output. Those that do not show the presence of hydrogen in their spectra are classified as type I SNe. A subclass of them, type Ia's are believed to arise from thermonuclear explosions in the electron degenerate cores of stars less massive than $8M_{\odot}$ and are very useful to map the geometry of our universe, because they serve as calibratable "standard" candles. These "thermonuclear supernovae" are usually more luminous in the optical bands than the core-collapse varieties, but while the former class put out several MeVs of energy per nucleon, the core-collapse SNe or ccSNe, emit several *hundreds* of MeVs per nucleon from their *central engines* (mostly in down-scattered neutrinos). Apart from the missing hydrogen lines, the type Ia SNe show an absorption "trough" in their spectra around maximum light at 6150 Å, due to blue shifted Si II lines [18]. Of the other type I SNe which show no hydrogen and no Si II trough, some show helium in their early spectra (type Ib) and others have little or no helium and have a composition enhanced in oxygen (type Ic) [19]. These, (Ib and Ic) together with the type IIs constitute the core collapse SNe.

[8] There are however examples to the contrary when thermonuclear reactions take place in an explosive manner, e.g. when a whole white dwarf (resulting from an evolved intermediate mass star) undergoes merger with another and explodes, as nuclear fuel (carbon) is ignited under degenerate conditions, such as in a type Ia supernova; explosive thermonuclear reactions also take place in post-bounce core

The recognition of the quantum mechanical tunneling effect prompted Atkinson and Houtermans [22] to work out the qualitative treatment of energy production in stars. They suggested that the nucleus serves as both a cooking pot and a trap. Binding energy difference of four protons and two electrons (the nuclear fuel) and their ash, the helium nucleus, some 26.7 MeV is converted to heat and light that we receive from the sun[9]. The photons in the interior are scattered many a times, for tens of millions of years, before they reach the surface of the sun. Some of the reactions produce neutrinos, which because of their small cross-section for interaction, are not stopped or scattered by overlying matter, – but stream out straight from the core. Neutrinos are thus the best probes of the stellar core [23, 106], while the photons bear information from their surface of last scattering – the photosphere.

2.1 Why do the stars burn slowly: a look at Gamow peaks

The sun has been burning for at least 4.6 billion years[10]. How does it manage to burn so slowly[11]? Under the ambient conditions in the core, the relevant

---

collapse supernovae, when the hydrodynamic shock ploughs through unburnt Si- or O-layers in the mantle.

[9] Lord Kelvin surmised in the nineteenth century that the solar luminosity is supplied by the gravitational contraction of the sun. Given the solar luminosity, this immediately defined a solar lifetime (the so-called Kelvin-Helmholtz time): $\tau_{KH} = GM_\odot^2/R_\odot L_\odot \sim$ few $\times\, 10^7$yr. This turned out to be much shorter than the estimated age of the earth at that time known from fossil records, and led to a famous debate between Lord Kelvin and the fossil geologist Cuvier. In fact, as noted above modern estimates of earth's age are much longer and therefore the need to maintain sunshine for such a long time requires that the amount of energy radiated by the sun during its lifetime is much larger than its gravitational energy or its internal (thermal) energy: $L_\odot \times t_{\rm life} \gg GM_\odot^2/R_\odot$. This puzzle was resolved only later with the realization that the star can tap its much larger nuclear energy reservoir in its core through thermonuclear reactions. The luminosity of the sun however is determined by an interplay of atomic and gravitational physics that controls the opacity, chemical composition, the balance of pressure forces against gravity, etc. Nuclear physics determines how fast nuclear reactions go under feedback control determined by the ambient conditions.

[10] Lord Rutherford [24] determined the age of a sample of pitchblende, to be 700 million years, by measuring the amount of uranium and radium and helium retained in the rock and by calculating the annual output of alpha particles. The oldest rock found is from Southwest Greenland: ≈ 3.8 Gyr old [15]. Radioactive dating of meteorites point to their formation and the solidification of the earth about 4.55 ± 0.07 years ago [25]. Since the sun and the solar system formed only slightly before, their age at isolation and condensation from the interstellar medium is taken to be 4.6 Gyr [26].

[11] The Nobel prize citation of Hans Bethe (1967) who solved this problem, noted that this "concerns an old riddle. How has it been possible for the sun to emit light and heat without exhausting its source not only during the thousands of

thermonuclear reaction cross sections are very small[12]. For reactions involving charged particles, nuclear physicists often encounter cross-sections near the Coulomb barrier of the order of millibarns. One can obtain a characteristic luminosity L_C based on this cross section and the nuclear energy released per reaction [15]:

$$L_C \sim \epsilon N \Delta E/\tau_C \tag{1}$$

where $\epsilon \approx 10^{-2}$ is the fraction of total number of solar nuclei $N \sim 10^{57}$ that take part in nuclear fusion reactions generating typically $\Delta E \sim 25$ MeV in hydrogen to helium conversion. Here, the τ_C is the characteristic time scale for reactions, which becomes minuscule for the cross-sections at the Coulomb barrier and the ambient density and relative speed of the reactants etc:

$$\tau_C \sim \frac{1}{n\sigma v} = \frac{10^{-8} s}{[n/(10^{26}\ \mathrm{cm}^{-3})][\sigma/1\ \mathrm{mbarn}][\mathrm{v}/10^9\ \mathrm{cm\ s}^{-1}]} \tag{2}$$

This would imply a characteristic luminosity of $L_c \approx 10^{20} L_\odot$, even for a small fraction of the solar material taking part in the reactions (i.e. $\epsilon \sim 10^{-2}$). If this was really the appropriate cross-section for the reaction, the sun would have been gone very quickly indeed. Instead the cross-sections are much less than that at the Coulomb barrier penetration energy (say at proton energies of 1 MeV), to allow for a long lifetime of the sun (in addition, weak-interaction process gives a smaller cross-section for some reactions than electromagnetic process, – see Sect. 3.1).

Stellar nuclear reactions can be either: a) charged particle reactions (both target and projectile are nuclei) or b) neutral particle (neutron) induced reactions. Both sets of reactions can go through either a resonant state of an intermediate nucleus or can be a non-resonant reaction. In the former reaction, the intermediate state could be a narrow unstable state, which decays into other particles or nuclei. In general, a given reaction can involve both

---

centuries the human race has existed but also during the enormously long time when living beings needing the sun for their nourishment have developed and flourished on our earth thanks to this source? The solution of this problem seemed even more hopeless when better knowledge of the age of the earth was gained. None of the energy sources known of old could come under consideration. A very important part of his work resulted in eliminating a great number of thinkable nuclear processes under the conditions at the center of the sun, after which only two possible processes remained..... (Bethe) attempted a thorough analysis of these and other thinkable processes necessary to make it reasonably certain that these processes, and only these, are responsible for the energy generation in the sun and similar stars."

[12] This makes the experimental verification of the reaction cross-sections a very challenging task, requiring in some cases, extremely high purity targets and projectiles so that relevant small event rates are not swamped by other reaction channels and products (see Rolfs and Rodney, Chapter 5 [13]).

types of reaction channels. In charged particle induced reactions, the cross-section for both reaction mechanisms drops rapidly with decreasing energy, due to the effect of the Coulomb barrier (and thus it becomes more difficult to measure stellar reaction cross-sections accurately). In contrast, the neutron induced reaction cross-section is very large and increases with decreasing energy (here, resonances may be superposed on a smooth non-resonant yield which follows the $1/v \sim 1/\sqrt{E}$ dependence). These reaction rates and cross-sections can be then directly measured at stellar energies that are relevant (if such nuclei are long lived or can be generated). The schematic dependence of the cross-sections are shown in Fig. 1. We shall not discuss the neutron capture elements in these notes.

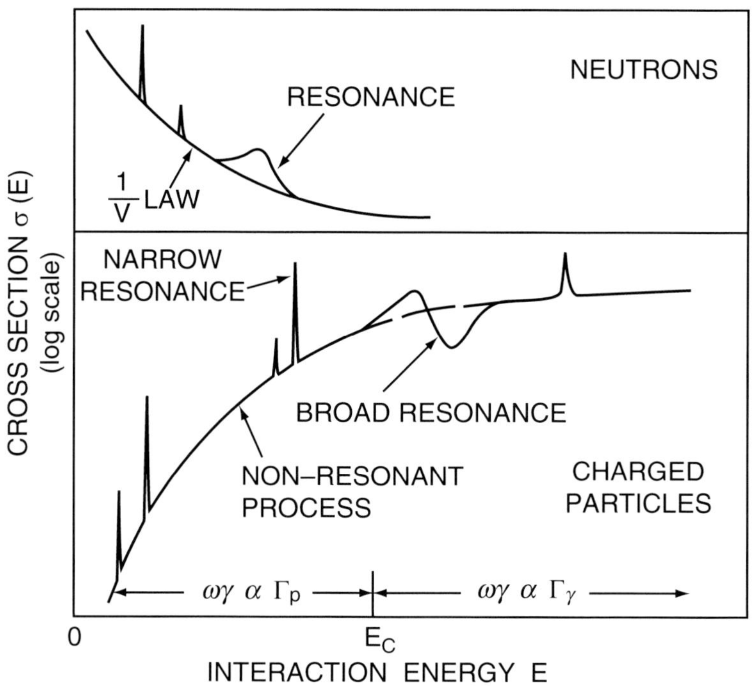

Fig. 1. Dependence of total cross-sections on the interaction energy for neutrons (top panel) and charged particles (bottom panel). Note the presence of resonances (narrow or broad) superimposed on a slowly varying non-resonant cross-section (after [13])

2.2 Gamow peak and the astrophysical S-factor

The sun and other "main-sequence" stars (burning hydrogen in their core quiescently) evolve very slowly by adjusting their central temperature such that the average thermal energy of a nucleus is small compared to the Coulomb repulsion an ion-ion pair encounters. This is how stars can live long for astronomically long times. A central temperature $T \geq 10^7$K (or $T_7 \geq 1$, hereafter a subscript x to a quantity, indicates that quantity in units of 10^x) is required for sufficient kinetic energy of the reactants to overcome the Coulomb barrier and for thermonuclear reactions involving hydrogen to proceed at an effective rate, even though fusion reactions have positive Q values i.e. net energy is liberated out of the reactions. The classical turning point radius for a projectile of charge Z_2 and kinetic energy E_p (in a Coulomb potential $V_C = Z_1Z_2e^2/r$, and effective height of the Coulomb barrier $E_C = Z_1Z_2e^2/R_n = 550$ keV for a p + p reaction), is: $r_{cl} = Z_1Z_2e^2/E_p$. Thus, classically a p + p reaction would proceed only when the kinetic energy exceeds 550 keV. Since the number of particles traveling at a given speed is given by the Maxwell Boltzmann (MB) distribution $\phi(E)$, only the tail of the MB distribution above 550 keV is effective when the typical thermal energy is 0.86 keV ($T_9 = 0.01$). The ratio of the tails of the MB distributions: $\phi(550 \text{ keV})/\phi(0.86 \text{ keV})$ is quite minuscule, and thus classically at typical stellar temperatures this reaction will be virtually absent.

Although classically a particle with projectile energy E_p cannot penetrate beyond the classical turning point, quantum mechanically, one has a finite value of the squared wave function at the nuclear radius R_n : $|\psi(R_n)|^2$. The probability that the incoming particle penetrates the barrier is:

$$P = \frac{|\psi(R_n)|^2}{|\psi(R_c)|^2} \tag{3}$$

where $\psi(r)$ are the wave-functions at corresponding points. Bethe [27] solved the Schrodinger equation for the Coulomb potential and obtained the transmission probability:-

$$P = \exp\Big(- 2\mathrm{KR_c}\big[\frac{\tan^{-1}(\mathrm{R_c/R_n} - 1)^{1/2}}{(\mathrm{R_c/R_n} - 1)^{1/2}} - \frac{\mathrm{R_n}}{\mathrm{R_c}}\big]\Big) \tag{4}$$

with $K = [2\mu/\hbar^2(E_c - E)]^{1/2}$. This probability reduces to a much simpler relation at the low energy limit: $E \ll E_c$, which is equivalent to the classical turning point R_c being much larger than the nuclear radius R_n. The probability is:

$$P = \exp(-2\pi\eta) = \exp[-2\pi \mathrm{Z_1Z_2e^2/(\hbar v)}] = \exp[-31.3\mathrm{Z_1Z_2}(\frac{\mu}{\mathrm{E}})^{1/2}] \tag{5}$$

where in the second equality, μ is the reduced mass in Atomic Mass Units and E is the center of mass energy in keV. The exponential quantity involving the square brackets in the second expression is called the "Gamow factor". The reaction cross-section between particles of charge Z_1 and Z_2 has this exponential dependence due to the Gamow factor. In addition, because the cross-sections are essentially "areas": proportional to $\pi(\lambda/2\pi\hbar)^2 \propto 1/E$, it is customary to write the cross-section, with these two energy dependences filtered out:

$$\sigma(E) = \frac{\exp(-2\pi\eta)}{\mathrm{E}} S(E) \tag{6}$$

where the factor $S(E)$ is called the astrophysical S-factor. The S-factor may contain degeneracy factors due to spin, e.g. $[(2J+1)/(2J_1+1)(2J_2+1)]$ as reaction cross-sections are summed over final states and averaged over initial states. Because the rapidly varying parts of the cross-section (with energy) are thus filtered out, the S-factor is a slowly varying function of center of mass energy, at least for the non-resonant reactions. It is thus much safer to extrapolate $S(E)$ to the energies relevant for astrophysical environments from the laboratory data, which is usually generated at higher energies (due to difficulties of measuring small cross-sections), than directly extrapolating the $\sigma(E)$, which contains the Gamow transmission factor (see Fig. 2). Additionally, in order to relate $\sigma(E)$ and $S(E)$, quantities measured in the laboratory to these relevant quantities in the solar interior, a correction factor f_0 due to the effects of electron screening needs to be taken into account [28].

In the stellar core with a temperature T, reacting particles have many different velocities (energies) according to a Maxwell Boltzmann distribution:-

$$\phi(v) = 4\pi v^2 \left(\frac{\mu}{2\pi kT}\right)^{3/2} \exp\left[-\frac{\mu \mathrm{v}^2}{2\mathrm{kT}}\right] \propto \mathrm{E}^{1/2} \exp[-\mathrm{E/kT}] \tag{7}$$

Nuclear cross-section or the reaction rates which also depend upon the relative velocity (or equivalently the center of mass energy) therefore need to be averaged over the thermal velocity (energy) distribution. Therefore, the thermally averaged reaction rate per particle pair is:

$$< \sigma v >= \int_0^\infty \phi(v)\sigma(v)v dv = \left(\frac{8}{\pi\mu}\right)^{1/2} \frac{1}{(kT)^{3/2}} \int_0^\infty \sigma(E) E \exp(-\mathrm{E/kT})\mathrm{dE} \tag{8}$$

The thermally averaged reaction rate per pair is, utilizing the astrophysical S-factor and the energy dependence of the Gamow-factor:

$$< \sigma v >= \left(\frac{8}{\pi\mu}\right)^{1/2} \frac{1}{(kT)^{3/2}} \int_0^\infty \mathrm{S(E)}\exp\left[-\frac{\mathrm{E}}{\mathrm{kT}} - \frac{\mathrm{b}}{\sqrt{\mathrm{E}}}\right]\mathrm{dE} \tag{9}$$

with $b^2 = E_G = 2\mu(\pi e^2 Z_1 Z_2/\hbar)^2 = 0.978\mu Z_1^2 Z_2^2$ MeV, E_G being called the Gamow energy. Note that in the expression for the reaction rate above, at low energies, the exponential term $\exp(-\mathrm{b}/\sqrt{\mathrm{E}}) = \exp(-\sqrt{(}\mathrm{E_G/E}))$ becomes very small whereas at high energies the Maxwell Boltzmann factor $\mathrm{E}^{1/2}\exp(-\mathrm{E/kT})$ vanishes. Hence there would be a peak (at energy, say, E_0) of the integrand for the thermally averaged reaction rate per pair (see Fig. 3). The exponential part of the energy integrand can be approximated as:

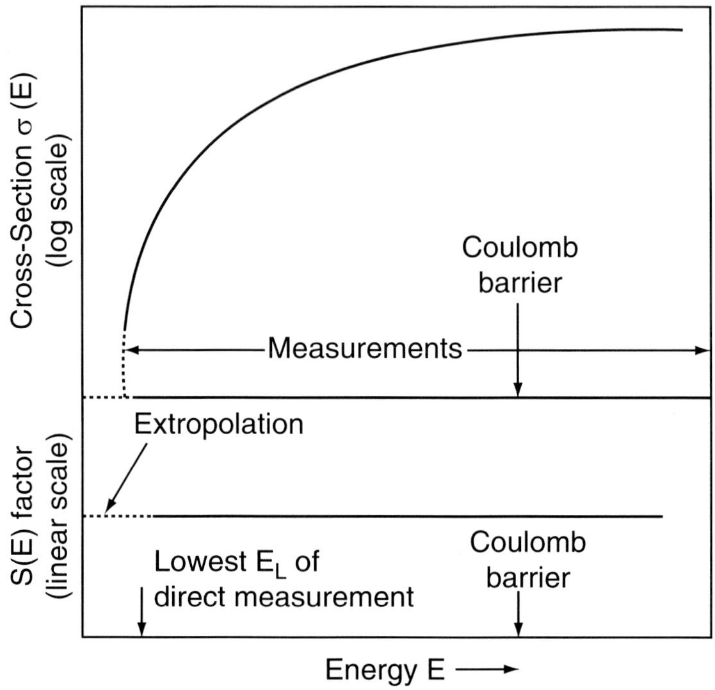

Fig. 2. Cross-section and astrophysical S-factor for charged particle reactions as a function of beam energy. The effective range of energy in stellar interiors is usually far less than the Coulomb barrier energy E_C or the lower limit E_L of laboratory measurements. The y-scale is logarithmic for cross-section but linear for S-factor; thus the cross section drops sharply in regions of astrophysical interest, while the change is much less severe for the S-factor. The extrapolation of laboratory data to lower energies relevant for astrophysical situations is more reliable for S-factor

$$\exp\Big[-\frac{\mathrm{E}}{\mathrm{kT}}-\mathrm{bE}^{-1/2}\Big] \sim \mathrm{C}\exp\Big[-\big(\frac{\mathrm{E}-\mathrm{E}_0}{\Delta/2}\big)^2\Big] \tag{10}$$

where

$$C=\exp(-\mathrm{E}_0/\mathrm{kT}-\mathrm{bE}_0^{-1/2})=\exp(-3\mathrm{E}_0/\mathrm{kT})=\exp(-\tau)$$

$$E_0=(bkT/2)^{\frac{2}{3}}=1.22\mathrm{keV}(\mathrm{Z}_1^2\mathrm{Z}_2^2\mu\mathrm{T}_6^2)^{\frac{1}{3}}$$

$$\varDelta=4(E_0kT/3)^{\frac{1}{2}}=0.75\mathrm{keV}(\mathrm{Z}_1^2\mathrm{Z}_2^2\mathrm{AT}_6^5)^{\frac{1}{6}}$$

Since most stellar reactions happen in a fairly narrow band of energies, S(E) will have a nearly constant value over this band averaging to S_0. With this, the reaction rate per pair of particles, turns out to be:

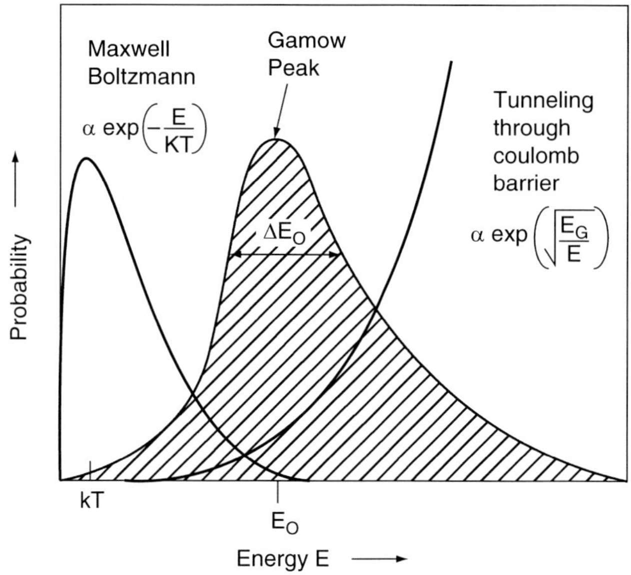

Fig. 3. The Gamow peak is a convolution of the energy distribution of the Maxwell Boltzmann probability and the quantum mechanical Coulomb barrier transmission probability. The peak in the shaded region near energy E_0 is the Gamow peak that gives the highest probability for charged particle reactions to take place. Usually the Gamow peak is at a much higher energy than kT, and in the figure the ordinate scale (for the Gamow peak) is magnified with respect to those of the M-B and barrier penetration factors. See also Table 1

$$< \sigma v >= [\frac{2}{kT}]^{\frac{3}{2}} \frac{S_0}{\pi\mu} \int_0^{\infty} e^{-\tau-4(\frac{E-E_0}{\Delta})^2} dE = 4.510^{14} \frac{S_0}{AZ_1Z_2} \tau^2 e^{-\tau} \mathrm{cm}^3\mathrm{s}^{-1} \tag{11}$$

Here,

$$\tau = 3E_0/kT = 42.5(Z_1^2 Z_2^2 \mu/T_6)^{\frac{1}{3}} \tag{12}$$

The maximum value of the integrand in the above equation is:

$$\mathrm{I}_{\mathrm{max}} = \exp(-\tau)$$

The values of $\mathrm{E}_0, \mathrm{I}_{\mathrm{max}}, \Delta$, etc., apart from the Coulomb barrier for several reactions are tabulated in Table 1 for $T_6 = 15$.

As the nuclear charge increases, the Coulomb barrier increases, and the Gamow peak E_0 also shifts towards higher energies. Note how rapidly the maximum of the integrand I_{max} decreases with the nuclear charge and the Coulomb barriers. The effective width Δ is a geometric mean of E_0 and kT, and $\Delta/2$ is much less rapidly varying between reactions (for $kT \ll E_0$). The rapid variation of I_{max} indicates that of several nuclei present in the stellar core, those nuclear pairs will have the largest reaction rates, which have the smallest Coulomb barrier. The relevant nuclei will be consumed most rapidly at that stage. (Note however that for the p+p reaction, apart from the Coulomb barrier, the strength of the weak force, which transforms a proton to a neutron also comes into play).

When nuclei of the smallest Coulomb barrier are consumed, there is a temporary dip in the nuclear generation rate, and the star contracts gravitationally until the temperature rises to a point where nuclei with the next lowest Coulomb barrier will start burning. At that stage, further contraction is halted. The star goes through well defined stages of different nuclear burning phases in its core dictated by the height of the Coulomb barriers of the fuels. Note also from the Table 1, how far E_0, the effective mean energy of reaction is below the Coulomb barrier at the relevant temperature. The stellar burning is so slow because the reactions are taking place at such a far sub-Coulomb region, and this is why the stars can last so long.

Table 1. Parameters of the thermally averaged reaction rates at $T_6 = 15$

Reaction	Coulomb Barrier (MeV)	Gamow Peak (E_0) (keV)	I_{max} ($e^{-3E_0/kT}$)	Δ (keV)	$(\Delta)I_{max}$
p + p	0.55	5.9	1.1×10^{-6}	6.4	7×10^{-6}
p + N	2.27	26.5	1.8×10^{-27}	13.6	2.5×10^{-26}
$\alpha + C^{12}$	3.43	56	3×10^{-57}	19.4	5.9×10^{-56}
$O^{16} + O^{16}$	14.07	237	6.2×10^{-239}	40.4	2.5×10^{-237}

The above discussion assumes that a bare nuclear Coulomb potential is seen by the charged projectile. For nuclear reactions measured in the laboratory, the target nuclei are in the form of atoms with electron cloud surrounding the nucleus and giving rise to a screened potential – the total potential then goes to zero outside the atomic radius. The effect of the screening is to reduce the effective height of the Coulomb barrier. Atoms in the stellar interiors are in most cases in highly stripped state, and nuclei are immersed in a sea of free electrons which tend to cluster near the nucleus. When the stellar density increases, the so called Debye-Huckel radius $R_D = (kT/4\pi e^2 \rho N_A \xi)^{1/2}$, (here: $\xi = \sum_i (Z_i^2 + Z_i) X_i / A_i$) which is a measure of this cluster "radius", decreases, and the effect of shielding upon the reaction cross-section becomes more important. This shielding effect enhances thermonuclear reactions inside the star. The enhancement factor $f_0 = \exp(0.188 Z_1 Z_2 \xi \rho^{1/2} T_6^{-3/2}$, varies between 1 and 2 for typical densities and compositions [28] but can be large at high densities.

3 Hydrogen burning: the pp chain

The quantitative aspects of the problem of solar energy production with details of known nuclear physics of converting hydrogen into helium was first worked out by von Weizsäcker (1937-38) [32], [33] and Bethe & Critchfield (1938-1939) [34], which made it clear that two different sets of reactions: the p-p chains and the CN cycle can do this conversion. This happens in the core of the star initially (at the "main sequence" stage), and then later in the life of a star in a shell of burning hydrogen around an inert core of He.

In the first generation of stars in the galaxy only the p-p cycle may have operated. In second generation, heavier elements like C, N from the ashes of burning in previous stars are available and they too can act as catalysts to have thermonuclear fusion of hydrogen to helium. Since in the very first generation, the heavier elements like C, N were practically absent, all stars including the massive stars, burnt hydrogen through the p-p cycle. [A recent discovery ([35]) of a low-mass star with an iron abundance as low as 1/200,000 of the solar value (compare the previous record of lowest iron abundance less than 1/10,000 that of the sun), suggests that such first generation stars are still around].

The sun with a central temperature of 15.7 million degrees, ($T^c_{6\odot} = 15.7$) burns by p-p chains. Slightly more massive star (with central temperature $T_6 \geq 20$) burns H by the CNO cycle also. Davis et al.s' solar neutrino experiment [23], which in 1968 had only an upper limit of the neutrino flux, itself put a limit of less than 9% of the sun's energy is produced by the carbon-nitrogen cycle (the more recent upper limit [36] is 7.3%, from an analysis of several solar neutrino experiments, including the Kamland measurements). Note however that for the standard solar model, the actual contribution of CNO cycle to solar luminosity is $\sim 1.5\%$ [15]). In CNO cycle, nuclei such as C, N, O serve as "catalysts" do in a chemical reaction. The pp-chain and the CNO cycle reaction sequences are illustrated in Figs. 4 and 10.

The pp-chain begins with the reaction $\mathrm{p} + \mathrm{p} \rightarrow \mathrm{d} + \mathrm{e}^+ + \nu_\mathrm{e}$. Bethe and Critchfield [34] showed that weak nuclear reaction is capable of converting a proton into a neutron during the brief encounter of a scattering event. (This reaction overcomes the impasse posed by the instability of ^{2}He in the $\mathrm{p} + \mathrm{p} \rightarrow^2$ He reaction via the strong or electromagnetic interactions, and as the next nucleus ^{3}Li reachable via these interactions is also unstable as a final product). Since a hydrogen atom is less massive than a neutron, such a conversion would be endothermic (requiring energy), except for the fact that a neutron in a deuterium nucleus ^{2}D can form a bound state with the proton with a binding energy of 2.224 MeV – thus making the reaction exothermic with an available kinetic energy of 0.42 MeV. The released positron soon pair annihilates into photons making the total energy released to be 1.442 MeV.

Because of the low Coulomb barrier, in the p+p reaction ($E_c = 0.55$ MeV), a star like the sun would have consumed all its hydrogen quickly (note the relatively large value of $(\Delta)I_{max}$ in Table 1), were it not slowed down by the

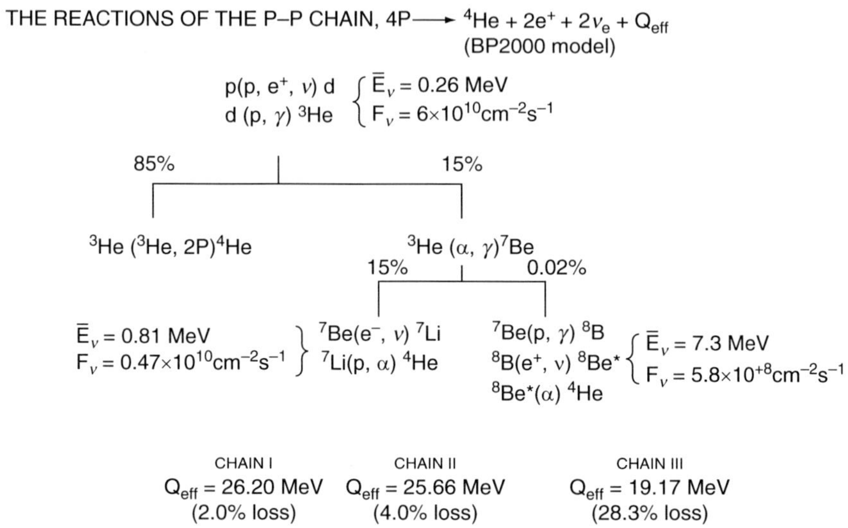

Fig. 4. The p-p chain starts with the formation of deuterium and ^3He. Thereafter, ^3He is consumed in the sun 85% of the time through ppI chain, whereas pp II and pp III chains together account for 15% of the time in the Bahcall Pinsonneault 2000 solar model. The pp III chain occurs only 0.02% of the time, but the ^8B β^+-decay provides the higher energy neutrinos (average $\bar{E}_\nu = 7.3$ MeV). The net result of the chains is the conversion of four protons to a helium, with the effective Q-values (reduced from 26.73 MeV) as shown, due to loss of energy in escaping neutrinos. See [38, 37] for updated branching ratios and neutrino fluxes for BPS2008(AGS) model

weakness of the weak interactions. The calculation of probability of deuteron formation consists of two separate considerations: 1) the penetration of a mutual potential barrier in a collision of two protons in a thermal bath and 2) the probability of the β-decay and positron and neutrino emission. Bethe and Critchfield used the original Fermi theory (point interaction) for the second part, which is adequate for the low energy process.

3.1 Cross-section for deuterium formation

The total Hamiltonian H for the p-p interaction can be written as a sum of nuclear term H_n and a weak-interaction term H_w. As the weak interaction term is small compared to the nuclear term, first order perturbation theory can be applied and Fermi's "Golden rule", gives the differential cross-section as:

$$d\sigma = \frac{2\pi\rho(E)}{\hbar v_i}| < f|H_w|i > |^2 \tag{13}$$

here $\rho(E) = dN/dE$, is the density of final states in the interval dE and v_i is the relative velocity of the incoming particles. For a given volume V, the number of states dn between p and p+dp is:-

$$dN = dn_e dn_\nu = (V\frac{4\pi p_e^2 dp_e}{h^3})(V\frac{4\pi p_\nu^2 dp_\nu}{h^3}) \tag{14}$$

By neglecting the recoil energy of deuterium (since this is much heavier than the outgoing positron in the final state) and neglecting the mass of the electron neutrino, we have:

$$E = E_e + E_\nu = E_e + cp_\nu \tag{15}$$

and $dE = dE_\nu = cp_\nu$, for a given E_e and,

$$\rho(E) = dN(E)/dE = dn_e(dn_\nu/dE) = \frac{16\pi^2 V^2}{c^3 h^6} p_e^2(E - E_e)^2 dp_e = \rho(E_e)dp_e \tag{16}$$

The matrix element that appears in the differential cross section, may be written in terms of the initial state wave function Ψ_i of the two protons in the entrance channel and the final state wave function Ψ_f as:

$$H_{if} = \int [\Psi_d \Psi_e \Psi_\nu]^* H_\beta \Psi_i d\tau \tag{17}$$

If the energy of the electron is large compared to $Z \times$ Rydberg (Rydberg $R_\infty = 2\pi^2 me^4/ch^3$), then a plane wave approximation is a good one: $\Psi_e = 1/(\sqrt{V})\exp(\mathrm{i}\mathbf{k_e}.\mathbf{r})$ where the wave-function is normalized over volume V. (For lower energies, typically 200 keV or less, the electron wave-function could be strongly affected by nuclear charge (see [39])). Apart from this, the final state wave function: $[\Psi_d \Psi_e \Psi_\nu]$ has a deuteron part Ψ_d whose radial part rapidly vanishes outside the nuclear domain (R_0), so that the integration need not extend much beyond $r \simeq R_0$ (for example, the deuteron radius $R_d = 1.7$ fm). Note that because of the Q-value of 0.42 MeV for the reaction, the kinetic energy of the electron ($K_e \leq 0.42$ MeV) and the average energy of the neutrinos ($\bar{E}_\nu = 0.26$ MeV) are low enough so that for both electrons and neutrino wave-functions, the product $kR_0 \leq 2.2 \times 10^{-3}$ and the exponential can be approximated by the first term of the Taylor expansion:

$$\Psi_e = 1/(\sqrt{V})[1 + \mathrm{i}(\mathbf{k_e}.\mathbf{r})] \sim 1/(\sqrt{\mathrm{V}}) \tag{18}$$

and

$$\Psi_\nu \sim 1/(\sqrt{V})$$

Then the expectation value of the Hamiltonian, for a coupling constant g is:

$$H_{if} = \int [\Psi_\mathrm{d}\Psi_\mathrm{e}\Psi_\nu]^* \mathrm{H}_\beta \Psi_\mathrm{i} \mathrm{d}\tau = \frac{\mathrm{g}}{\mathrm{V}} \int [\Psi_\mathrm{d}]^* \Psi_\mathrm{i} \mathrm{d}\tau \tag{19}$$

The $d\tau$ integration can be broken into space and spin parts M_{space} and M_{spin}:

$$d\sigma = \frac{2\pi}{\hbar v_i} \frac{16\pi^2}{c^3 h^6} g^2 M_{spin}^2 M_{space}^2 p_e^2 (E - E_e)^2 dp_e \tag{20}$$

The total cross-section upto an electron energy of E is proportional to:

$$\int_0^E p_e^2 (E - E_e)^2 dp_e = \frac{(m_e c^2)^5}{c^3} \int_1^W (W_e^2 - 1)^{1/2} (W - W_e)^2 W_e dW_e \tag{21}$$

where $W = (E + m_e c^2)/m_e c^2$. The integral over W can be shown as:

$$f(W) = (W^2 - 1)^{1/2} [\frac{W^4}{30} - \frac{3W^2}{20} - \frac{2}{15}] + \frac{W}{4} \ln[\mathrm{W} + (\mathrm{W}^2 - 1)^{1/2}] \tag{22}$$

so that:

$$\sigma = \frac{m_e^5 c^4}{2\pi^3 \hbar^7} f(W) g^2 M_{space}^2 M_{spin}^2 \tag{23}$$

At large energies, the factor $f(W)$ behaves as:

$$f(W) \propto W^5 \propto \frac{1}{30} E^5 \tag{24}$$

The final state nucleus (deuterium in its ground state) in the reaction: $\mathrm{p} + \mathrm{p} \rightarrow \mathrm{d} + \mathrm{e}^+ + \nu_\mathrm{e}$, has $J_f^\pi = 1^+$, with a predominant relative orbital angular momentum $l_f = 0$ and $S_f = 1$ (triplet S-state). For a maximally probable super-allowed transition, there is no change in the *orbital* angular momentum between the initial and final states of the nuclei. Hence for such transitions, the initial state two protons in the $\mathrm{p} + \mathrm{p}$ reaction must have $l_i = 0$. Since the protons are identical particles, Pauli principle requires $S_i = 0$, so that the total wave-function will be antisymmetric in space and spin coordinates. Thus, we have a process: $|S_i = 0, l_i = 0 > \rightarrow |S_f = 1, l_f = 0 >$. This is a pure Gamow-Teller[13] transition with coupling constant $g = C_A$ (the axial vector coupling component can be obtained, from the pure GT decay $^6He(0^+) \rightarrow ^6 Li(1^+)$).

[13] In the beta-decay *allowed* approximation, we neglect the variation of the lepton wave-functions over the nuclear volume and the nuclear momentum (this is equivalent to neglecting all total lepton orbital angular momenta $L > 0$). The total angular momentum carried off by the leptons is their total spin: i.e. $S = 1$ or 0, since each lepton has $\frac{s=1}{2}$. When the lepton spins in the final state are antiparallel, $s_e + s_\nu = s_{tot} = 0$ the process is the Fermi transition with Vector coupling constant $g = C_V$ (e.g. a pure Fermi decay: $^{14}O(J_i^\pi = 0^+) \rightarrow ^{14} N(J_f^\pi = 0^+)$). When the final state lepton spins are parallel, $s_e + s_\nu = s_{tot} = 1$, the process is

The spin matrix element in the above expression for energy integrated cross-section σ, is obtained from summing over the final states and averaging over the initial states *and* dividing by 2 to take into account that we have two identical particles in the initial state. Thus,

$$\lambda = \frac{1}{\tau} = \frac{m^5 c^4}{2\pi^3 \hbar^7 v_i} f(W) g^2 \frac{M^2_{space} M^2_{spin}}{2} \tag{25}$$

where, $M^2_{spin} = \frac{(2J+1)}{(2J_1+1)(2J_2+1)} = 3$. And the space matrix element is:

$$M_{space} = \int_0^\infty \chi_f(r)\chi_i(r) r^2 dr \tag{26}$$

in units of $\mathrm{cm}^{3/2}$. The above integral contains the radial parts of the nuclear wave-functions $\chi(r)$, and involves Coulomb wave-functions for barrier penetration at (low) stellar energies. The integral has been evaluated by numerical methods ([42]), and Fig. 5 shows schematically how the M_{space} is evaluated for the overlap of the deuterium ground state wave-function with the initial pair of protons state. (See also [43], [44] for details of calculations of the overlap integral and writing the astrophysical S-factor in terms of the beta decay rate of the neutron [43] which takes into account of radiative corrections to the axial-vector part of the neutron decay through an effective matrix

Gamow-Teller with $g = C_A$. For Fermi coupling, there is no change in the (total) angular momentum between the initial and final states of the nuclei ($\Delta J = 0$). For the Gamow-Teller coupling, the selection rules are: $\Delta J = 0$ or ± 1 (but the possibility $\Delta J = 0$ is excluded between two states of zero angular momentum). The size of the matrix element for a transition depends on the overlap of the wave-functions in the initial and final states. In the case of "mirror pair" of nuclei (the nucleus $A_Z = (2Z+1)_Z$ is the mirror of the nucleus $(2Z+1)_{Z+1}$), the wave-functions are very much alike as shown through simple heuristic arguments [40]. For these nuclei, ft-values range from $\sim 1000 - 5000$ and are called super-allowed transitions. For super-allowed transitions, which have maximum decay probabilities, there are no changes in the *orbital* angular momentum between the initial and final states of the nuclei. In the p + p $\rightarrow$ D + e^+ + ν_e reaction, the initial proton state is antisymmetric to an interchange of space and spin coordinates and the final deuteron is symmetric in this respect. (In fact when the two protons are in the S state (which is most favorable for their close approach), their spins will be anti-parallel (a singlet state) whereas the ground state of the deuteron is a triplet S state). If this were the complete description of the exchange symmetry properties of the Gamow-Teller transition (permitting a change of spin direction of the proton as it transforms to a neutron, changing the total spin by one unit) advocated here this would actually be forbidden. However in the use of configuration space in beta-decay process one must include isotopic spin as well. The ^1S state of the two protons is symmetric to exchange of this coordinate, whereas the deuteron (consisting of both a proton and a neutron) function is antisymmetric in this coordinate. In the complete coordinate system the transition is from an initial antisymmetric state to another antisymmetric final state accompanied by a positron emission ([41]).

element, the assumption being that these are the same as that for the proton beta decay in the pp reaction above). In the overlap integral one needs only the S-wave part for the wave-function of the deuteron ψ_d, as the D-wave part makes no contribution to the matrix element [42], although its contribution to the normalization has to be accounted for. The wave-function of the initial two-proton system ψ_p is normalized to a plane wave of unit amplitude, and again only the S-wave part is needed. The asymptotic form of ψ_p (well outside the range of nuclear forces) is given in terms of regular and irregular Coulomb functions and has to be defined through quantities related to the S-wave phase shifts in p-p scattering data). The result is a minuscule total cross-section of $\sigma = 10^{-47}\mathrm{cm}^2$ at a laboratory beam energy of $E_p = 1$ MeV, which cannot be measured experimentally even with milliampere beam currents.

The reaction $\mathrm{p} + \mathrm{p} \rightarrow \mathrm{d} + \mathrm{e}^+ + \nu_\mathrm{e}$ is a non-resonant reaction and at all energies the rate varies smoothly with energy (and with stellar temperatures), with $S(0) = 3.8 \times 10^{-22}$ keV barn and $dS(0)/dE = 4.2 \times 10^{-24}$ barn. As for example, the central temperature of the sun $T_6 = 15$, this gives: $< \sigma v >_{pp} = 1.2 \times 10^{-43}$ cm^3 s^{-1}. For density in the center of the sun $\rho = 100$ gm cm^{-3} and equal mixture of hydrogen and helium ($X_H = X_{He} = 0.5$), the mean life of a hydrogen nucleus against conversion to deuterium is $\tau_H(H) = 1/N_H < \sigma v >_{pp} \sim 10^{10}$yr. This is comparable to the age of the old stars. The reaction is so slow primarily because of weak interactions and to a lesser extent due to the smallness of the Coulomb barrier penetration factor (which contributes a factor $\sim 10^{-2}$ in the rate), and is the primary reason why stars consume their nuclear fuel of hydrogen so slowly. For a calculation of the weak capture of protons on protons using calculated wavefunctions obtained from modern, realistic high precision interactions, see [29].

3.2 Deuterium burning

Once deuterium is produced in the weak interaction mediated $\mathrm{p} + \mathrm{p}$ reaction, the main way this is burnt in the sun turns out to be:

$$d + p \rightarrow \; ^3He \; + \gamma \tag{27}$$

This is a non-resonant direct capture reaction to the ^3He ground state with a Q-value of 5.497 MeV and $S(0) = 2.5 \times 10^{-3}$keV barn. The angle averaged cross-sections measured as a function of proton + deuterium center of mass energy, where the capture transitions were observed in gamma-ray detectors at several angles to the incident proton beam direction, are well explained by the direct capture model (see Fig. 6 after [13]). The LUNA collaboration [31] has measured the cross section down to a 2.5 keV c.m. energy, well below the solar Gamow peak using a segmented Bismuth germanate (BGO) gamma-ray detector and found the S(E) factor to be in fair agreement with extrapolation of data at higher energies.

The reactions comprising the rest of the (three) pp-chains start out with the predominant product of deuterium burning: ^3He (manufactured from d+p

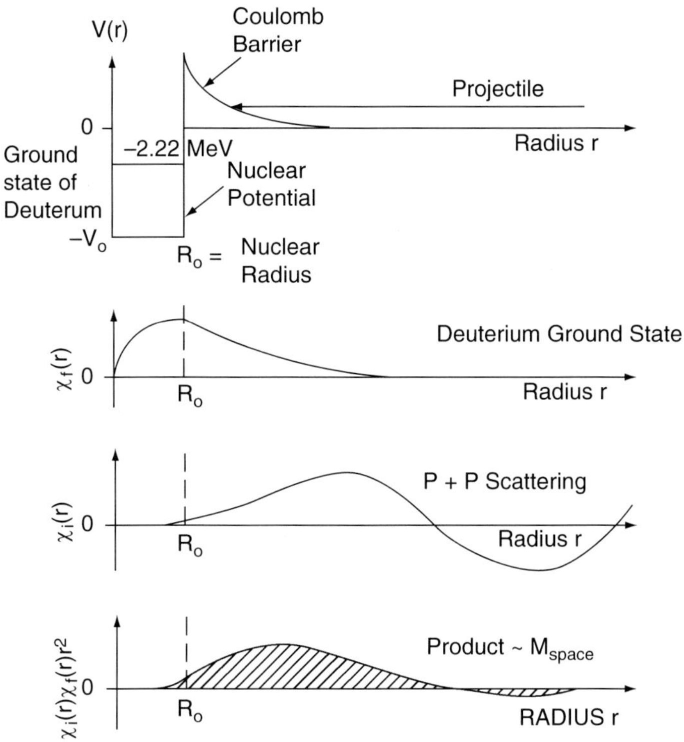

Fig. 5. Schematic representation (after [13]) of the numerical calculation of the spatial part of the matrix element M_{space} in the $\mathrm{p} + \mathrm{p} \to \mathrm{d} + \mathrm{e}^+ + \nu_\mathrm{e}$ reaction. The top part shows the potential well of depth V_0 and nuclear radius R_0 of deuterium with binding energy of −2.22 MeV. The next part shows the radius dependence of the deuterium radial wave function $\chi_d(r)$. The wave-function extends far outside the nuclear radius with appreciable amplitude due to the loose binding of deuterium ground state. The p-p wave-function $\chi_{pp}(r)$ which comprise the $l_i = 0$ initial state has small amplitude inside the final nuclear radius. The radial part of the integrand entering into the calculation of M_{space} is a convolution of both χ_d and χ_{pp} in the second and third panels and is given with the hatched shading in the bottom panel. It has the major contribution far outside the nuclear radius

reaction) as the starting point. The only other reactions with a $S(0)$ greater than the above are: $\mathrm{d(d,p)t}$, $\mathrm{d(d,n)^3He}$, $\mathrm{d(^3He,p)^4He}$, and $d(^3He,\gamma)^5Li$. However, because of the overwhelmingly large number of protons in the stellar thermonuclear reactors, the process involving protons on deuterium dominates. The rate of this reaction is so fast compared to its precursor:

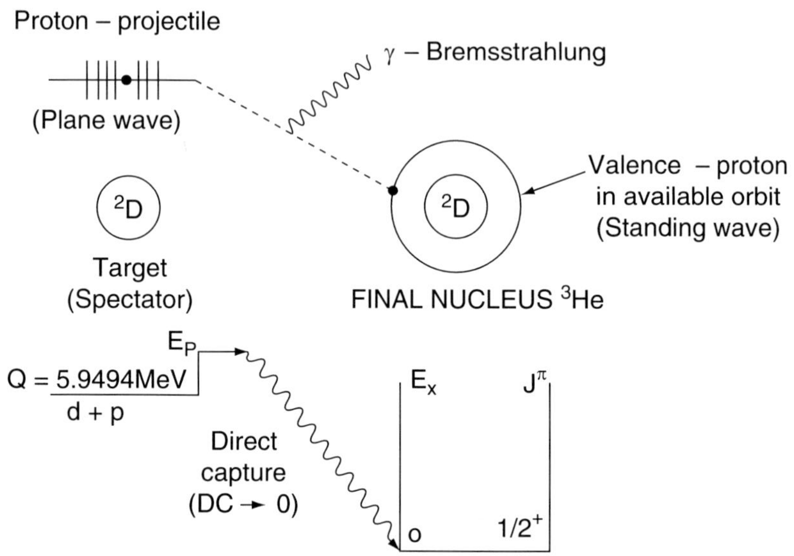

Fig. 6. The Direct Capture reaction $d(p,\gamma)^3He$ to form ^3He in its ground state. The proton projectile (shown as a plane wave) radiates away a bremsstrahlung photon to be captured in a "valence" orbital around the ^2D

$p + p \rightarrow d + e^+\nu_e$, that the overall rate of the pp-chain is not determined by this reaction.

One can show that the abundance ratio of deuterium to hydrogen in a quasi-equilibrium has an extremely small value, signifying that deuterium is destroyed in thermonuclear burning. The time dependence of deuterium abundance D is:

$$\frac{dD}{dt} = r_{pp} - r_{pd} = \frac{H^2}{2} < \sigma v >_{pp} - HD < \sigma v >_{pd} \tag{28}$$

The self regulating system eventually reaches a state of quasi-equilibrium and has:

$$(D/H) =< \sigma v >_{pp} /(2 < \sigma v >_{pd}) = 5.6 \times 10^{-18} \tag{29}$$

at $T_6 = 5$ and 1.7×10^{-18} at $T_6 = 40$. For the solar system however, this ratio is 1.5×10^{-4} and the observed $(D/H)_{obs}$ ratio in the cosmos is $\sim 10^{-5}$. The higher cosmic ratio is due to primordial nucleosynthesis in the early phase of the universe before the stars formed. (The primordial deuterium abundance is a key quantity used to determine the baryon density in the universe). Stars only destroy the deuterium in their core due to the above reaction.

3.3 ^{3}He burning

The pp-chain-I is completed (see Fig. 4) through the burning of ^3He via the reaction:

$$^3\text{He} + {}^3\text{He} \rightarrow \text{p} + \text{p} + {}^4\text{He} \tag{30}$$

with an S-factor: $S(0) = 5500$ keV barn and Q-value = 12.86 MeV. In addition, the reaction:

$$^3\text{He} + \text{D} \rightarrow {}^4\text{He} + \text{p} \tag{31}$$

has an S-factor: $S(0) = 6240$ keV barn, but since the deuterium concentration is very small as argued above, the first reaction dominates the destruction of ^3He even though both reactions have comparable $S(0)$ factors. The cross section for the former reaction near the Gamow energy for the sun, has been measured in [30].

^3He can also be consumed by reactions with ^4He (the latter is pre-existing from the gas cloud from which the star formed and is synthesized in the early universe and in Pop III objects). These reactions proceed through Direct Captures and lead to the pp II and pp III parts of the chain (happening 15% of the time). Note that the reaction $^3He(\alpha,\gamma)^7Be$ together with the subsequent reaction $^7Be(p,\gamma)^8B$ control the production of high energy neutrinos in the sun and are particularly important for the ^{37}Cl solar neutrino detector constructed by Ray Davis and collaborators.

3.4 Reactions involving ^7Be

As shown in Fig. 4, about 15% of the time, ^3He is burned with ^4He radiatively to ^7Be. Subsequent reactions involving ^7Be as a first step in alternate ways complete the fusion process: $4H \rightarrow {}^4He$ in the pp II and pp III chains.

Electron capture process

The first step of the pp II chain is the electron capture reaction on ^7Be : $^7Be + e^- \rightarrow {}^7Li + \nu_e$ (see Fig 7). This decay goes both to the ground state of ^7Li as well as to its first excited state at $E_X = 0.478$ keV, $J^\pi = \frac{1}{2}^-$) – the percentage of decays to the excited state being 10.4 % in the laboratory. The energy released in the reaction with a Q-value of 0.862 keV is carried away by escaping mono-energetic neutrinos with energies: $E_\nu = 862$ and 384 keV. The measured laboratory mean life of the decay is $\tau = 76.9$d. The capture rate in the laboratory can be obtained from Fermi's Golden Rule and utilizing the fact that the wave-functions of both the initial nucleus and the final one vanish rapidly outside the nuclear domain and the electron wave-function in that domain can be approximated as its value at $r = 0$ and the neutrino wave-function by a plane wave normalizes to volume V, so that

$H_{if} = \Psi_e(0)g/\sqrt{V}\int \Psi^*_{^7Li}\Psi_{^7Be}d\tau = \Psi_e(0)gM_n/\sqrt{V}$, where M_n represents the nuclear matrix element and the resultant capture rate is:

$$\lambda_{EC} = 1/\tau_{EC} = (g^2 M_n^2/\pi c^3\hbar^4)E_\nu^2|\Psi_e(0)|^2 \tag{32}$$

In the laboratory capture process, any of the various electron shells contribute to the capture rate; however the K-shell gives the dominant contribution. At temperatures inside the sun, e.g. $T_6 = 15$, nuclei such as ^7Be are largely ionized. The nuclei however are immersed in a sea of free electrons resulting from the ionization process and therefore electron capture from continuum states is possible (see e.g., [45], [46]). Since all factors in the capture of continuum electrons in the sun are approximately the same as those in the case of atomic electron capture, except for the respective electron densities, the ^7Be lifetime in a star, τ_s is related to the terrestrial lifetime τ_t by:

$$\frac{\tau_{fr}}{\tau_t} \sim \frac{2|\Psi_t(0)|^2}{|\Psi_{fr}(0)|^2} \tag{33}$$

where $|\Psi_{fr}(0)|^2$ is the density of the free electrons $n_e = \rho/m_H$ at the nucleus, ρ being the stellar density. The factor of 2 in the denominator takes care of the two spin states of calculation of the λ_t whereas the corresponding λ_{fr} is calculated by averaging over these two orientations. Taking account of distortions of the electron wave-functions due to the thermally averaged Coulomb interaction with nuclei of charge Z and contribution due to hydrogen (of mass fraction X_H) and heavier nuclei, one gets the continuum capture rate as:

$$\tau_{fr} = \frac{2|\Psi_t(0)|^2\tau_t}{(\rho/M_H)[(1+X_H)/2]2\pi Z\alpha(m_e c^2/3kT)^{1/2}} \tag{34}$$

with $|\Psi_e(0)|^2 \sim (Z/a_0)^3/\pi$. Bahcall et al. [44] obtained for the ^7Be nucleus a lifetime:

$$\tau_{fr}(^7Be) = 4.72 \times 10^8 \frac{T_6^{1/2}}{\rho(1+X_H)}\mathrm{s}$$

The temperature dependence comes from the nuclear Coulomb field corrections to the electron wave-function which are thermally averaged. For solar condition the above rate [46] gives a continuum capture lifetime of $\tau_{fr}(^7Be) = 140d$ as compared to the terrestrial mean life of $\tau_t = 76.9d$. Actually, under stellar conditions, there is a partial contribution from some ^7Be atoms which are only partially ionized, leaving electrons in the inner K-shell. So the contributions of such partially ionized atoms have to be taken into account. Under solar conditions the K-shell electrons from partially ionized atoms give another 21% increase in the total decay rate. Including this, gives the solar lifetime of a ^7Be nucleus as: $\tau_\odot(^7Be) = 120d$. In addition, the solar fusion reactions have to be corrected for plasma electrostatic screening enhancement effects. For a recent discussion of the issues see [47,48].

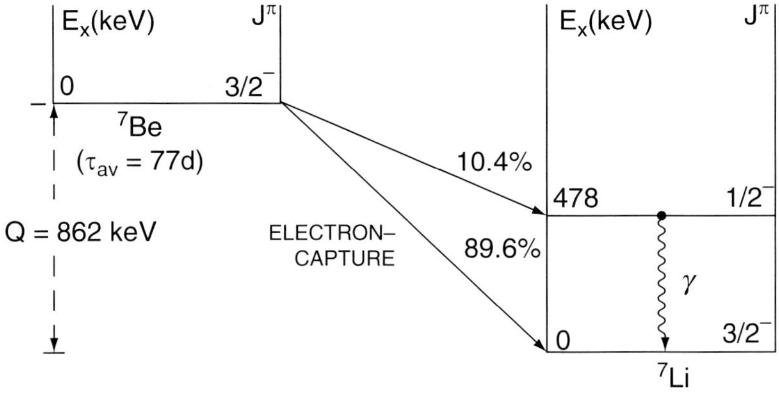

Fig. 7. Electron capture on ^7Be nucleus. The capture proceeds 10.4% of the time to the first excited state of ^7Li at 478 keV, followed by a decay to the ground state by the emission of a photon. The average energy of the escaping neutrinos (which are from the pp II chain) is 814 keV

Capture reaction leading to ^8B

Apart from the electron capture reaction, the ^7Be that is produced is partly consumed by proton capture via: $^7Be(p,\alpha)^8B$ reaction. Under solar conditions, this reaction happens only 0.02% of the time. The proton capture on ^7Be proceeds at energies away from the 640 keV resonance via the direct capture process. Since the product ^7Li nucleus emits an intense γ-ray flux of 478 keV, this prevents the direct measurement of the direct capture to ground state γ-ray yield. The process is studied indirectly by either the delayed positron or the breakup of the product ^8B nucleus into two alpha particles. This reaction has a weighted average $S(0) = 0.0238$ keVbarn [49]. The $^7Be(p,\alpha)^8B$ reaction cross section measurement has been attempted both by direct capture reactions as well as by the Coulomb dissociation of ^8B. For a comparison of the $S_{17}(0)$ factors determined by the two methods and a critical review of the differences of direct and indirect methods, see [50].

The product ^8B is a radioactive nucleus with a lifetime $\tau = 1.1$ s:

$$^8B \rightarrow {}^8Be + e^+ + \nu_e \tag{35}$$

The positron decay of $^8B(J^\pi = 2^+)$ goes mainly to the $\Gamma = 1.6$ MeV broad excited state in ^8Be at excitation energy $E_x = 2.94$ MeV ($J^\pi = 2^+$) due to the selection rules (see Fig. 8). This excited state has very short lifetime and quickly decays into two α-particles. This completes the pp III part of the pp-chain. The average energy of the neutrinos from ^8B reactions is: $\bar{E}_\nu(^8B) = 7.3$ MeV. These neutrinos, having relatively high energy, play an important role in several solar neutrino experiments. The neutrino spectrum is not the same as that obtained from allowed β-decay approximation and is affected by the broad α-unstable ^{8}Be final state. Winter et al. [51] measured the total energy of the α-particles emitted following the β-decay of ^8B and determined the neutrino energy spectrum corrected for recoil order effects in the ^{8}Be final state and constructed from the decay strength function.

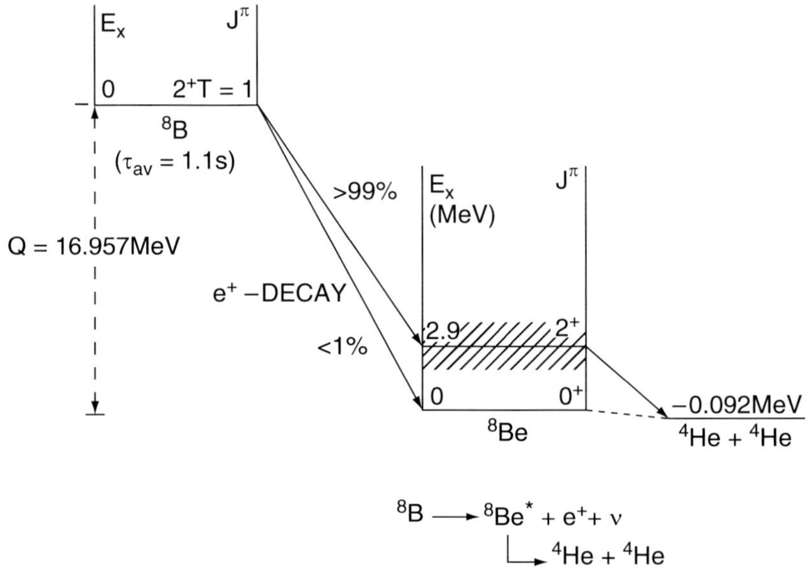

Fig. 8. The decay scheme of ^8B with positron emission, which goes to the first excited state of ^{8}Be at $E_X = 2.9$ MeV with a width of $\Gamma = 1.6$ MeV. The ^8Be nucleus itself fissions into two alpha particles. The neutrinos accompanying the positron decay of ^8B are the higher energy solar neutrinos with $\bar{E}_\nu = 7.3$ MeV

4 The CNO cycle and hot CNO

The sun gets most of its energy generation through the pp-chain reactions (see Fig. 9). However, as the central temperature (in stars more massive than the sun) gets higher, the CNO cycle (see below for reaction sequence) comes to dominate over the pp-chain at T_6 near 20 (this changeover assumes the solar CNO abundance, the transition temperature depends upon CNO abundance in the star; in fact, if one is able to isolate the neutrino flux from the Sun's weak CN cycle, as in say future extended Borexino or SNO+ experiments with the capability to detect low energy neutrinos, this could in turn directly constrain the metallicity of the Sun's core[14] [38]. The early generation of stars (usually referred to as the Population II (Pop II) stars, although there is an even earlier generation of Pop III metal poor massive stars[15]) generated energy primarily through the pp-chain. The Pop II stars are still shining in globular clusters, and being of mass lower than that of the sun, are very old. Most other stars that we see today are later generation stars formed from the debris of heavier stars that contained heavy elements apart from (the most abundant) hydrogen. Thus in the second and third generation stars (which are slightly heavier than the sun) where higher central temperatures are possible because of higher gravity, hydrogen burning can take place through faster chain of reactions involving heavy elements C, N, and O which have some reasonable abundance (exceeding 1%)) compared to other heavy elements like Li, Be, B which are extremely low in abundance. The favored reactions involve heavier elements (than those of the pp-chain) which have the smallest Coulomb barriers but with reasonably high abundance. Even though the Coulomb barriers of Li, Be, B are smaller than those of C, N, O (when protons are the lighter reactants (projectiles)), they lose out due to their lower abundance.

In 1937-1938, Bethe and von Weizsäcker independently suggested the CN part of the cycle, which goes as:

$$^{12}C(p,\gamma)^{13}N(e^{+}\nu_{e})^{13}C(p,\gamma)^{14}N(p,\gamma)^{15}O(e^{+}\nu)^{15}N(p,\alpha)^{12}C \tag{36}$$

This has the net result, as before: $4p \rightarrow^{4} He + 2e^{+} + 2\nu_{e}$ with a $Q = 26.73$. In these reactions, the ^{12}C and ^{14}N act merely as catalysts as their nuclei are "returned" at the end of the cycle. Therefore the ^{12}C nuclei act as seeds that can be used over and over again, even though the abundance of the seed material is minuscule compared to the hydrogen. But note that there is a loss of the catalytic material from the CN cycle that takes place through the $^{15}N(p,\gamma)^{16}O$ reactions. However, the catalytic material is subsequently returned to the CN cycle by the reaction: $^{16}O(p,\gamma)^{17}F(e^{+}\nu_{e})^{17}O(p,\alpha)^{14}N$.

In the CN cycle (see Fig 10), the two neutrinos involved in the beta decays (of ^{13}N ($t_{1/2} = 9.97$min) and ^{15}O ($t_{1/2} = 122.24$s)) are of relatively low energy and most of the total energy $Q = 26.73$ MeV from the conversion of four protons into helium is deposited in the stellar thermonuclear reactor. The rate of the energy production is governed by the slowest thermonuclear reaction in the cycle. Here nitrogen isotopes have the highest Coulomb barriers in charged particle reactions, because of their $Z = 7$. Among them $^{14}N(p,\gamma)^{15}O$ is the

---

[14] A measurement of CN-cycle neutrino flux (with an expected total flux of about 5×10^{8} cm^{-2} s^{-1}) would test an assumption of the Standard Solar Model that during the early pre-main-sequence Hayashi phase the Sun became homogeneous due to convective mixing and that subsequent evolution has not appreciably altered the distribution of metals [38].

[15] Though subsequently when Carbon has been synthesized by triple-α process, these Pop III stars turn on their CN cycle.

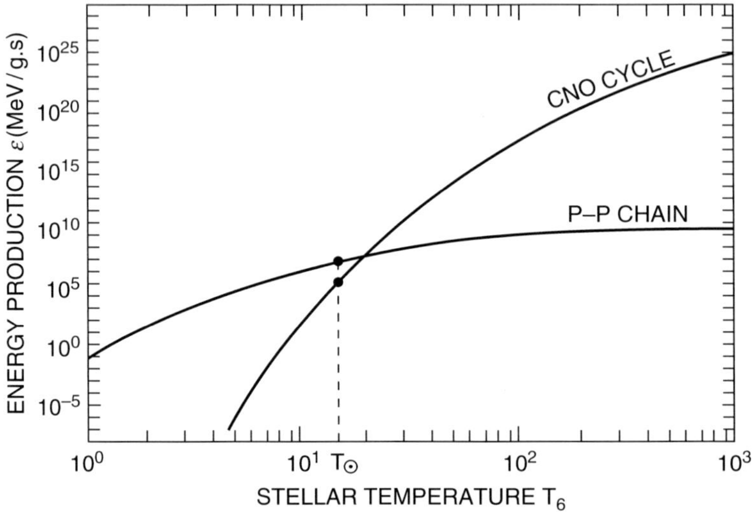

Fig. 9. Comparison of the temperature dependence of the p-p chain and the CNO cycle energy production. The points marked for the solar central temperature $T_{\odot} = T_6 = 15.7$ are shown on both graphs. The CNO cycle generation dominates over the pp-chain at temperatures higher than $T_6 = 20$, so that for sun like stars, the pp-chain dominates. For more massive stars, the CNO cycle dominates as long as one of the catalysts: C, N, or O have initial mass concentration at least 1%. Note the logarithmic scales of the graph and how both rates drop sharply with decreasing temperature, with that of CNO cycle even more drastic due to higher Coulomb barriers

slowest because this reaction having a final state photon is governed by electromagnetic forces while that involving the other nitrogen isotope: $^{15}N(p,\alpha)^{12}C$ is governed by strong forces and is therefore faster.

From the CN cycle, there is actually a branching off from ^{15}N by the reaction $^{15}N(p,\gamma)^{16}O$ mentioned above. This involves isotopes of oxygen, and is called the ON cycle; finally the nitrogen is returned to the CN cycle through ^{14}N. Together the CN and the ON cycles, constitutes the CNO bi-cycle. The two cycles differ considerably in their relative cycle-rates: the ON cycle operates only once for every 1000 cycles of the main CN cycle. This can be gauged from the S(0) factors of the two sets of reactions branching off from ^{15}N: for the $^{15}N(p,\alpha)^{12}C$ reaction $S(0) = 65$ MeV b, whereas for $^{15}N(p,\gamma)^{16}O$, it is 64 keV b, i.e. a factor of 1000 smaller.

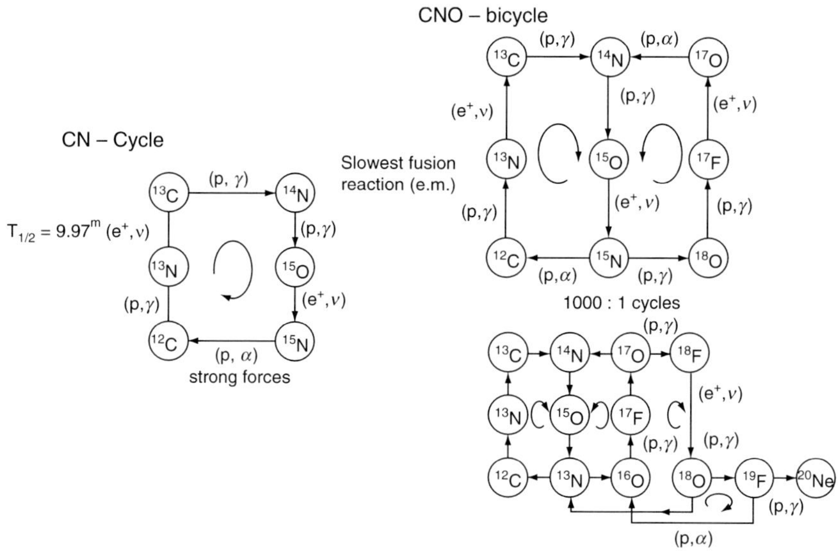

Fig. 10. The various CNO cycles. The left part is the CN cycle where only C and N serve as catalysts for the conversion of four protons into ^4He. Here the slowest fusion reaction is (p,γ) reaction on ^{14}N whereas the slower β-decay has a half-life of 9.97m. In the CNO bi-cycle (right part), there is leakage from the CN cycle to the ON cycle through the branching at ^{15}N. The flow is returned to the CN cycle (which cycles 1000 times for each ON cycle) through $^{17}O(p,\alpha)^{14}N$. The right bottom part represents additional cycles linking into the CNO cycle through the $^{17}O(p,\gamma)^{18}F$ reaction [13]

4.1 Hot CNO and rp-process

The above discussion of CNO cycle is relevant for typical temperatures $T_6 \leq 80$. These are found in quiescently hydrogen burning stars with solar composition which are only slightly more massive than the sun. There are situations where the hydrogen burning takes place at temperatures ($T \sim 10^8 - 10^9$ K) which are in excess of those found in the interiors of the ordinary "main sequence" stars. Examples of these are: hydrogen burning at the accreting surface of a neutron star or in the explosive burning on the surface of a white dwarf, i.e. novae, or the outer layers of a supernova shock heated material in the stellar mantle. These hot CNO cycles operate under such conditions on a rapid enough time scale (few seconds) so that even "normally" β-unstable nuclei like ^{13}N will live long enough to be burned by thermonuclear charged particle reactions, before they are able to β-decay [52], [53]. So, unlike the normal CNO the amount of hydrogen to helium conversion in hot CNO is limited by the β-decay lifetimes of the proton-rich nuclei like: ^{14}O and ^{15}O rather than the proton capture rate of ^{14}N. Wallace and Woosley [54] showed that for temperatures, $T \geq 5 \times 10^8 K$, nucleosynthesised material can leak out of the cycles. This leads to a diversion from lighter to heavier nuclei and is known as the rapid proton capture or rp-process. The flow between the hot CNO cycle and the rp capture process in X-ray bursts from the atmosphere of a neutron star is regulated by the $^{15}O(\alpha,\gamma)^{19}Ne$ reaction. Another alpha capture reaction $^{18}Ne(\alpha,\gamma)^{21}Na$, continuously processes the available ^4He nuclei flowing towards heavier elements. For a recent discussion on the hot CNO and the rp process on accreting neutron stars, see [55].

The nucleosynthesis path of rp-process of rapid proton addition is analogous to the r-process of neutron addition (for neutron capture processes in the early galaxy see [56]). The hot hydrogen bath converts CNO nuclei into isotopes near the region of proton unbound nuclei (the proton drip line). For each neutron number, a maximum mass number A is reached where the proton capture must wait until β^+-decay takes place before the buildup of heavier nuclei (for an increased neutron number) can take place. Unlike the r-process the rate of the rp-process is increasingly hindered due to the increasing Coulomb barrier of heavier and higher-Z nuclei to proton projectiles. Thus the rp-process does not extend all the way to the proton drip line but runs close to the beta-stability valley and runs through where the β^+-decay rate compares favorably with the proton captures.

5 Helium burning and the triple-α reaction

After hydrogen burning in the core of the star has exhausted its fuel, the helium core contracts slowly. Its density and temperature goes up as gravitational energy released is converted to internal kinetic energy. The contraction also heats hydrogen at the edge of the helium core, igniting the hydrogen to burn in a shell. At a still later stage in the star's evolution, the core has contracted enough to reach central temperature density conditions: $T_6 = 100-200$ and $\rho_c = 10^2 - 10^5$ gm cm^{-3} when the stellar core settles down to burn ^4He in a stable manner. The product of helium burning is ^{12}C. Since in nature, the $A = 5$ and $A = 8$ nuclei are not stable, the question arises as to how helium burning bridges this gap. A direct interaction of three α particles to produce a ^{12}C nucleus would seem at first sight, to be too improbable (as was mentioned, for example, in Bethe's 1939 paper [57], which was primarily on the CN cycle). However, Öpik [58] and Salpeter [43], [59] independently proposed a two step process where in the first step, two α particles interact to produce a ^8Be nucleus in its ground state (which is unstable to α-breakup), followed by the unstable nucleus interacting with another α-particle process to produce a ^{12}C nucleus.

Thus the triple alpha reaction begins with the formation of ^8Be that has a lifetime of only 1×10^{-16} s (this is found from the width $\Gamma = 6.8$ eV of the ground state and is the cause of the $A = 8$ mass gap). This is however long compared to the transit time 1×10^{-19} s of two α-particles to scatter past each other non-resonantly with kinetic energies comparable to the Q-value of the reaction namely, $Q = -92.1$ keV. So it is possible to have an equilibrium build-up of a small quantity of ^8Be in equilibrium with its decay or reaction products: $\alpha + \alpha \rightarrow {}^8Be$. The equilibrium concentration of the ^8Be nucleus can be calculated through the Saha equation

$$N_{12} = \frac{N_1 N_2}{2} \left(\frac{2\pi}{\mu kT}\right)^{3/2} \hbar^3 \frac{(2J+1)}{(2J_1+1)(2J_2+1)} \exp\left(-\frac{\mathrm{E_R}}{\mathrm{kT}}\right) \tag{37}$$

at the relevant temperature $T_6 = 11$ and $\rho = 10^5$ gm cm^{-3} to be:

$$\frac{N(^8Be)}{N(^4He)} = 5.2 \times 10^{-10} \tag{38}$$

Salpeter suggested that this small quantity of ^8Be serves as the seed for the second stage of the triple α-capture into the ^{12}C nucleus. It was however shown by Hoyle [61] that the amount of ^{12}C produced for the conditions inside a star at the tip of the red-giant branch is insufficient to explain the observed abundance, *unless* the reaction proceeds through a resonance process [60]. The presence of such a resonance greatly speeds up the rate of the triple-α process which then proceeds through an s-wave ($l = 0$) resonance in ^{12}C near the threshold of $^8Be + \alpha$ reaction. Since ^8Be and ^4He both have $J^\pi = 0^+$, an s-wave resonance would imply that the resonant state in question has to be 0^+ in the ^{12}C nucleus. Hoyle suggested the excitation energy to be: $E_X \sim 7.68$ MeV in the ^{12}C nucleus and this state was experimentally found by W.A. Fowler's group ([62]) with spin-parity: $J^\pi = 0^+$. This state has a total width ([13]) $\Gamma = 8.9 \pm 1.08$ eV, most of which lies in Γ_α, due to the major propensity of the ^{12}C nucleus to break-up through α-decay. (The decay of the excited state of ^{12}C by γ-rays cannot go directly to the ground state, since the resonance state as well as the ground state of the ^{12}C nucleus have both $J^\pi = 0^+$ and $0^+ \rightarrow 0^+$ decays are forbidden. This partial width due to gamma-decay is several thousand times smaller than that due to α-decay). So, $\Gamma = \Gamma_\alpha + \Gamma_{rad} \sim \Gamma_\alpha$ and $\Gamma_{rad} = \Gamma_\gamma + \Gamma_{e^+e^-} = 3.67 \pm 0.50$ meV. Again the radiative width Γ_{rad} is dominated by the width due to photon width deexcitation: $\Gamma_\gamma = 3.58 \pm 0.46$ meV. (Note the scales of *milli*electron Volts). The reaction scheme for the first and the second parts of the triple-alpha reaction is given in Fig. 11. The locations of the Gamow energy regions near the above resonance state (for several stellar temperatures) are shown only schematically.

The reaction rate for the ^{12}C formation can be calculated by using the properties of the resonant state and the thermally averaged cross-section:

$$r_{3\alpha} = N_{^8Be} N_\alpha < \sigma v >_{^8Be+\alpha} \tag{39}$$

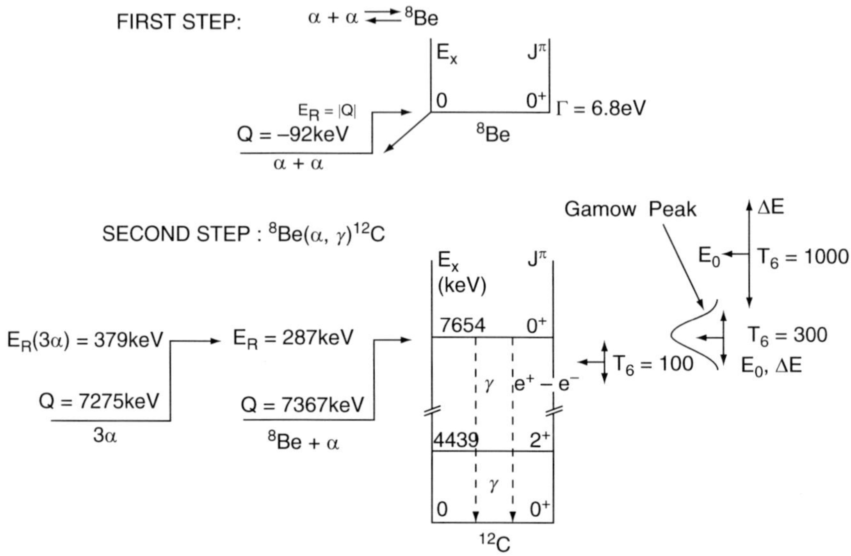

Fig. 11. Triple alpha process of ^{12}C synthesis. In the first step a small amount of ^{8}Be nuclei builds up in equilibrium with its decay products (forward and backward reactions involve alpha particles). The second step involves a capture of another alpha particle by the unstable ^{8}Be nucleus which proceeds via an s-wave resonance state in the product nucleus ^{12}C located close to the Gamow energy window for temperatures indicated schematically by the three-way arrows on the right

Here $N_{^8Be}$ and N_α are the number densities of interacting ^8Be and ^4He nuclei and the angular brackets denote thermal averaging over a Maxwell Boltzmann distribution $\psi(E)$. This averaging leads to:

$$r_{3\alpha} = N_{^8Be} N_\alpha \int_0^\infty \psi(E) v(E) \sigma(E) dE \tag{40}$$

with

$$\psi(E) = \frac{2}{\sqrt{\pi}} \frac{E}{kT} \exp(-\mathrm{E/kT}) \frac{\mathrm{dE}}{(\mathrm{kTE})^{1/2}}$$

and

$$\sigma(E) = \pi \left(\frac{\lambda}{2\pi}\right)^2 \frac{2J+1}{(2J_1+1)(2J_2+1)} \frac{\Gamma_1 \Gamma_2}{(E-E_R)^2 + (\Gamma/2)^2}$$

is the Breit-Wigner resonant reaction cross section with the resonant energy centroid at $E = E_R$. The total width Γ is a sum of all decay channel widths such as $\Gamma_1 = \Gamma_\alpha$ and $\Gamma_2 = \Gamma_\gamma$. If the width Γ is only a few eVs then the functions $\psi(E)$ and $v(E)$ can be pulled out of the integral. Then, the reaction rate will contain an integral like: $\int_0^\infty \sigma_{BW}(E)dE = 2\pi(\lambda/2\pi\hbar)^2\omega\Gamma_1\Gamma_2/\Gamma$, where $\omega = (2J+1)/[(2J_1+1)(2J_2+1)]$ and the functions pulled out of the integral need to be evaluated at $E = E_R$. Since most of the time the excited state of the $^{12}C^*$ breaks-up into α-particles, we have $\Gamma_1 = \Gamma_\alpha$ dominating over Γ_γ and $(\Gamma_1\Gamma_2/\Gamma) \sim \Gamma_2$. This limit usually holds for resonances of energy sufficiently high so that the incident particle width (Γ_1) dominates the natural width of the state (Γ_2). In that case, we can use the number density of the ^8Be nuclei in equilibrium with the α-particle nuclei bath as described by Saha equilibrium condition:

$$N(^8Be) = N_\alpha^2 \omega f \frac{h^3}{(2\pi\mu kT)^{3/2}} \exp(-\mathrm{E_r/kT}) \tag{41}$$

where f is the screening factor. It is possible to get the overall triple-alpha reaction rate by calculating the equilibrium concentration of the excited (resonant) state of ^{12}C reached by the $^8Be + \alpha \rightarrow^{12} C^*$ reaction and then multiplying that concentration by the gamma-decay rate $\Gamma_\gamma/\hbar$ which leads to the final product of ^{12}C. So, the reaction rate for the final step of the triple-alpha reaction turns out to be:

$$r_{3\alpha} = N_{^8Be} N_\alpha \hbar^2 \left(\frac{2\pi}{\mu kT}\right)^{3/2} \omega f \Gamma_2 \exp(-\mathrm{E_r^{'}/kT}) \tag{42}$$

where μ is the reduced mass of the reactants ^8Be and α particle. This further reduces by the above argument to:

$$r_{3\alpha \rightarrow ^{12}C} = \frac{N_\alpha^3}{2} 3^{3/2} \left(\frac{2\pi\hbar^2}{M_\alpha kT}\right)^3 f \frac{\Gamma_\alpha \Gamma_\gamma}{\Gamma \hbar} \exp(-\mathrm{Q/kT}) \tag{43}$$

The Q-value of the reaction is the sum of $E_R(^8Be+\alpha) = 287$ keV and $E_R(\alpha+\alpha) = |Q| = 92$ keV and turns out to be: $Q_{3\alpha} = (M_{^{12}C^*} - 3M_\alpha)c^2 = 379.38 \pm 0.20$keV [63]. Numerically, the energy generation rate for the triple-alpha reaction is:

$$\epsilon_{3\alpha} = \frac{r_{3\alpha} Q_{3\alpha}}{\rho} = 3.9 \times 10^{11} \frac{\rho^2 X_\alpha^3}{T_8^3} f \exp(-42.94/\mathrm{T_8}) \text{ erg gm}^{-1} \text{ s}^{-1} \tag{44}$$

The triple alpha reaction has a very strong temperature dependence: near a value of temperature T_0, one can show that the energy generation rate is:

$$\epsilon(T) = \epsilon(T_0)(\frac{T}{T_0})^n \tag{45}$$

where, $n = 42.9/T_8 - 3$. Thus at a sufficiently high temperature and density, the helium gas is very highly explosive, so that a small temperature rise gives rise to greatly accelerated reaction rate and energy liberation. When helium thermonuclear burning is ignited in the stellar core under degenerate conditions, an unstable and sometimes an explosive condition develops.

6 Survival of ^{12}C in red giant stars and $^{12}C(\alpha,\gamma)^{16}O$ reaction

The product of the triple-alpha reactions ^{12}C, is burned into ^{16}O by α-capture reactions:

$$^{12}C + \alpha \rightarrow ^{16} O + \gamma \tag{46}$$

If this reaction proceeds too efficiently in helium burning Red giant stars, then all the carbon will be burned up to oxygen. Carbon is however the most abundant element in the universe after hydrogen, helium and oxygen, and the cosmic C/O ratio is about 0.6. In fact, the O and C burning reactions and the conversion of He into C and O take place in similar stellar core temperature and density conditions. Major ashes of He burning in Red Giant stars are C and O. Red Giants are the source of the galactic supply of ^{12}C and ^{16}O. Fortuitous circumstances of the energy level structures of these alpha-particle nuclei are in fact important for the observed abundance of oxygen and carbon.

For example, if as in the case of the 3α reaction, there was a resonance in the $^{12}C(\alpha,\gamma)^{16}O$ reaction near the Gamow window for He burning conditions (i.e. $T_9 \sim 0.1 - 0.2$), then the conversion of $^{12}C \rightarrow \ ^{16}O$ would have proceeded at a very rapid rate. However, the energy level diagram of ^{16}O shows that for temperatures upto about $T_9 \sim 2$, there is no level available in ^{16}O to foster a resonant reaction behavior (Fig. 12). But since this nucleus is found in nature, its production must go through either: 1) a non-resonant direct capture reaction or 2) non-resonant captures into the tails of nearby resonances (i.e. sub-threshold reactions). In Fig. 12, also shown on the left of the ^{16}O energy levels, is the threshold for the $^{12}C +^4 He$ reaction, drawn at the appropriate level with respect to the ground state of the ^{16}O nucleus. The Gamow energy regions drawn on the extreme right for temperatures $T_9 = 0.1$ and above, indicates that for the expected central temperatures, the effective stellar (center of mass) energy region is near $E_0 = 0.3$ MeV. This energy region is reached by the low energy tail of a broad resonance centered at $E_{CM} = 2.42$ MeV above the threshold (the $J^\pi = 1^-$ state at 9.58 MeV above the ground state of ^{16}O) with a (relatively large) resonance width of 400 keV. On the other hand, there are two sub-threshold resonances in ^{16}O (at $E_X = 7.12$ MeV and $E_X = 6.92$ MeV), i.e. -45 keV and -245 keV *below* the α-particle threshold that have $J^\pi = 1^-$ and $J^\pi = 2^+$, that contribute to stellar burning rate by their high energy tails. However, electric dipole (E1) γ-decay of the 7.12 MeV state is inhibited by isospin selection rules. Had this not been the case, the $^{12}C(\alpha,\gamma)^{16}O$ reaction would have proceeded fast and ^{12}C would have been consumed during helium burning itself. The two sub-threshold states at -45 keV and -245 keV give contributions to the astrophysical S-factor of: $S_{1^-}(E_0) = 0.1$ MeV barn and $S_{2^+}(E_0) = 0.2$ MeV barn respectively at the relevant stellar energy $E_0 = 0.3$ MeV. The state at $E_{CM} = 2.42$ MeV ($J^\pi = 1^-$ state at 9.58 MeV) gives a contribution: $S_{1^-}(E_0) = 1.5 \times 10^{-3}$MeV barn. The total S-factor at $E_0 = 0.3$ MeV is therefore close to 0.3 MeV barn. These then provide low enough S or cross-section to not burn away the ^{12}C entirely to ^{16}O, so that $C/O \sim 0.1$ at the least.

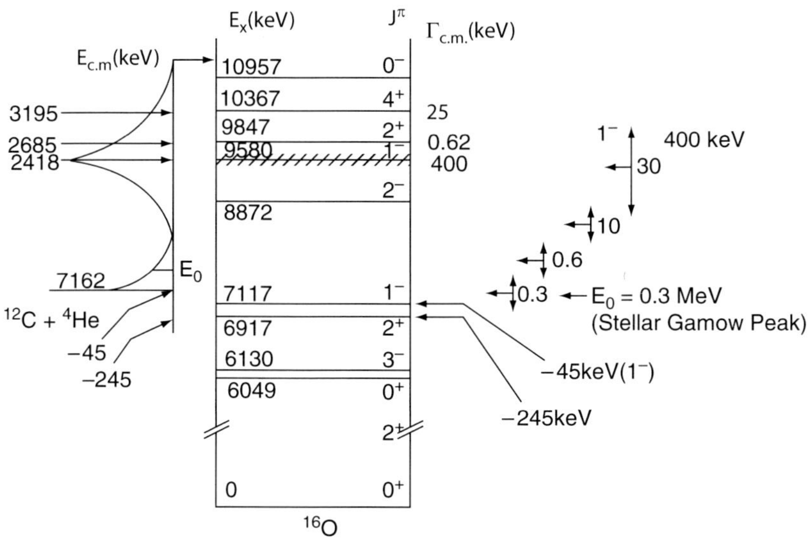

Fig. 12. Energy levels of ^{16}O nucleus near and above the alpha-particle threshold of capture on ^{12}C. Shown on the right are effective stellar energy regions corresponding to the temperatures given near the three-way arrows. The reaction rate is influenced mainly by the high energy tails of two sub-threshold resonances in ^{16}O at $E_R = -45$ keV and $E_R = -245$ keV, plus the low energy tail of another high-lying broad resonance at 9580 keV

Additionally, ^{16}O nuclei are not burnt away by further α-capture in the reaction:

$$^{16}O +^4 He \rightarrow \, ^{20}Ne + \gamma \tag{47}$$

A look at the level schemes of ^{20}Ne (see Fig. 13) shows the existence of a $E_X = 4.97$ MeV state ($J^\pi = 2^-$) in the Gamow window. However, this state cannot form in the resonance reaction due to considerations of parity conservation (unnatural parity of the resonant state)[16]. The lower 4.25 MeV state ($J^\pi = 4^+$) in ^{20}Ne also cannot act as a sub-threshold resonance as it

[16] Whether or not a resonant state can be formed or accessed via a given reaction channel depends upon the angular momentum and parity conservation laws. The spins of the particles in the entrance channel, j_1, j_2 and relative angular

lies too far below threshold and is formed in the g-wave state. Therefore only direct capture reactions seem to be operative, which for (α, γ) reactions lead to cross-sections in the range of nanobarns or below. Thus the destruction of ^{16}O via: $^{16}O(\alpha, \gamma)^{20}Ne$ reaction proceeds at a very slow rate during the stage of helium burning in Red Giant stars, for which the major ashes are carbon and oxygen and these elements have their galactic origin in the Red Giants.

To summarize, the synthesis of two important elements for the evolution of life as we know on the earth have depended upon fortuitous circumstances of nuclear properties and selection rules for nuclear reactions. These are: 1) the mass of the unstable lowest (ground) state of ^{8}Be being close to the combined mass of two α-particles; 2) there is a resonance in ^{12}C at 7.65 MeV which enhances the alpha addition reaction (the second step); and 3) parity conservation has protected ^{16}O from being destroyed in the $^{16}O(\alpha, \gamma)^{20}Ne$ reactions by making the 4.97 MeV excited state in ^{20}Ne of unnatural parity.

The experimental determination of the reaction rate $^{12}C(\alpha, \gamma)^{16}O$ has been an important goal in nuclear astrophysics for several decades. Its cross section at the position of the Gamow window for a typical stellar temperature of 2.5×10^{8}K is comparable to that of weak interaction cross-sections. At those energies, this reaction is practically a non-resonant reaction and its cross-section is determined by the tails of interfering resonance and sub-threshold states [64]. The low cross section and the complexity of low energy contributions to the reaction rate makes a reliable prediction difficult [65, 66].

7 Advanced stages of thermonuclear burning

As the helium burning progresses, the stellar core is increasingly made up of C and O. At the end of helium burning, all hydrogen and helium is converted into a mixture[17] of C and O, and since H, He are most abundant elements in the original gas from which the star formed, the amount of C and O are far more in the core than the traces of heavy elements in the gas cloud. Between these two products, the Coulomb barrier for further thermonuclear reaction involving the products is lower for C nuclei. At first the C+O rich

momentum l adds upto the angular momentum of the resonant state $J = j_1 + j_2 + l$. Therefore, for spin-less particles like the closed shell nuclei $^{4}He, ^{16}O$ $(j_1 = 0, j_2 = 0)$, we have $J = l$. In the entrance channel of the reacting particles, the parity would be: $(-1)^{l}\pi(j_1)\pi(j_2) = (-1)^{l=0}(1)(1)$. If the parity of the resonance state were the same as that of the entrance channel, then the resultant state would have been a "natural parity" state. However, since the 4.97 MeV state in ^{20}Ne has an assignment: $J^{\pi} = 2^{-}$, this is an "unnatural parity" state.

[17] Note however the caveat: if the amount of ^{12}C is little (either due to a long stellar lifetime of He burning or due to a larger rate of the $^{12}C + \alpha \rightarrow ^{16}O + \gamma$ reaction whose estimate outlined in the earlier section is somewhat uncertain), then the star may directly go from He-burning stage to the O-burning or Ne-burning stage skipping C-burning altogether ([67]).

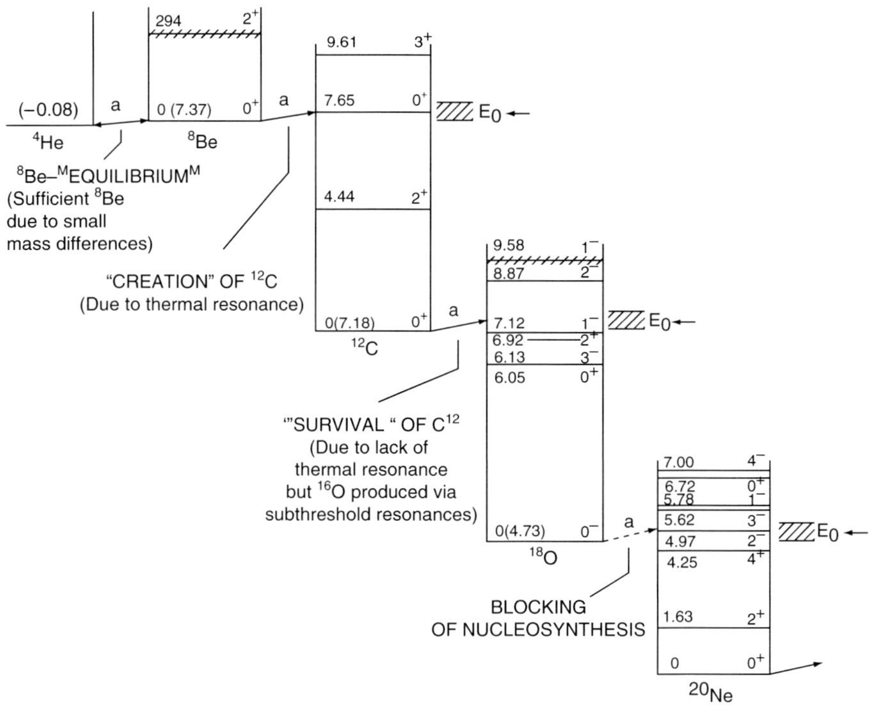

Fig. 13. Energy levels of nuclei participating in thermonuclear reactions during the helium burning stage in red giant stars (after [13]). Survival of both ^{12}C and ^{16}O in red giants, from which terrestrial abundances result, depends upon the fortuitous circumstances of nuclear level structures and other properties of these nuclei

core is surrounded by He burning shells and a helium rich layer, which in turn may be surrounded by hydrogen burning shell and the unignited hydrogen rich envelope. When the helium burning ceases to provide sufficient power, the star begins to contract again under its own gravity and as implied by the Virial theorem the temperature of the helium exhausted core rises. The contraction continues until either the next nuclear fuel begins to burn at rapid enough rate or until electron degeneracy pressure halts the infall.

7.1 Carbon burning

Stars somewhat more massive than about 3 $M_\odot$ contract until the temperature is large enough for carbon to interact with itself (stars less massive on the main sequence may settle as degenerate helium white dwarfs). For stars which are more massive than $M \geq 8-10\ M_\odot$ (mass on the main sequence, - *not* the mass of the C+O core), the contracting C+O core remains non-degenerate until C starts burning at $T \sim 5 \times 10^8 K$ and $\rho = 3 \times 10^6$ gcm^{-3}. Thereafter sufficient power is generated and the contraction stops and quiescent (hydrostatic, not explosive) C-burning proceeds (see Fig. 14).

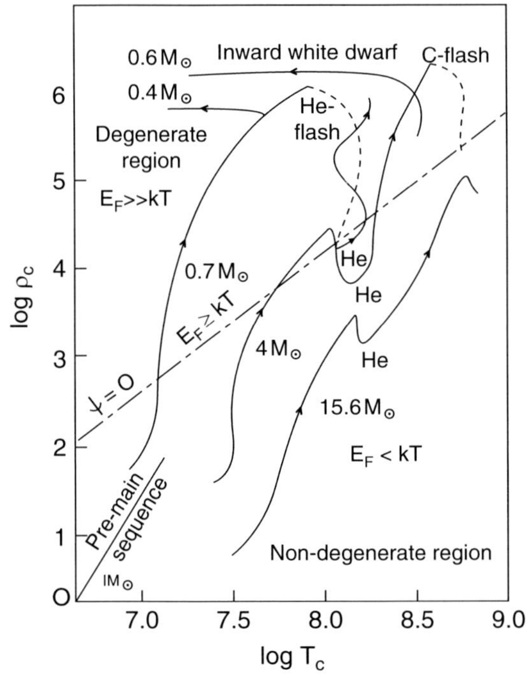

Fig. 14. Tracks in the core temperature, density plane of stars of various masses (at the start of hydrogen burning i.e. main sequence masses). Note that a star of mass $M \sim 15M_\odot$ ignites all its fuels in non-degenerate conditions, whereas a star of mass $M \sim 4M_\odot$ ignites carbon under strongly degenerate conditions. (After [68])

The combined mass of two reacting ^{12}C nuclei falls at an excitation energy of 14 MeV in the compound nucleus of ^{24}Mg. At this energy there are many compound nuclear states, and the most effective range of stellar energies (the Gamow window) at the relevant temperature is about 1 MeV; hence a number of resonant states can contribute to the decay of the compound nucleus, and even the large angular momentum resonances may be important because the penetration factors in massive nuclei are not affected by centrifugal barriers. The carbon on carbon burning can proceed through multiple, energetically allowed reaction channels, listed below:

$$^{12}C + ^{12}C \rightarrow \ ^{20}Ne + ^{4}He \ \ (Q = 4.62 \text{ MeV}) \tag{48}$$

$$\rightarrow \ ^{23}Na + p \ \ (Q = 2.24 \text{ MeV}) \tag{49}$$

$$\rightarrow \ ^{23}Mg + n \ \ (Q = -2.62 \text{ MeV}) \tag{50}$$

At the temperatures where carbon burning starts, the neutron liberating reactions requires too much particle kinetic energy to be effective. In addition, based on laboratory measurements at higher energies compared to the stellar energies, the electromagnetic decay channel ($^{24}Mg + \gamma$) and the three particle channel ($^{16}O + 2\alpha$) have lower probability compared to the two particle channels: $^{23}Na + p$ and $^{20}Ne + \alpha$. The latter two channels have nearly equal probabilities (see [14]; at the lowest center of mass energies for which cross-sections are measured in the laboratory for the proton and α channels, (i.e. about 2.45 MeV [69]), the branching ratios were $b_p \sim 0.6$ and $b_\alpha \sim 0.4$), and therefore the direct products of carbon burning are likely to be ^{23}Na, ^{20}Ne, protons and alpha particles. The rate for this reaction per pair of ^{12}C nuclei is ([70]):

$$\log\lambda_{12,12} = \log f_{12,12} + 4.3 - \frac{36.55(1 + 0.1\mathrm{T}_9)^{1/3}}{\mathrm{T}_9^{1/3}} - \frac{2}{3}\log\mathrm{T}_9 \tag{51}$$

the factor $f_{12,12}$ is a screening factor. Now, at the temperatures of ^{12}C burning, the liberated protons and alpha particles can be quickly consumed through the reaction chain: $^{12}C(p,\gamma)^{13}N(e^+\nu_e)^{13}C(\alpha,n)^{16}O$. Thus, the net effect is that the free proton is converted into a free neutron (which may be further captured) and the α-particle is consumed with ^{12}C into ^{16}O. The α-particles are also captured by other alpha-particle nuclei, resulting in, at the end of carbon burning in nuclei like: ^{16}O, ^{20}Ne, ^{24}Mg and ^{28}Si. These secondary reactions augment the energy released by the initial carbon reaction and Reeves (1959) estimated that each pair of ^{12}C nuclei release about 13 MeV of energy. Towards the end of carbon burning phase there are also other reactions such as: $^{12}C +^{16} O$ and $^{12}C +^{20} Ne$ which take place. But these are less rapid and are not expected to play major roles compared to the $^{12}C+^{12}C$ reactions, due to their increased Coulomb barriers. A recent discussion of the heavy ion reactions involving C and O is contained in [1] Sect. 3.6.

During the carbon-burning and subsequent stages, the dominant energy loss from the star is due to neutrinos streaming out directly from the stellar thermonuclear furnace, rather than by photons from the surface. The neutrino luminosity is a sensitive function of core temperature and quickly outshines the surface photon luminosity of the star at carbon burning stage. The (thermal) evolutionary time scale of the star, due to the neutrino emission becomes very short and the core evolves rapidly, – so rapidly (compared to the "cooling" time scale Kelvin-Helmholtz time: $\tau_{KH} \sim GM^2/RL_{ph}$) that the conditions in the core are "uncommunicated" to the surface, since this happens by photon diffusion. The surface conditions (e.g. the temperature) then does not change much as the core goes beyond the carbon burning stage, and it may not be possible just by looking at a star's surface conditions whether the core is close to a supernova stage or has many thousands of years of hydrostatic thermonuclear burning to go.

7.2 Neon burning

The result of carbon burning is mainly neon, sodium and magnesium, but aluminum and silicon are also produced in small quantities by the capture of α, p and n released during carbon burning. When carbon fuel is exhausted, again the core contracts and its temperature T_c goes up. At approximately $T_9 \sim 1$, energetic photons from the high energy tail of the Planck distribution function can begin to disintegrate the ^{20}Ne ash (see Fig. 13) so that one has the reaction: $^{20}Ne + \gamma \rightarrow ^{16}O + ^4He$.

Nucleons in a nucleus are bound with typical binding energy of several to 8 MeV. An energetic γ-ray photon is required to photo-eject a nucleon. Two nucleon ejection requires more energy. Alpha particles are however released at approximately the same energy as a nucleon due to the low separation energy of an alpha particle in the nucleus. For example, the alpha separation energy in ^{20}Ne is 4.73 MeV. Thus, the major photo-nuclear reactions are: (γ, n), (γ, p) and (γ, α) processes. For a photo-disintegration reaction to proceed through an excited state E_X in the mother, the decay rate is:-

$$\lambda(\gamma,\alpha) = \left[\exp\left(-\frac{\mathrm{E_X}}{\mathrm{kT}}\right)\frac{2\mathrm{J_R}+1}{2\mathrm{J_0}+1}\frac{\Gamma_\gamma}{\Gamma}\right] \times \frac{\Gamma_\alpha}{\hbar} \tag{52}$$

In the above equation, the first factor in square brackets on the RHS is the probability of finding the nucleus in the excited state E_X and spin J_R (with J_0 being the ground state spin), while the second factor $\Gamma_\alpha/\hbar$ is the decay rate of the excited state with an alpha particle emission. Now since $E_X = E_R + Q$, we have:

$$\lambda(\gamma,\alpha) = \frac{\exp(-\mathrm{Q/kT})}{\hbar(2\mathrm{J_0}+1)}(2J_R+1)\frac{\Gamma_\alpha\Gamma_\gamma}{\Gamma}\exp(-\mathrm{E_R/kT}) \tag{53}$$

At $T_9 \geq 1$, the photo-disintegration is dominated by the 5.63 MeV level in ^{20}Ne (see Fig. 13). At approximately $T_9 \sim 1.5$, the photo-dissociation rate becomes greater than the rate for alpha capture on ^{16}O to produce ^{20}Ne (i.e. the reverse reaction), thus leading effectively to the net dissociation of ^{20}Ne. The released ^4He reacts with the unspent ^{20}Ne and leads to: $^4He + {}^{20}Ne \rightarrow$

$^{24}Mg + \gamma$. Thus the net result of the photo-dissociation of two ^{20}Ne nuclei is: $2 \times {}^{20}Ne \rightarrow {}^{16}O + {}^{24}Mg$ with a net Q-value of 4.58 MeV. The brief neon burning phase concludes at T_9 close to ~ 1.5.

7.3 Oxygen burning

At the end of the neon burning the core is left with a mixture of alpha particle nuclei: ^{16}O and ^{24}Mg. After this another core contraction phase ensues and the core heats up, until at $T_9 \sim 2$, ^{16}O begins to react with itself:

$$^{16}O + {}^{16}O \rightarrow {}^{28}Si + {}^{4}He \quad (54)$$

$$\rightarrow {}^{32}S + \gamma \quad (55)$$

The first reaction takes place approximately 45% of the time with a Q-value of 9.593 MeV. In addition to Si and S, the oxygen burning phase also produces, Ar, Ca and trace amounts of Cl, K, etc. upto Sc. Then at $T_9 \sim 3$, the produced ^{28}Si begins to burn in what is known as the Si burning phase.

7.4 Silicon burning

As we have seen, most of the stages of stellar burning involve thermonuclear fusion of nuclei to produce higher Z and A nuclei. The first exception to this is neon burning where the photon field is sufficiently energetic to photo-dissociate neon, before the temperature rises sufficiently to allow fusion reactions among oxygen nuclei to overcome their Coulomb repulsion. Processing in the neon burning phase takes place with the addition of helium nuclei to the undissociated neon rather than overcoming the Coulomb barrier of two neon nuclei. This trend continues in the silicon burning phase. In general, a photo-disintegration channel becomes important when the temperature rises to the point that the Q-value, i.e. the energy difference between the fuel and the products is smaller than approximately $30k_BT$ ([71]).

With typical Q-values for reactions among stable nuclei above silicon being 8-12 MeV, photo-disintegration of the nuclear products of neon and oxygen burning begins to play an important role once the temperature exceeds: $T_9 \geq 3$. Then nuclei with smaller binding energies are destroyed by photo-dissociation in favor of their more tightly bound neighbors, and many nuclear reactions involving α-particles, protons and neutrons interacting with all the nuclei in the mass range $A = 28 - 65$ take place. In contrast to the previous burning stages where only a few nuclei underwent thermonuclear reactions upon themselves, here the nuclear reactions are primarily of a rearrangement type, in which a particle is photo-ejected from one nucleus and captured by another and a given fuel nucleus is linked to a product nucleus by a multitude of reaction chains and cycles and it is necessary to keep track of many more nuclei (and many reaction processes involving these) than for previous burning stages. More and more stable nuclei form in a nuclear reaction network as the rearrangement proceeds. Since there exists a maximum in the binding energy per nucleon at the ^{56}Fe nucleus, the rearrangements lead to nuclei in the vicinity of this nucleus (i.e. iron-group nuclei).

In the mass range $A = 28 - 65$, the levels in the compound nuclei that form during silicon burning are so dense that they overlap. Moreover, at the high temperatures that are involved ($T_9 = 3 - 5$), the net reaction flux may be small compared to the large forward and backward reactions involving a particular nucleus and a quasi-equilibrium may ensue between groups of nuclei which are connected between separate groups by a few, slow, rate-limiting reactions ("bottlenecks"). However, as the available nuclear fuel(s) are consumed and thermal energy is removed due to escaping neutrinos, various nuclear reactions may no longer occur substantially rapidly ("freeze-out"). Thielemann and Arnett [72] found that for cores of massive stars in hydrostatic cases, the bottlenecks between quasi-equilibrium (QSE) groups coincided with Z=21 nuclei whereas for lower mass stars, lower temperatures and Y_e and higher density this bridge involved neutron rich isotopes of Ca capturing protons. Hix and Thielemann [71] discussed and contrasted these results with those of earlier workers; in general the reaction flow across the boundary of the QSE groups are influenced by the neutronization of the material, i.e. the overall Y_e. It is in this context that weak interaction processes such as electron capture and beta decay of nuclei are important, by influencing the Y_e and thereby the reaction flow. These ultimately affect both the stellar core density and entropy structures, and it is important to track and include the changing Y_e of the core material not only in the silicon burning phase, but even from earlier oxygen burning phases. The calculation of stellar weak processes on nuclei has spawned extensive literature (see [73], [74], [75] etc., and for a review [76]).

Iron, nickel and neighboring nuclei which are the products of the hydrostatic Si burning in the core are mostly trapped in the collapsing core that ends up as the compact remnant and little of this may reach the interstellar medium. However, the blast wave shock launched after the core bounce impacts through the onion-like outer shells that the star's earlier evolution has left behind. The nearest part of the Si shell that is still unburned is heated to sufficient temperatures to form iron peak nuclei. Regions which are not as close can undergo incomplete Si burning and be left with substantial amounts of Si, S, Ar, Ca and Ti. Three separate outcomes may result depending upon the initial density and the peak temperature: 1) incomplete Si burning; 2) a "normal freezeout" and 3) an "α- rich freezeout". In the first case, with initial temperatures of about 5×10^9 K and peak density of $\rho = 10^9$ g cm^{-3}, significant amounts of Si and other intermediate-mass elements remain after the charged particle reactions freezeout in the expanding ejecta. In the normal freezeout, an initial condition of 7×10^9 K and peak density of $\rho = 10^9$ g cm^{-3} leads to complete Si burning. The "α-rich freezeout" however takes place at lower peak densities $\rho = 10^7$ g cm^{-3} though similar peak temperatures: 7×10^9 K. This ends up with an abundance of ^4He nuclei which produces a significant flow through the triple alpha reaction into intermediate mass nuclei, in addition to the iron group nuclei products. The feedback between the rate of nuclear recombination and that of temperature evolution in the expansion critically controls the "α-richness" of matter and production of important radioactive nuclei e.g. ^{44}Ti and is one of the challenges of computational simulation of silicon burning. For a recent discussion of the computational aspects of the nuclear evolution during silicon burning, see [77].

In summary, a few key points concerning the thermonuclear burning of ^{28}Si are as follows:-

• Direct thermonuclear fusion of two ^{28}Si nuclei does not take place because their Coulomb barrier is too high. Instead thermonuclear fusion takes place by successive additions of α-particles, neutrons and protons.

• Although this is actually a large network of nuclear reactions it is called "silicon burning" because ^{28}Si offers the largest resistance to photo-dissociation because of its highest binding energy among intermediate mass nuclei.

• The source of the α-particles which are captured by ^{28}Si and higher nuclei is ^{28}Si itself. Silicon, sulphur etc. partially melt-down into α-particles, neutrons and protons by photo-dissociation. These then participate in reaction networks involving quasi-equilibrium clusters linked by "bottleneck" links.

• Although beta decay and electron captures on stellar core nuclei do not produce energy in major ways they nevertheless play a crucial role in shifting the pathways of nuclear and thermodynamic evolution in the core conditions. These ultimately determine the mass of the core and its entropy structure which finally collapses in a supernova explosion.

8 Core collapse SNe: electron capture and neutrinos

At the end of nuclear burning in the core of a massive star, inert Iron group nuclei are left in the innermost region, – inert because nuclear burning cannot extract any further energy from the most tightly bound nuclei. At the last stages of nuclear burning, neutrino cooling which is far more efficient than the radiation of photons from the surface, would leave the core compact and degenerate [78] (where electron Fermi energy is much larger than the thermal energy kT) due to a number of processes such as e^+/e^- pair annihilation process, photoneutrino process and plasmon decay into neutrinos. These neutrinos escape from the star freely at this stage since their interaction cross section is so small, much smaller than that of the photons. In the picture of the late stages of a spherical star that $\mathrm{B^2FH}$ [79] proposed one had an onion-skin model: nested shells of progressively heavier elements, with the densest iron core at the center. Successively high-Z elements are ignited in increasingly central parts of the core in decreasingly lower specific entropy environments, and are contained within entropy barriers left behind by previous generations of nuclear burning. The degenerate core becomes unstable to gravitational collapse at roughly the same Chandrasekhar mass (modulo relatively small variations in the electron fraction or entropy profiles in the core) practically independent of the total mass of the star, which for example could range from ~ 8 M$_\odot$ to ~ 60 M$_\odot$ [67]. This "core convergence" establishes the important connection between the Chandrasekhar mass and the masses of neutron stars that are formed from core-collapse supernovae (though the details of the boundaries of the main sequence masses that lead to neutron stars, as well as the composition of the cores, e.g. O-Mg-Ne cores vs iron cores etc., have evolved as further research was undertaken). The supernova explosion itself is triggered by the collapse of the core [80], followed by the sudden release of gravitational energy. When the iron core grows in size, its temperature increases, and when $T > 7 \times 10^9 K$, iron nuclei photo-dissociate in endothermic reactions (consuming energy, even though the free energy F = U -TS decreases due to increased entropy) into alphas and neutrons. This leads to the effective adiabatic index decrease below 4/3 and the iron core becomes unstable to collapse. Collapse initiation through photo-dissociation happens in relatively massive cores of more massive stars whereas in less massive stars the collapse may initiate due to reduction of electrons (which provide the bulk of the supporting pressure due to their high Fermi momenta) as they are captured by nuclei in "inverse beta decay" in regions where the electron Fermi energy exceed the capture thresholds.

The core collapses on a dynamical scale until the infall is suddenly halted when the central density overshoots that of the nuclear matter, at $4 - 5 \times 10^{14}$gm cm^{-3}. In between, the iron core collapsed onto itself nearly freely at about a quarter of the speed of light. Initially it had the size of the Earth, but towards the end it is a hot, dense, neutron rich sphere about 30 km in radius (see [81] for a review). A static accretion shock wave forms initially at the edge of the quasi-free falling core, which soon starts propagating outward through the outer core and mantle as more kinetic energy is brought into it by infalling matter. The shock wave soon stalls (at least for most of the range of main sequence masses beyond about 10 $M_\odot$) which may however be helped after about 500 milliseconds ("after a pause that refreshes" – according to Hans Bethe) [82] by neutrinos which are freely streaming out from the inner core, but manage to deposit enough energy (albeit a small fraction of their total energy) "reviving" the shock. During this post-bounce phase the proto-neutron star radiates away most of its energy (3×10^{53} erg) in the form of neutrinos and antineutrinos of various flavors and this accounts for the larger intrinsic energy per unit mass of the central engines of core collapse SNe.

8.1 Electron capture on nuclei and protons: a core thermometer

During the gravitational collapse, the entropy of the core stays low, which permit the nuclei of various elements present in the core to (largely) survive thermal disintegration and coexist with a small fraction of "dripped" nucleons ([83] hereafter BBAL). Around the density 5×10^{11}gm cm^{-3}, neutrinos produced through electron capture on these nuclei and free protons are trapped in the core. Much of the information pertaining to the conditions in which the neutrinos are originally produced, such as the nuclear and thermodynamic properties of the core of the supernova are altered because these neutrinos undergo *inelastic* scattering with the overlying stellar matter in the post neutrino trapping phase. Neutrinos which are emitted through electron captures on the nuclei present in the pre-supernova and collapsing core *before* it reaches neutrino trapping density ($\simeq 10^{12}$ gm cm^{-3}) [84], [85], however, stream freely through the stellar matter without any further interactions. These pre-trapping neutrinos carry with them information on both physical conditions within the core, as well as it's nuclear configuration e.g. the ratio of the number density of free protons to that of heavy nuclei. The last quantity can depend on the nuclear equation of state relevant to collapse. Since neutrinos act as probes of the dynamic, thermodynamic and nuclear properties of the pre-supernova and collapsing core, their detection and measurement of energy spectrum can have significant implications. The time evolution of the detected spectrum could also reveal the dynamical time scale – a clue to the average density and mass of the stellar core which may have implications for neutron star vs black hole formation[18]. The reduction of lepton fraction during stellar collapse has implications for shock formation stages and the overall dynamics - even in the delayed explosion stage, since it determines, through the original energy of the bounce shock, and the entropy profile in the outer core, the position of the stalled shock.

The loss of neutrinos at low and intermediate densities is important in determining the saturation ratio of the lepton to baryon ratio at the time of core bounce (the leptons determine the pressure in the core whereas the baryons, mainly nucleons, determine the mass of the homologous collapsing core). Brown et al. [87] argued that the hydrodynamic collapse is nearly homologous, i.e. the density structure of the collapsing core remains self-similar throughout the collapse until the time of bounce. This greatly simplifies the study of various processes during the collapse and quantities in the core can be calculated (largely semi-analytically) through the evolution of a "mean" "one-zone" symbolizing the core properties (see e.g. [88], [89]).

The core of a massive star collapses under its own gravity when the pressure support from degenerate electrons is reduced through the capture of electrons in the stellar material. The electron capture on neutron rich heavy nuclei initially proceeds primarily through the allowed type ($\Delta l = 0$) Gamow Teller transitions. As core density exceeds $\simeq 10^{11}$gm/cm^3, the nuclei become more and more massive and too neutron rich to allow e$^-$-capture to take place through allowed Gamow-Teller transitions from the ground state. This is because the allowed states for p to n transition within the nucleus are already filled by the neutrons (neutrons shell blocked) and the transition strength for typical captures like ^{56}Fe $\rightarrow$ ^{56}Mn used earlier (as in [90])

[18] The collapse of a star could in principle also lead to a "naked singularity" [86]

is no longer representative of typical nuclear e$^-$-capture rates. It was shown [91],[92] that the dominant unique first forbidden transition strength was actually negligible compared to the thermally unblocked strength under the typical core collapse conditions. Therefore, after neutron-shell blocking sets in, (when (A,Z) > ^{74}Ge) the sum rule for the Gamow Teller transition operator $|M_{GT}|^2$ decreases from a typical value of 2.5 [90], [88] to about 0.1. The e$^-$-capture rate on a single nucleus $X(A,Z)$ in the initial state i to the final state j is:

$$\lambda_{ij} = \ln 2 \frac{f_i(T, \mu_e, Q_{ij})}{ft_{ij}} \tag{56}$$

where ft_{ij} is related to $|M_{GT}|^2$ by $ft_{ij} = 3.596 - \log |M_{GT}|^2$ for allowed Gamow-Teller type transitions (for free protons, log ft$_{f.p.}$ = 3.035). The factor $f_i(T, \mu_e, Q_{ij})$ is the phase space factor for allowed transition, which is a function of the ambient temperature T, the Fermi-energy of the electron μ_e and the Q-value for the reaction. The neutrino energy is $E_{\nu_e} = E_e - Q_{ij}$.

The change in entropy during collapse controls the fraction of the dripped protons with respect to that of the heavy nuclei and this influences the overall neutrino spectrum received on earth as the spectrum of neutrinos generated by electron capture on protons are different from captures on heavy nuclei. The received neutrino spectrum depends not only upon the initial conditions from which the collapse started, but also on the details of the electron capture properties of the stellar matter. Properties of nuclei at finite temperatures and density during this phase of the collapse, where shell and pairing corrections are relevant were computed in [94] and utilized to evolve self-consistently with the electron capture physics and the consequent changes in nuclear and thermodynamic variables.

The rate of generation of neutrinos per nucleon within energy band E_ν to $E_\nu + dE_\nu$ after accounting for the relative abundance of free protons and nuclei, is:

$$dY_\nu(E_\nu) = d\lambda_{fp}(E_\nu) X_p + d\lambda_H(E_\nu)(1 - X_n - X_p)/A \tag{57}$$

here A represents the Atomic weight of the ensemble of nuclei present in the core, taken to be represented by a single "mean" nucleus as in [90]. The differential neutrino production rates for free protons and heavy nuclei are:

$$d\lambda_{fp,H} = \frac{\log 2}{(ft)_{fp,H}} \frac{< G >}{(m_e c^2)^5} \frac{E_\nu^2 (E_\nu + Q_{fp,H}) \sqrt{(E_\nu + Q_{fp,H})^2 - (m_e c^2)^2)}}{(1 + \exp(E_\nu + Q_{fp,H} - \mu_e))} dE_\nu \tag{58}$$

where the Coulomb correction factor $< G >$ has been taken as ≈ 2 for heavy nuclei and 1 for free protons. The Q-value is given as: $Q = (\hat{\mu} + 1.297 + E_{GT})$ assuming that the strength is concentrated in a single state and here $\hat{\mu}$ $(= \mu_n - \mu_p)$ is the difference in the neutron and proton chemical potentials when free nucleons coexist with a distribution of neutron rich nuclei in nuclear statistical equilibrium, and E_{GT} is the energy of the Gamow-Teller Resonance centroid. The centroids in fp-shell nuclei, found from experimental data from (n,p) reactions have been used for characterizing GT transitions in these nuclei [93] and are close to the value (3 MeV) used here. The GT centroid in the e-capture direction is itself a function of the ambient temperatures as a quasi-particle random phase approximation (QRPA) calculation shows [95]. At the relevant high densities, when neutron rich nuclei with $A > 65$, are abundant and the electron chemical potential is noticeably larger than typical nuclear Q-values, electron capture rates are mainly sensitive to the centroid and total strength of the GT+ distributions – these are reasonably well described within the random phase approximation [96]. We note that the temperature dependence of the nuclear symmetry energy can also affect the neutronization of the stellar core prior to neutrino trapping during gravitational collapse [97], [98] since ambient temperatures can reach upto several MeV. Not only the reaction Q-values but also the equation of state of bulk dense matter, the free nucleon abundances, the degree of dissociation into alpha-particles and the nuclear internal excitations are modified by changes in the symmetry energy.

The Fermi-energy $\mu_e = 11.1(\rho_{10}Y_e)^{1/3}$ MeV. The difference in chemical potentials $\hat{\mu}$, and the relative fraction of free protons are obtained from a low density analytic equation of state similar to that in [90] with modifications noted in [88]. Shown in Fig. 15 is a typical "snapshot" spectrum of neutrinos from a 15 $M_\odot$ star's core collapse within a narrow range of stellar core density around 10^{11} gm cm^{-3}. Note the two separate peaks from captures on free protons and heavy nuclei which have non-thermal spectra.

8.2 Number of neutrinos emitted and predictions of detections

The number of neutrinos and their energy distributions that could have been detected by large underground neutrino experiments like the Super-Kamioka experiment and the Sudbury Neutrino Observatory (SNO)[19] has been calculated [99]. The spectrally integrated fluence of ν_e at a distance of 1 kpc, as Y_e changes from 0.42 to 0.39 in a 1.4 $M_\odot$ stellar core (of a 15 $M_\odot$ star) is: $F_\nu = 4.2 \times 10^{11}$cm^{-2}. The energy of the infall neutrino burst up to this stage is: $E_{\nu_e} = 7.2 \times 10^{50}$erg. The flux, direction and the spectra of the neutrinos could have been measured by the charge current dissociation of the deuterium nucleus ($\nu_e(d,pp)e^-$) in the ("classic") SNO [100] with a fiducial mass of 1 Kt of high purity D_2O. It would have been also possible to detect

[19] The original SNO experiment with heavy water has finished taking data. Thus the heavy water based predictions if a nearby supernova took place are only indicative. However there are plans of an extended SNO (called SNO+) which will use 1 kTon of liquid scintillator which will greatly reduce the lower energy threshold of neutrinos and will remain sensitive to neutrinos from a nearby SN. Moreover, after an accident the photomultiplier tubes in Super-Kamioka, the low energy thresholds originally designed for are no longer operative.

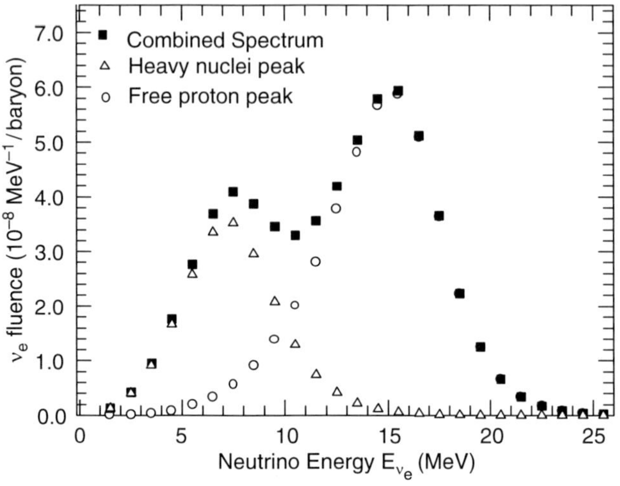

Fig. 15. "Snapshot" neutrino fluence (MeV^{-1} baryon^{-1}) in a density interval $\Delta\rho_{10} = 0.0002$ around $\rho_{10} = 9.8668$ gm cm^{-3} for a 15 M$_\odot$ star (D = 1 kpc) [99]

ν_e and obtain spectral information by means of the neutrino-electron scattering reaction ($\nu_e + e^- \rightarrow \nu_e + e^-$) in SNO as well as in the light water detector Super-Kamioka. Apart from the reactions mentioned above which can perform spectroscopy (i.e. measure the incoming neutrino energy), the neutral current dissociation of d by the reaction $\nu_e(d, pn)\nu_e$ could have obtained the total neutrino flux (of all flavors). The number of ν_e events which could have been detected in the SNO detector through neutrino-induced c.c. reaction on the target d nuclei is given by: $n_{\nu_e} = F_\nu \sigma_{cc}(\epsilon_{\nu_e}) N_d$ where N_d ($= 6.02 \times 10^{31}$) is the total number of target nuclei present in the 1Kt detector. The charge current and neutral current cross-sections ($\sigma_{cc}(\epsilon_{\nu_e})$ and $\sigma_{nc}(\epsilon_{\nu_x})$ respectively), have been computed for the ν-d process by Bahcall et al. [101] and accurate fits to these cross-sections between 5 to 40 MeV are given as [102]: $\sigma_i = \alpha_i(\epsilon_\nu - \epsilon_{th,i})^{2.3}$ where $i = cc$ and nc, $\alpha_{cc} = 1.7 \times 10^{-44}$ cm^2, $\alpha_{nc} = 0.85 \times 10^{-44}$ cm^2, $\epsilon_{th,cc} = 2.2$ MeV and $\epsilon_{th,nc} = 1.44$ MeV.

For the H_2O based Cerenkov detector (Super-Kamioka) the (ν_e, e^-) scattering events would be the main source of ν_e spectral information since the corresponding energy thresholds for charge current and neutral current interactions for ordinary water are much higher. (During the collapse stage, the neutrino flux is almost entirely in neutrinos of the electron type; anti-neutrinos of various kinds, as well as neutrinos of the mu or tau type are generated in copious numbers only in the hot post core bounce phase). The relevant (ν_e, e^-) scattering cross-section is [103]: $\sigma_e = (1/2)(4G^2 m_e^2 \hbar^2/\pi c^2)(7/12)(\epsilon_\nu/m_e c^2)$.

The number of detections by Super-Kamioka (mass 32 kt) and SNO for a supernova explosion 1 kpc away, for several possible scenarios of stellar core collapse are reported in Table 2 [99]. The 15 $M_\odot$ star collapse is initiated from thermodynamic conditions as in [88] ($Y_{ei} = 0.420$, $S_i/k_B = 1.00$, $T_i = 0.7178$, $\rho_{10} = 0.37$), while the 25 $M_\odot$ star's single zone initial conditions are similarly derived from the data reported in [104] and an expression for the core averaged entropy ($Y_{ei} = 0.423$, $S_i/k_B = 1.14$, $T_i = 0.6756$, $\rho_{10} = 0.15$).

Table 2. Pre-trapping neutrino detections predicted in SNO (heavy water) and Super Kamioka [99] with hardness ratios up to $\rho_{10} = 24.16$ for indicated heavy nuclear e-capture matrix elements for 15 $M_\odot$ Fuller (1982) and 25 $M_\odot$ WWF presupernova stars. Note the caveat in footnote 20

Star Mass	$\|M_{GT}\|^2$	$t_{collapse}$ (ms)	Pre-trapping Variables Y_{ef}	S_f/k_B	No. Detected SNO	S-K	Hardness Ratio† SNO	S-K
15 $M_\odot$	1.2/0.1	120	0.3909	1.0021	82	394	0.2786	0.8540
	2.5/0.1	120	0.3896	1.0085	66	344	0.2876	0.9537
25 $M_\odot$	1.2/0.1	190	0.3828	1.1080	120	566	0.2878	0.8319
	2.5/0.1	190	0.3813	1.1204	99	499	0.2916	0.9190

† The hardness ratio denotes the number of neutrino events in the 5 MeV $\leq E_{\nu_e} \leq$ 12 MeV and 12 MeV $\leq E_{\nu_e} \leq$ 25 MeV bands.

Neutrino spectroscopy of the final state of a star would be possible provided that the event occurs at a relatively close distance. Although, a priori, these may be rare events, there have been a number of historical Supernovae, as well as detected radio pulsars within a distance of about 2 kpc. There are a number of star forming regions nearby (such as the Orion complex - about 440 pc away), which are sites of potential supernova progenitor. A detection by underground neutrino experiments would constrain features of theoretical calculations of both collapse and explosion era of type II Supernovae as well as shed light on the characteristics of the stellar core. The 19 neutrinos detected from SN1987A were most likely to have been emitted during the post-bounce phase as their total fluence during the proto neutron star cooling phase (at $\simeq 10^{58}$) is much larger than that during the collapse phase ($\simeq 10^{56}$).

9 Detected neutrinos from SN 1987A and future neutrino watch

Gravitational collapse of the core of the massive star under its own gravity leads to a supernova explosion. These are extremely energetic explosions where the observable energy in the kinetic energy of the exploded debris and electromagnetic radiation add up to several times 10^{51} erg. The actual energy scale is typically 3×10^{53} erg or higher, but most of this is radiated away neutrinos. Although the full understanding of the process of explosion in a gravitational collapse leading to a supernova has not been achieved despite several decades of theoretical and computational work, a watershed in the field was achieved observationally when a supernova exploded close by in a satellite galaxy of our own, namely SN1987A in the Large Magellanic Cloud (LMC). A few neutrinos were detected from this supernova [105], [106], which were the first detections of the astrophysical neutrinos from outside of our solar system. By using the energetics of the neutrinos, their approximate spectral distribution, the distance to the LMC it was possible to show that the overall energy of the explosion was indeed $E_T \sim 2 - 3 \times 10^{53}$ erg.

In addition, the duration of the neutrino burst was of the order of a few seconds as opposed to a few milliseconds, signifying that the observed neutrinos had to diffuse out of the dense and opaque stellar matter as predicted theoretically, instead of directly streaming out of the core. The spectral characteristics also indicated that the object that is radiating the neutrinos was indeed very compact, roughly of the same dimensions as that of a protoneutron star. Thus SN1987A provided the observational confirmation of the broad aspects of the theoretical investigation of stellar collapse and explosion. For a review of the understanding of the astrophysics of SN1987A, see [107].

Physicists are now gearing up to detect not only another supernova in our own galaxy, but by hoping to build very large neutrino detectors, they aim to detect supernova neutrinos from the local group of galaxies ([108], [109]). As neutrinos from the supernova travel directly out from the core, they arrive a few hours ahead of the light flash from the exploding star, since light can only be radiated from the surface of the star, and it takes the supernova shock launched at the deep core several hours to reach the surface. In the case of SN1987A, this time delay allowed the estimation of the size of the star that exploded. Thus some advance warning of the supernova in the optical band can be gotten from a "neutrino watch". Physicists have now connected a worldwide array of neutrino detectors through the Internet (SN Early Warning System or SNEWS[20]) which will notify astronomers to turn their optical, UV and other telescopes to the right direction when they find a burst of neutrinos characteristic of a supernova explosion.

10 What X-ray spectroscopy reveals about nucleosynthesis in SNe and SNRs

X-rays from a supernova explosion arise from the interaction of the supersonic ejecta with the circumstellar medium (CSM). The CSM typically consists of a slow-moving wind. When the ejecta collides with the CSM, it creates two shocks: a high-temperature, low-density, forward-shock ploughing through

[20] See the site: http://hep.bu.edu/~snnet/

the CSM (known as blast-wave shock) and a low-temperature, high-density, reverse-shock moving into the expanding ejecta. Initially the X-rays come from the forward-shocked shell dominated by continuum radiation, but after a few days X-rays arise also from the reverse-shock, which can have substantial line emission, thus providing nucleosynthetic fingerprints of the ejecta. The temporal evolution of the X-ray luminosity of a supernova can yield information on the density distribution in the outer parts of the exploding star ($\rho \propto r^{-n}$, here n can be in the range $7 - 12$, typically $n \sim 10$ for a Blue Supergiant (BSG) and $n \sim 12$ for a Red Supergiant (RSG)- see [110]). These studies are therefore of interest to stellar structure and evolution.

To date thirty-six supernovae have been detected in the X-ray bands[21]. Among these the most extensively studied is SN 1987A because it was so bright. Another extensively studied object, of the supernova remnant (SNR) kind is Cassiopeia A. A SNR is just an older supernova after it has picked up enough material from the surrounding medium, which slows it down. The debris of the explosion, or the ejecta, radiates when it is reheated after an initial cooling, having been hit by a inward propagating reverse shock.

10.1 Supernova Remnant Cassiopeia A

Cassiopeia A (Cas A for short and also known as 3C461 or G111.7-2.1) is the second youngest SNR in our Milky Way galaxy. It was believed to be the product of a SN explosion in $\sim$ 1672 [112] only about 3.4 kpc away [113]. The British astronomer John Flamsteed may have recorded it as a sixth magnitude star in 1680, but it may have faded rapidly after explosion which could have acted against its widespread reportage. It is a shell type SNR which was rediscovered by radio astronomers Ryle and Smith [114] and is the brightest extra-solar radio object. Cas A is an oxygen-rich SNR with heavy element distribution typical of a core-collapse SN. Its closeness, young age and high brightness across the whole electromagnetic spectrum have underscored its importance for studying supernovae[22]. A recent optical spectrum of the original supernova near maximum brightness, obtained from scattered light echo more than three centuries after the direct light of the explosion was received on Earth, showed it to be a type IIb SN [116][23], somewhat like the well studied

[21] See S. Immler's X-ray supernova page at http://lheawww.gsfc.nasa.gov/users/immler/supernovae_list.html and [111].

[22] Here we discuss thermal emission from Cas A, which directly connects to nucleosynthetic products. Many shell-type SNRs, including Cas A, also show non-thermal synchrotron X-rays - see [115].

[23] Krause et al. note that even with the overall lack of hydrogen emission in most knots and the nitrogen enrichment in the remnant which is widely interpreted as signatures of the collapse of a Wolf-Rayet star, i.e. a type Ib SN [117], Cas A cannot be classified as arising out of a type Ib SN, since Cas A light echo spectrum does not match well the spectrum of the prototype SN 2005bf.

prototype SN 1993J[24] [123]. The estimated mass loss rate of 2×10^{-5} $M_\odot$ yr^{-1} and a wind velocity of 10 km s^{-1} consistent with the hydrodynamical state of the Cas A remnant [124] is also similar to what has been interpreted from radio and X-ray observations of SN 1993J [125], [126].

SNR Cassiopeia A has a bright clumpy emission ring with a diameter of about $3'$. This is associated with the SNR ejecta, whereas at an outer diameter of about $5'$, there is fainter emission in a filamentary ring which is due to the forward ("blast-wave") shock in the circumstellar medium. The SNR has a nearly circular appearance, though the ejecta shows a bipolar structure, and jets (mainly in the NE, i.e. upper left of the image and a fainter extension in the SW direction, called the "counterjet") that were formed by the explosion (detected both in the optical [127] and in the Si line X-ray emission [129], [128]). The remnant has been extensively observed in the optical and more recently with a million second exposure of the Chandra X-ray Observatory [128] (see also XMM-Newton [130]) and includes high resolution grating X-ray spectroscopy with Chandra [131] as also the Spitzer Space Telescope using the Infrared Array Camera [132] and the IR Spectrograph [133].

Many of Cas A's optically bright knots have been identified as shocked ejecta which is still visible due to its young age. The optical radiation emitting regions have been classified into two groups: the so called fast moving knots (or 'FMK's with speeds 4000 km s^{-1} < v < 15000 km s^{-1}) and the slow moving quasi-stationery flocculi (or 'QSF's with speeds $v < 300$ km s^{-1}). The FMKs believed to comprise of SN ejecta, have H-deficient emission (lacks in $H\alpha$) which is dominated by forbidden lines of O and S emission [134] while the QSF emission is rich in N, and is believed to originate from the gas lost from the star before its explosion (circumstellar envelope) which is now hit by the blast wave shock [131].

Cas A shows the nucleosynthesis products of both hydrostatic and explosive nucleosynthesis. C and O are produced in He burning, Ne and Mg first appears through C burning, and O and Al are added with Ne burning [135]. Spectral lines of heavier elements: Si, S, Ar, Ca etc. are seen. These are produced in O burning, with alpha reactions on Mg also contributing, – they are from zones where explosive O-burning and incomplete explosive Si-burning occurs. Fe group elements are produced in the "Si burning" chain of reactions and result from complete and incomplete Si-burning. Much of the layered

[24] The collapse of a red supergiant [118] in a binary system [119] with the larger star with a mass on the main sequence of $\sim$ 10 to 20 led to SN 1993J and could explain its light curve and other characteristics [120], [121]. Photometry and spectroscopy at the position of the fading SN with Hubble Space Telescope and Keck Telescope, a decade after its explosion showed the signature of a massive star, – the companion to the progenitor that exploded. While the binary system initially consisted of e.g. a $15M_\odot$ star and $14M_\odot$ star in an orbit of 5.8 yr, at the time of the explosion, the primary had a mass of $5.4M_\odot$ (with a helium exhausted core of $5.1M_\odot$) and a secondary which gained mass in transfer and which we still see today, ended up with $22M_\odot$ [122].

nucleosynthetic structure has been preserved during the explosion process in Cas A, with layers of N-, S-, O-rich ejecta seen beyond the outer shock [134]. However, in the optical bands, "mixed emission knots" showing N and S lines suggest that clumps of high speed S ejecta have penetrated through outer N-rich layers [134]. X-ray line data show that for example, in the southeast region of the SNR, Fe is farther out and is moving faster than the Si/O region. Hughes et al. [136] consider the high surface brightness knots enriched in Si and S are consistent with nucleosynthetic products of explosive O burning, whereas elsewhere, the more diffuse, lower surface brightness features with enhanced Fe could be the result of explosive Si burning.

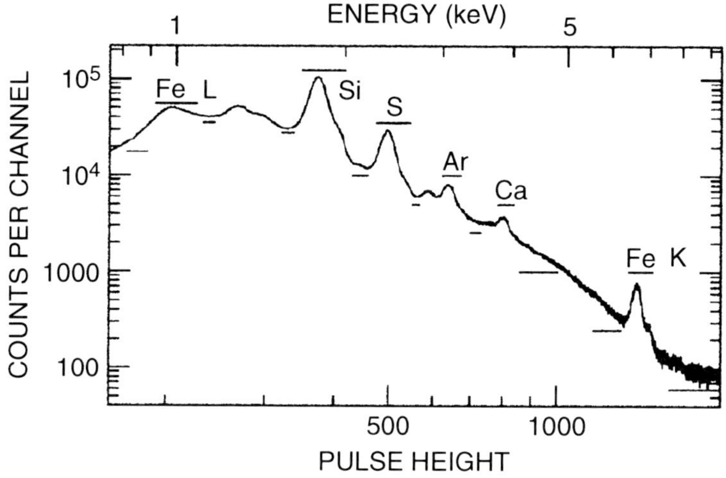

Fig. 16. Spectrum of the entire supernova remnant Cassiopeia A taken by Chandra (AXAF CCD Imaging Spectrometer (ACIS)) on January 30-31, 2000 against pulse height (lower x-axis) in the spectral channel and photon energy (upper x-axis) using a gain value of 4.8 eV per channel (Fig. reproduced by permission of the AAS and courtesy of U. Hwang [129])

The "non-dispersed"[25] X-ray CCD spectrum for the entire Cas A remnant is shown in Fig. 16. A 50 ks exposure in January 2000 collected approximately 16 million photons during the observation of the SNR. It shows prominent lines of Si, S, Ar, and Ca, in their Helium-like ionic state and undergoing ($n = 2 \to n = 1$) transitions as also the L and K transitions of Fe.

Maps were constructed of the entire remnant in the spectral line energies of several specific elements. The Si and S X-ray line maps and the optical

[25] This means the X-ray beam did not pass through any dispersive element like a grating, but the spectrum is obtained by measuring the energy of the incident X-ray by the intrinsic spectral sensitivity of the CCD.

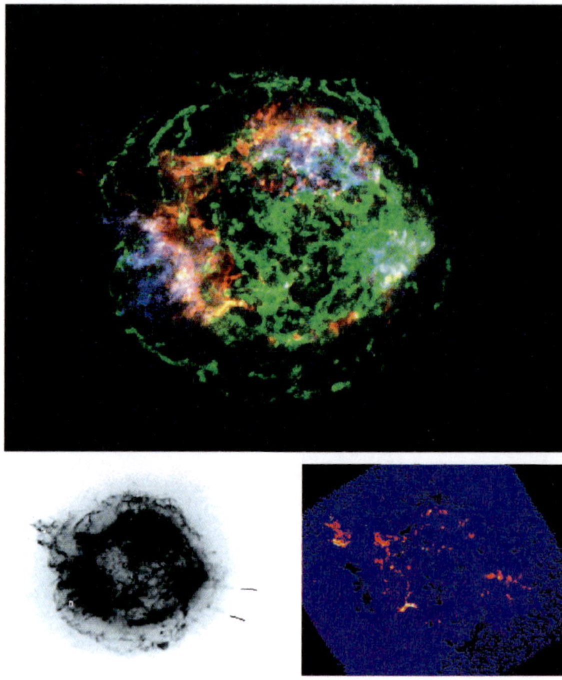

Fig. 17. Top: Three-color image of Cas A with red = Si Heα (1.78–2.0 keV), blue = Fe K (6.52–6.95 keV), and green = 4.2–6.4 keV continuum. The remnant is roughly 5' across. Bottom left: Overexposed broadband image showing faint features. The spectral regions are indicated (top left of this box: northeast jet; bottom left of this box: Fe-rich region; lines at bottom right point to two southwest jet filaments). Bottom right: On the same scale, the ratio image of the Si Heα (1.78–2.0 keV) and 1.3–1.6 keV (Mg Heα, Fe L), without subtraction of the continuum contribution. The image highlights the jet and counterjet traced by Si emission, although features at the lowest intensity levels are uncertain. (Figure reproduced by permission of the AAS and courtesy of U. Hwang [128])

maps resemble each other and this demonstrated that the X-ray and optically emitting ejecta are largely spatially coexistent. In fact the X-ray Si and S line emission, expected to include a significant contribution from the ejecta, shows asymmetric Doppler shifts corresponding to bulk velocities of ~ 2000 km s^{-1} which are comparable to those of the optical ejecta knots [129]. Si, S, Ar, and Ca have similar general morphologies, but examination of their images in their line energies also show significant variations. At the positions of the brightest knots in the Si image at the inner boundary of the shell to the northeast and southeast the corresponding Ar and Ca X-ray images are much weaker in the remnant. Hughes et al. [136] using the Chandra first light observation of Cas A ($\sim 5000s$ on 1999 August 20) showed that the Fe-rich ejecta lie outside the S-rich material, and claimed that this is due to extensive, energetic bulk motions which caused a spatial inversion of a part of the supernova core. However, on the basis of Spitzer Space Telescope imaging and spectral data Rudnick and collaborators [137], [133] claim that there are two roughly spherical shocks, (the blast wave and the reverse shocks) but "the wide spatial variations in composition seen in infrared, optical and X-ray studies are not due to local differences in ionization age or temperature, but instead reflect very specific asymmetries in the geometry of the underlying explosion. These asymmetries are ones in which the nucleosynthetic layers have remained mostly intact along each radial direction, but the velocity profiles along different radial directions vary over a range of approximately five" (see also [132]). In some directions, only the upper C-burning layers have been probed by the reverse shock, while in other directions, deeper O- and Si-burning layers have reached the reverse shock.

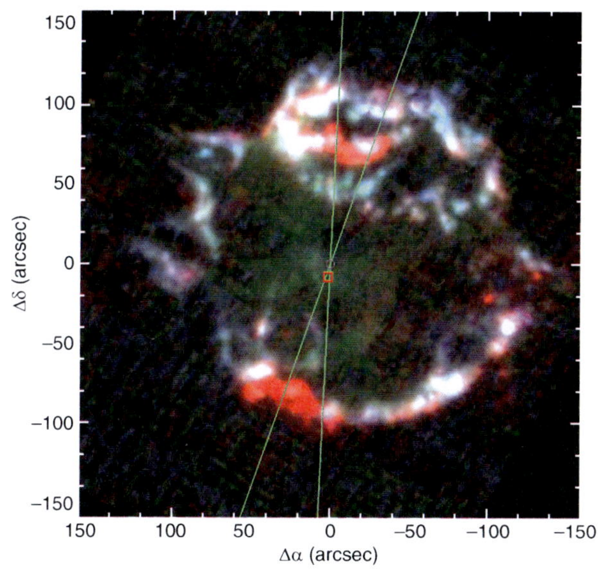

Fig. 18. Composite Spitzer image of Cas A in [Ar III] (blue), [S IV] (green), and [Ne II] (red), showing two pronounced neon-rich crescent-shaped regions to the N and S. The size of the image is 320" with north to the top and east to the left. Shown are the kinematic center of the remnant at the magenta circle and X-ray localization of the remnant's compact object (red square, 7" to the south). Green lines indicate the 1σ range of the "kick vector" direction of the compact object from the ejecta's expansion center: $169^o \pm 8.4^o$ [112]. The two regions of enhanced neon abundance lie very close to this projected direction. Several much smaller neon-enhanced regions lie in the West along the X-ray jet direction (Fig reproduced by permission of AAS and courtesy of J. Smith [133])

The gas infrared lines of Ar, Ne, O, Si, S and Fe are seen at different locations, along with higher ionization states of the same elements visible in the optical and X-ray lines. For example, strong [Ar II] regions in the IR match very closely the places where helium- and hydrogen-like ionization states of Si and S are seen in the X-ray and [S III] seen in the optical [134]. These are the same regions with multiple temperatures and ionization states but show products of oxygen burning instead of carbon burning. Other regions, for example the crescent-shaped green feature slightly east of south in the Spitzer IRAC line and continuum images depicting gas in Fig 2 of [132] show strong [Ne II] emission in the IR (see also the red regions in the image Fig 18 reproduced from [133]), increasing [O III] emission in the optical and a gap in the silicon dominated X-ray emission, are interpreted as the locations where carbon-burning layers are presently encountering the reverse shock. Different layers containing various types of nucleosynthetic products seems to show the presence of characteristic types of dust, e.g. the deep layers contain silicates while the upper layers contain dust dominated by Al_2O_3 and carbon grains. They also find evidence for circumstellar dust heated by the blast wave shock [137].

Smith et al. [133] identify IR line emission from ejecta materials in the interior, prior to their encounter with the reverse shock, as well as from the post-shock bright ring. There is a dramatic increase in the electron density (by a factor ≥ 100 to $\sim 10^4\text{cm}^{-3}$) as well as a concomitant change in the ionization state of the ejecta as it encounters the reverse shock. There is a clear layering of ionization state, from low ionization species in the interior, e.g. [Si II] (8.2 eV), higher energies on the IR bright ring [S IV] (34.8 eV) where optical emission is also seen, and very high energies traced by X-ray line emission from H-like and He-like K-alpha resonance lines of Si XIII (0.5 keV) and Si XIV (2.4 keV), extending beyond the IR-bright rim. In addition, they find two compact, crescent shaped clumps with high neon abundance (mentioned above) which are arranged symmetrically around the central neutron star, and the crescent regions are closely aligned with the kick direction of the neutron star from the remnant's expansion center. These regions contain a huge amount of ionized neon (dominated by Ne^+ and Ne^{++}, and excluding neutral neon), almost 1.8×10^{-4} $\text{M}_\odot$, flowing outwards 20 degrees from the plane of the sky at roughly -5500 km s^{-1} in the south and $+4200$ km s^{-1} in the north, while the entire SNR may contain an ionized neon mass $\sim 8.6 \times 10^{-4}$ $\text{M}_\odot$. Smith et al. comment that the *apparent* macroscopic elemental mixing mentioned in [136], may actually arise from different compositional layers of ejecta passing through the reverse shock at present along different directions.

X-ray grating spectra of Cassiopeia A and SN 1987A

In the optical bands, thousands of individual knots have been observed to yield kinematic information, whereas the X-rays probe more dynamically important information since a much larger fraction of the ejecta mass is probed by X-rays

($\sim 4M_\odot$ as compared to $< 0.1M_\odot$ in the optical). The Chandra High Energy Grating Spectroscopy (HETGS) [131] provided for the first time the high spatial and spectral resolution X-ray map of Cas A, comparable to what is achievable in the optical. Because of Cas A's bright emission lines and narrow bright filaments and small bright clumps that stand out against the diffuse, continuum emission, this grating X-ray spectroscopy was possible and yielded rich information about the kinematic and plasma properties of the emitting knots. The HETGS has two grating arms with different dispersion directions, with the medium energy range (MEG) covering the range 0.4 – 5.0 keV and spectral FWHM of 0.023 Å, while the high energy range (HEG) covers 0.9 – 10.0 keV and spectral FWHM 0.012 Å. The grating spectra were obtained by Lazendic et al. on some 17 different positions on the Cas A image, mainly near the reverse shock regions which are X-ray bright.

As mentioned already, Cas A image in the HETGS both in the non-dispersed (zeroth-order) as well as dispersed images in different energy bands contain H- and He-like ionization states of O+Ne+Fe (0.65-1.2 keV), Mg (1.25-1.55 keV), Si (1.72-2.25 keV), S (2.28-2.93 keV), Ar (2.96-3.20 keV), Ca (3.75-4.0 keV). In addition images in the Li- and Be-like states of Fe K lines (6.3-6.85 keV) are also obtained. A key advantage of high resolution grating spectroscopy over the low resolution "non-dispersive" CCD spectroscopy is that the former can be used to resolve individual lines and thereby interpret conditions in the radiating plasma in various parts of the SNR. The He-like ions of Si and S and other elements present in Cas A are the dominant ion species for each element over a wide range of temperatures and they emit strong K-shell lines of these ionic stages. For typical plasma densities in the SNR, the He-like triplet of Si and S lines shows strong forbidden (f) and resonance lines (r) and a comparatively weaker intercombination line (i) (see Sect. 4.5 of [138] for an illustration of these transitions in O-ions). The G-ratio $= (f+i)/r$ is, for example, a good diagnostic of the electron temperature [139]. Thus the measured line ratios and abundance ratios of H- and He-like ions of the same element (say Si) were used [131] to measure the electron temperature kT_e and the ionization time scale $\tau = n_e t$ (n_e being the electron density and t being for example the time since the region was hit by the shock). The results of [131] show that for most of the selected regions, resolved spectroscopy of Si He-like triplet lines and Si H-like lines yields temperatures around ~ 1 keV, consistent with reverse shocked ejecta. However, two regions, (designated R8 and R10) had significantly higher temperatures ~ 4 keV, and could be part of the circumstellar material. The dominant element in Cas A seems to be Oxygen, and the electron density in the X-ray line emitting regions seem to vary between 20 to 200 cm^{-3}, which is more typical of the unshocked interior of the IR line-emitting plasma, than that at the shock front as shown in [133] (see above). Lazendic et al. had derived unambiguous Doppler shifts for their selected 17 regions. While the SE region of the SNR show mostly blue shifted velocities reaching upto -2500 km s^{-1} (compare the IR line velocities in the south -5500 km s^{-1}), the NW side of the SNR had extreme red shifted values up to $+4000$ km s^{-1} (compare IR line velocity in the north: $+4200$ km s^{-1}).

Deep high resolution X-ray spectroscopy of SN 1987A was undertaken with Chandra HETG, 20 years after its explosion [140]. The impact of the SN ejecta with the circumstellar medium dominates the observed luminosity, the X-ray luminosity having brightened by a factor of 25 from that in 1999 and is presently increasing at the rate of 40 % per year [141]. An expanding elliptical ring is seen in the X-ray image whose brightness distribution seems to correlate with the rapidly brightening optical hot spots on the inner circumstellar ring observed with Hubble Space Telescope. The HETG spectrum for a total life time of 355 ksec shows H-like and He-like lines of Si, Mg, and Ne, and O VIII lines and bright Fe XVII lines. Fig. 19 taken from Dewey et al. [140] shows the resolved Si and Mg triplets of (r, i, and f lines); the data have similarity with model G-ratios $(f + i)/r$ mentioned earlier. Since the dispersed X-ray spectrum is a convolution of the spatial structure of the X-ray image and the motion of the X-ray emitting gas, Chandra's sub-arc second resolution was important to resolve the circumstellar ring of dimensions 1.2" × 1.6".

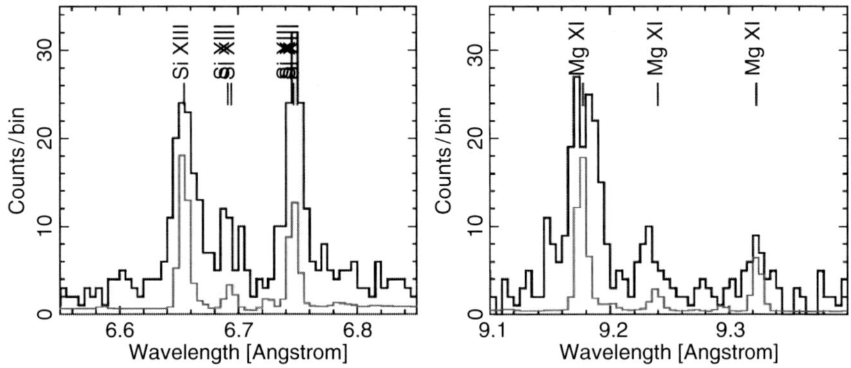

Fig. 19. Si (left panel) and Mg (right panel) triplets (resonance, intercombination and forbidden lines) resolved by Chandra High Energy transmission Grating Spectroscopy. The data are shown by the solid black histogram and an arbitrarily scaled point-source version of the two-shock model (no spatial-velocity effects) is shown in red to suggest the relative similarity of the data and model G-ratios [(f+i)/r]. (Fig reproduced by permission of the AAS and courtesy of D. Dewey [140])

The global fit to the HETGS data with a two shock model yielded element abundances and absorption column densities which are consistent within the same 90% confidence limits that were derived from LETGS by [142]. The lower temperature shock seems to give the same kT_{low} for all the LETG data sets (in 2004 and 2007) and the HETG observation, whereas the kT_{high} values seem to show a general evolution towards lower values among the grating datasets and similar trend seen in the ACIS monitoring of SN 1987A.

The HETG data implies a relatively low bulk radial velocity of the shocked gas in the ring compared to model expectation from a plane parallel strong shock entering stationary gas and the temperature range inferred from the spectral modeling of the emitting gas. Since the X-ray image is correlated with the optical hot spots, it appears that the blast wave ahead of the SN ejecta is overtaking dense clumps of circumstellar gas in the hot spots. The blast wave may be encountering the dense clumps either at normal incidence or at oblique incidence. In the former case, the reflected shock would give rise to gas that has been shocked twice and having nearly stationary bilk velocity but further elevated temperature. With more and more X-ray emission coming from gas behind reflected shocks, there would be an increase of the fraction of higher temperature X-ray emission, consistent with what is seen. If the blast wave encounters dense clumps at an oblique incidence, the shocked gas will have significant velocity component parallel to the shock surface, and a significantly fast transmitted component emitting X-rays can result. The Doppler broadening seen in the line profiles of the Chandra data seems to suggest that both transmitted and reflected shocks encountering the circumstellar ring at normal and oblique angles may be at play [140].

10.2 Live radioactive decays in Cas A, SN 1987A

Gamma-ray, X-ray, optical and infrared line spectroscopy of SNe and SNRs have been used to observe nucleosynthesis and the abundance distribution of elements freshly synthesized (both radioactive isotopes and stable decay products) and to extract dynamical information about the explosion. In particular the radioactive isotopes provide unique tracers of nucleosynthetic processes (what, where and how much) and its related dynamics. The best examples are the observations of gamma-ray lines in supernovae, but X-ray line spectroscopy of decay products of radioactive nuclei have also been attempted, and specific elements in numerous SNRs and a few SNe identified.

After a few days, the main energy input to the SN ejecta comes from radioactive decay. For SN 1987A, at first, the important isotopes are ^{56}Ni ($\tau_{1/2} = 6.1d$) followed by ^{56}Co ($\tau_{1/2} = 78.8d$). Beyond $\sim$ 1100 days, ^{57}Co is more important and at $\geq 2000d$, the dominant role is played by ^{44}Ti. Using bolometric and broad-band UBVRI light curves [143] estimated the masses of the three most important radio-isotopes in SN 1987A to be $^{56}Ni(0.069 \pm 0.003 M_\odot)$, $^{57}Ni(0.003 M_\odot)$ and $^{44}Ti(1 \pm (0.5-2) \times 10^{-4} M_\odot)$.

Among the radioactive isotopes accessible to gamma-ray astronomy, ^{44}Ti is a key isotope for the investigation of the inner regions of core collapse SNe and their young SNRs. It has a half-life of 58.9 $\pm$ 0.3 yr [144] and a decay scheme: $^{44}Ti \rightarrow ^{44}Sc \rightarrow ^{44}Ca$. The discovery [145] of the 1157 keV gamma-ray line emission[26] from Cas A with the Compton Gamma Ray Observatory was the first direct proof of synthesis of this short lived, freshly

[26] $J^\pi = 2^+ \rightarrow 0^+$ transition in ^{44}Ca reached from ^{44}Sc following electron capture.

made radio-isotope in SNe. Renaud et al. [146] using the INTEGRAL spacecraft recently reported the detection of both 67.9 and 78.4 keV gamma-ray lines of ^{44}Sc in Cas A. There was a clear separation of the two lines due to an improved detection of the hard X-ray continuum up to 100 keV. The line flux of $(2.5 \pm 0.3) \times 10^{-5}$ cm^{-2} s^{-1} leads to a tightly constrained ^{44}Ti mass of $(1.6^{+0.6}_{-0.3}) \times 10^{-4} M_{\odot})$. This is actually high compared to the predictions of the standard models [135], [147], or their improved versions [148], [149]. Since the production of ^{44}Ti is sensitive to the explosion energy and asymmetries and Cas A is known to be asymmetric and energetic (2×10^{51} erg instead of the standard 1×10^{51} erg), this could be a factor in its apparent overproduction compared to models. At the same time, the recent revision of the $^{40}Ca(\alpha, \gamma)^{44}Ti$ reaction rate [150] has led to an increase of ^{44}Ti production by a factor of ~ 2.

Live radioactive isotopes freshly synthesized in the explosion were detected from SN 1987A by directly detecting 1238 keV and 847 keV gamma-ray lines [151], which provided detailed diagnostics of the nuclear burning conditions and the explosion dynamics. Another radio-nuclide ^{57}Co, which is a decay product of ^{57}Ni made in the supernova explosion, was also detected directly by measuring gamma-ray lines [152]. The calculated gamma-ray light curves for 847 keV and 1238 keV and other lines [153] could be made consistent with the Solar Maximum Mission (SMM) measurement only if the ^{56}Co was mixed up to a mass coordinate of $M_r \sim 13.5 M_{\odot}$. The calculated fluxes were still slightly smaller than the observed fluxes at $t \sim 200d$, but this discrepancy could be removed by taking account of the effect of clumpiness in the ejecta which is more significant in the earlier phases. Also, the observed flux ratio between the two gamma-ray lines at 847 keV and 1238 keV is close to unity at early stages because of the smaller cross section for the 1238 keV lines than for 847 keV. As the column depth of the overlying matter decreased due to dilution with time in an expanding envelope, the observed flux ratio approached the expected value of 0.68 based on branching ratios [153].

X-ray fluorescence spectrometry technique has been used by archaeologists to determine the major and trace elements in ceramics, glass etc. Here, a sample is irradiated with X-rays and the wavelengths of the released (fluorescent) X-rays are measured. Since different elements have characteristic wavelengths, they can be identified and their concentrations estimated from the intensity of the released X-rays. Trace elements analysis may, for example, help in identifying the (geographical) source of a material. A similar method has been proposed [154] and attempted for SN 1987A [155]. It utilizes electron capture decays of freshly synthesized radio-nuclei. Following the electron capture, the K-shell vacancies are in most cases filled in by downward transitions from other bound shells. As the fluorescence X-ray yields are well known, measuring the X-ray line fluxes can then estimate the number of freshly synthesized nuclei. For example, attempts were made to detect the 5.9 keV Kα line from the stable nucleus ^{55}Mn in the Chandra X-ray spectrum of SN 1987A, which is due to the decay of radioactive ^{55}Fe. Both $^{55}Fe(\tau_{1/2} = 2.7$ yr) and

$^{55}Co(\tau_{1/2} = 18$ hr) are produced in explosive, incomplete Si burning as well as in normal freezeout of nuclear statistical equilibrium, in the inner ejecta of core collapse supernovae. However, no evidence of the 5.9 keV line emission from Mn could be found in 400 ks of Chandra ACIS data and the upper limit to the mean flux was $< 3 \times 10^{-7}$ cm^{-2}s^{-1}. Rauscher et al. [148] calculated the ejected mass of $A = 55$ radioactive nuclei to be 7.7×10^{-4} $M_\odot$ for $20 M_\odot$ models of which most was ^{55}Co. If only about half this mass of ^{55}Fe were ejected, the reduced flux would be consistent with the observed upper limit. On the other hand, even if the total mass inside were as much as 1×10^{-3} $M_\odot$, but the ^{55}Fe abundance was zero outside the radial velocity shells at 1500 kms^{-1}, the line flux would be still consistent with data, as at late times the emerging flux depends sensitively on the presence of ^{55}Fe in the outer zones.

10.3 Other X-ray supernovae

Apart from SN 1987A, which occurred close by and was therefore easy to detect, there are supernovae with relatively high intrinsic X-ray luminosity. SN 1993J and SN 1995N are two of them and they are at the high end of the X-ray luminosity [156], [157], which makes it suitable for study in the X-ray wave bands even at late stages [159]. Nymark et al. [158] have shown that SNe with strong X-ray emission are likely to have radiative shocks and that in these shocks a large range of temperatures contribute to the spectrum. The cooling of the shock also affects the hydrodynamics of the flow structure and the volume of the emitting gas and thus the total luminosity from the interaction region. A shock is radiative depending on whether the cooling time scale t_{cool} of the shock is short compared to its expansion time t. Since the cooling function (from electron bremsstrahlung or free-free scattering and X-ray line emission from bound-bound transitions of ionized atomic species) is a function of the shock temperature this condition translates to a condition on the shock temperature for radiative vs. adiabatic shock. The boundary of the two regimes depends upon the shock velocity V_s and the ejecta density profile (given by the index η in $\rho \propto v^{-\eta}$) and the separation in turn depends upon the chemical composition of the ejecta (see Fig. 20 reproduced from [160]). For a given shock velocity (and a composition and mass loss parameter of the progenitor stellar wind and the ejecta velocity scale), the shock is likely to be radiative for steeper density gradients.

As already mentioned, the spectroscopic type of SN 1993J from a type II-like SN at early times due to detected Balmer lines soon weakened and it came to resemble a type Ib SN, leading to a reclassification as a type IIb SN [161]. This transition is best explained if the progenitor star prior to its explosion, had lost its hydrogen envelope due to interaction with a binary companion [120], [121], [162], [163], [122]. If the H-envelope was indeed already thin, then the reverse shock would transit the H-rich region quickly, and soon the emission may be dominated by material dredged up from the nuclear processed parts of the star in the interior. In the 4H47 model of Shigeyama et al. [164], by

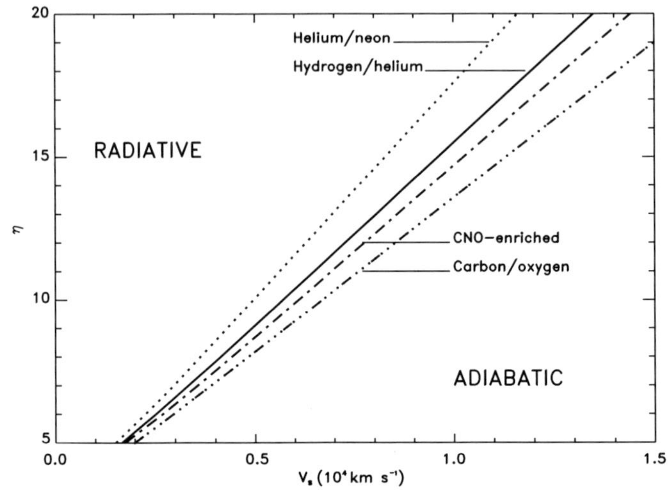

Fig. 20. The border between radiative and adiabatic regions of the shocks in supernovae as a function of shock velocity and ejecta density profile index η ($\rho \propto v^{-\eta}$). The boundaries depend upon the composition of the ejecta through which the shock may be propagating as also the mass loss parameter of the progenitor stellar wind and the ejecta velocity scale (Fig. courtesy T. Nymark [160])

2600 days after explosion, the hydrogen-rich envelope had $\sim 0.47 M_{\odot}$. Inside this envelope is the outer part of the He-zone which is enriched in N (about $\sim 0.3 M_{\odot}$), while its inner part is C-rich (has $\sim 1.5 M_{\odot}$). If the circumstellar medium or the ejecta is clumpy, they may be hit by the reverse shock obliquely and the shock in the clump will be slower and the shock temperature lower. The presence of both hard X-rays and optical emission from the shocked CSM shows the importance of a clumpy CSM, for example in SN 1987A where a range of shock speeds are necessary to explain the observed spectrum [142]. SN 1993J is not as bright as SN 1987A or Cas A due to its larger distance, and both Chandra and XMM-Newton data is comparatively of a moderate signal to noise, at the low (CCD) resolution. The XMM spectrum of SN 1993J is dominated by the Fe L emission at 0.7 - 1.0 keV and is blended with strong Ne IX-X emission. Above 1.0 keV, emission lines which may be present are from Mg XI-XII, Si XIII-XIV, as well as S XVI. Nymark et al. find that the spectrum of SN 1993J is fit best by a combination of an adiabatic shock ($kT_{rev} = 2.1$ keV) propagating through a zone with CNO burning products, with He dominant and N and Ne being the most abundant metals, and a second radiative shock ($kT_{rev} = 1.0$ keV). The latter could have been caused by instabilities at the reverse shock front or by a clumpy CSM. The XMM spectrum of SN 1993J and the model fits used [160] are shown in Fig 21.

SN 1995N is a type IIn supernova, a type that shows unusual optical characteristics and spans a very broad range of photometric properties such as

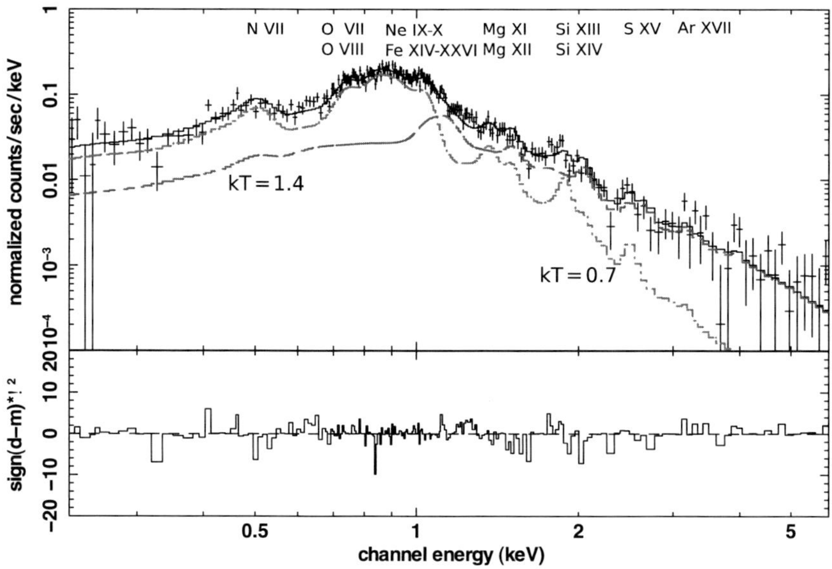

Fig. 21. XMM spectrum of SN 1993J (shown by data-points with crosses from April 2001). The data is fitted with an adiabatic shock at $kT_{rev} = 2.1$ keV (long dashed in blue) and a radiative shock at $kT_{rev} = 1.0$ keV (short dashed in red) for a He/N dominated composition from model s15s7b [163]. (Fig. courtesy T. Nymark [160])

decline rates at late times [165]. It is likely that these differences are related to their progenitor's structure, mass, composition as well as the composition and the density profile of the CSM [166]. These supernovae show the presence of strong, narrow Balmer line emission on top of the broader emission lines in their early spectra. The narrow emission lines may originate in the dense and ionized circumstellar (CS) gas [167, 168]. The presence of strong Hα emission line, the high bolometric luminosity and the broad Hα emission base powered by the interaction of the supernova shock with the CSM, all point towards a very dense circumstellar environment [169]. When the stellar core collapses and explodes, the supernova lights up the slow-moving gas into narrow emission lines leading to the type IIn supernova classification (n for narrow emission line). This interaction of the supernova shock with the dense CS gas is indicated by strong radio and X-ray emission detected from several type IIn supernovae and in particular SN 1995N.

Chandra ACIS observations of SN 1995N show a best-fit line energy at 1.02 keV which is ascribed to Ne X [170]. There is also the possibility of another line with best-fit line energy at 0.85 keV with identification as Ne IX. The ionization potentials of all ionized Neon species upto Ne VIII are less than 240 eV, whereas those of Ne IX and Ne X are above 1195 eV. Hence at temperatures found by the Chandra observations, the predominant species of Neon are expected to be Ne X and Ne XI. The mass of Neon was estimated to be about $5 \times 10^{-3} M_{\odot} - 1 \times 10^{-2} M_{\odot}$. This is consistent with Neon being in the Helium layer for a $15 M_{\odot}$ star where it was co-synthesized with C, N, and He. For other stellar masses and other composition zones, the required Neon mass is much larger than observed. Therefore, Neon in the Helium core of the $15 M_{\odot}$ star is the probable site where it was co-synthesized with C, N.

Conclusions

Much insight about how stars burn nuclear fuel, evolve and ultimately explode has been obtained by calculations and computer simulations. Theoretical developments have been augmented by crucial astronomical observations over decades in all bands of electromagnetic radiation and through other signal channels like neutrinos. Most important have been the wider availability of powerful telescopes both from the ground and space platforms and state of the art computers. Crucial inputs to the field of nuclear astrophysics are also coming from laboratory experiments involving radioactive ion beams (RIB) and intense beams of energetic nucleons and nuclei. Short lived nuclei can only be studied close to their sites of formation in the laboratory before they decay. Such facilities will further advance the future of nuclear astrophysics.

Acknowledgments

I thank Aruna Goswami and the organizers of the School for the invitation to Kodaikanal Observatory. Discussions with Katherina Lodders and Bruce Fegley at the School and Claus Rolfs and Richard McCray in other meetings are thankfully acknowledged. I thank Poonam Chandra and Firoza Sutaria for comments on this manuscript and for long term collaboration. Nuclear astrophysics research at Tata Institute is a part of the Plan Project: 11P-409.

References

1. W. D. Arnett, *Supernovae and Nucleosynthesis*, Princeton University Press (1996)
2. M. N. Saha: Phil. Mag. **40**, 472 (1920)
3. M. N. Saha: Proc. Roy. Soc. Ser. A **99**, 135 (1921)
4. H. Reeves, *Stellar Evolution and Nucleosynthesis*, Gordon and Breach, New York, (1968)
5. B. J. Carr, J. Bond, W. D. Arnett, ApJ **277**, 445 (1984)
6. B. J. Carr, ARAA **419**, 904 (1994)
7. W. Haxton, *Nuclear Astrophysics Course* (1999) http://ewiserver.npl.washington.edu/phys554/phys554.html
8. Oak Ridge National Lab, *Nuclear Data for Nuclear Astrophysics* http://www.phy.ornl.gov/astrophysics/data

9. Lawrence Berkeley National Lab, *Nuclear Astrophysics Reaction Rates* http://ie.lbl.gov/astro/astrorate.html
10. Joint Institute for Nuclear Astrophysics, *REACLIB library* http://www.nscl.msu.edu/nero/db
11. Computational Infrastructure for Nuclear Astrophysics http://www.nucastrodata.org
12. M. S. Smith et al., in *CP1016, Origin of Matter and Evolution of Galaxies*, ed. by T. Suda et al., (AIP) p.466 (2008)
13. C. E. Rolfs, W. S. Rodney, *Cauldrons in the Cosmos*, University of Chicago Press (1988)
14. D. D. Clayton, *Principles of Stellar Evolution and Nucleosynthesis*, University of Chicago Press (1984)
15. J. N. Bahcall: *Neutrino Astrophysics*, Chapter 3, Cambridge University Press (1989); see also: http://www.sns.ias.edu/~jnb/.
16. R. N. Boyd *An Introduction to Nuclear Astrophysics*, University of Chicago Press (2008)
17. A. Ray, in *5th SERC School on Radioactive Ion Beams and Physics of Nuclei Away from the Line of Stability*, Chandigarh, ed. by I. M. Govil, R. K. Puri (Elite publishers, New Delhi), p. 99 (2003); arXiv:astro-ph/0405568.
18. A. V. Filippenko, in *Supernovae and Stellar Evolution, proceedings of the school and workshop in Goa*, ed. by A. Ray, T. Velusamy, World Scientific, Singapore, p 34 (1991)
19. R. P. Harkness, J. C., Wheeler, in *Supernovae*, ed. by A. Petschek, Springer, Berlin, p 1 (1990)
20. A. Goswami, N. Prantzos, A&A **359**, 191 (2000)
21. A. S. Eddington, Observatory **43**, 341 (1920)
22. R. d'E. Atkinson, F. G. Houtermans, Z. Phys. **54**, 656 (1929)
23. R. Davis, D. S. Harmer, K. C. Hoffman, Phys. Rev. Lett. **20**, 1205 (1968)
24. E. Rutherford: Nature **123**, 313 (1929)
25. T. Kirsten, in *The origin of the solar system*, ed. by S. F. Dermott, New York: Wiley, p 267 (1978)
26. W. A. Fowler, in *Proc. Welch Found. Conf. on Chemical Research*, ed. by W.D. Milligan, Houston Univ. Press, p 61 (1977)
27. H. A. Bethe, Rev. Mod. Phys. **9**, 69 (1937)
28. E. E. Salpeter, Australian J. Phys. **7**, 373 (1954)
29. R. Schiavilla, Phys. Rev. **C 58**, 1263 (1998)
30. T. Itahashi et al., Nucl. Phys. **A 718**, 466c (2003)
31. LUNA Collaboration, C. Casella, et al., Nucl. Phys. **A 706**, 203 (2002)
32. C. F. von Weizsäcker, Phys. Z. **38**, 176 (1937)
33. C. F. von Weizsäcker, Phys. Z. **39**, 633 (1938)
34. H. A. Bethe, C. L. Critchfield, Phys. Rev. **54**, 248 (1938)
35. N. Chriestlieb et al., Nature **419**, 904 (2002)
36. J. N. Bahcall, M. C. Gonzalez-Garcia, C. Pefia-Garay, Phys. Rev. Lett. **90**, 131301 (2003)
37. C. Pena-Garay, A. Serenelli, arXiv: 0811.2424 (2008)
38. W. Haxton, A. Serenelli, ApJ **687**, 678 (2008)
39. G. K. Schenter, P. Vogel, Nucl. Sci. Engg. **83**, 393 (1983)
40. E. Fermi, *Nuclear Physics, Course Notes*, University of Chicago Press, p 83 (1951)

41. H. A. Bethe, C. L. Critchfield, Phys. Rev. **54**, 862 (1938)
42. E. Frieman, L. Motz, Phys. Rev **89**, 648 (1951)
43. E. E. Salpeter, Phys. Rev. **88**, 547 (1952); also ApJ **115**, 326 (1952)
44. J. N. Bahcall, R. M. May, ApJ **155**, 501 (1969)
45. H. A. Bethe, R. F. Bacher, Rev. Mod. Phys. **8**, 82 (1936)
46. J. N. Bahcall, C. P. Moeller, ApJ **155**, 511 (1969)
47. J. N. Bahcall, L. S. Brow, A. Gruzinov, R. F. Sawyer, A&A **383**, 291 (2002)
48. P. Quarati, A. M. Scarfone, J. Phys. G. **36**, 025203 (2009)
49. B. W. Filippone, et al., Phys. Rev. **C 28**, 2222 (1983)
50. M. Gai, Phys. Rev. **C 74**, 025810 (2006)
51. W. T. Winter et al., Phys. Rev. **C 73**, 025503 (2006)
52. L. van Wormer et al., ApJ **432**, 326 (1994)
53. A. E. Champagne, M. Wiescher, Ann Rev. Nucl. Part. Sci. **42**, 39 (1992)
54. R. K. Wallace, S. E. Woosley, ApJS **45**, 389 (1981)
55. M. Wiescher et al., Prog in part & Nucl Phys. **59**, 51 (2007)
56. C. Sneden, J. J. Cowan, R. Gallino, ARAA **48**, 241 (2008)
57. H. A. Bethe, Phys. Rev. **55**, 103 and 434 (1939)
58. E. J. Öpik: Proc. Roy. Irish Acad. **A 54**, 49 (1951)
59. E. E. Salpeter, Phys. Rev. **107**, 516 (1957)
60. F. Hoyle, ApJS **1**, 121 (1954)
61. F. Hoyle, D. N. F. Dunbar, W. A. Wenzel, W. Whaling, Phys. Rev., **92**, 1095 (1953)
62. C. W. Cook, W. A. Fowler, C. C. Lauritsen, T. Lauritsen, Phys. Rev. **107**, 508 (1957)
63. J. A. Nolen, S. M. Austin, Phys. Rev. **C 13**, 1773 (1976)
64. C. Angulo, *Exp Tools for Nuclear Astrophysics*, Lecture Notes in Physics **764**, 253 (2009)
65. L. Buchmann, C. A. Barnes: Nucl. Phys. **A 777**, 254 (2006)
66. A. Lefebvre-Schuhl, in *Frontiers in Nuclear Structure, Astrophysics and Reactions: FINUSTAR 2*, ed. by P. Demetriou et al., (AIP), p 150 (2008)
67. S. E. Woosley, T. A. Weaver, ARAA **24**, 205 (1986)
68. C. Hayashi, R. Hoshi, D. Sugimoto, Prog. Theor. Phys. Suppl. **22**, (1951)
69. M. Mazarkis, W. Stephens, ApJ **171**, L97 (1972)
70. H. Reeves, E. E. Salpeter, Phys. Rev. **116**, 1505 (1959)
71. W. R. Hix, F. K. Thielemann, ApJ **460**, 869 (1996)
72. F. K. Thielemann, W. D. Arnett, ApJ **295**, 264 (1985)
73. G. M. Fuller, W. A. Fowler, M. J. Newman, ApJ **252**, 715 (1982)
74. K. Kar, A. Ray, S. Sarkar, ApJ **434**, 662 (1994)
75. J. -U. Nabi, M. Sajjad, Phys. Rev. **C 77**, 055802 (2008)
76. K. Langanke, G. Martinez-Pinedo, *nucl-th/0203071*, Rev. Mod. Phys. **75**, 819 (2003)
77. W. R. Hix et al., ApJ **667**, 476 (2007)
78. H. Y. Chiu, Ann of Phys **15**, 1 (1961)
79. E. M. Burbidge, G. R. Burbidge, W. A. Fowler, F. Hoyle, Rev. Mod. Phys. **29**, 547 (1957)
80. G. Gamow, M. Schoenberg, Phys. Rev. **59**, 539 (1941)
81. S. E. Woosley, H. -T. Janka, Nature Physics **1**, 147 (2005)
82. H. A. Bethe, J. Wilson, ApJ **295**, 14 (1985)

83. H. A. Bethe, G. E. Brown, J. Applegate, J. Lattimer, Nucl. Phys. **A 324**, 487 (1979)
84. W. D. Arnett, ApJ **218**, 815 (1977)
85. S. W. Bruenn, ApJS **58**, 771 (1985)
86. P. S. Joshi, Scientific American **300**, 36 (2009)
87. G. E. Brown, H. A. Bethe, G. Baym, Nucl. Phys. A **375**, 481 (1982) (BBB)
88. G. M. Fuller, ApJ **252**, 741 (1982)
89. A. Ray, S. M. Chitre, K. Kar, ApJ **285**, 766 (1984)
90. H. A. Bethe, G. E. Brown, J. Applegate, J. M. Lattimer, Nucl. Phys. **A 324**, 487 (1979)
91. K. Kar, A. Ray, Phys Lett **A 96**, 322 (1983)
92. A. Zaringhalam, Nucl. Phys. **A 404**, 599 (1983)
93. F. K. Sutaria, A. Ray: Phy. Rev. **C 52**, 3460 (1995)
94. F. K. Sutaria, A. Ray, J. A. Sheikh, P. Ring, A&A **349**, 135 (1999)
95. O. Civitarese, A. Ray, Physica Scripta **59**, 352 (1999)
96. H.-T. Janka, K. Langanke, A. Marek, G. Martnez-Pinedo, B. Mller, Phys. Rept. **442**, 38 (2007)
97. A. F. Fantina, P. Donati, P. M. Pizzochero, Phys. Lett. **B 676**, 140 (2009)
98. D. J. Dean, K. Langanke, J. M. Sampaio, Phys Rev. **C 66**, 45802 (2002)
99. F. K. Sutaria, A. Ray, Phys. Rev. Lett. **79**, 1599 (1997)
100. Sudbury Neutrino Observatory Proposal, SNO-87-12, October (1987)
101. J. N. Bahcall, K. Kubodera, S. Nozawa, Phys. Rev. **D 38** 1030 (1988)
102. A. S. Burrows, in *Supernovae*, ed. by A. G. Petschek, Springer-Verlag (1990)
103. L. M. Segal, Nucl. Phys. **B 70**, 61 (1974)
104. T. Weaver, S. E. Woosley, G. M. Fuller, Nuclear Astrophysics: Essays in honor of J. Wilson, ed. by J. Centrella et al., 374 (1985)
105. R. M. Bionta et al., Phys. Rev. Lett., **58**, 1494 (1987) (IMB collaboration)
106. K. Hirata, et al., Phys. Rev. Lett. **58**, 1490 (1987) (K II collaboration)
107. W. D. Arnett, J. N. Bahcall, R. P. Kirshner, S. E. Woosley, ARAA **27**, 629 (1989)
108. D. Casper, *UNO A Next Generation Detector for Nucleon Decay and Neutrino Physics* meco.ps.uci.edu/lepton_workshop/talks/casper/uno.pdf (2000).
109. K. J. Chang, *hep-ex/0005046* (2000)
110. R. A. Chevalier, C. Fransson, in *Supernovae and Gamma-Ray Bursters*, ed. by Kurt Weiler, Lecture Notes in Physics, 598, 171 (2003)
111. S. Immler, W. H. G. Lewin, in *Supernovae and Gamma-Ray Bursters*, ed. by Kurt Weiler., Lecture Notes in Physics, vol. 598, p. 91 (2003)
112. R. A. Fesen et al., ApJ **645**, 283 (2006)
113. J. E. Reed, J. J. Hester, A. C. Fabian, P. F. Winkler, ApJ **440**, 706 (1995)
114. M. Ryle, F. G. Smith, Nature **162**, 462 (1948)
115. S. P. Reynolds, ARAA **48**, 89 (2008)
116. O. Krause et al., Science **320**, 1195 (2008)
117. S. E. Woosley, N. Langer, T. A. Weaver, ApJ **411**, 823 (1993)
118. G. Aldering, R. M. Humphreys, M. Richmond, AJ **107**, 662 (1994)
119. N. Rathnasree, A. Ray, JAA **13**, 3 (1992)
120. A. Ray, K. P. Singh, F. K. Sutaria, JAA **14**, 53 (1993)

121. K. Nomoto et al., Nature **364**, 507 (1993)
122. J. R. Maund, S. J. Smartt, R. P. Kudritzki, P. Podsiadlowski, G. F. Gilmore, Nature **427**, 129 (2004)
123. T. Matheson et al., AJ **120**, 1487 (2000)
124. R. A. Chevalier, J. Oishi, ApJ **593**, L23 (2003)
125. C. Fransson, P. Lundqvist, R. A. Chevalier, ApJ **461**, 993 (1996)
126. P. Chandra, A. Ray, S. Bhatnagar, ApJ **612**, 974 (2004)
127. R. A. Fesen, K. S. Gunderson, ApJ **470**, 967 (1996)
128. U. Hwang et al., ApJ **615**, L117 (2004)
129. U. Hwang, S. S. Holt, R. Petre, ApJ **537**, L119 (2000)
130. R. Willingale et al., A&A **381**, 1039 (2002)
131. J. S. Lazendic, D. Dewey, N. S. Schulz, C. R. Canizares, ApJ **651**, 250 (2006)
132. J. A. Ennis et al., ApJ **652**, 376 (2006)
133. J. D. T. Smith et al., ApJ **693**, 713 (2009)
134. R. A. Fesen et al., AJ **122**, 2644 (2001)
135. S. E. Woosley, T. A. Weaver, ApJS **101**, 181 (1995)
136. J. P. Hughes et al., ApJ **528**, L113 (2000)
137. L. Rudnick, in *The 10th International Symposium on Origin of Matter and Evolution of Galaxies, Sapporo, Japan*, ed. by Takuma Suda et al., AIP Conference Proceedings, Volume 1016, p. 353 (2008)
138. D. A. Liedahl, in *X-ray Spectroscopy in Astrophysics, X EADN School Amsterdam*, ed. by J. van Paradijs, J. A. M. Bleeker, Springer, Berlin, 189 (1999)
139. D. Porquet et al., A&A **376**, 1113 (2001)
140. D. Dewey, S. A. Zhekov, R. McCray, R. Canizares, ApJ **676**, L131 (2008)
141. S. Park et al., in *Supernova 1987A: 20 Years After*, ed. by S. Immler, K. W. Weiler, R. McCray (New York: AIP) 43 (2007)
142. S. A. Zhekov et al., ApJ **645**, 293 (2006)
143. C. Fransson, C. Kozma, New Astron Rev. **46**, 487 (2002)
144. I. Ahmad et al., Phys. Rev. **C 74**, 065803 (2006)
145. A. F. Iyudin et al., A&A **284**, L1 (1994)
146. M. Renaud et al., ApJ **647**, L41 (2006)
147. F. K. Thielemann, K. Nomoto, M. Hashimoto, ApJ **460**, 408 (1996)
148. T. Rauscher et al., ApJ **576**, 323 (2002)
149. M. Limongi, A. Chieffi, ApJ **592**, 404 (2003)
150. H. Nassar et al., Phys. Rev. Lett. **96**, 041102 (2006)
151. S. M. Matz et al., Nature **331**, 416 (1988)
152. J. D. Kurfess et al., ApJ **399**, L137 (1992)
153. K. Nomoto et al., in *Supernovae and Stellar Evolution, proceedings of the school and workshop in Goa*, ed. by A. Ray, T. Velusamy, World Scientific, Singapore p. 116 (1991)
154. M. D. Leising, New Astr. Reviews **46**, 529 (2002)
155. M. D. Leising, ApJ **651**, 1019 (2006)
156. H. U. Zimmermann, B. Aschenbach, A&A **406**, 969 (2003)
157. D. W. Fox et al., MNRAS **319**, 1154 (2000)
158. T. Nymark, C. Fransson, C. Kozma, A&A **449**, 171 (2006)
159. P. Chandra et al., ApJ **699**, 388 (2009)
160. T. Nymark, P. Chandra, C. Fransson, A&A **494**, 179 (2009)

161. A. V. Filippenko, T. Matheson, L. C. Ho, ApJ **415**, L103 (1993)
162. Ph. Podsiadlowski et al., Nature **364**, 509 (1993)
163. S. E. Woosley et al., ApJ **429**, 300 (1994)
164. T. Shigeyama et al., ApJ **420**, 341 (1994)
165. A. V. Filippenko, ARAA **35**, 309 (1997)
166. W. Li, A. V. Filippenko, S. D. Van Dyk, J. Hu, PASP **114**, 403 (2002)
167. R. B. C. Henry, D. Branch, PASP **99**, 112 (1987)
168. A. V. Filippenko, in *Supernova 1987A and other supernova, ESO Conference and Workshop*, ed. by J. Danziger and Kurt Kjaer, p 343 (1991)
169. N. N. Chugai, I. J. Danziger, Astron. Lett. **29**, 649 (2003)
170. P. Chandra et al., ApJ **629**, 933 (2005)

The Evolution of Massive Stars and the Concomitant Non-explosive and Explosive Nucleosynthesis

Marcel Arnould

Institut d'Astronomie et d'Astrophysique, Université Libre de Bruxelles, CP-226, B-1050 Brussels, Belgium

Summary. These lectures are concerned with some aspects of the evolution of massive stars and of the concomitant nucleosynthesis. They complement other lectures in this volume. Special emphasis is put on the production of the nuclides heavier than iron by the r- and p-processes.

Keywords: Stars: massive – Stars: evolution – Stars: nucleosynthesis – Solar System: composition – r-process – p-process

1 Introduction

Massive stars classically refer to those stars that are able to experience all possible nuclear burning stages from hydrogen to the so-called silicon burning. This definition limits the relevant mass range to $M \gtrsim 10M_{\odot}$, with some uncertainty on this boundary value.

These stars have been of central concern in a myriad of observational and theoretical works. No wonder! They indeed play a key role in many chapters of astrophysics. In particular, they influence the physical and chemical states of their circumstellar environments or of the interstellar medium through their intense radiation and mass losses during their non-explosive phases of evolution, and even more so, as a result of their final supernova explosions. They may act as triggers of star formation, are essential agents of the evolution of the nuclidic content of the galaxies, accelerate particles to cosmic ray energies, and leave neutron stars or black holes at the end of their evolution. They are also the progenitors of certain γ-ray bursts.

Chapter 5 in this volume discusses some aspects of the physics and astrophysics of massive stars. We concentrate here on the production of the nuclides heavier than iron by the r- and p-processes (the third mode of production of the heavy nuclides, that is the s-process, is discussed by Karakas in this volume). As an introduction to these topics, we briefly sketch in Sect. 2

A. Goswami and B.E. Reddy (eds.), *Principles and Perspectives in Cosmochemistry*, Astrophysics and Space Science Proceedings,
DOI 10.1007/978-3-642-10352-0_6,

the evolution of massive stars before dwelling into some details of their non-explosive nucleosynthesis (Sect. 3). The explosion of massive stars that leads to the supernova phenomenon is the subject of Sect. 4, and the concomitant synthesis of the nuclides up to the iron peak is summarised in Sect. 5. The following sections are devoted to the r- and p-processes. The separation between the s-, r- and p-process contributions to the bulk Solar System composition is discussed in Sect. 6.2, with special emphasis on the largely neglected uncertainties that affect this splitting procedure. Deviations to the bulk composition in the form of isotopic anomalies are briefly mentioned in Sect. 6.3, while Sect. 6.4 deals with the evolution of the r-nuclide content of the Galaxy. The topics of the actinide abundances in the Solar System, in the local interstellar medium or in stars, and of the r-nuclide content of the galactic cosmic rays are covered in Sects. 6.6 and 6.7. Site-free parametric approaches of the r-process are reviewed in Sect. 7. Section 7.1 is concerned with the so-called canonical and multi-event models, and dynamical high-temperature r-process approaches are presented in Sect. 7.2. Section 7.3 deals instead with a high-density scenario. Attempts to relate the r-process to specific astrophysical situations follow. Sections 8 and 9 review the possibility of development of the r-process in massive star explosions and as a result of the coalescence of two neutron stars. The review of the r-process is concluded in Sect. 10 with some comments on the difficulties to model the evolution of the r-nuclide content of the Galaxy and on nucleo-cosmochronology. The review of the p-process starts with some generalities (Sect. 11). Sections 11.1 and 11.2 discuss the possibility of production of the p-nuclides in Type II and in Type Ia supernovae, while the development of the p-process in sub-Chandrasekhar white dwarf explosions is examined in Sect. 11.3. Section 11.4 is devoted to a brief account of the isotopic anomalies attributed to the p-process, and to attempts to build a chronometry based on this process. Some general conclusions are drawn in Sect. 12.

2 Some generalities about the evolution of massive stars

The structure and evolution of the stars of various masses and initial compositions are discussed in some detail in chapters 3 and 4 in this volume. We limit ourselves here to some general considerations regarding the evolution of massive stars ($M \gtrsim 10 M_{\odot}$) only, with special emphasis on the accompanying nucleosynthesis. In short, and as pictured very schematically in Fig. 1, the evolution of the central regions of a massive star is made of successive 'controlled' thermonuclear burning stages and of phases of gravitational contraction. The latter phases are responsible for a temperature increase, while the former ones produce nuclear energy through charged-particle induced reactions. Of course, composition changes also result from these very same reactions, as well as, at some stages at least, from neutron-induced reactions, which in contrast do not play any significant role in the stellar energy budget. The nuclear reactions involved in the different non-explosive nuclear burning stages are briefly discussed in Sect. 3. Let us simply say here that such a sequence develops in time with nuclear fuels of increasing charge number Z and at temperatures increasing from several tens of 10^6 K to about 4×10^9 K. Concomitantly the duration of the successive nuclear-burning phases decreases in a dramatic way. This situation results from the combination of (i) a decreasing energy production when going from H burning to the later burning stages, and (ii) an increasing neutrino production, and the consequent energy losses, with temperatures exceeding about 5×10^8 K. Figure 1 also depicts schematically that a nuclear burning phase, once completed in the central regions, migrates into a thin peripheral shell. As a consequence the deep stellar regions look like an onion with various 'skins' of different compositions.

It is quite important to stress that the true stellar structure is certainly much more complicated than sketched in Fig. 1, even when effects like deviations from spherical symmetry (induced by rotation or certain mechanisms of transport of matter) are neglected. This spherically symmetric picture of a star may break down, especially during the advanced stages of the evolution of massive stars, and would lead to a dramatic growing of the complication of the stellar structure and evolution (e.g. [2,3], and references therein). This increased complexity is demonstrated by multi-dimensional simulations of the structure of massive stars. The consideration of rotation of course brings additional difficulties. Steady mass loss from a star may also affect its evolution in various ways. Finally, binarity may lead to specific evolutionary patterns resulting for the largest part from episodic mass transfers from one component to the other.

After having experienced all the burning phases terminating with Si burning (Fig. 1), massive stars are seen to develop an Fe core that is lacking further nuclear fuels, as any transformation of the strongly-bound Fe nuclei is endothermic. In fact this core becomes dynamically unstable and implodes as a result of free-electron captures and Fe photodisintegration, the former transformation playing an especially important role at the low end of the relevant stellar-mass range. Through a very complex chain of physical events the implosion can, in certain cases at least, turn into an explosion referred to as a supernova (SN). It is accompanied with the ejection into the interstellar medium (ISM) of a substantial amount of material with kinetic energies typically of the order of 10^{51} ergs. Outliers in the form of more energetic (hypernovae) or less energetic (faint) SNe have been identified, however. The innermost parts of the star are predicted to escape explosion. They remain bound in a 'residue' which may be a neutron star (observable as a pulsar if it is magnetized and rapidly rotating), or even a black hole. Some more details on the modelling of SN explosions can be found in Sect. 4. A more thorough review of the SN phenomenon can be found in [4].

Let us just note here that, observationally, a supernova is classified as being of Type II (SN II) if it shows H-lines in its spectrum. It is likely that most, if not all, of the exploding massive stars still have some H-envelope

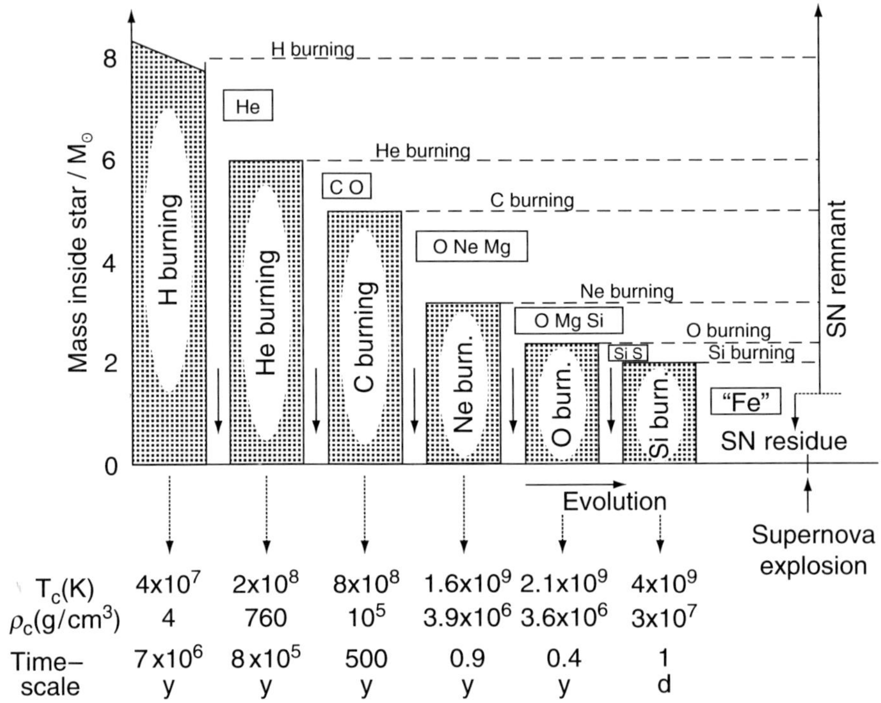

Fig. 1. Schematic representation of the evolution of the internal structure of a spherically-symmetric massive ($M \approx 25M_{\odot}$) star. The shaded zones correspond to nuclear burning stages. A given burning phase starts in the central regions and then migrates into thin peripheral burning shells. Typical central temperatures T_c, central densities ρ_c and durations of the core burning phases are indicated at the bottom of the figure (adapted from [1]). In between the central nuclear burning phases are episodes of gravitational contraction (downward arrows). The chemical symbols represent the most abundant nuclear species left after each nuclear-burning mode ("Fe" symbolises the iron-peak nuclei with mass numbers A in the $50 \lesssim A \lesssim 60$ range). If the star eventually explodes as a supernova, the most central parts may leave a 'residue,' while the rest of the stellar material is ejected into the ISM, where it is observed as a supernova 'remnant'

left, and thus exhibit such a feature. In contrast, Type I SNe (SNI) lack H in their ejecta. Specific spectral features have led to the identification of different SNI subclasses, among which are the SNIa and SNIb/c supernovae. The latter concern the fate of massive stars suffering pre-explosion mass losses that are strong enough for ejecting their H-rich envelopes. A situation of this type is typically encountered in stars that are massive enough to evolve through the Wolf-Rayet phase. In contrast, SNIa are classically viewed as the explosion of a white dwarf following its accretion of matter from a companion star in a binary system.

3 Non-explosive stellar evolution and concomitant nucleosynthesis

In the following we present a very brief account of the nucleosynthesis developing during the various non-explosive stages in the evolution of massive stars depicted in Fig. 1.

3.1 Hydrogen burning

In massive stars, hydrogen burns essentially through the cold CNO cycles represented in Fig. 2 which develop at temperatures in excess of some 10^7 K.[1] They produce fresh ^{4}He, while the initial C, N and O are mostly transformed into ^{14}N. Some H is also consumed by the NeNa and MgAl chains (Fig 3), which play only a minor role in the stellar energy budget, but are of significance in the production of the Na to Al isotopes. Most important, the MgAl chain might synthesize ^{26}Al, which is a very interesting radio-active nuclide for γ-ray astronomy and cosmochemistry. The p-p chains of H burning play only a very minor energetic or nucleosynthesis role in massive stars, in contrast to the situation encountered in low-mass stars (see e.g. [5]).

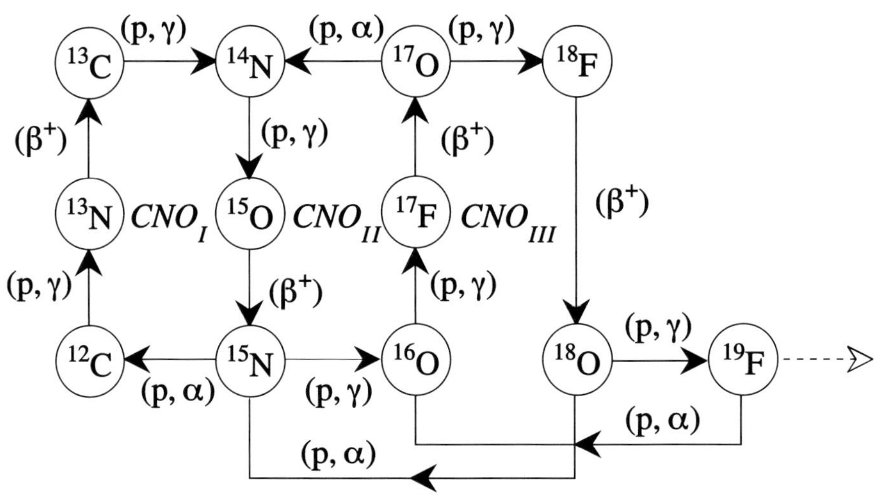

Fig. 2. Reactions involved in the cold CNO mode of hydrogen burning

[1] A burning mode is referred to as ‘cold’ if an unstable nucleus, once produced, has time to β-decay before suffering a particle capture. The reverse situation characterises ‘hot’ modes instead.

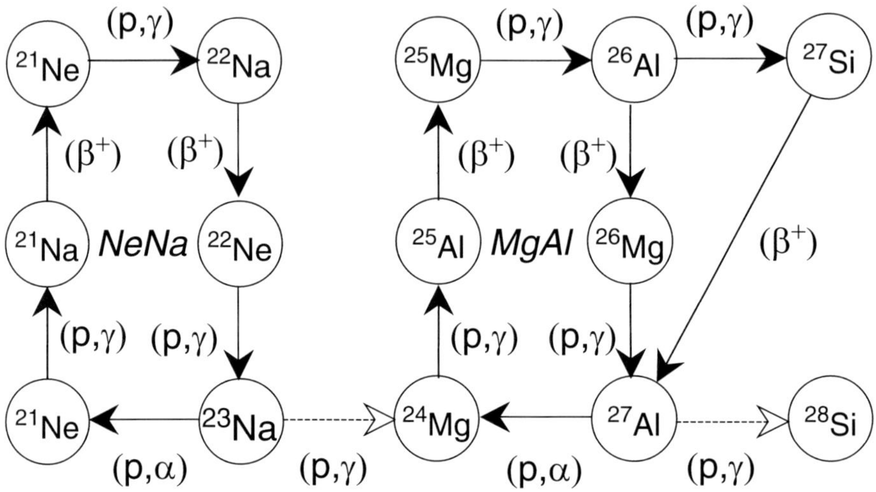

Fig. 3. Reactions involved in the cold NeNa and MgAl modes of hydrogen burning

Much experimental and theoretical effort has been devoted to the reactions involved in the H-burning modes, as summarized in NACRE [6], which also provides typical uncertainties still affecting the relevant reactions (see also [7]). The yields from the cold CNO, NeNa and MgAl burning modes based on the NACRE adopted rates and their uncertainties have been analyzed by [8] based on the NACRE data and on schematic stellar conditions. These predictions, as well as others based on specific stellar models (see e.g. [9, 10] for stars with masses $M < 10M_{\odot}$) demonstrate that better determinations of certain reaction rates would be desirable in order to set up meaningful comparisons between certain abundance predictions and observations.

3.2 Helium burning and the s-process

The reactions involved in the He-burning stage have been discussed in many places (e.g. [5], NACRE [6]). The main ones are displayed in Fig 4. They develop at temperatures in excess of 10^8 K, and mainly transform ^{4}He into ^{12}C and ^{16}O, with some limited contribution to the abundance of some heavier α-particle nuclei (esp. ^{20}Ne,....), at least in massive enough stars. Of very special and dramatic importance for the theories of stellar evolution and of nucleosynthesis is the famed $^{12}\mathrm{C}\,(\alpha\,,\gamma)\,^{16}\mathrm{O}$ reaction, which has been the subject of a flurry of experimental investigations, as well as of theoretical efforts (for a review, see [11]; see also [12] for a recent re-analysis of the case). In spite of that, uncertainties remain, and preclude certain nuclear astrophysics predictions to be made at a satisfactory level (see e.g. [13]).

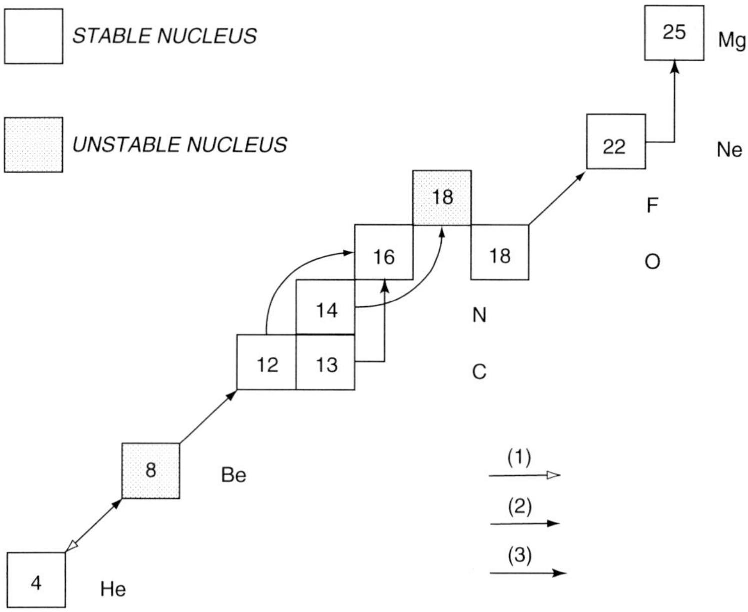

Fig. 4. Main reactions involved in the burning of helium. Arrows of types (1), (2) and (3) refer to (γ, α), (α, γ) and $(\alpha.n)$ reactions

Other α-particle induced reactions are of special importance, and have been the subject of many dedicated experimental and theoretical works. This is particularly the case for $^{13}\mathrm{C}\,(\alpha\,,\mathrm{n})\,^{16}\mathrm{O}$ and $^{22}\mathrm{Ne}\,(\alpha\,,\mathrm{n})\,^{25}\mathrm{Mg}$. These reactions are considered as the main sources of neutrons for the s-process nucleosynthesis, the latter one being of special significance in massive stars.

The proper treatment of the s-process also requires the knowledge of a host of neutron-capture cross sections at typical energies from about 10 to 100 keV on targets in the whole $12 \leq A \leq 210$ mass range. Much dedicated experimental work has led to a substantial improvement in our knowledge of relevant (n,γ), as well as (n,p) and (n,α) cross sections. Some data are still lacking, especially for neutron-captures by unstable targets close to the valley of nuclear stability that can be involved in the s-process. In these cases, the astrophysical rates of the competing β-decays have also to be known.

3.3 Carbon burning

Carbon burning develops at temperatures in excess of about 5×10^8 K, and produces mainly nuclides in the $20 \lesssim A \lesssim 27$ mass range through reactions the most important of which are displayed in Fig. 5. The C-burning phase raises the very interesting question of the fusion of light heavy ions below the Coulomb barrier, and in particular of the origin of the very pronounced structures observed in the $^{12}\mathrm{C} + {}^{12}\mathrm{C}$ fusion cross section at low energies. The devoted experimental (e.g. [14], and references therein) and theoretical works do not provide fully satisfactory solutions.

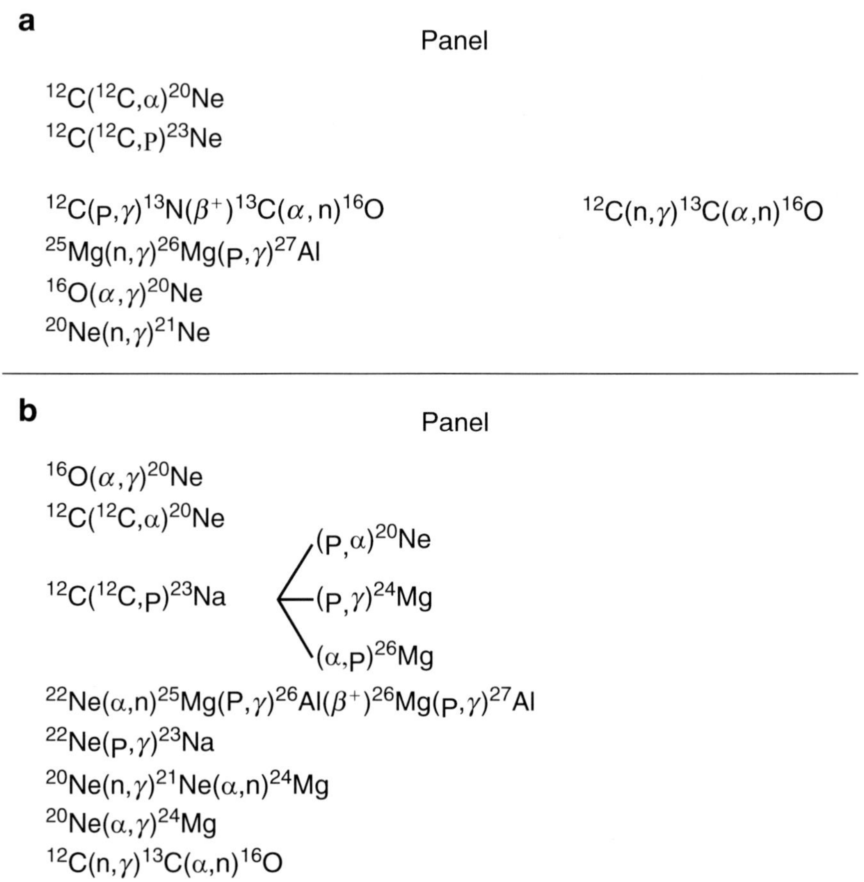

Fig. 5. Most important reactions involved in the carbon burning core of massive stars. Panels (a) and (b) concern the first and later stages of the burning in a Population I 25 $M_\odot$ star (from [15])

3.4 Neon, oxygen, and silicon burning

As shown in Fig. 6, the Ne-burning phase is initiated at temperatures somewhat in excess of 10^9 K by $^{20}\mathrm{Ne}\,(\gamma\,,\alpha)\,^{16}\mathrm{O}$, the first major energetically significant photodisintegration reaction experienced by a star in the course of its evolution. A complementary ^{20}Ne destruction channel is $^{20}\mathrm{Ne}\,(\alpha\,,\gamma)\,^{24}\mathrm{Mg}$. Both α-capture reactions have been studied experimentally and theoretically [6]. The main ashes of Ne burning are ^{16}O. A variety of other nuclides up to ^{28}Si are also synthesised.

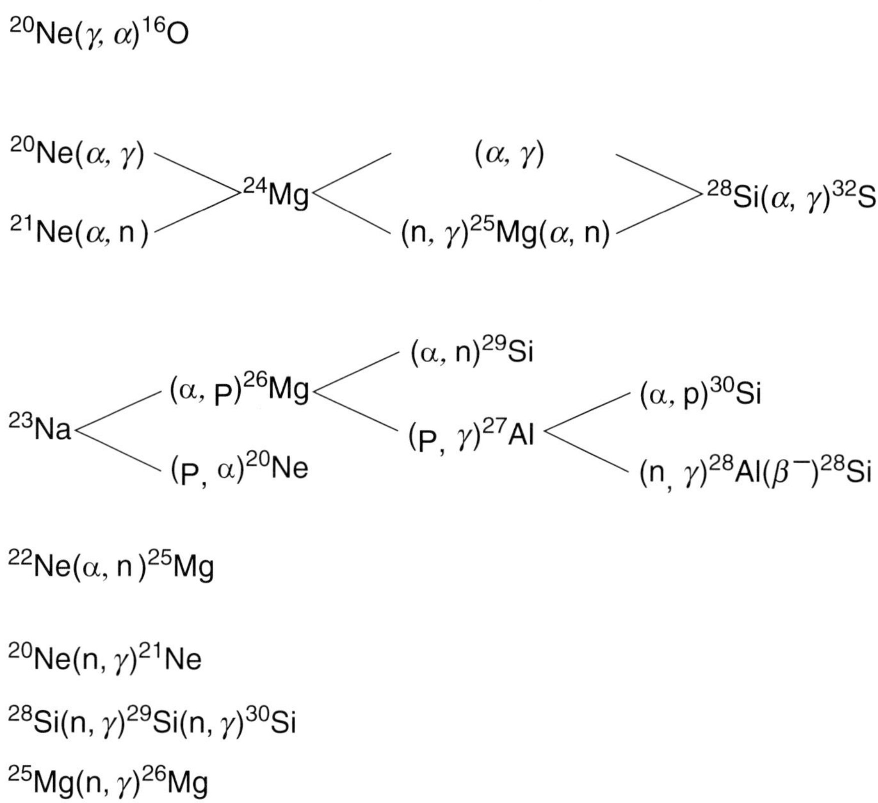

Fig. 6. Most important reactions involved in the central Ne burning of massive stars (here a Population I 25 $M_{\odot}$ star (from [15])

The main reactions involved in the central burning of oxygen are displayed in Fig. 7. They develop at temperatures in excess of about 2×10^9 K and produce mainly ^{28}Si, as well as a variety of nuclides in the approximate $30 \lesssim A \simeq 40$ mass range. It is important to emphasize that free electron captures also start playing the important role of changing the neutron excess during the course of the burning. It is defined as $\sum_i (N_i - Z_i) Y_i$, where $N_i(Z_i)$ is the number of (bound plus free) neutrons (protons), and Y_i is the mole fraction of nuclide i ($Y_i = X_i/A_i$ with X_i the mass fraction of i and A_i its atomic weight). This change has an important impact on the composition of the central region of a massive star at its presupernova stage (see below). Also note that, on the nuclear side, the ^{16}O+^{16}O fusion reaction that governs the O-burning phase does not exhibit intricacies comparable to those encountered with ^{12}C+^{12}C.

Silicon burning comprises a very complex pattern of nuclear reactions that has been analysed in particular detail by [16]. It develops at temperatures in excess of about 4×10^9 K, and is mainly made of a nuclear flow starting at ^{28}Si and leading to the production of lighter species through a series of photodisintegrations. The nucleons and α-particles released by these reactions can be captured by the remaining ^{28}Si, which generates a nuclear flow moving upward. The reactions involved in the downward and upward moving flows progressively lead to a situation where limited ensembles of nuclei reach a situation of 'Nuclear Statistical Equilibrium (NSE)'. These ensembles are referred to as 'Quasi-Statistical Equilibrium (QSE)' clusters. In this situation, the relative abundances of nuclei belonging to the same QSE cluster are obtained from the equations of equilibrium statistical mechanics (the so-called nuclear Saha equations; see below). In such a regime, it is not required to solve a full nuclear reaction network. In contrast, the relative abundances of species belonging to different QSE clusters can be obtained only from the consideration of a reaction network. Figure 8 displays a snapshot of the Si burning pattern of nuclear flows at $T = 3.5 \times 10^9$ K in a Population I 25 $M_\odot$ star. Two QSE clusters are seen to have developed, corresponding to two ensembles of nuclei connected by a large number of reactions (which translates in the figure by grey patches). These QSE clusters are interconnected by a smaller number of reactions that are too slow for allowing these clusters to be in equilibrium at the selected time. Note that the boundaries of the QSE clusters and abundances are time-dependent, as a result of changes in temperature and in neutron excess, the latter being due to the operation of free electron captures. The general trend is that the two QSE clusters displayed in Fig. 8 are more fragmented at earlier epochs. At later times, a QSE cluster involving lighter nuclei develops, and all the clusters finally merge to lead to a Nuclear Statistical Equilibrium (NSE) state.[2] This regime has been described in detail by e.g. [17], and is not reviewed here. Let us simply say that the

[2] Note that weak interaction processes remain out of equilibrium as long as neutrinos are not equilibrated with matter and radiation. This is the case as long as the density remains lower than about 10^{11} g/cm^3. At higher densities, a state of so-called complete equilibrium is obtained.

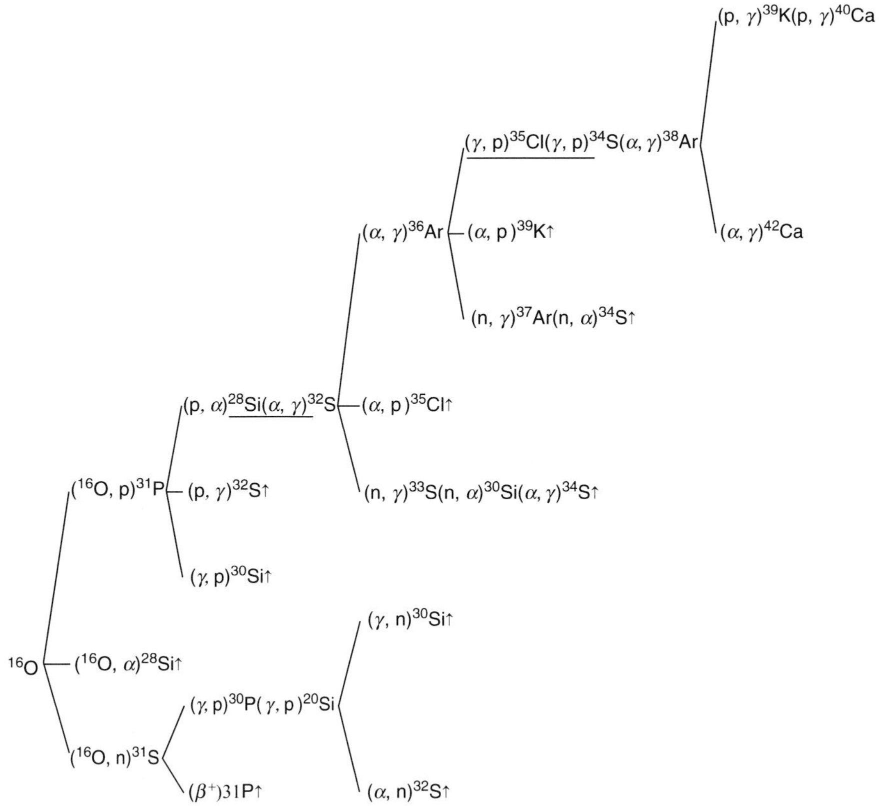

Fig. 7. The main reactions involved in the O burning in the core of massive stars (here a Population I 25 $M_\odot$ star). The reverse reactions of the underlined ones may be activated at some point during the burning (from [15])

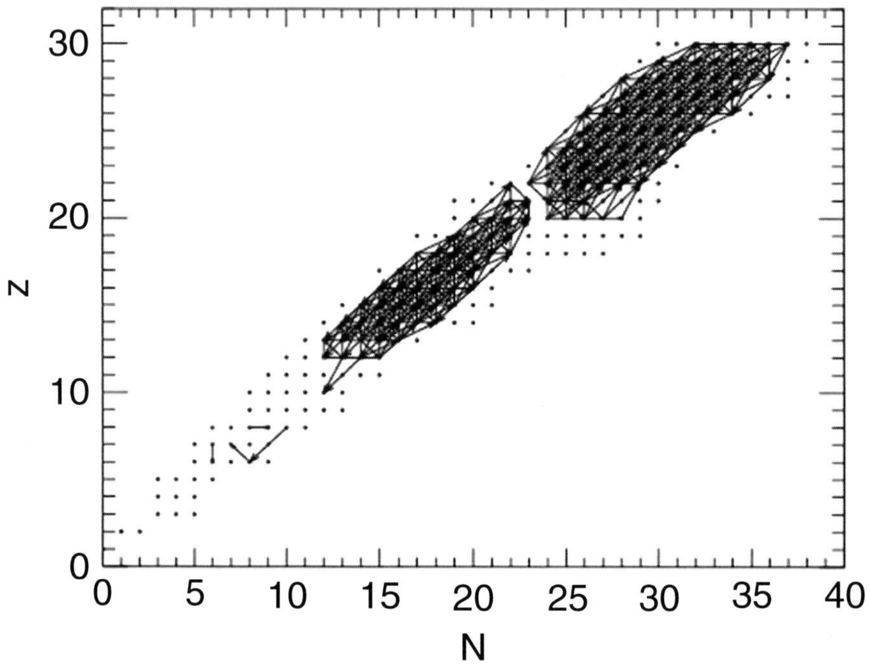

Fig. 8. Illustration of the development of two QSE clusters during Si burning at 3.5×10^9 K in a Population I 25 $M_\odot$ star (from [15])

abundances are obtained simply by solving a set of nuclear Saha equations under the constraint of mass conservation and electrical neutrality. The resulting abundances do not depend on reaction rates, but only on temperature, nuclear binding energies, nuclear partition functions and neutron excess. As in the case of the QSE regime, the NSE abundances are subjected to time variations as a result of temperature and neutron excess changes. In any case, the NSE conditions favour the production of the most stable, that is the iron peak, nuclides. The iron core represented in Fig. 1 results.

In the framework of stellar evolution models, Si burning has often been approximated in different ways relying on the QSE cluster concept. These approximations are not fully appropriate, particularly in view of the time dependence of the QSE cluster boundaries. This has been stressed by [15] who adopt instead for the QSE and NSE regimes a detailed Si-burning network coupled to the stellar evolution equations.

4 The explosive fate of massive stars

As said in Sect. 2, the iron core left over following Si burning suffers a dynamical instability as a result of endothermic electron captures and Fe photodidintegration. To a first approximation, this gravitational instability sets in near the classical Chandrasekhar mass limit for cold white dwarfs, $M_{\rm Ch} = 5.83\,Y_{\rm e}^2$, $Y_{\rm e}$ being the electron mole fraction. In the real situation of a hot stellar core, collapse may start at masses that differ somewhat from this value, depending on the details of the core equation of state. The reader is referred to [18] (especially Chaps. 12 and 13) for a detailed discussion of the implosion mechanism and for its theoretical outcome and observable consequences. Here, we just briefly summarise the situation.

The gravitational collapse of the iron core does not stop before the central densities exceed the nuclear matter density $\rho_0 \approx 2.5 \times 10^{14}\,{\rm g\,cm^{-3}}$ by about a factor of two. At this point, the innermost ($M \lesssim 0.5\ M_\odot$) material forms an incompressible, hot and still lepton-rich 'proto-neutron' star (PNS) whose collapse is stopped abruptly, and which eventually bounces. A shock wave powered by the gravitational binding energy released in the collapse propagates supersonically into the infalling outer layers. For many years, it has been hoped that this shock could be sufficiently strong for ejecting explosively most of the material outside the core, producing a so-called 'prompt core collapse supernova' (PCCSN) with a typical kinetic energy of 1–2 $\times\ 10^{51}$ ergs, as observed. The problem is that the shock is formed roughly half-way inside the iron core, and looses a substantial fraction of its energy in the endothermic photodisintegrations of the iron-group nuclei located in the outermost portion of the core. The shock energy loss is aggravated further by the escape of the neutrinos produced by electron captures on the abundant free protons in the shock-heated material. Detailed one-dimensional hydrodynamic simulations conclude that the initially outgoing shock wave transforms within a few milliseconds after bounce into an accretion shock. The matter behind the shock continues to accrete on the PNS. No recent simulation is able to predict a successful PCCSN for a Fe-core progenitor star ($M \gtrsim\ 10\ M_\odot$). This failure is illustrated in Fig. 9 for a 15 $M_\odot$ star.

Even so, some hope to get a CCSN of a non-prompt type has been expressed if there is a way to 'rejuvenate' the shock efficiently enough to obtain an explosive ejection of the material outside the PNS. This rejuvenation remains a matter of intensive research. Neutrinos might well play a pivotal role in this matter. They are produced in profusion from the internal energy reservoir of the PNS that cools and deleptonises hundreds of milliseconds after bounce, and their total energy might amount to several 10^{53} ergs, that is about 100 times the typical explosion energy of a SN II. The deposition of a few percent of this energy would thus be sufficient to unbind the stellar mantle and envelope, and provoke a 'delayed' CCSN (DCCSN) (these qualitative statements assume that a black hole is not formed instead of a PNS; see below). Many attempts to evaluate the precise level of neutrino energy

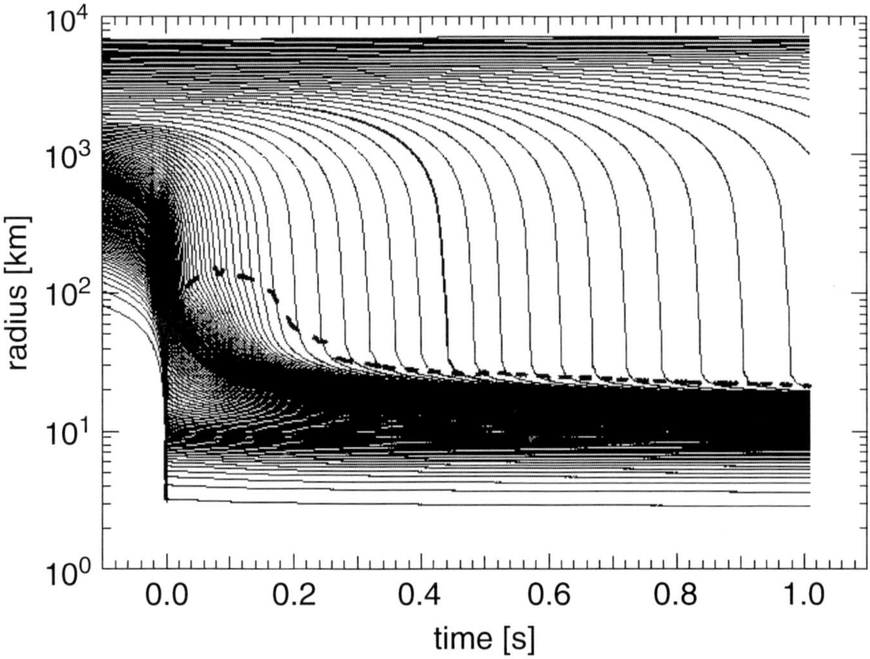

Fig. 9. Radial trajectories of several mass elements of the core of a 15 $M_\odot$ star versus time after bounce. The trajectories are plotted for each 0.02 $M_\odot$ up to 1 $M_\odot$, and for each 0.01 $M_\odot$ outside this mass. The thick dashed line indicates the location of the shock wave. The prompt shock stalls within 100 ms after reaching 150 km, and recedes down to below 100 km. No sign of a revival of the shock that possibly leads to a successful D(elayed-)CCSN is seen either, even after 300 ms. Instead, a stationary accretion shock forms at several tens of km. A PNS is seen to form, reaching 1.6 $M_\odot$ around 1 s after bounce (from [19])

deposition have been conducted over the last decades, based on more or less controversial simplifications of the treatment of the neutrino transport (e.g. [20] for a recent re-analysis of the problem, which is made even more complex by the due consideration of neutrino flavor mixing). In fact, theoretical investigations and numerical simulations performed with increasing sophistication over the past two decades have not been able to come up with a clearly successful CCSN for a Fe-core progenitor ($M \gtrsim 10\ M_\odot$). This conclusion is apparently robust to changes in the highly complex physical ingredients (like the neutrino interactions, or the equation of state), and in the numerical techniques (e.g. [20]). In fact, the neutrino-energy deposition should have to be significantly enhanced over the current model values in order to trigger an explosion. An illustration of a failed DCCSN is shown in Fig. 9.

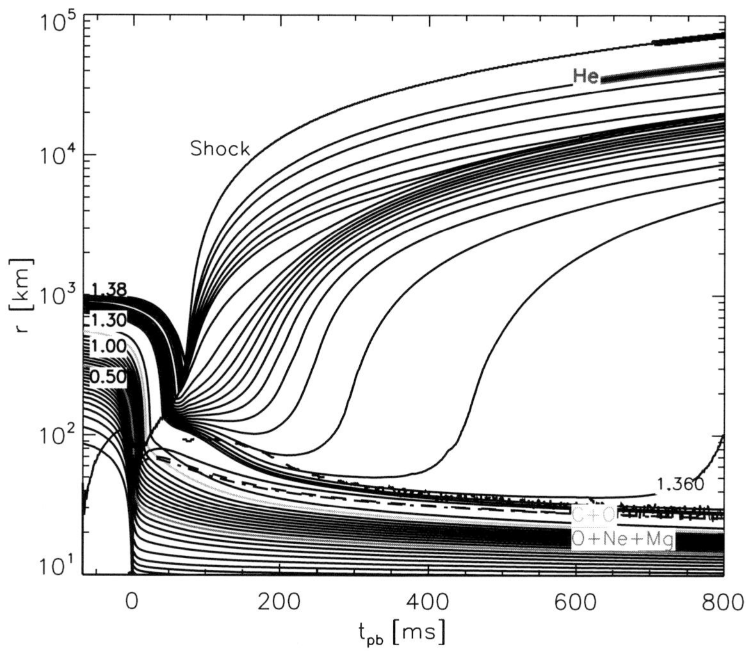

Fig. 10. Simulation of an electron-capture supernova following the collapse of an O-Ne core. The time evolution of the radius of various mass shells is displayed with the inner boundaries of the O+Ne, C+O and He shells marked by thick lines. The inner core of about 0.8 $M_{\odot}$ is mainly made of Ne at the onset of collapse ([21], and references therein). The explosion is driven by the baryonic wind caused by neutrino heating around the PNS. The thick solid, dashed, and dash-dotted lines mark the neutrino spheres of $\nu_{\rm e}$, $\bar{\nu}_{\rm e}$, and heavy-lepton neutrinos, respectively. The thin dashed line indicates the gain radius which separates the layers cooled from those heated by the neutrino flow. The thick line starting at $t = 0$ is the outward moving supernova shock (from [22])

This adverse circumstance may not mark the end of any hope to get a DCCSN, however. In the case of the single stars considered here, one might just have to limit the considerations to the stars in the approximate 9 to 10 $M_{\odot}$ range that possibly develop O-Ne cores instead of iron cores at the termination of their hydrostatic evolution. Efficient endothermic electron captures could trigger the collapse of that core, which could eventually transform into a so-called electron-capture supernova that may be of the SN Ia or SN II type, depending upon the extent of the pre-explosion wind mass losses.[3] In these situations, illustrated in Fig. 10, the neutrino heating is indeed efficient enough for rejuvenating the shock wave about 150 ms after bounce, and mass shells start being ablated from the PNS surface about 50 ms later, leading to a so-called 'neutrino-driven wind'.[4] No information is provided by the current simulations on the conditions at times much later than a second after bounce. Note that the predicted successful delayed electron-capture supernova is characterised by a low final explosion energy (of the order of 0.1×10^{51} ergs, which is roughly ten times lower than typical SN values), and by just a small amount of ejected material (only about 0.015 $M_\odot$). These features might suggest a possible connection with some sub-luminous (faint) SN II events. Let us stress, however, that the structure of the progenitors of the electron-capture supernovae remains especially uncertain (e.g. [23]), which endangers any conclusion one may draw on these SN types.

A major effort has been put recently in the development of simulations of explosions that go beyond the one-dimensional approximation. This is motivated not only by the difficulty of obtaining successful CCSNe in one-dimensional simulations, but also by the mounting observational evidence that SN explosions deviate from spherical symmetry, not to mention the possible connection between the so-called soft long-duration gamma-ray bursts, and grossly asymmetric explosions accompanied with narrow jets of relativistic particles, referred to as JetSNe. The multi-dimensional extension of the simulations opens the potentiality to treat in a proper way different effects that may turn out to be essential in the CCSN or JetSNe process. As briefly reviewed by e.g. [26], they include fluid instabilities, or rotation and magnetic fields on top of the neutrino transport already built into the one-dimensional models. Acoustic power may be another potential trigger of CCSNe [27] (see also [24] for a brief review of multidimensional simulations).

In summary, there are obviously many crucial questions that remain to be answered before one can hope putting together a clear and coherent picture of the CCSN fate of massive stars. The structure of the pre-supernova stars remains uncertain in many important aspects which may have a significant impact on the properties, and even the very existence, of the explosive fate of the massive stars. This concerns in particular the mass loss rates,

---

[3] The range of initial masses of single stars which could experience an electron-capture instability is still quite uncertain, and depends in particular on a subtle competition between the growth of the stellar cores resulting from thermal pulses developing during the Asymptotic Giant Branch evolution and their erosion resulting from steady mass losses. Other stars in the approximate 8 to 12 $M_\odot$ mass range might end up as O-Ne white dwarfs or experience of Fe core collapse instead of experiencing an electron-capture supernova resulting from the collapse of the O-Ne core ([23] for a review). Binary systems might offer additional opportunities of obtaining electron-capture supernovae (see [24] or [25])

[4] Unless otherwise stated, neutrino-driven winds refer to transonic as well as subsonic winds, the latter being referred to as a breeze regime (e.g. [24])

angular-momentum distributions, couplings to magnetic fields, chemical mixing, not to mention multi-dimensional effects. The simulations of CCSNe and of JetSNe face crucial problems of micro- and macro-physics nature. Aborted model explosions are currently commonplace. PCCSNe appear to be excluded, and so are 1D DCCSNe. Multi-dimensional simulations leave some hope through the interplay between fluid instabilities, acoustic waves, rotation, magnetic fields and neutrinos. Mild or weak explosions of stars developing O-Ne cores or of accreting and rotating O-Ne white dwarfs have been obtained thus far, sometimes at the expense of an artificial enhancement of the neutrino luminosity. Detailed three-dimensional simulations are most needed in order to clarify the role of various mechanisms listed above, and their precise couplings.

5 Nucleosynthesis associated with CCSN events

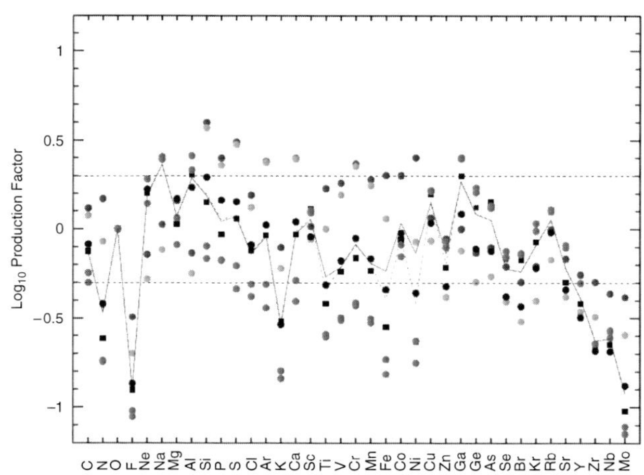

Fig. 11. Production factors for the elements between C and Mo following the SN explosion of stars with metallicity $Z = 0.02$ and with different masses (13 $M_\odot$: triangles; 15 $M_\odot$: squares; 20 $M_\odot$: open circles; 25 $M_\odot$: filled circles; 30 $M_\odot$: pentagons; 35 $M_\odot$: asterisks in open circles). In all cases, the production factors have been normalised to an oxygen production factor of unity. The lines refer to production factors obtained by integrating over a Salpeter IMF ($dn/dM \propto M^{-2.35}$). The solid line refers to an assumed relation between the mass of the SN residue, which is equivalent to the ^{56}Ni yields, and the progenitor mass (0.15, 0.10, 0.08, 0.07, 0.05 and 0.05 $M_\odot$ for the 13, 15, 20, 25, 30 and 35 $M_\odot$ stars). The dotted line is obtained by assuming that the ^{56}Ni yield of each SN is 0.05 $M_\odot$ (from [28])

Detailed nucleosynthesis predictions are in general available for 1D explosion models only, under the assumption that a shock wave propagates outward through most of the supernova structure, compresses the various traversed layers, heats them up before pushing them successfully outward until their ejection into the ISM. This expansion is of course accompanied by the cooling of the material. This heating and cooling process of the layers hit by the supernova shock wave allows some nuclear transformations to take place during a quite brief time, modifying more or less significantly the pre-explosion composition of the concerned layers. The study of the composition of the ejected material that makes up the supernova remnant is one of the main chapters of the theory of 'explosive nucleosynthesis.'

These lectures are mainly concerned with the explosive production of the r- and p-nuclides heavier than iron. The explosive yields of the lighter nuclides have been the subject of many calculations, and are not reviewed in any detail here. Let us just emphasize that the absence of fully self-consistent CCSN explosions imposes a parametric approach of the associated nucleosynthesis. It involves in particular the choice of a suitable final kinetic energy of the ejecta compatible with the observations (of the order of 10^{51} ergs), or the selection of the ejected amount of ^{56}Ni that can influence the SN light curve. This amount relates to the mass of the residue left from the explosion, as ^{56}Ni is the result of the explosive burning of Si in the deepest stellar layers. The mass of the residue is indeed a free parameter in absence of self-consistent SN simulations. The propagation of the outward moving shock wave and the properties of the shocked layers are computed in various ways (e.g. [28]).

Roughly speaking, the most abundant species are the result of the pre-explosion nucleosynthesis, while less abundant ones can be more or less substantially produced explosively. Figure 11 provides an example of the composition of the ejecta from various stars with metallicity $Z = 0.02$. The result of an integration of the yields from individual stars over an Initial Mass Function (IMF) is also displayed. It appears that most of the resulting yields are compatible (within a factor of about 2) with the oxygen one. Deviations may be due to the neglect of the contribution from stars with masses lower than 13 $M_\odot$ stars, from SNIa explosions, or from stars with metallicities different from $Z = 0.02$. It has also to be stressed that many uncertainties of different natures remain. At the level of individual stars, they concern in particular the pre-SN and SN evolutionary stages (including multi-dimensional effects), or the rate of certain key reactions, like $^{12}\mathrm{C}\,(\alpha\,,\gamma)\,^{16}\mathrm{O}$. Many more large uncertainties remain at the level of the evolution of the nuclidic content of the galaxies predicted by so-called 'chemical evolution models'.

6 The synthesis of the nuclides heavier than iron: generalities

Since the early days of the development of the theory of nucleosynthesis (e.g. [29]), it has proved operationally rewarding to introduce three categories of heavy nuclides referred to as s-, p-, and r-nuclides. This splitting is not a mere game. It corresponds instead to the 'topology' of the chart of the nuclides, which exhibits three categories of stable heavy nuclides: those located at the bottom of the valley of nuclear stability, called the s-nuclides, and those situated on the neutron-deficient or neutron-rich side of the valley, named the p- or r-nuclides, respectively. Three different mechanisms are called for to account for the production of these three types of stable nuclides. They are naturally referred to as the s-, r-, and p-processes.

Over the years, the development of models for the production of these three types of nuclides has largely relied on the bulk Solar System abundances, and on the decomposition of these abundances into the contributions from the s-, r- and p-processes. These data have been complemented with a myriad of spectroscopic observations from which heavy element abundances have been derived in a large sample of stars in different galactic locations and with different metallicities.

6.1 The bulk Solar System composition

Since the fifties, much effort has been devoted to the derivation of a meaningful set of elemental abundances representative of the composition of the bulk material from which the Solar System (referred to in the following as SoS) formed some 4.6 Gyr ago.

Early in the development of the theory of nucleosynthesis, it has been recognised that this bulk material is made of a well-mixed blend of many nucleosynthesis contributions to the SoS composition over the approximate 10 Gyr that have elapsed between the formations of the Galaxy and of the SoS. The latest detailed analysis of the SoS is due to [30]. As in previous compilations, the selected abundances are largely based on the analysis of a special class of rare meteorites, the CI1 carbonaceous chondrites, which are considered as the least-altered samples of available primitive SoS matter. Solar abundances derived from spectroscopic data now come in quite good agreement with the CI1 data for a large variety of elements. Some noticeable exceptions exist, however (e.g. [31]).

The bulk SoS isotopic composition shows a high level of homogeneity. This is why it is mostly based on the terrestrial data. For H and the noble gases, as well as for Sr, Nd, Hf, Os, and Pb, some adjustments are required [30]. There are exceptions, however, to this high bulk isotopic homogeneity. One is due to the decay of relatively short-lived radionuclides that existed in the early SoS, and decayed in early formed solids in the solar nebula. Also interplanetary dust particles contain isotopic signatures apparently caused by chemical processes. Additional isotopic 'anomalies' are observed in some meteoritic inclusions or grains (see Sect. 6.3).

As it is well known, the SoS nuclidic abundance distribution exhibits a high 'iron peak' centered around ^{56}Fe followed by a broad peak in the $A \approx 80 - 90$ mass region, whereas double peaks show up at $A = 130 \sim 138$ and $195 \sim 208$. These peaks are superimposed on a curve decreasing rapidly with increasing mass number. It has been realized very early that these peaks demonstrate the existence of a tight correlation between SoS abundances and nuclear neutron shell closures.

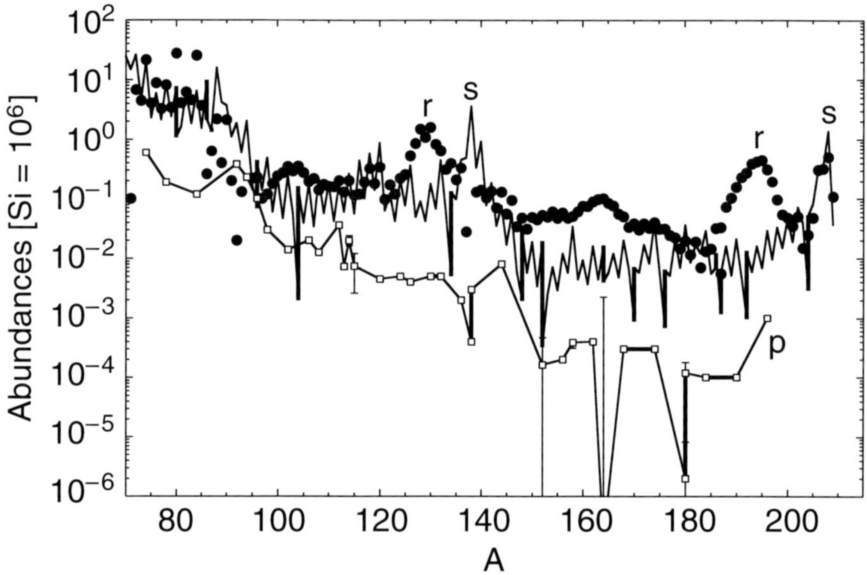

Fig. 12. Decomposition of the solar abundances of heavy nuclides into s-process (*solid line*), r-process (*black dots*) and p-process (*open squares*) contributions. The uncertainties on the abundances of some p-nuclides due to a possible s-process contamination are represented by vertical bars (from [32]). See Figs. 13 - 15 for the uncertainties in the s- and r-nuclide data

6.2 The s-, r- and p-nuclides in the Solar System

As mentioned above, it is very useful to split the abundance distribution of the nuclides heavier than iron into three separate distributions giving the image of the SoS content of the p-, s- and r-nuclides. A rough representation of this splitting is displayed in Fig. 12. In its details, the procedure of decomposition is not as obvious as it might be thought from the very definition of the different types of nuclides, and is to some extent dependent on the models for the synthesis of the heavy nuclides. These models predict in particular that the stable nuclides located on the neutron-rich (deficient) side of the valley of nuclear stability are produced, to a first good approximation, only by the r-(p-)process. These stable nuclides are naturally called 'r-only' and 'p-only' nuclides, and their abundances are deduced directly from the SoS abundances. The situation is more intricate for the nuclides situated at the bottom of the valley of nuclear stability. Some of them are produced solely by the s-process, the typical flow of which is located very close to the valley (see the discussion on s-process in this volume). They are referred to as 's-only' nuclides, and are encountered only when a stable r- or p-isobar exists, which 'shields' the s-isobar from the r- and p-processes. As a result, only even-Z heavy elements possess an s-only isotope. In general, a phenomenological model of the s-process is used to fit at best the abundances of all the s-only nuclides. Such a model is described in e.g. [24]. Once the parameters of this model have been selected in such a way, it is used to predict the s-process contributions to the other s-nuclides. The subtraction of these s-process contributions from the observed SoS abundances leaves for each isotope a residual abundance that represents the contribution to it of the r-process (if neutron-rich) or p-process (if neutron-deficient). These nuclides of mixed origins are called 'sr' or 'sp' nuclides.

Figure 12 shows that about half of the heavy nuclei in the SoS material come from the s-process, and the other half from the r-process, whereas the p-process is responsible for the production of about 0.01 to 0.001 of the abundances of the s- and r-isobars, except in the Mo-Ru region. It also appears that some elements have their abundances dominated by an s- or r-nuclide. They are naturally referred to as s- or r-elements. Clearly, p-elements do not exist. If these global abundance patterns remain valid in other locations than the SoS, stellar spectroscopy can provide information on the s- or r- (but not on p-) abundances outside the SoS. Even if the dominance of the s- or r-processes on a given element remains true in all astrophysical locations, a wealth of observations demonstrate departures from the SoS s- and r-element mixtures. As mentioned above, such departures exist in the SoS itself in the form of isotopic anomalies, or in stars with different ages, galactic locations, or evolutionary stages. The SoS abundances and their s-, r- and p-process contributions do not have any 'universal' character.

From the above short description of the splitting strategy between s-, r- and p-nuclides, it is easily understood that uncertainties affect the relative s- and r-(p-)process contributions to the SoS abundances of the sr(p)-nuclides. Even so, they are quite systematically put under the rug. This question clearly deserves a careful study, especially in view of the sometimes very detailed and far-reaching considerations that have the s-r SoS splitting as an essential starting point.

The splitting of the SoS s-, r- and p-nuclide abundances has been reviewed in some detail by [32, 24]. In view of its importance, we repeat here most aspects of the procedure.

As recalled above, the SoS r-nuclide abundance distribution is obtained by subtracting from the observed SoS abundances those predicted to originate from the s-process. These predictions are classically based on a parametric model, referred to as the canonical exponential model initially developed by [33], and which has received some refinements over the years (e.g. [34]). This model assumes that stellar material composed only of iron nuclei is subjected to neutron densities and temperatures that remain constant over the whole period of the neutron irradiation. In addition, the SoS s-abundance pattern is viewed as originating from a superposition of two exponential distributions of the time-integrated neutron exposure, $\tau_{\rm n} = \int_0^t N_{\rm n} v_T {\rm d}t$ (where $N_{\rm n}$ is the neutron number density, and v_T is the most probable relative neutron-nucleus velocity at temperature T). These distributions are traditionally held responsible for the so-called weak ($70 \lesssim A \lesssim 90$) and main ($A \gtrsim 90$) components of the s-process. A third exponential distribution is sometimes added in order to account for the $204 < A \leq 209$ s-nuclides. Through an adequate fitting of the parameters of the three τ-distributions, the superposition of the two or three resulting abundance components reproduces quite successfully the abundance distribution of the s-only nuclides in the SoS, from which it is concluded that the s-contribution to the sr-nuclides can be predicted reliably. It has to be stressed that this result is rooted only in the nuclear properties of the species involved in the s-process, and does not rely at all on specific astrophysics scenarios. Many s-process calculations have been performed in the framework of models for stars of various masses and initial compositions (e.g. [24], and other contributions to this volume). Some model calculations along the line have been used to obtain the contributions of the s- and r-processes to the SoS abundances [35,36]. This procedure is currently not advisable. Large uncertainties remain in the s-abundances predicted from individual model stars. In addition, the SoS s-nuclide abundances result from a long evolution of the galactic composition that cannot be mimicked reliably enough.

Despite the success of the canonical model in fitting the solar s-nuclide distribution, some of its basic assumptions deserve questioning. This concerns in particular a presumed exponential form for the distribution of the neutron exposures τ, which has been introduced by [33] in view of their mathematical ease in abundance calculations. In addition, the canonical model makes it difficult in the s-nuclide abundance predictions to evaluate uncertainties of nuclear or observational nature. As a result, the concomitant uncertainties in the solar r-abundances are traditionally not evaluated. The shortcomings of the canonical model are cured to a large extent by the so-called multi-event s-process model (MES) [37]. In view of the importance to evaluate the uncertainties affecting the solar distribution of the abundances of the r-nuclides, we review the MES in some detail. A similar multi-event model has also been developed for the r-process (MER), and is presented in [38].

The MES relies on a superposition of a given number of canonical events, each of them being defined by a neutron irradiation on the ^{56}Fe seed nuclei during a time $t_{\rm irr}$ at a constant temperature T and a constant neutron density $N_{\rm n}$. In contrast to the canonical model, no hypothesis is made concerning any particular distribution of the neutron exposures. Only a set of canonical events that are considered as astrophysically plausible is selected a priori. We adopt here about 500 s-process canonical events covering ranges of astrophysical conditions that are identified as relevant by the canonical model, that is, $1.5 \times 10^8 \leq T \leq 4 \times 10^8$ K, $7.5 \leq \log N_{\rm n}[{\rm cm}^{-3}] \leq 10$, and 40 chosen $t_{\rm irr}$-values, corresponding to evenly distributed values of $n_{\rm cap}$ in the $5 \leq n_{\rm cap} \leq 150$ range, where

$$n_{\rm cap} = \sum_{Z,A} A\ N_{Z,A}(t = t_{\rm irr}) - \sum_{Z,A} A\ N_{Z,A}(t = 0) \qquad (1)$$

is the number of neutrons captured per seed nucleus (^{56}Fe) on the timescale $t_{\rm irr}$, the summation extending over all the nuclides involved in the s-process. For each of the selected canonical events, the abundances $N_{Z,A}$ are obtained by solving a reaction network including 640 nuclear species between Cr and Po. Based on these calculated abundances, an iterative inversion procedure described in [39] allows us to identify a combination of events from the considered set that provides the best fit to the solar abundances of a selected ensemble of nuclides. This set includes 35 nuclides comprising the s-only nuclides, complemented with ^{86}Kr and ^{96}Zr (largely produced by the s-process in the canonical model), ^{152}Gd and ^{164}Er (unable in the p-process and able in the s-process to be produced in solar abundances [32]), and ^{208}Pb (possibly produced by the strong s-process component in the canonical model).

On grounds of the solar abundances of [40], it has been demonstrated in [37] that the derived MES distribution of neutron irradiation agrees qualitatively with the exponential distributions assumed in the canonical model, even though some deviations are noticed with respect to the canonical weak and strong components.[5] The MES provides an excellent fit to the abundances of the 35 nuclides included in the considered set of species, and in fact performs to a quite-similar overall quality as that of the exponential canonical model predictions of [40]. Even a better fit than in the canonical framework is obtained for the s-only nuclides (see [37] for details). The MES model is therefore expected to provide a decomposition of the solar abundances into their s- and r-components that is likely to be more reliable than the one derived from the canonical approach for the absence of the fundamental assumption of exponential distributions of neutron exposures.

Compared with the canonical approach, the MES model has the major advantage of allowing a systematic study of the various uncertainties affecting the abundances derived from the parametric s-process model, and consequently the residual r-nuclide abundances. The uncertainties in these residuals have been evaluated in detail by [37] from due consideration of the uncertainties in (i) the observed SoS abundances as given by [40] (see footnote[1]), (ii) the experimental and theoretical radiative neutron-capture rates involved in the s-process network, and in (iii) the relevant β-decay and electron-capture rates. Total uncertainties resulting from a combination of (i) to (iii) have finally been evaluated. The results of such a study for the elements with $Z \geq 30$ are displayed in Figs. 13 and 14. The corresponding SoS isotopic r-residuals

[5] A MES calculation with the revised solar abundances [30, 31] has not been done, but is expected not to give significantly different results from those reported here.

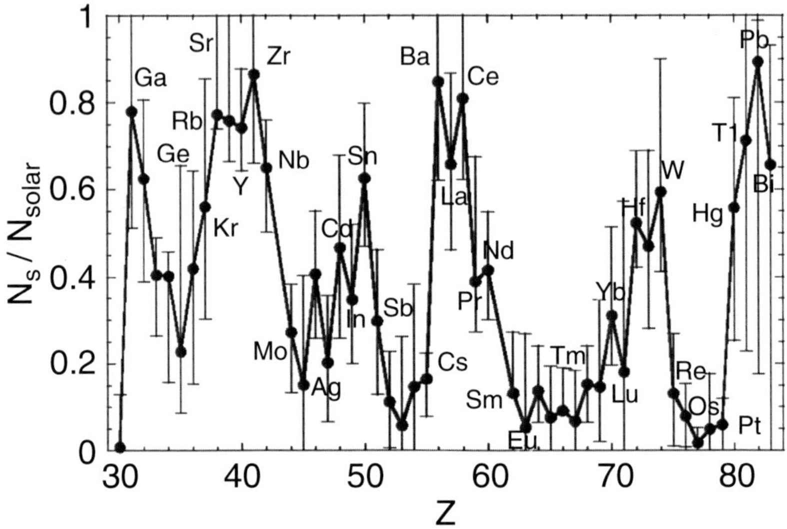

Fig. 13. MES predictions of the s-process contribution to the SoS abundances N_{solar} [40] of the elements with $Z \geq 30$. Uncertainties are represented by vertical bars (from the calculations of [37])

and their uncertainties are shown in Fig. 15. They are tabulated in [24]. Different situations can be identified concerning the uncertainties affecting the r-residuals. Many sr-nuclides are predicted to have a small s-process component only. The r-process contribution to these species, referred to as r-dominant, is clearly quite insensitive to the s-process uncertainties. The situation is just the opposite in the case of s-dominant nuclides.

Some r-process residuals are seen to suffer from remarkably large uncertainties, which quite clearly cannot be ignored when discussing the r-process and the virtues of one or another model for this process. This concerns in particular the elements Rb, Sr, Y, Zr, Ba, La, Ce and Pb. Some of them, and in particular Ba or La, are often used as tracers of the levels of s- or r-processing during the galactic history (see Sect. 6.4). Lead has also a special status in the studies of the s-process (e.g. [89] for references), as well as of the r-process (see Sect. 6.6). It could well be of pure s-nature if a strong s-process component can indeed develop in some stars, but a pure r-process origin cannot be excluded. These uncertainties largely blur any picture one

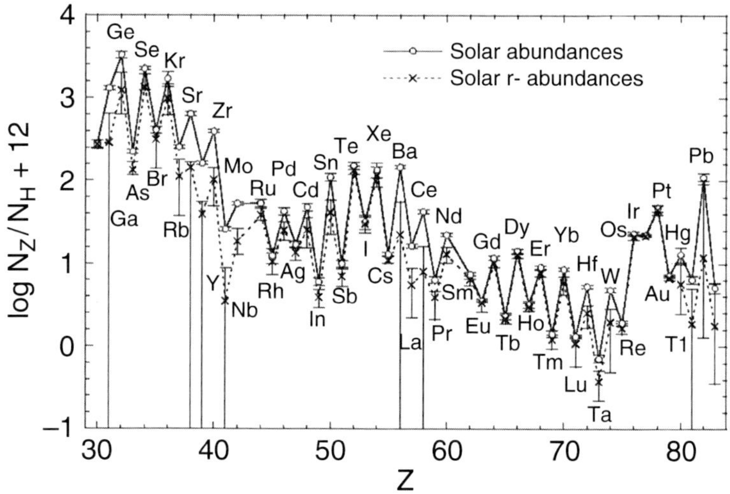

Fig. 14. SoS r-residuals and their uncertainties for the $Z \geq 30$ elements based on the s-abundances of Fig. 13. The abundances N_{Z} and N_{H} refer to element Z and to H. The ordinate used here is about 1.55 dex larger than the [Si=10^6] unit used elsewhere

might try to draw from spectroscopic observations and from simplistic theoretical considerations.

6.3 Isotopic anomalies in the solar composition

The bulk SoS composition has been of focal interest since the very beginning of the development of the theory of nucleosynthesis. Further astrophysical interest and excitement have developed with the discovery of the fact that a minute fraction of the SoS material has an isotopic composition deviating from that of the bulk. Such 'isotopic anomalies' are observed in quite a large suite of elements ranging from C to Nd (including the rare gases), and are now known to be carried by high-temperature inclusions of primitive meteorites, as well as by various types of meteoritic grains. The inclusions are formed from SoS material out of equilibrium with the rest of the solar nebula. The grains are considered to be of circumstellar origin, and to have survived the process of incorporation into the SoS.

Isotopic anomalies contradict the canonical model of an homogeneous and gaseous protosolar nebula, and provide new clues to many astrophysical

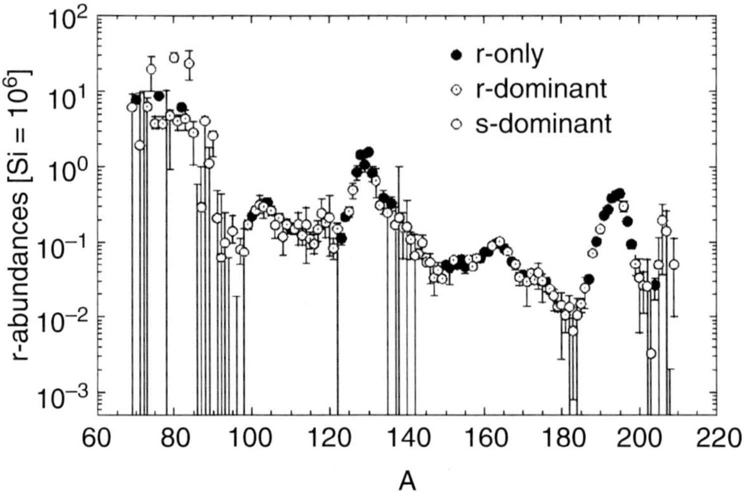

Fig. 15. SoS isotopic r-residuals corresponding to the elemental abundances of Fig. 14. Different symbols identify different relative levels of r-process contribution. The s-dominant nuclides are defined here as those predicted by MES to have more than 50% of their abundances produced by the s-process. The s-process contribution varies between 10 and 50% in the case of the r-dominant species, and does not exceed 10% for the r-only nuclides

problems, like the physics and chemistry of interstellar dust grains, the formation and growth of grains in the vicinity of objects with active nucleosynthesis, the circumstances under which stars (and in particular SoS-type structures) can form, as well as the early history of the Sun (in the so-called 'T-Tauri' phase) and of the SoS solid bodies. Last but not least, they raise the question of their nucleosynthesis origin and offer the exciting perspective of complementing the spectroscopic data for chemically peculiar stars in the confrontation between abundance observations and nucleosynthesis models for a very limited number of stellar sources, even possibly a single one. This situation is in marked contrast with the one encountered when trying to understand the bulk SoS composition, which results from the mixture of a large variety of nucleosynthesis events, and consequently requires the modelling of the chemical evolution of the Galaxy.

Among the identified anomalies, several concern the p-, s- and r-nuclides. Those attributed to the r- and p-processes are discussed in some detail by [32, 24], and are not reviewed here. As a very brief summary, let us just state that various blends of p, s-, and r-nuclides that differ more or less markedly from the bulk SoS mixture depicted in Sect. 6.2 are identified in a variety of meteorites at various scales, including bulk samples, refractory inclusions or grains interpreted from their many highly anomalous isotopic signatures as grains of circumstellar origins. This is generally interpreted in terms of the decoupling between the three mechanisms producing these nuclides, and of a non-uniform mixing of their products. One of the surprises of main relevance to these lectures is that those grains that are generally interpreted in terms of supernova condensates do not carry the unambiguous signature of the r-process that would be expected if indeed supernovae are the privileged r-process providers (see Sect. 8).

6.4 Evolution of the r-nuclide content of the Galaxy

In practice, the question of the evolution of the galactic content of the nuclides heavier than iron concerns the s- and r-nuclides only. It is traditionally assumed indeed that the p-nuclides are just as rare in all galactic locations as in the SoS. In such conditions, the p-nuclide abundances outside the SoS are out of reach of spectroscopic studies. On the other hand, the s-process and its contribution to the galactic abundances at different epochs are discussed elsewhere in this volume, so that we focus here only on the evolution of the r-nuclide galactic content.

A substantial observational work has been conducted in recent years on this subject. This effort largely relies on the abundance evolution of Eu, classified as an r-process element on grounds of SoS abundance analyses. The main conclusions derived from this observational work may be summarized as follows (see [24] for details and references).

(i) The Eu data summarised in Fig. 16 are classically used to support the idea that the r-process has contributed very early to the heavy element content of the Galaxy. However, the observed Eu abundance scatter introduces some confusion when one tries to establish a clear trend of the Eu enrichment with metallicity. It is also difficult to identify the value of [Fe/H] at which the signature of the s-process becomes identifiable. This conclusion relies in particular on La abundances derived from observation (see Fig. 17), La being classically considered as an s-element in the SoS (even if a non-negligible r-process contribution cannot be excluded). A most useful information on the relative evolution of the s- and r-process efficiencies in the Galaxy would be provided by the knowledge of the isotopic composition of the neutron-capture elements (see Sect. 6.5).
(ii) Much excitement has been raised by the observation that the patterns of abundances of heavy neutron-capture elements between Ba and Pb in r-process-rich metal-poor stars are remarkably similar to the SoS one. This claimed ‘convergence’ or ‘universality’ illustrated in Fig. 18 has to be taken

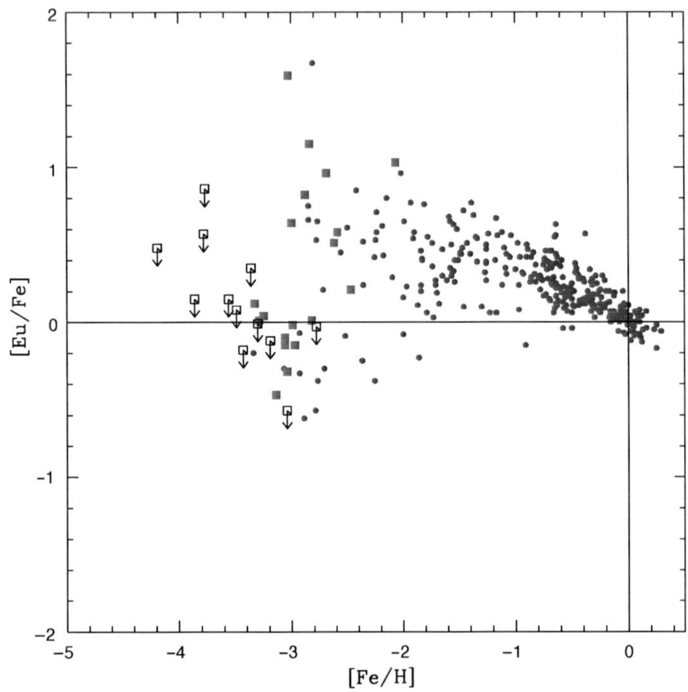

Fig. 16. Ratio [Eu/Fe] versus [Fe/H] for a large stellar sample. The data are from various sources, the details of which can be found in [42]

with some care, however, as it largely relies on the assumption that the decomposition between s- and r-process contributions in metal-poor stars is identical to the SoS one, which has yet to be demonstrated. An interpretation of this universality is proposed in Sect. 7.1.

(iii) No universality appears to hold for $Z \lesssim 58$ and for $Z \gtrsim 76$, where this concerns in particular the Pb-peak elements and the actinides. This situation has far-reaching consequences, particularly regarding the possibility of building galactic chronologies on the actinide content of very metal-poor stars (see Sect. 10). No universality is observed either for the elements lighter than Ba (like Sr, Y or Zr). This is illustrated in Fig. 19, which shows in particular that Sr has a quite different behaviour than the heavier Eu, with no clear identifiable trend of [Sr/Eu], in contrast to [La/Eu].

(iv) The different behaviours of the abundance patterns of the elements below and above Ba have laid the ground for speculations on possible different sites of the r-process. It is discussed by [24] that the interest of these speculations is rather limited at the present stage of desperate search for a single suitable site for the r-process (see Sects. 8 and 9).

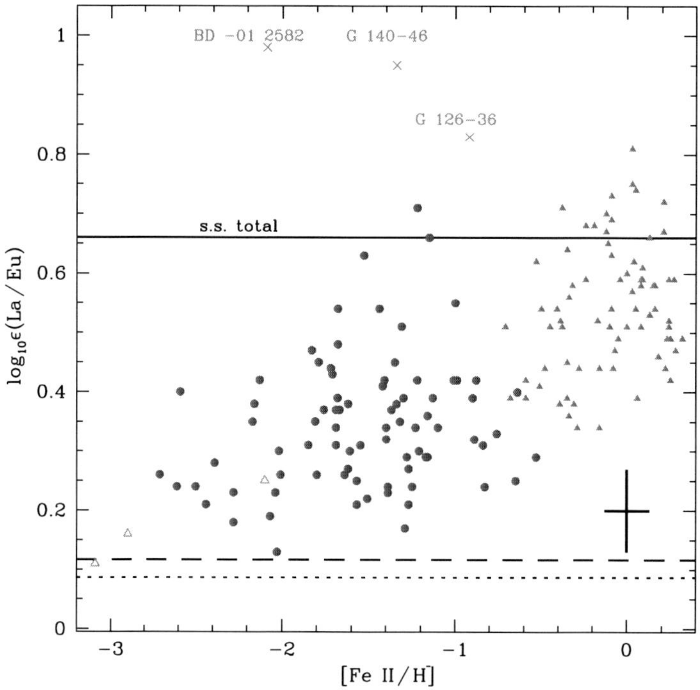

Fig. 17. Values of $\log_{10}\epsilon(\mathrm{La/Eu}) = \log_{10}(\mathrm{N_{La}/N_H}) - \log_{10}(\mathrm{N_{Eu}/N_H})$ versus [Fe/H] from various studies (different symbols). A typical error is shown as a dagger. The three labelled points correspond to metal-poor La-rich stars. The solid line is the total SoS La/Eu ratio, and the broken lines indicate the SoS s- and r-process abundance breakdowns (from [36], where the references to the observations are provided)

6.5 Can the available isotopic data tell something about the prevalence of the s- or of the r-process at early galactic times?

The considerations above rely on elemental abundance data. A most useful information regarding the relative evolution of the s- and r-process efficiencies in the Galaxy would be provided by the knowledge of the isotopic composition of the neutron-capture elements. Such data are unfortunately very scarce, and largely concern Ba only. The hyperfine splitting of its spectral lines has been used to evaluate the fractional abundance $f_{\mathrm{odd}} = [N(^{135}\mathrm{Ba}) + N(^{137}\mathrm{Ba})]/N(\mathrm{Ba})$ of the odd Ba isotopes. This ratio has indeed been considered as a measure of the relative contributions of the r- and s-processes to Ba. The SoS f_{odd} is about 0.18. The r-process fraction to Ba is given by $r/(r+s) = [f_{\mathrm{odd}} - f^{\mathrm{s}}_{\mathrm{odd}}]/[f^{\mathrm{r}}_{\mathrm{odd}} - f^{\mathrm{s}}_{\mathrm{odd}}]$. Here, the fractions $f^{\mathrm{s}}_{\mathrm{odd}}$ and $f^{\mathrm{r}}_{\mathrm{odd}}$ are of the odd-mass Ba isotopes in the cases of *pure* s- and *pure* r-processes, respectively, which for the time being cannot be evaluated in any other way than by analysing the SoS Ba isotopic compositions. Table 1 of [24] gives $0.06 \lesssim f^{\mathrm{s}}_{\mathrm{odd}} \lesssim 0.23$ with the 'standard' value being $f^{\mathrm{s}}_{\mathrm{odd}} \approx 0.10$. On the

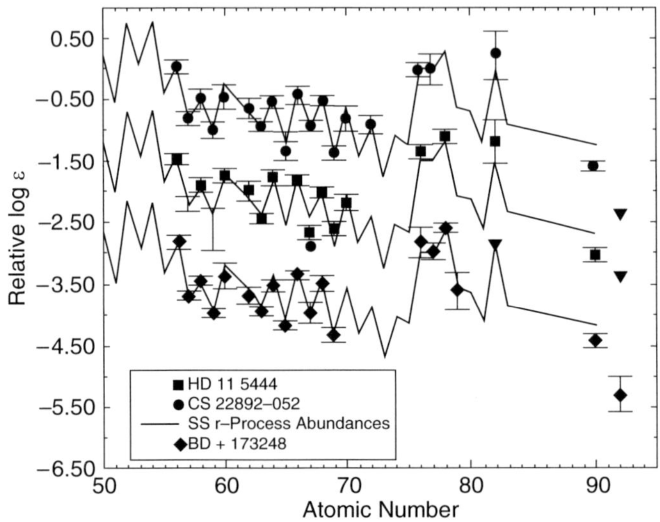

Fig. 18. Heavy-element abundance patterns for three heavy-element-rich metal-poor stars, the [Eu/Fe] ratios of which are included in Fig. 16. The solid lines represent a scaled SoS r-nuclide distribution. Inverted triangles indicate upper limits (from [43])

other hand, any value of $f^{\rm r}_{\rm odd}$ ($0 \lesssim f^{\rm r}_{\rm odd} \lesssim 1$) is permissible in consideration of the largely uncorrelated uncertainties, whilst its standard value is 0.66. On top of the uncertainties in $f^s_{\rm odd}$ and $f^r_{\rm odd}$, the derivation of $f_{\rm odd}$ from observation is also hampered with substantial difficulties, as illustrated by the diverging conclusions concerning HD 140283 by [46] who concludes that Ba in this star is of typical SoS s-r mix, and by [47] who instead claims that it is of pure r-process origin. A solar mixture of Ba isotopes is also preferred by [48] for a sample of metal-poor main sequence or close to main sequence cool stars. Note that all these statements are based on roughly the same $f_{\rm odd}$ values. The Ba isotopic composition has been analysed recently in 25 cool thick and thin disc dwarf stars with $-1.35 \leq$ [Fe/H] ≤ 0.25 [45]. The derived $f_{\rm odd}$ values are displayed in Fig. 20. They are seen to be smaller in the thin disc than in the thick disc, whose age is estimated to be comparable to the one of the halo. Note that the $f^s_{\rm odd}$ and $f^r_{\rm odd}$ values (0.10 and 0.46) indicated in Fig. 20 are those derived from the predictions of certain AGB models, and thus may not be reliable to be applied in the problem at hand because of the large intrinsic uncertainties of the models themselves. In particular, to combine [45]

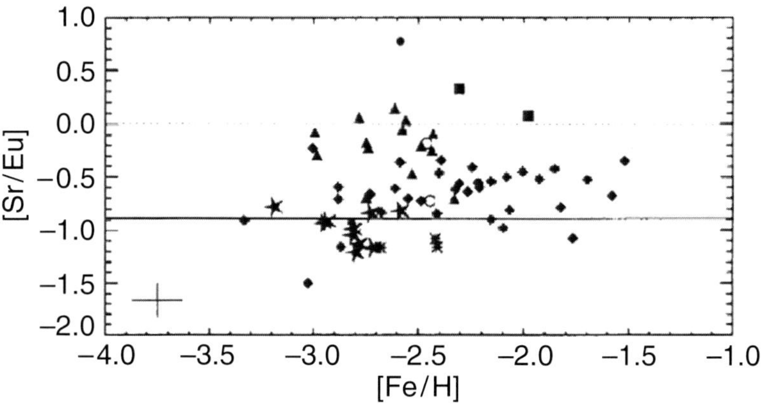

Fig. 19. Values of [Sr/Eu] versus [Fe/H] for an ensemble of stars with different metallicities. The average relative error is shown at the bottom left. The horizontal line represents an estimate of the pure r-process SoS value (from [44])

the metallicity-dependent AGB s-process models along the chemical evolution history of the Galaxy with the SoS isotopic composition is quite a dubious practice because it is then effectively dictating the possible metallicity dependence of the r-process. Nonetheless, with these or our 'standard' $f^s_{\rm odd}$ and $f^r_{\rm odd}$ values taken for granted as a measure, the observed $f_{\rm odd}$ trend seemingly suggests that the r-process contribution to Ba decreased during the galactic evolution. However, before deriving firm conclusions, the uncertainties in the $f_{\rm odd}$ values surely do not have to be swept under the rug.

6.6 Actinides in the Solar System, in the Local Interstellar Medium, and in stars

Actinides have a very special status in the theory of nucleosynthesis, as they are the only ones of clear and unique r-process origin. In addition, their radioactivity, and in some cases their lifetimes commensurable with the presumed age of the Galaxy, makes them potentially suited for chronological considerations.

For long, the abundances of the actinides have been known only in the SoS essentially through meteorite analyses [30]. The astrophysical importance of Th and U has been enhanced further with the first observation of Th in

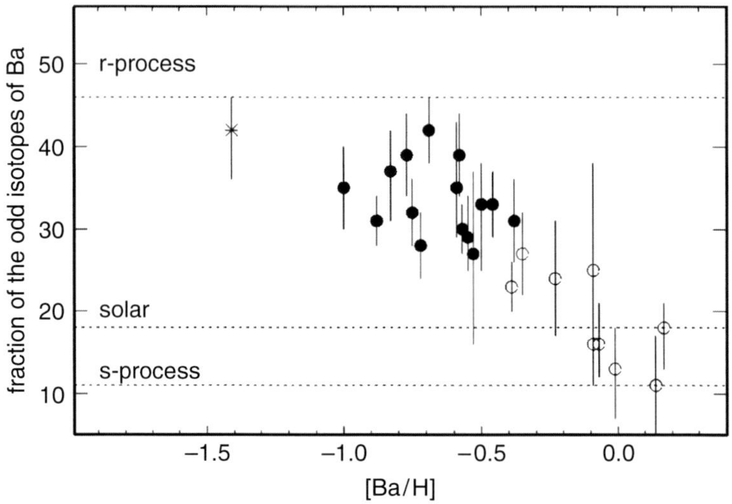

Fig. 20. Values of $f_{\rm odd}$ (in %) versus [Ba/H] for a sample of thick disc (black circles) and thin disc (open circles) stars with $-1.35 \leq$ [Fe/H] ≤ 0.25. The displayed s-process ($f^{\rm s}_{\rm odd}$) and r-process ($f^{\rm r}_{\rm odd}$) values (horizontal lines) are derived from an AGB star model (from [45])

some stars with different metallicities (Fig. 21). The Th data have been used by several authors in combination with the corresponding Eu abundances to evaluate the ages of the concerned stars. In this respect, it would clearly be desirable to rely instead on the Th/U abundance ratio in stars, as it is done in the SoS case. Unfortunately, detection of U is difficult because of the weakness of its spectral lines, combined with its present low abundance in the studied stars. In spite of these difficulties, U has been successfully measured in the [Fe/H] = -2.9 giant CS 31082-001 ([50], and references therein). Only upper limits have been obtained by [49] for the 7 stars in which they have derived Th abundances. The reliability of the age determination of specific stars based on the use of the Th/Eu and Th/U derived from observations is discussed in Sect. 10.

It has to be noted that ^{232}Th, ^{235}U and ^{238}U all decay to Pb. The Pb abundance has been measured in CS 31082-001 [50]. From this observation, it is concluded [50] that more than 50% of the total Pb in this star are the actinide progeny. This does not provide any strong constraint on the fraction of the Pb in CS 31082-001 that is a direct (instead of an actinides decay) product of the r-process. Lead in very metal-poor stars can indeed originate from the s-process as well [51]. It has also to be remarked that the SoS r-process Pb is highly uncertain, the fractional contribution of this process derived from

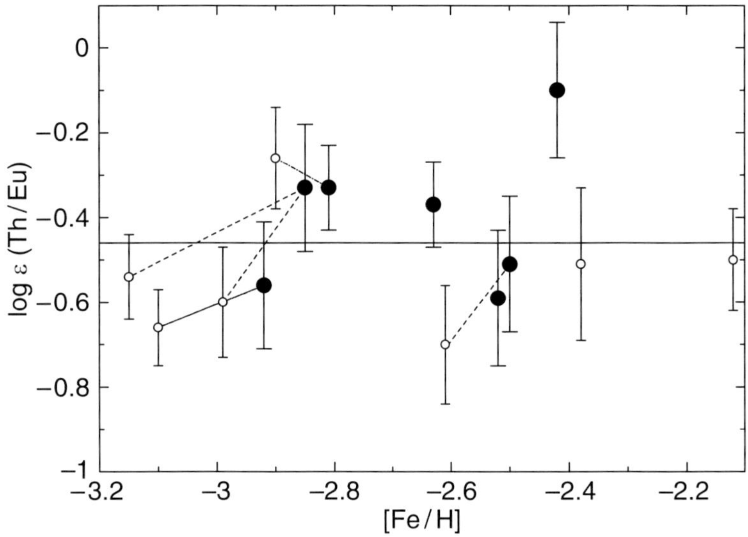

Fig. 21. Values of Th/Eu (in the ϵ scale defined in Fig. 17) versus [Fe/H] obtained from different observations (from [49]). The black circles are from [49], whereas the open circles are earlier observations with sometimes different [Fe/H] assignments. The solid horizontal line corresponds to the SoS value

Table 1 of [24] lying between 1 and 80%! In such conditions, Pb data in low-metallicity stars or in the SoS can hardly provide useful information on the r-process.

Finally, let us recall the attempts to measure the ^{244}Pu content in the local ISM, which may have some interesting astrophysical implications. At present, this can be done through the analysis of dust grains of identified interstellar origin recovered in deep-sea sediments (e.g. [52]). In a near future, the determination of elemental and isotopic composition of the ISM grains will be a major goal of research with their recovery to Earth by the Stardust mission [53].

6.7 The r-nuclide content of Galactic Cosmic Rays

The measured abundances in the Galactic Cosmic Rays (GCRs) of all the elements in the approximate $30 \lesssim Z \lesssim 60$ range of atomic numbers are roughly consistent with a solar composition at the source of the GCRs (the very nature of which remains quite mysterious) once corrections for atomic selection effects have been duly taken into account. In particular, no clear trend is identified in this range of atomic numbers for a specific enhancement or deficiency of either r- or s-nuclides. The elements with $Z \gtrsim 40$ might be overabundant *relative to* Fe. This conclusion is very sensitive, however, to the modelling of the GCR propagation conditions (see e.g. [54]).

The situation appears to be different in the $Z \gtrsim 60$ range, where observations suggest that the abundances in the 'Pt group' (Pt, Ir and Os) are in excess to those around Pb (the 'Pb group') relative to the SoS composition (e.g. [55]), this conclusion being relatively insensitive to the propagation conditions. Various interpretations of this relative excess in the Pt group have been proposed. One of them [54] relates to the high volatility of Pb and of other Pb-group elements, which contrasts with the refractory character of the Pt-group elements. In these views, the observations could well be consistent with a mixture of the SoS type. Another interpretation (e.g. [55]) calls for an r-process enrichment of the material to be accelerated to GCR energies. A note of caution is in order at this point. As in the analysis of stellar spectroscopic data, the discussion concerning the Pt group to the Pb group abundance ratio relies heavily on the SoS splitting between s- and r-nuclides. As already stressed before, these two contributions to Pb are very uncertain (these uncertainties are much lower in the Pt group case, the r-process contribution to which varies from about 85 to 100%; see Table 1 of [24]). Recall that GCRs are made of much younger material than the SoS (i.e., 20 $\sim$ 30 Myr old). There is no proof at this time that the r-process(es) that has(have) contributed to this recent sample of galactic material is(are) similar to the one(s) contained in the SoS material.

The identification of actinides in the GCRs has been made possible quite recently by the use of the Trek detector [55]. An accurate measurement of their abundances relative to each other and to the Pt group is within the reach of the planned Extremely Heavy Cosmic Ray Composition Observer (ECCO) [56]. Such data would in particular help discriminating between various GCR sources that have been proposed, including fresh supernova ejecta, supper-bubble material, or old galactic material (see [24] for some details and references).

As in the stellar case, information on the isotopic composition of the GCRs would be of prime interest in helping to evaluate the fractional contribution of freshly synthesised r-process material to this quite recent sample of galactic material. In fact, GCR composition measurements with isotopic resolution up to at least $Z \approx 40$ are within the reach of present detector technology, as exemplified by the R-process Isotope Observer RIO project currently under study [57].

7 The astrophysics of the r-process: parametrized site-free scenarios

7.1 Canonical and 'multi-event r-process (MER)' high-temperature models

It has already been proposed by [29] that the r-process results from the availability of neutron concentrations that are so high that neutron captures (especially of the radiative type) are faster than β-decays, at least for a substantial number of neutron-rich nuclides that are not located too far from the valley of nuclear stability. This is in marked contrast to the situation envisioned for the s-process. Such conditions clearly provide a natural way to transform any pre-existing material into very neutron-rich species, viewed as the progenitors of the r-nuclides. A classical additional hypothesis has been that the otherwise unspecified stellar location is hot enough to allow (γ,n) photodisintegrations to counteract to a more or less large extent the action of the inverse radiative neutron captures. Finally, it is supposed that a decrease of temperature that allows the 'freezing-out' of the photodisintegrations occurs concomitantly with a decrease of the neutron density to values that are low enough to freeze the neutron captures.[6]

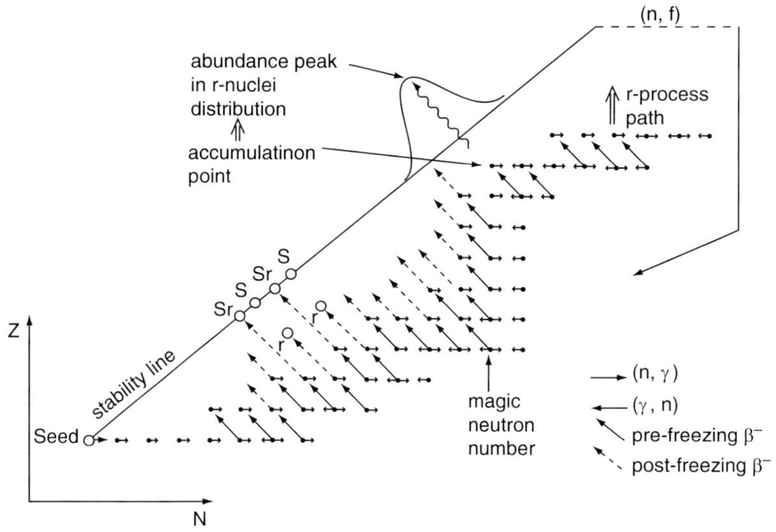

Fig. 22. Schematic representation of the r-process in the (N, Z)-plane

The aforementioned requirements on neutron concentration and temperature suffice to fix qualitatively several of the main features of the nuclear flow associated with the r-process and to identify the involved nuclear physics. Figure 22 depicts the situation very schematically. In the course of the transformation of a given seed into more neutron-rich isotopes by a series of (n,γ) reactions, (γ,n) photodisintegrations have a rate increasing with the neutron excess or, equivalently, with the associated decrease of the neutron separation energy $S_{\rm n}$. For low enough $S_{\rm n}$, the (γ,n) reactions counteract efficiently the radiative neutron captures. At this point, the nuclear flow may proceed to

[6] Let us recall that a transformation is said to be 'frozen' if its typical mean lifetime gets longer than a typical evolutionary timescale of the considered astrophysical site.

higher Z elements through β-decays. In this picture, the flow takes a special character at neutron closed shells. The especially low S_n values just past a magic neutron number indeed hinder the flow to proceed to more neutron-rich species, so that β-decays drive the material closer to the valley of stability following a path with increasing Z at practically constant N. The β-decays of the corresponding relatively less exotic nuclei become less probable and tend to slow down the nuclear flow. As a consequence, some material accumulates at nuclei with a magic neutron number. However, when the path gets close enough to the stability line, S_n is finally large enough to allow (n,γ) reactions to become more rapid than the β-decays, and to proceed without being largely counteracted by the (γ,n) photodisintegrations. The flow then resumes normally until a new neutron magic number is reached. In this picture, the accumulation of matter at neutron closed shell nuclei due to the relatively slow β-decay bottlenecks provides a natural explanation of the SoS r-process peaks (Fig. 15), as suggested in Fig. 22.

If the nuclear flow towards increasing Z values reaches the actinide or transactinide region, it is stopped by neutron-induced or β-delayed fissions which lead to a recycling of a portion of the material to lower Z values. At freezing of the neutron captures or inverse photodisintegrations, mainly β-decays, but also spontaneous or β-delayed fissions and single or multiple β-delayed neutron emissions, drive the neutron-rich matter towards the valley of stability. These post-freezing transformations are shown schematically in Fig. 22.

The hypothesized high neutron fluence and temperature has been the framework adopted by the vast majority of studies of the r-process. In many cases, the consequences of such an assumption have been scrutinized only from a purely nuclear physics point of view, just considering that one astrophysical site or the other, and in particular the inner regions of massive star supernova explosions, could be the required neutron provider. The early works of [29] and of [58] have proposed the simplest and most widely used form of the r-process scenario, referred to as the canonical r-process model (see e.g. [24] for details). It assumes constant thermodynamic conditions (temperatures, densities, neutron concentrations) during a given irradiation time. Over this period, pre-existing iron-peak material is driven by neutron captures into a location of the neutron-rich region determined by the neutron fluence and by the highly temperature-sensitive reverse photodisintegrations. When the irradiation stops, all neutron captures and photodisintegrations are abruptly frozen (this implies that the capture of neutrons produced in β-delayed processes is neglected). Even if this canonical model does not make reference to any specific astrophysics scenario, but builds on nuclear properties only, it has helped greatly paving the way to more sophisticated approaches of the r-process.

A parametric approach of the r-process referred to as the 'multi-event r-process (MER)' has been developed recently (see [24] for details). It drops some of the basic assumptions of the canonical model, but keeps the

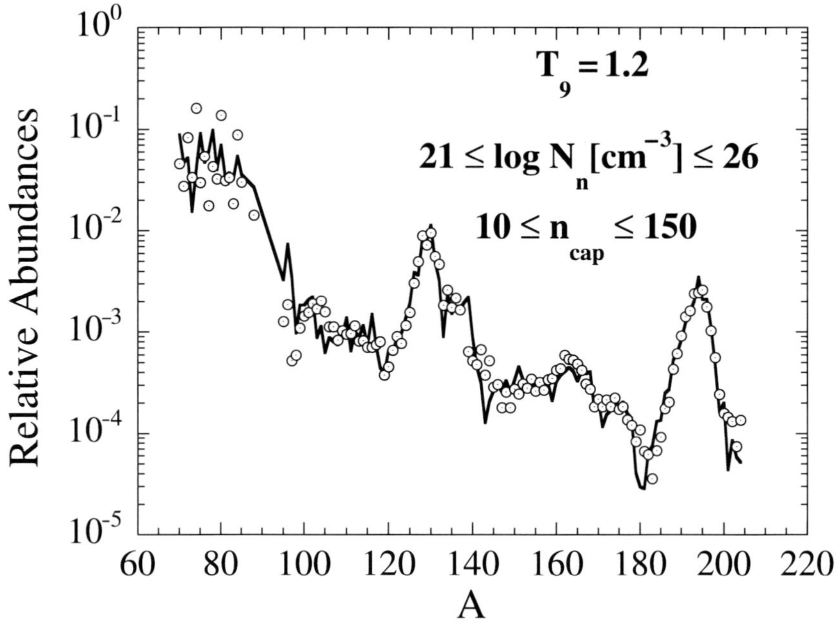

Fig. 23. Comparison between the SoS r-abundances (Fig. 15; the uncertainties are not shown) and a MER fit obtained assuming a constant temperature $T_9 = 1.2$ (in billion K) and a superposition of events characterized by free neutron concentrations $N_{\rm n}$ and numbers of neutrons captured by iron seed $n_{\rm cap}$ (see Eq. (1)) located in the indicated ranges. Within these prescribed limits, the characteristics of the events involved and their relative contributions to the displayed yields are derived from an iterative inversion procedure (see [24] for details and references)

simplification of constant thermodynamic conditions and neutron concentrations during the irradiation time, after which all nuclear reactions are frozen. It is an especially efficient tool to carry out systematic studies of the impact of uncertainties of nuclear physics nature on yield predictions, which, in this approach, necessitate the solution of huge nuclear reaction networks involving thousand of nuclear species, and the availability of an even larger body of input nuclear data (reaction and β-decay rates, fission probabilities,...). Figure 23 provides an example of fit to the SoS r-abundance distribution obtained with the use of MER.

Let us also note that MER has been used to interpret the high similarity of the SoS r-abundances between Ba and Os and those observed in r-process-rich metal-poor stars. This situation is generally interpreted as the signature of a 'universality' of the r-process (see Sect. 6.4). As reviewed by [24], the main conclusions drawn from the MER results are that (1) the pattern of abundances in the Ba to Os range is mainly governed by nuclear physics properties (and in particular by the fact that even Z elements have more stable isotopes that can be fed by the r-process). If this is indeed true, a possible universality in this Z range does not tell much about specific astrophysical conditions, and (2) the convergence of abundances in the above mentioned range does not provide any demonstration of any sort of a more global universality involving lighter and heavier elements. With time, these reservations have received mounting support from observation, as noted in Sect. 6.4.

7.2 Dynamical high-temperature r-process approaches (DYR)

In associating the r-process with supernova explosions, several attempts to go beyond the canonical and MER models have been made by taking into account some evolution of the characteristics of the sites of the r-process during its development. These models are coined 'dynamical' (DYR) in the following in order to remind of the time variations of the thermodynamic state of the r-process environment (see [24] for references). These models do not rely on any specific explosion scenario. They just assume that a material that is initially hot enough for allowing a nuclear statistical equilibrium (NSE) to be achieved expands and cools in a prescribed way on some selected timescale. This evolution is fully parameterized.

With the requirement of charge and mass conservation, and if the relevant nuclear binding energies are known, the initial NSE composition is determined from the application of the nuclear Saha equation (e.g. Sect. 7-2 of [17] for a general presentation) for an initial temperature and density (or, equivalently, entropy), and electron fraction (net electron number per baryon) Y_e. These three quantities are of course free parameters in a site-free r-process approach. The evolution of the abundances during expansion and cooling of the material from the NSE state is derived by solving an appropriate nuclear reaction network. The freeze-out of the charged-particle induced reactions might be followed by an r-process.

With temperature, density and Y_e as free parameters, many choices of initial NSE compositions may clearly be made, involving a dominance of light or heavy nuclides, as illustrated in Fig. 24. However, in view of its relevance to the supernova models, an initial NSE at temperatures of the order of 10^{10} K is generally considered. It favours the recombination of essentially all the available protons into α-particles (the region noted NSE [n,α] in Fig. 24). The evolution of this initial composition to the stage of charged-particle induced reaction freeze-out has been analyzed in detail by [60], and we just summarize here some of its most important features that are of relevance to a possible subsequent r-process:

(1) at some point in the course of the expansion and cooling of the initially α-rich material, full NSE breaks down as a result of the slowness of a fraction of the charged-particle reactions relative to the expansion timescale. The formation of quasi-equilibrium (QSE) clusters results. In this state, the intra-QSE composition still follows the NSE Saha equation, but the relative inter-cluster abundances do not, and depend on the kinetics of the nuclear flows into and out of the QSE clusters. To be more specific, the QSE phase is dominated in

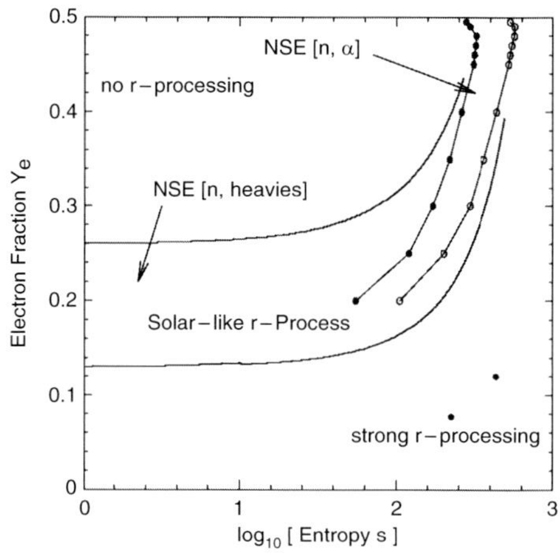

Fig. 24. The likelihood of a DYR r-process for given combinations of the electron fraction Y_e and the entropy per baryon s. A SoS-like r-process is expected for a suitable superposition of conditions between the black lines. The results inferred from an initial NSE phase at low s are smoothly connected to those of various nuclear network calculations for high s values. In the latter cases, the assumed expansion timescales imply that the freeze-out of the charged-particle induced reactions is reached after dynamical timescales τ_{dyn} in excess of about 50 - 100 ms. The two dotted lines represent the contours of successful r-processing for $\tau_{dyn} = 50$ ms (left line) and 100 ms (right line) (see [59] for details)

its early stages by a light cluster made of neutrons, α-particles and traces of protons, and by a cluster made of ^{12}C and heavier species. The population of the latter is determined mainly by the $\alpha + \alpha + n$ reaction, followed by $^9\text{Be}(\alpha, n)^{12}\text{C}(n, \gamma)^{13}\text{C}(\alpha, n)^{16}\text{O}$, as first noticed by [61].
(2) As the temperature decreases further, the QSE clusters fragment more and more into smaller clusters until total breakdown of the QSE approximation, at which point the abundances of all nuclides have to be calculated from a full nuclear reaction network. In the relevant α-particle-rich environment, the reaction flows are dominated by (α, γ) and (α, n) reactions with the addition of radiative neutron captures. Nuclei as heavy as Fe or even beyond may result. For a low enough temperature, all charged-particle-induced reactions freeze-out, only neutron captures being still possible. This freeze-out is made even more efficient if the temperature decrease is accompanied with a drop of the density ρ, which is especially efficient in bringing the operation of the ρ^2-dependent $\alpha + \alpha + n$ reaction to an end.

In the following, the neutron-rich α-rich process summarized above will be referred to as the α-process for simplicity, and for keeping the terminology introduced by [62]) (the α-process of [29] refers to a different nuclear process).

The composition of the material at the time of freeze-out depends on the initial Y_e, on the entropy s (see [60] for a detailed discussion), as well as on the dynamical timescale τ_{dyn}. The heavy nuclei synthesized at that moment may have on average neutron numbers close to the $N = 50$ closed shell, and an average mass number around $A = 100$. These nuclei can be envisioned to be the seeds for a subsequent r-process, in replacement of the iron peak assumed in the canonical and MER models. For a robust r-process to develop, favourable conditions have to be fulfilled at the time of the α-process freeze-out. In particular, the ratio at that time of the neutron concentration to the abundance of heavy neutron-rich seeds has to be high enough for allowing even the heaviest r-nuclides to be produced. As an example, $A = 200$ nuclei can be produced if an average of 100 neutrons are available per $A = 90$ nucleus that could emerge from the α-process, as it is the case if Y_e is close to 0.5 [62], (ii) at lower entropies if Y_e is low enough, and/or (iii) if the temperature decrease is fast enough for avoiding a too prolific production of heavy seeds. Figure 24 sketches in a semi-quantitative way the conclusions of the discussion above concerning the likelihood of development of a successful r-process in terms of entropy and Y_e.

7.3 A high-density r-process scenario (HIDER)

Early in the development of the theory of nucleosynthesis, an alternative to the high-T r-process canonical model (Sects. 7.1 and 7.2) has been proposed [63]. It relies on the fact that very high densities (say $\rho > 10^{10}$ gcm^{-3}) can lead material deep inside the neutron-rich side of the valley of nuclear stability as a result of the operation of endothermic free-electron captures, this so-called 'neutronisation' of the material being possible even at the $T = 0$ limit. The astrophysical plausibility of this scenario in accounting for the production of the r-nuclides has long been questioned, and has remained largely unexplored until the study of the composition of the outer and inner crusts of neutron stars and of the decompression of cold neutronised matter resulting from tidal effects of a black hole on a neutron-star companion ([24] for references). The decompression of cold neutron star matter has recently been studied further (Sect. 9).

In view of the renewed interest for a high-density r-process, a simple steady flow model, referred to in the following as HIDER, may be developed. Irrespective of the specific details of a given astrophysical scenario, it allows to follow in a very simple and approximate way the evolution of the composition of an initial cold (say $T = 0$) highly neutronized matter under the combined effect of β-decays and of the captures of free neutrons that are an important initial component of the considered material. These are the only two types of transformations that have to be considered if fissions are disregarded, and if any heating of the material resulting from the β-decay energy deposition is neglected, so that photodisintegrations can be ignored.

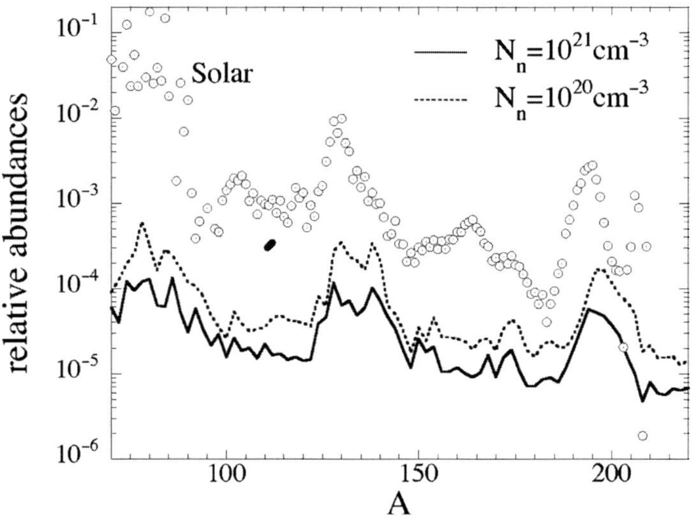

Fig. 25. Abundance distributions predicted by the steady flow HIDER for $N_n = 10^{20}$ and 10^{21} cm^{-3}. The details of the adopted nuclear physics can be found in [24]. The SoS abundances are shown for illustrative purposes

The predictions of HIDER under the additional assumption of a steady flow ($dN(A)/dt = 0$, $N(A)$ being the total abundance of all the isobars with mass number A; see [24] for details) are illustrated in Fig. 25. This model does roughly as well in reproducing the three SoS abundance peaks as a steady state high-T canonical model for comparable neutron densities. In other words, a high-T environment is not a necessary condition to account either for the location, or for the width of the observed SoS r-abundance peaks.

8 The neutrino-driven DCCSNe: a high-temperature site for the r-process?

The neutrino-driven winds that may accompany successful DCCSNe are certainly interesting from a purely hydrodynamical point of view, even if they may turn out not to be the key triggering agents of DCCSNe. In addition, their nucleosynthesis has been scrutinized in detail, especially following the excitement raised by the hope that they could provide a natural site for an α-process and for a subsequent dynamical r-process (Sect. 7.2, and e.g. [64] for early calculations). This hope has gained support from a one-dimensional DCCSN simulation of an iron-core progenitor predicting that entropies as high as about 400 could be attained in the wind more than 10 seconds after bounce (e.g. [64]). Such a high entropy allows the development of a robust r-process for a large variety of values of the neutron excess or Y_e and dynamical timescale τ_{dyn} (see Sect. 7.2 and Fig. 24). However, another one-dimensional iron-core DCCSN model has predicted about five times lower entropies, so that the development of an extended r-process is severely endangered [65]. The subsequent studies have confirmed that this r-process scenario could only be recovered at the expense of some twists that are difficult to justify in general (e.g. [66]).

In such an unsatisfactory state of affairs, the best one can do is to try to understand better the physics of neutrino-driven winds through the development of (semi-)analytical models some aspects of which may be inspired by (failed) explosion simulations, and to try to delineate on such grounds favourable conditions for the development of the r-process. These analytical models confirm that the wind nucleosynthesis depends on Y_e, entropy s, and τ_{dyn}, as in the α-process discussed in Sect. 7.2. The wind mass-loss rate $\dot{M}$ is influential as well. Ultimately, the quantities acting upon the synthesis in the neutrino-driven DCCSN model depend crucially on the details of the interaction of neutrinos with the innermost supernova layers, as well as on the mechanisms that might aid to get a successful DCCSN, and whose relative importance remains to be quantified in detail.

Several wind models of analytical nature exist. They differ in their level of physical sophistication and in their way to parametrize the wind characteristics. In all cases, the wind is assumed to be spherically symmetric, which appears to be a reasonable first approximation even in two-dimensional simulations, at least late enough after core bounce. In addition, the wind is generally treated as a stationary flow, meaning no explicit time dependence of any physical quantity at a given radial position. Newtonian and post-Newtonian descriptions of a spherically symmetric stationary neutrino-driven (supersonic) wind or (subsonic) breeze emerging from the surface of a PNS have been developed. The reader is referred to [24] for the presentation of a Newtonian, adiabatic and steady-state model for the wind and breeze regimes, and for a general-relativistic steady-state wind solution.

A comment is in order here concerning the relevance of the wind or breeze regimes. Is one of the two favoured by the DCCSN physics? This question is far from being just academic, as it is likely that its answer may have some impact on the predicted development of the r-process. It is quite intricate as well. One difficulty arises as the neutrino-driven material is likely not to flow unperturbed to infinity in a variety of DCCSN situations. The wind may in particular interact with matter and radiation in that portion of the star through which the SN shock has already passed. This interaction is likely to depend, among other things, on the pre-SN structure. As an example, it is clearly more limited as the mass of the outer layers decreases when going from massive SN II progenitors to SN Ib/c events whose progenitors (Wolf-Rayet stars) have lost their extended H-rich envelope prior to the explosion. The interaction of the material ablated from the PNS and the outer SN layers has several important consequences. It may give rise to a reverse shock responsible for the fallback of a more or less large amount of material onto the PNS, and whose properties (location and strength) alter more or less deeply the characteristics of the neutrino-ejected material. For large enough energies of the reverse shock, the wind may indeed transform into a breeze.

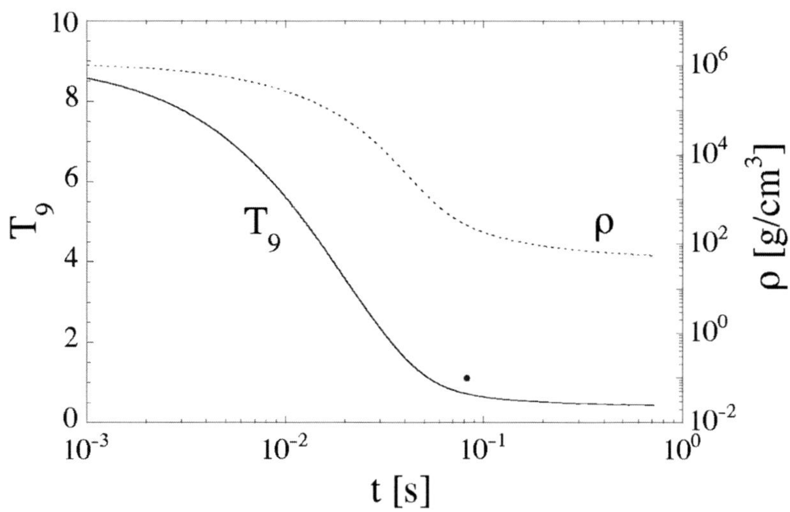

Fig. 26. Evolution of density ρ and of temperature T_9 (in 10^9 K) calculated in the Newtonian approximation for a breeze starting about 14 km away from a $1.5M_\odot$ PNS with a temperature $T_9 = 9$. The displayed solution depends on other parameters (including entropy and mass loss rate) that are discussed in [24]

Figure 26 displays the evolution of temperature and density for a particular breeze solution discussed in [24]. Some snapshots of the corresponding progressive build-up of heavy nuclei by the α-process, followed by the production of r-nuclides are shown in Fig. 27. The r-nuclide abundance distributions calculated for a breeze solution that is slightly different from the one displayed in Fig. 26 (even if the PNS mass and entropy are the same; see [24] for details) and three different mass loss rates (leading to three different cooling timescales) are shown in Fig. 28. The influence on the r-abundances of different

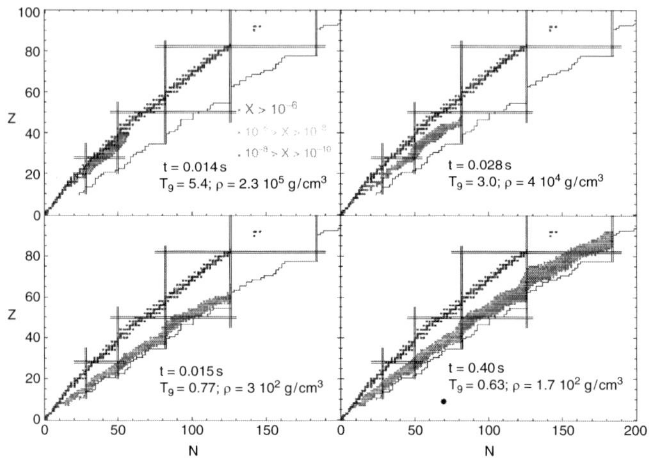

Fig. 27. Snapshots of the nuclear flows calculated for the breeze solution of Fig. 26. The first three panels ($t \leq 0.15$ s) describe the progressive build-up of heavy nuclei by the α-process. These act as the seeds for the r-process that develops after the α-process freeze-out, as shown in the last panel

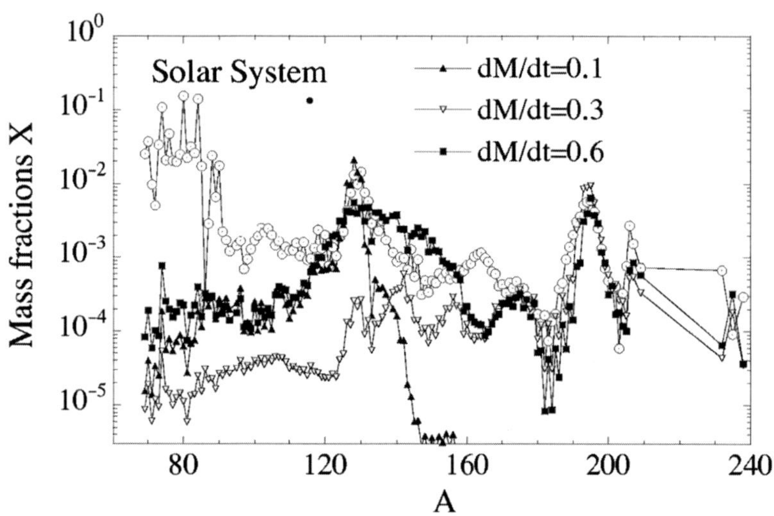

Fig. 28. Distribution of the r-nuclide abundances derived for a breeze solution obtained with a PNS mass of $1.5M_\odot$, an initial electron fraction $Y_e = 0.48$, and different values of the mass loss rate (in units of 10^{-5} $M_\odot$/s). The upper curve corresponds to the SoS r-nuclide abundances normalized to $\sum_i X_i = 1$ (see [24] for details, particularly on the selected breeze solution)

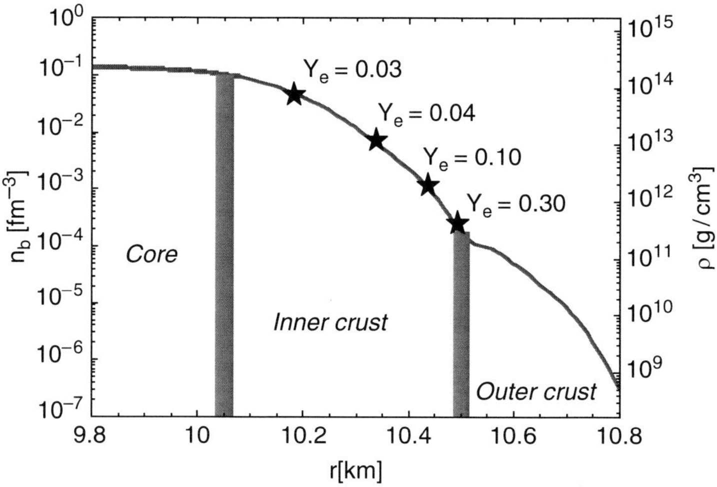

Fig. 29. Density profile (expressed in baryon number density n_b and matter density ρ in a typical 1.4 $M_\odot$ NS showing the structure of the crust. The values of the electron fraction Y_e are given at $\rho = 4\ 10^{11}, 2 \times 10^{12}, 10^{13}$ and 10^{14} g/cm^3

values of other breeze parameters is discussed by [24], as is the influence of the neutrino luminosity.

A discussion of the prospects of development of an r-process in a variety of multi-dimensional simulations including rotation, magnetic fields, or accretion in binary systems can also be found in [24]. In short, it is fair to say that no firm conclusion can be drawn on the possibility of development of the r-process in CCSNe.

9 Compact objects: a site for the high-density r-process scenario?

As reminded in Sect. 7.3, the decompression of the crust of cold neutron stars (NSs) made of a lattice of very neutron-rich nuclei immersed in a gas of neutrons and degenerate electrons has long been envisioned as a possible site for the development of a high-density r-process (HIDER). This decompression could result from the coalescence of two NSs or of a NS and a black hole (BH) in a binary system. It could also result from the ejection of material from magnetars.

The modelling of the coalescence of two NSs or of a NS and a BH has recently attracted a flurry of interest (see the review by [24]). These events are indeed considered as being among the strongest known sources of gravitational wave radiation, this emission being in fact responsible for the coalescence after

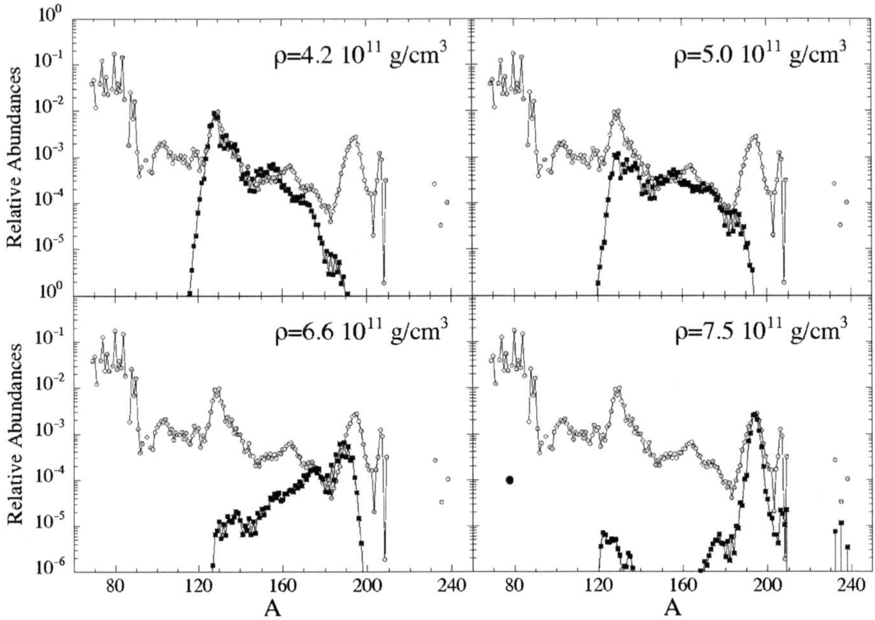

Fig. 30. Final composition of the ejected NS inner crust material with four different initial densities lower than 10^{12} g/cm^3 (black squares). The SoS r-abundance distribution is also shown by open dots (from [24])

typical times of the order of tens of millions to billions of years. They are also viewed as the likely progenitors of the class of short hard gamma-ray bursts.

Magnetars may result from certain JetSNe. Their properties are reviewed in several places (see [24] for references). They are observed as Soft Gamma Repeaters emitting sporadically bright bursts of energy over a period that is estimated to be of the order of 10000 years. Their most remarkable characteristic is their magnetic fields whose values are typically of the order of 10^{14-15} G, that is 100 to 1000 times larger than classical pulsar values.

The matter ejected from magnetars or from a NS coalescence can be highly neutron-rich, the distribution of Y_e in this material being in fact reminiscent of the neutron crust values (Fig. 29). Figure 30 shows the final composition of the material ejected from the NS inner crust at densities $\rho < 10^{12}$ g/cm^3. Although the composition is seen to be density-dependent, it is characterized by peaks with location and width similar to the SoS ones. For inner crust layers at higher densities ($\rho > 10^{12}$ g/cm^{-3}), large neutron-to-seed ratios bring the nuclear flow into the superheavy region, from which fission recycles material to lighter nuclides. Figure 31 shows the abundance distribution resulting from the decompression of a clump of material with initial density $\rho = 10^{14}$g/cm^3.

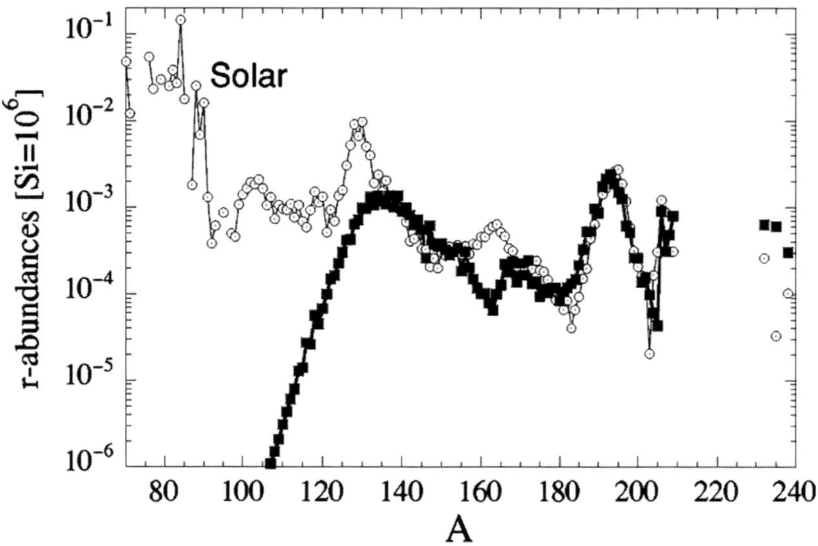

Fig. 31. Final r-abundance distribution (black squares) for a clump of material with initial density $\rho = 10^{14}\mathrm{g/cm}^3$ expanding on a timescale $\tau_{exp} = 6.5$ ms. The SoS r-abundance distribution is also shown

10 Some brief comments on the modelling of the evolution of the r-nuclide content of the Galaxy and on nucleo-cosmochronology

A quite natural astrophysicists' dream is to understand the wealth of data on the evolution of the r-nuclide content of the Galaxy that are accumulating from very many spectroscopic observations (Sect. 6.4). These observations clearly demonstrate a huge complexity that will probably keep rising as new observations become available. The best one can thus imagine at this stage is to explain broad trends which may be identified through the analysis of the r-nuclide abundance information. In this exercise, one has always to keep in mind that, at best, more or less reliable *elemental* abundances are derived from spectral analyses that often rely on approximate classically used techniques. This necessitates to disentangle the s- and r-process contributions to a given elemental abundance. It is generally done by assuming that these two nucleosynthesis contributions are at a largely metallicity-independent relative level, and thus do not differ widely from the SoS case. This assumption cannot yet be ascertained in any quantitative way, and is in fact not expected to hold for the s-process.

Various galactic chemical evolution toy models have been constructed, which often focus on the evolution of the abundances of two representative elements, Ba (a s-process element in the SoS) and Eu (a r-process element in the SoS). They adopt different schematic descriptions of the galactic halo and disk, and different prescriptions for the physical input quantities to these models. In particular, fully ad-hoc assumptions are made or free parameters are chosen concerning the r-process yields from stars of different masses and metallicities.

Some predictions for [Eu/Fe] have been made recently by [67]. They are based on a homogeneous one zone model in which it is assumed that stars in prescribed mass ranges produce an artificially selected amount of r-nuclides through the neutrino wind or prompt explosion mechanisms (Sects. 4 and 8). As expected, the predicted [Eu/Fe] ratio is very sensitive to the selected stellar mass ranges. This result might be optimistically considered as providing a way to constrain the site(s) of the r-process from observation. Reality is most likely less rosy, as very many uncertainties and severe approximations drastically blur the picture.

The assumption of the homogeneity of the interstellar medium at all times is dropped by [68]. The resulting inhomogeneous model might increase the plausibility of the predictions especially at early times in the galactic history. With the granularity of the nucleosynthesis events duly considered, one might hope to better account for the large observed scatter of the r-nuclide abundances at very low metallicities. In addition, the model of [68] takes into account the r-process contribution from NS mergers (Sect. 9) on top of the one from supernovae in selected mass ranges. The many other simplifications generally made in other chemical evolution models are also adopted by [68]. This concerns in particular the r-process yields from supernovae, as well as from NS mergers, that are just taken to be SoS-like. From their predicted [Eu/Fe] ratio, [68] conclude that the scenario assuming the predominance of SN II events in the 20 to 50 $M_{\odot}$ range allows the best fit to the observations. This result is obtained for total masses of r-nuclides per supernova varying from about $10^{-4} M_{\odot}$ down to about $10^{-7} M_{\odot}$ when going from 20 to 50 $M_{\odot}$ stars. Again, this conclusion has to be taken with great care in view of the many uncertainties and approximations involved in the chemical evolution model. Within the same model, it is also claimed that NS mergers are ruled out as the major source of r-nuclides in the Galaxy. This conclusion relies on a very approximate and highly uncertain time-dependent frequency of the events. In order to cope at best with observational constraints, coalescence timescales and amount of r-nuclides ejected per merger are adopted by [68] to vary from about 0.1 to $10^{-4} M_{\odot}$, depending upon other parameters of the NS merging model.

All in all, we consider that the galactic chemical evolution models devised up to now are by far too schematic and uncertain to provide a reliable tool to account for the observed evolution of the r-nuclide content of the Galaxy, or for constraining the possible sites of the r-process. A priority would clearly be at least to identify with a reasonable confidence a single site for the r-process before dwelling on the grand project of constructing models for the evolution of the r-nuclide content of a whole galaxy, and of the SoS in particular.

As far as nucleo-cosmochronology is concerned, the actinides produced by the r-process enter in particular attempts to estimate the age of the Galaxy through their present SoS content, or through their abundances evaluated at the surface of very metal-poor stars. These attempts face some severe problems related to the nuclear physics and astrophysics uncertainties that affect the predictions of the actinide production. The situation is worsened further by the especially large uncertainties in the contribution of the r-process to the solar-system Pb and Bi content (see [24] for a brief review and references). Concerning the ^{232}Th - ^{238}U and ^{235}U - ^{238}U pairs classically used to date the Galaxy from their present meteoritic abundances, the opinion has been expressed more than 20 years ago by [69] that they have just limited chronometric virtues. This is in marked contrast to a widely-spread and repeated claim following [70], and based on simple analytic models for the evolution of Th and U in the Galaxy. In addition, the chronometric predictions based on the observations of Th and U in very metal-poor stars have to be considered with great care (see [24] for a brief review and references). In order for them to be reliable, it is not only required that the production of the actinides by the r-process is well known, but also, and very decisively, that the production of Th with respect to U and to the Pt peak is universal. Observation demonstrates now that this is not the case.

The bottom line of this brief review of the r-process is that unanswered questions are by far more numerous than solved problems when one is dealing with this nucleosynthesis mechanism. They concern especially the astrophysics of the process, as no single site has been identified decisively yet. It also raises many nuclear physics questions. In such conditions, the modelling of the evolution of the r-nuclide content of the Galaxy and actinide-based chronometric evaluations cannot be based on solid grounds yet. This is in fact a very pleasing situation, as hope for many exciting discoveries is still ahead of us!

11 The p-process: Some generalities

The nuclear physics and astrophysics aspects of the p-process have been reviewed in substantial detail by [32]. We limit ourselves here to a summary of this review.

Quite clearly, the neutron-deficiency of the p-nuclides forbids their production in neutron capture chains of the s- or r-types. In contrast, and as already proposed very early in the development of the theory of nucleosynthesis [29], they could well be synthesized from the destruction of pre-existing s- or r-nuclides by different combinations of (p,γ) captures, (γ,n), (γ,p) or (γ, α) reactions. Some β-decays, electron captures or (n,γ) reactions can possibly complete the nuclear flow. These reactions may lead directly to the production of a p-nuclide. In most cases, however, they are synthesized through an unstable progenitor that transforms into the stable p-isobar through (a chain of) β-decay(s). Figure 32 displays in a very schematic way some possible nuclear routes through which seed s-(r-) nuclides can be transformed into p-nuclides.

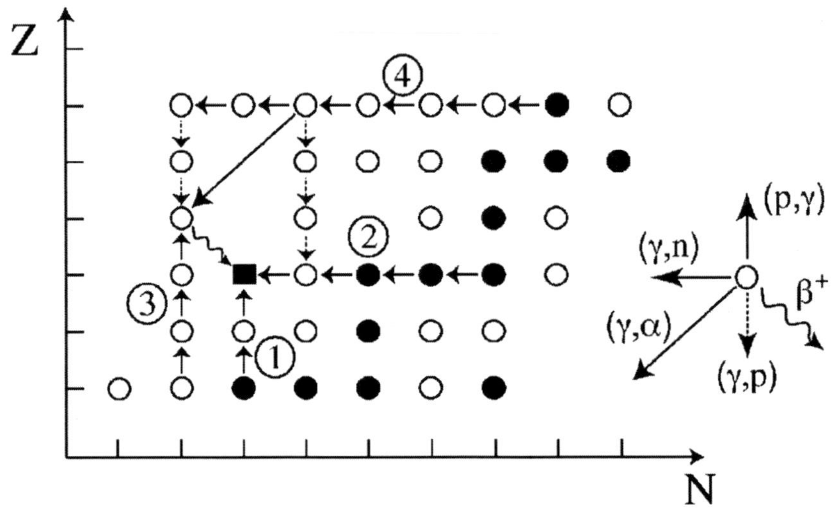

Fig. 32. Schematic representation of some simple possible nuclear routes through which seed s- or r-nuclides (black dots) can be transformed into a p-nuclide (black square). Unstable nuclei are represented by open dots. Routes (1) and (2) are made of a succession of (p,γ) and (γ,n) reactions leading directly to the p-nuclide. Slightly more complicated chains involve (p,γ) reactions followed by β-decays (route (3)), or a combination of (γ,n) and (γ,p) or (γ,α) and β-decays (route (4)). More complicated flow patterns involving combinations of the represented ones can also be envisioned. The p-nuclide destruction channels are not represented

The relative importance of the various ways to make p-nuclides from more neutron-rich seeds depends critically on temperature. This relates largely to the properties of the photodisintegration rates which are known to be strongly increasing functions of temperature. In fact, values typically in excess of about 1.5×10^9 K are required for the photodisintegrations to take place on timescales commensurable with stellar evolution ones. It is also mandatory for these reactions to freeze-out before a too strong photoerosion of the heavy nuclides, which would leave iron peak nuclei as the main end products. In practice, this forces temperatures not to exceed about 3.5×10^9 K during a short enough time only. These three constraints (abundant enough seed nuclei, high enough temperatures, short enough timescales for the hot phases) are obviously complemented by the necessity of considering proton-rich layers if (p,γ) captures are indeed envisioned.[7]

The details of the nuclear physics for the p-process have been discussed by [32], and are not reviewed here. Static properties have to be known for a large variety of neutron-deficient nuclei from about iron to lead located between the valley of nuclear stability and the vicinity of the proton-drip line, As made clear in Fig. 32, nucleon and α-particle capture rates by these nuclei are needed as well. Free electron captures may also be requested.

In spite of the general consensus that the p-nuclides largely originate from photo-eroded s- or r-nuclides, the identification of the stellar sites in which the p-process has a chance to develop has changed most dramatically since [29], where it has been considered that the constraints on the operation of the p-process mentioned above could be best satisfied in the H-rich envelopes of SN II explosions. This scenario has been put to its limits by [74]. At the end of the sixties already, it was realized that the required high temperatures are very unlikely to be reached in SN II envelopes. This is why it has been suggested by [75, 76] to locate the p-process in the deep O-Ne-rich layers of massive stars either in their pre-SN or SN phases. Subsequently, other sites have been considered as well, like SN Ia or exploding sub-Chandrasekhar mass white dwarfs following the accretion of He-rich material. Other possibilities are reviewed in detail by [32].

11.1 The p-process in SN IIe

Following the suggestion by [75,76], most of the investigations of the p-process have been concerned with the explosion of the O-Ne-rich layers associated with SN II events. The first calculations of this type [77] made use of parametrized conditions for the thermodynamic description of the relevant supernova layers, of solar s-nuclide seeds, and of a simplified reaction network. The most detailed studies performed so far [78, 79] are conducted in the framework of one-dimensional explosions of non-rotating solar-metallicity stars in the 13 to $15 \leq M \leq 25 M_{\odot}$ mass range.

These computations demonstrate that the p-process can develop in the O-Ne layers of the considered massive stars explosively heated to peak temperatures T_m comprised approximately between 1.7 and 3.3×10^9 K. These zones, referred to in the following as the p-process layers (PPLs), are located far enough from the base of the supposedly ejected material for their physical properties not to be affected drastically by the very complicated physics of the deeper SN II layers possibly of importance for the r-process (Sect. 8).

[7] Note that the above considerations concern only thermonuclear modes of p-nuclide synthesis. Some non-thermonuclear scenarios have in fact been proposed, like production by spallation reactions in the interstellar medium [71, 72], or photonuclear reactions triggered by non-thermalized photons [73]. These models suffer from either too low efficiencies, or from constraints that limit their astrophysical plausibility. They are not discussed further here.

In fact, what is basically required is just a successful explosion with a typical SN kinetic energy [32]. Another essential requirement is to have enough s- (or possibly r-) process seeds for the production of the p-nuclides. The s-process accompanying core He burning in massive stars can do the job. It strongly enhances the $60 \lesssim A \lesssim 90$ s-nuclei, while the abundances of the heavier nuclei are only very weakly increased.

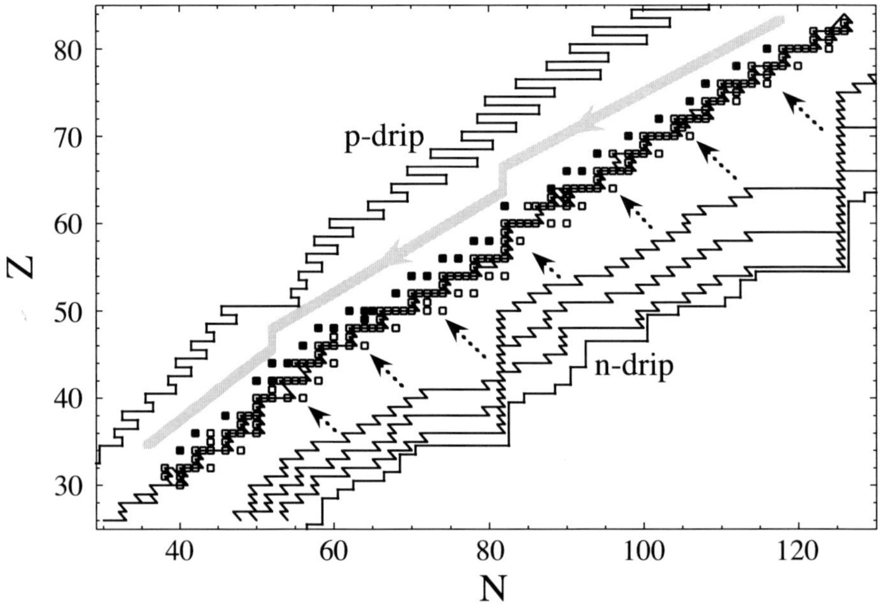

Fig. 33. Location in the (N, Z)-plane of the stable isotopes of the elements between Fe and Bi. The p-isotopes are represented by black squares, while both the s-, r-, sr- or sp-isotopes are identified with open squares (see Figs. 13 - 15 for details). The p-nuclides are the progeny of unstable neutron-deficient isobars located on the down-streaming p-process flow (thick black line; for more details on the p-process flow, see [32]). Some possible r-process flows derived from a high-temperature parametric model (Sect. 7) are also shown, as well as the up-streaming s-process flow (thin black line) confined at the bottom of the valley of nuclear stability. The proton and neutron drip lines correspond to the locations of zero proton and neutron separation energies

Figure 33 provides a schematic view of a typical SNII p-process flow. Its distinguishing feature is that it evolves from heavy to light nuclides as a result of the dominant action of photodisintegrations. Figure 34 displays p-nuclide yields in the form of normalized overproduction factors computed for a variety of SNII explosions of solar-metallicity stars (results for each star

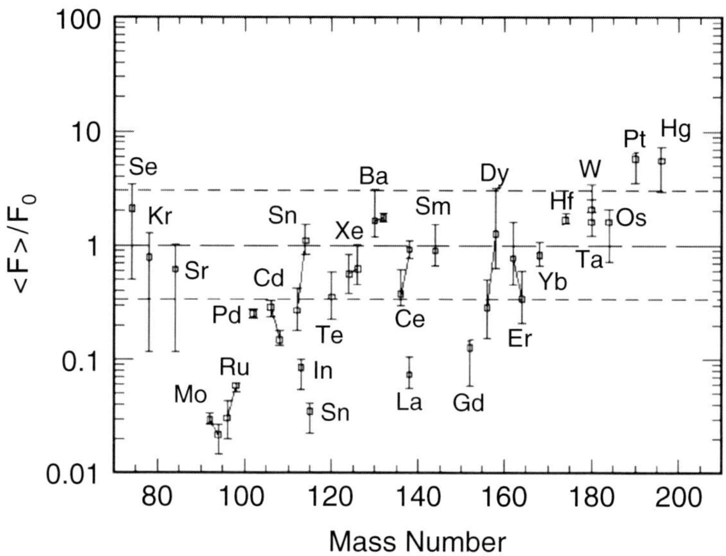

Fig. 34. Yields of the p-nuclides from SNIIe. They are expressed in terms of normalized overproduction factors $\langle F_i \rangle(M)/F_0(M)$. The mean overproduction factor $\langle F_i \rangle(M)$ for a star with mass M is defined as the total mass of the p-nuclide i produced in its PPLs divided by the corresponding mass if the PPLs had a solar composition. The normalizing factor $F_0(M)$ is the mean overproduction factor averaged over the 35 p-nuclides listed in [32]. With these definitions, all the normalized overproductions would be equal to unity if the derived abundance patterns were solar. Ranges of variations of the normalized p-nuclide overproductions represented by vertical bars cover the results obtained for individual SNII explosions of solar metallicity stars with masses in the 13 to 25 $M_\odot$ range. Open squares indicate the values of the normalized overproductions obtained by integration over the IMF proposed by [80]. Solid lines join different p-isotopes of the same element. The displayed ^{180}Ta data correspond to the sum of the ^{180}Ta$^{\rm g}$ and ^{180}Ta$^{\rm m}$ abundances. Roughly speaking, ^{180}Ta$^{\rm m}$ (the only Ta form to be present in the SoS) represents about half of the total ^{180}Ta SNII yields (see [32])

are not shown. Instead, only the ranges covered by the individual cases are represented). It is seen that, roughly speaking, the overproduction factors depend only relatively weakly on the star masses, this being less true, however, for the lightest p-nuclides ^{74}Se, ^{78}Kr and ^{84}Sr. As a zeroth-order approximation to the build-up of a full galactic chemical evolution model, the p-process yields have been averaged over an IMF, leading to the IMF-averaged normalized overproduction factors also displayed in Fig. 34. Of course, the build-up of a model for the temporal evolution of the galactic content of the p-nuclides not only requires the knowledge of the stellar mass dependence of their yields, but also their variations with the metallicity. This question has been first addressed qualitatively by [78] on grounds of calculations performed for the supernova SN 1987A. More details about these results can be found in [32].

From Fig. 34, it is concluded that about 60% of the produced p-nuclides fit the SoS composition within a factor of 3. Some discrepancies are also apparent. In particular, the Mo and Ru p-isotopes are severely underproduced. A possible remedy based on uncertainties in the seed s-process abundances due to an uncertain rate for the neutron production reaction ^{22}Ne(α, n)^{25}Mg is discussed by [32]. More specifically, an increase of this rate leads to an enhanced production of the Mo and Ru p-isotopes. The nuclides ^{113}In, ^{115}Sn, ^{138}La and ^{152}Gd are also underproduced. No other clear nucleosynthetic source has been identified yet for ^{113}In. In contrast, some ^{115}Sn and a substantial amount of the solar ^{152}Gd may owe their origin to the s-process. This process also contributes to ^{164}Er. From the level of p-process production of this nuclide shown in Fig. 34, it appears that ^{164}Er may well have a combined s- and p-origin without having to fear any insuperable overproduction with respect to the neighbouring p-nuclides. The rare odd-odd nuclide ^{180}Ta$^{\rm m}$ emerges naturally from all the considered SNII models in quantities consistent with its SoS abundance. This interesting case is discussed further by [32]. The other rare odd-odd p-nuclide ^{138}La is found to be of unlikely thermonuclear origin in SNII explosions, even if reasonable uncertainties on the production and destruction rates are taken into account. As reviewed by [32], ^{138}La could result instead from the captures by the initially relatively abundant ^{138}Ba of electron-neutrinos streaming out of the nascent PNS in a SNII explosion.

The impact on the p-nuclide yields displayed in Fig. 34 of a change of the SNII explosion energy, of metallicity, and of reaction rate uncertainties is also discussed by [32].

In order for SNII to be potentially important contributors to the galactic p-nuclides, it remains to make sure that they do not produce too much of other species, and in particular oxygen, the abundance of which is classically attributed to SNII explosions. This question has been examined in some detail by [78], who conclude that the p-nuclides are globally underproduced in a solar metallicity $25 M_\odot$ star by a factor of about 4 ± 2 relative to oxygen when all the abundances are normalized to the bulk SoS values. The problem of the oxygen overproduction may be eased in lower mass SNII explosions, while it is worsened with decreasing metallicity. It may thus be that the p-process enrichment of the Galaxy has been slower than the oxygen enrichment. There is at present no observational test of this prediction. The O overproduction may also be cured by a higher than nominal production of s-nuclides, which might also help solving the Mo-Ru underproduction problem, as discussed above. At this point, it must also be kept in mind that uncertainties in stellar convection introduce a large uncertainty (a factor of 2–3) in the oxygen yield of massive stars [81]. Likewise, the p-nuclide production might be affected as well. It has also to be stressed that the influence of multi-dimensional phenomena, and possibly of rotation, on the relative production of oxygen and the p-nuclides remains largely unknown.

To conclude this section, let us note the following: (1) it has been proposed that some p-nuclides could be produced under certain circumstances in $Y_e > 0.5$ neutrino-driven winds suggested by some SNII simulations [82], or through antineutrino absorption by intermediate-mass nuclei formed at some distance from the PNS in a SNII explosion [83]. It remains to be convincingly demonstrated that these scenarios can in particular help solving the Mo-Ru p-isotope puzzle, as sometimes claimed; (2) massive stars exploding as SNIb/c following a Wolf-Rayet phase might also provide a suitable site for the p-process in their Ne-O-rich layers. No detailed study of this scenario has been conducted to-date, and (3) massive enough stars might explode as so-called 'pair-creation supernovae'. The possibility of development of the p-process in such explosions is reviewed by [32].

11.2 The p-process in SNIa

The canonical model of SNIa explosions calls for a thermonuclear runaway of carbon in a C-O white dwarf (WD) that has reached a mass close to the Chandrasekhar limit ($M \approx 1.4\ M_\odot$) following the accretion of matter from a companion in a binary system. The nuclear burning in the course of the explosion is confined to very thin layers. These propagate either diffusively as subsonic deflagrations ('flames') or by shock compression as supersonic detonations, both modes being hydrodynamically unstable to spatial perturbations. A so-called 'delayed detonation' regime has also been proposed. It corresponds to the formation of a detonation wave when the outward moving deflagration flame is quenched by the expansion of the crossed layers. Some variants to the exploding C-O WD scenario have been proposed. They include the thermonuclear explosion of oxygen-neon WDs or the merging of two WDs. Large uncertainties of various sorts still affect the modelling of SNIa explosions (e.g. [84]). They influence more or less drastically the nucleosynthesis predictions, including those concerning the p-process.

Figure 35 presents the total p-nuclide yields from the SNIa deflagration model considered by [32], and with two different s-seed distributions, the proper one being largely unknown at this time. The conclusions one can derive from the inspection of this figure are quite similar to those resulting from the consideration of massive star explosions (Fig. 34), or from a delayed detonation scenario (see [32]). This concerns in particular the general pattern of the p-nuclide distribution, its high sensitivity to the assumed seed distribution, and the set of underproduced species (esp. ^{92,94}Mo, ^{96,98}Ru, ^{113}In, ^{115}Sn, and ^{138}La. Note that ^{180}Ta is also underproduced in the deflagration model adopted to construct Fig. 35, in contrast to the SNII predictions shown in Fig. 34). It has of course to be kept in mind that the simulation of deflagrations is extremely intricate, and model uncertainties certainly bring their share to the reliability of the yield predictions.

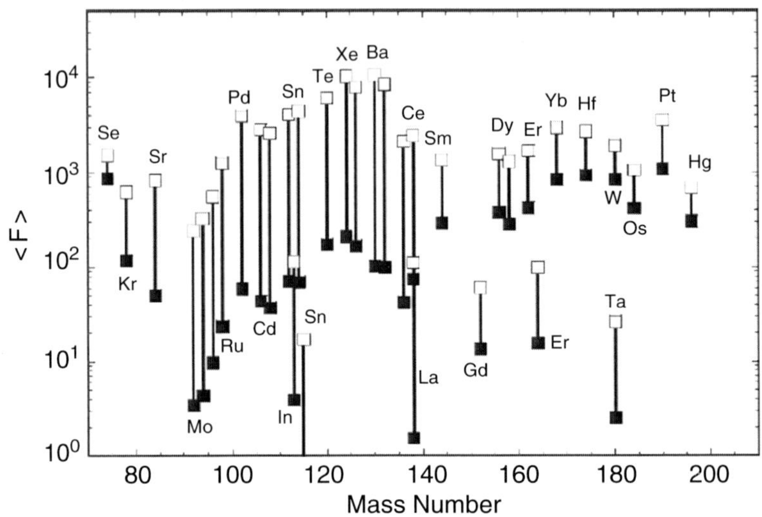

Fig. 35. Abundances of the p-nuclides in a SNIa deflagration model. They are calculated with solar s-seeds (black squares), or with a seed distribution which is representative of the s-process in solar-metallicity AGB stars (open squares) (see [32] for details)

11.3 The p-process in sub-Chandrasekhar white dwarf explosions

A scenario referred to as a 'sub-Chandrasekhar-mass' supernova envisions a C-O WD capped with a helium layer accreted by a companion, and which explodes as the result of a hydrodynamical burning before having reached the Chandrasekhar limit. This type of explosions may exhibit properties which do not match easily the observed properties of typical SNIa events. It cannot be excluded, however, that they are responsible for some special types of events, depending in particular on the He accretion rate and on the CO-sub-Chandrasekhar WD (SCWD) initial mass (e.g. [85]). Unidimensional simulations of He cataclysmics characterized by suitably selected values of these quantities reach the conclusion that the accreted He-rich layer can detonate. Most commonly, this explosion is predicted to be accompanied with the C-detonation of the CO-SCWD. In some specific cases, however, this explosive burning might not develop, so that a remnant would be left following the He detonation. Multidimensional calculations cast doubt on the nature, and even occurrence, of the C-detonation in CO-SCWD (e.g. [86]).

The possibility of development of the p-process following the detonation of the He-rich material assumed to be accreted onto a CO-SCWD has been explored by [88] in the framework of a 1D model of the He detonation of a non-rotating $0.8M_{\odot}$ WD. This work has been extended to a 3D simulation by [41]. It gives results that largely confirm the yield predictions of the 1D calculation, to which we limit the discussion in what follows.

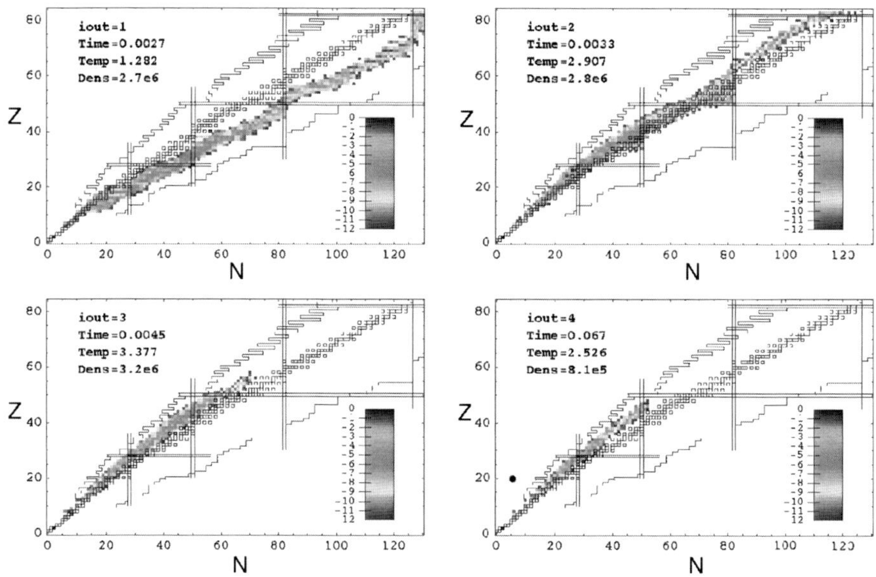

Fig. 36. Snapshots in the nuclidic chart of flow patterns in a 1D model of a detonating He layer accreted onto a $0.8 M_{\odot}$ WD. The selected times and corresponding temperatures or densities are given in different panels. The stable nuclides are indicated with open squares. The magic neutron and proton numbers are identified by vertical and horizontal double lines. The drip lines predicted by a microscopic mass model are also shown. The abundances are coded following the grey scales shown in each panel. At early times (bottom left panel), an r-process type of flow appears on the neutron-rich side of the valley of nuclear stability. At somewhat later times (top left panel), the material is pushed back to the neutron-deficient side rather close to the valley of β-stability. As time passes (two right panels), a pn-process [87] develops

A thermonuclear runaway develops near the base of the He envelope. The associated outward-moving detonation wave heats the matter to temperatures around 3×10^9 K, and leads to the ejection of about 0.2 $M_{\odot}$ into the interstellar medium. The associated nucleosynthesis is especially complex, and is reviewed in detail by [32]. The situation may be summarized by saying that a quite high neutron density builds up at the beginning of the detonation, and leads to a weak r-process. Subsequently, the nuclear flow is pushed back to the valley of nuclear stability, and eventually into the neutron-deficient region, by photodisintegrations that become more and more efficient as temperature rises. A process coined 'pn-process' by [87] then develops. The associated flow is displayed in Fig. 36. Its main path lies much further away from the proton-drip line than in the classical rp-process, which eases somewhat the nuclear physics problems. This results from the lower proton concentration than classically considered in the rp-process, and from the non-zero neutron concentrations encountered in the He detonation. Further details on the pn-process can be found in [32].

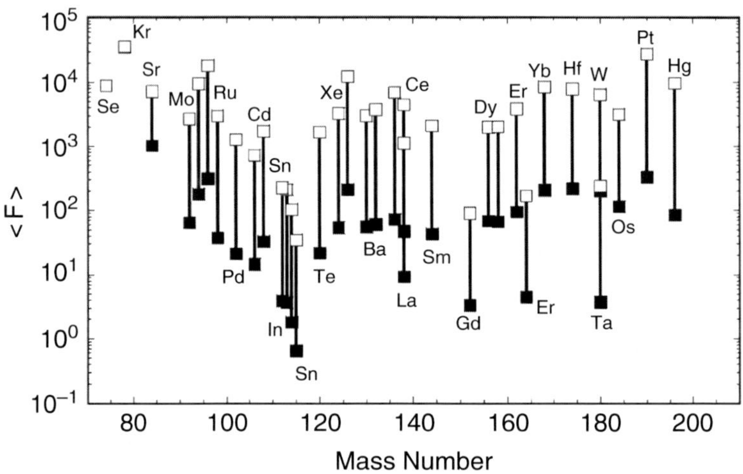

Fig. 37. Overproduction of the p-nuclides resulting from the He-detonation considered in Fig. 36. The full and open symbols correspond to SoS s-seed abundances, and to SoS seeds increased by a factor 100, respectively (from [32])

Figure 37 shows that almost all the p-nuclei are overproduced in solar proportions within a factor of 3 as a combined result of the p- and pn-processes. This includes the puzzling Mo and Ru p-isotopes, as well as ^{138}La, which appears to be overproduced at the same level as the neighbouring p-nuclides. This contrasts with the situation encountered in SNII (Sect. 11.1) or SNIa (Sect. 11.2) explosions.

An embarrassment with the SCWD scenario may come from the fact that the Ca-to-Fe nuclei are predicted to be overabundant with respect to the p-nuclei, but ^{78}Kr, by a factor of about 100. Taken at face value, this situation implies that the considered He detonation cannot be an efficient source of the bulk SoS p-nuclides (see [32] for details). In order to cure this problem, one may envision enhancing the initial abundance of the s-seeds, which has already been seen in Sect. 11.1 to be an essential factor determining the level of production of the p-nuclides. Figure 37 shows that an increase by a factor 100 of the s-nuclide abundances over their SoS values makes the overproduction of a substantial variety of p-nuclides comparable to the one of ^{78}Kr and of the Ca-to-Fe nuclei. The factor 100 enhancement would have to be increased somewhat if the material processed in the core of the CO-SCWD by a C-detonation were ejected along with the envelope. At this point, one essential question concerns the plausibility of the required s-nuclide enhancement.

There is no definite answer to this key question. This enrichment of the accreting WD might result from its past AGB history if indeed some of its outer s-process enriched layers could be mixed convectively with part at least of the accreted He-rich layers before the detonation. Alternatively, the He-rich layer accreted by the WD could be (or become) enriched in s-process elements. Such speculations need to be confirmed by detailed simulations.

An analysis of the impact of nuclear uncertainties on the p-abundances concludes that the results displayed in Fig. 37 are not drastically affected by other choices of nuclear reaction sets (see [32] for details).

11.4 Some comments on the p-process isotopic anomalies and chronometry

There is now observational evidence for the existence of isotopic anomalies involving the p-isotopes of Kr, Sr, Mo, Xe, Ba and Sm in various meteoritic materials [32]. These anomalies manifest themselves as excesses or deficits of the abundances of the p-nuclides with respect to the more neutron-rich isotopes, when comparison is made with the bulk SoS mix. In addition, two isotopic anomalies are attributable to the now extinct neutron-deficient radionuclides $^{92}Nb^{g}$ and ^{146}Sm which have decayed in the meteoritic material where excesses of ^{92}Zr and ^{142}Nd are observed.

Some of these anomalies, in particular in Xe, are puzzling, and cannot be explained by any p-process scenario [32]. As far as $^{92}Nb^{g}$ is concerned, there is some observational evidence for its presence in the early Solar System. With its half-live of 3.6×10^7 yr, some hope has been expressed that it could help building a chronometry for the p-process. However, the controversial early SoS $^{92}Nb^{g}$ abundance combined with the substantial uncertainties of nuclear physics and astrophysics natures which remain in the evaluated $^{92}Nb/\,^{92}Mo$ production ratios make by far premature any attempt to develop a meaningful ^{92}Nb-based p-process chronometry [32]. The radionuclide ^{146}Sm (α-decay half-live $t_{1/2} = 1.03 \times 10^8$ yr) has also been proposed as a p-process chronometer. Again, and particularly in view of uncertainties in its production by the p-process, it is without doubt highly risky, if not impossible, to construct a meaningful p-process chronometry based on ^{146}Sm [32].

By way of conclusion of this brief review of the p-process, one may say that substantial progress has been made in the understanding of this nucleosynthetic mechanism since [29]. A variety of astrophysical sites have been identified as possible active p-nuclide contributors. The relative success of the p-process calculations relying on detailed stellar models does not have to hide some puzzling problems raised by specific cases, the most publicized ones concerning ^{92,94}Mo, ^{96,98}Ru and ^{138}La. Some remedies have been proposed, but remain to be scrutinized in more detail. Some questions of nuclear physics or astrophysics nature the study of which would help putting the p-process modelling on safer grounds are identified by [32].

12 Summary and prospects

In view of their importance in many astrophysical questions, massive stars have been the subject of much work devoted to their evolution and nucleosynthesis. In spite of this, much obviously remains to be done, particularly on various astrophysics aspects of both the pre-supernova and supernova stages of their evolution. Among the questions that still call for further progress, let us mention the proper description of the transport of matter or of rotation, and of the structural effects of magnetic fields or of various instabilities that can develop prior to an eventual explosion. Uncertainties in the rates of some nuclear reaction and of weak interaction processes certainly do not have to be forgotten. Multi-dimensional simulations, especially at the late pre-supernova phases, are clearly in need.

The supernova explosions are still poorly understood, and it is still a long way to go to their successful simulations based on less uncertain pre-supernova models. Here again, the multi-dimensional treatment of a variety of physical effects, including rotation, magnetic fields, instabilities of different origins, and the transport of neutrinos, appears to be required.

As a direct result of the remaining structural and evolutionary uncertainties, the nucleosynthetic yields from massive stars have to be taken with some care, even if general trends seem to be understood. Of course, nuclear physics uncertainties also bring their share of difficulties to these predictions.

The r-process remains one of the most vexing nucleosynthetic puzzle. We briefly summarise here a selection of some r-process problems (the reader is referred to [24] for more details).

(1) *The r-nuclide content of the solar system.* Since [29], the bulk SoS content of r-nuclides has been a key source of information. Uncertainties unfortunately still affect a variety of SoS r-nuclide abundances, which weakens to some extent their constraining virtues on r-process models. These uncertainties stem from various origins, including meteoritic data or, more importantly, models for the SoS s-process content, from which r-abundances are traditionally derived. They need to be kept in mind in particular when interpreting spectroscopic data [see (2) below]. It is difficult to see how this situation can be improved in a decisive way. Isotopic anomalies involving some r-nuclides have been identified in meteorites. They do not bring much useful information on the identification of the site and characteristics of the r-process.

(2) *The evolution of the r-nuclide content of the Galaxy.* A substantial observational work has been conducted in recent years in order to identify the level of r-process contamination of the Galaxy with time. This effort largely relies on the abundance evolution of Eu, classified as an r-process element on grounds of SoS abundance analyses. The main conclusions derived from this observational work may be summarised as follows:
(i) the Eu data are classically used to support the idea that the r-process has contributed very early to the heavy-element content of the Galaxy. However, the scatter of the observed Eu abundances introduces some confusion when

one tries to establish a clear trend of the Eu enrichment with metallicity. It is also difficult to identify the value of [Fe/H] at which the signature of the s-process becomes identifiable. This conclusion relies in particular on La abundances derived from observation, La being classically considered as an s-element in the SoS (even if a non-negligible r-process contribution cannot be excluded). A most useful information on the relative evolution of the s- and r-process efficiencies in the Galaxy would be provided by the knowledge of the isotopic composition of the neutron-capture elements. Such data are unfortunately very scarce and mainly concern the Ba isotopic composition in a limited sample of stars. They are still under some debate, but suggest that the relative r-process contribution to Ba has decreased during the galactic evolution.
(ii) Much excitement has been raised by the observation that the patterns of abundances of heavy neutron-capture elements between Ba and Pb in r-process-rich metal-poor stars are remarkably similar to the SoS one. This claimed 'convergence' or 'universality' has to be taken with some care, however, as it largely relies on the assumption that the decomposition between s- and r-process contributions in metal-poor stars is identical to that in the SoS, a point that has yet to be demonstrated. A case is made that the convergence does not tell much about the astrophysics of the r-process, and is largely a signature of nuclear properties in the $58 \lesssim Z \lesssim 76$ range. This claim is in marked contrast with a statement that is often found in the literature. On the other hand, no 'universality' appears to hold for $Z \gtrsim 76$. This concerns in particular the Pb-peak elements and the actinides. This situation has far-reaching consequences. In particular, it invalidates the many attempts that have been made to build detailed galactic chronologies based on the actinides content of very metal-poor stars. At best, some lower limits to the age can be derived. No universality exists either for $Z \lesssim 58$ elements. In this mass range, some universality appears to be limited at best to stars with a large Eu overabundance.
(iii) The different behaviours of the abundance patterns of the elements below and above Ba have been the ground for the hypothesised existence of two so-called r-process 'components'. The question of the number of such components brings us more than forty years back, when it was discussed in the framework of the newly-constructed canonical r-process model. At a time when one is desperately trying to identify a suitable astrophysical site for the development of an (even limited) r-process [see (4) below], numbering the 'components' may not deserve a very high priority.

(3) *The nuclear physics for the r-process.* This has been reviewed in some detail by [24], and is not dealt with in these lectures. Let us just say that much effort has been put recently in the development of microscopic nuclear models aiming at reliable predictions of nuclear ground-state properties of thousands of nuclides located between the valley of nuclear stability and the neutron-drip line that may be involved in the r-process. Besides some remarkable achievements in the field, clearly much remains to be done.

(4) *The astrophysics aspects of the r-process.* This is clearly and by far the most unsatisfactorily understood facet of the r-process modelling, and the one that calls most desperately for progress. After some fifty years of research on this subject, the identification of a fully convincing r-process astrophysical site remains an elusive dream. The attempts conducted thus far may be briefly summarised as follows:
(i) some simplified site-free r-process models have been devised. They have the virtue of shedding some light on the conditions that are required in order for an r-process to develop. Broadly speaking, they can be divided into high-temperature scenarios and high-density scenarios. The former ones rely on the assumption that high neutron concentrations and high temperatures are both mandatory. They include, in order of increasing complexity (or decreasing number of simplifying assumptions), (a) the canonical waiting-point approximation model, (b) the multi-event model, and (c) dynamical approaches. The high-density models rely on the possibility of extensive neutronisation as a result of endothermic free-electron captures at high enough densities, followed by the decompression of this highly-neutronised material. A simple steady-flow model is constructed, and shows that the location and width of the SoS r-process peaks can be roughly accounted for.
(ii) Inspired by simulations that attempt to account for successful supernova explosions, (semi-)analytical models have been constructed in order to estimate the properties of the material ablated from a proto-neutron-star following the deposition of energy and momentum by the neutrinos streaming out of it. These models help identifying the physical quantities that have a significant impact on the nucleosynthesis in the ablated material, and the suitable ranges of their values for the development of a successful r-process. It appears that entropy (depending on temperature and density), expansion timescales (depending on the energy of the ablated material and on its mass-loss rate), and electron fraction are decisive quantities for the r-process nucleosynthesis, as illustrated by some network integrations. Recall also that the r-process efficiency tends to be reduced for increasing neutrino luminosities (which also affect the electron fraction).
(iii) The conclusions derived from the analytical models referred to above largely ruin the hope of having a successful r-process developing in the ablated material from a proto-neutron star, at least if one relies on realistic simulations of prompt or delayed iron core-collapse supernovae. The identified necessary conditions for the r-process are indeed not met in these models. True, one may get some hope that this conclusion will be invalidated when successful explosion simulations will at last be available!
(iv) Some hope might also come from a high-density r-process that could develop following the decompression of crust material ejected as a result of neutron-star coalescence. This is made plausible by some yield calculations based on numerical neutron-star merger simulations. Additional possibilities could be offered by the outflow of material from the discs forming around the coalesced neutron-stars. Considering the present status of the simulations of

core-collapse supernovae and of neutron-star coalescence, one may conclude that the latter scenario offers better potentialities for the r-process. Of course, uncertainties of different natures affect this scenario. They concern some modelling details and the calculated r-process yields. The limited efficiency of the scenario at early times in the galactic history based on the assumed low coalescence frequency in the young Galaxy is often considered as incompatible with the observed r-nuclide abundance at very low metallicity. This may well be the case. One has to acknowledge, however, that the frequency of the coalescence or magnetar events along the galactic history, as well as the amount of ejected matter per event, remain very uncertain.
(v) Attempts to interpret the observed trends in the galactic r-nuclide abundances have been conducted on grounds of models for the chemical evolution of the Galaxy. This is a task that is seemingly impossible to complete in the present state of affairs, given the absence of a reliable identification of an r-process site, and the schematic nature of the available galactic chemical evolution models, not to mention various intricacies of observational nature.
(vi) The actinides produced by the r-process enter in particular attempts to estimate the age of the Galaxy through their present SoS content, or through their abundances evaluated at the surface of very metal-poor stars. These attempts face some severe problems related to the nuclear and astrophysics uncertainties that affect the predictions of the actinides production. The situation is worsened further by the especially large uncertainties in the contribution of the r-process to the SoS Pb and Bi content. Concerning the ^{232}Th - ^{238}U and ^{235}U - ^{238}U pairs classically used to date the Galaxy from their present meteoritic abundances, it may be worth reiterating an opinion first expressed more than two decades ago that they have just limited chronometric virtues. The chronometric predictions based on the observations of Th and U in very metal-poor stars have to be considered with great care as well. In order for them to be reliable, it is not only required that the production of the actinides by the r-process is well known, but, and very decisively, that the production of Th with respect to U and to the Pt peak is 'universal.' Observation demonstrates now that this is not the case.

As far as the p-process is concerned, the calculations conducted in 'realistic' models for the various proposed sites, and in particular for the O-Ne-rich layers of SNeII, indicate, roughly speaking, that they may all provide yields enriched with a suite of p-nuclides at a level that is compatible (within a factor of 3 or so) with the solar values.

This relative success does not have to hide some puzzling problems raised by specific cases, the most publicized ones concerning ^{92,94}Mo, ^{96,98}Ru and ^{138}La. It has been proposed that the underproduction of the p-isotopes of Mo and Ru in SNII explosions is in fact due to some misrepresentation of the production in the He-burning core of massive stars of the s-nuclide seeds for the p-process. The modeling of the s-process in these conditions indeed faces some nuclear physics and astrophysics uncertainties which are not always appreciated at their true value. An increase of the $^{22}\mathrm{Ne}(\alpha, \mathrm{n})^{25}\mathrm{Mg}$ neutron production rate could help solving the underproduction of the Mo and Ru p-isotopes, but the validation of this possibility necessitates further nuclear physics laboratory efforts. Helium-detonating sub-Chandrasehkar-mass CO white dwarfs could be suitable ^{92,94}Mo and ^{96,98}Ru producers at least if the seed s-nuclides for the p-process are enhanced with respect to typical solar abundances in the material experiencing the detonation. As far as ^{138}La is concerned, SNII explosions could well be suitable sites for the production of this rare odd-odd nuclide mainly through ν_e-captures by ^{138}Ba. Exploding sub-Chandrasekhar-mass CO white dwarfs might also be significant ^{138}La producers. The other rare odd-odd nuclide ^{180}Ta$^{\rm m}$ has been shown years ago to be a natural product of the p-process in SNII explosions, a situation that is too often incorrectly appreciated in the literature. Charged current ν_e-captures on pre-existing ^{180}Hf could efficiently complement the thermonuclear ^{180}Ta$^{\rm m}$ production. Some calculations have also suggested that it could be made by the s-process in AGB stars. This possibility remains controversial at this time. Another quite puzzling case which does not attract much attention is ^{113}In. It does not appear to be efficiently produced in the p-process scenarios reviewed here, and no other source has been identified to-date.

A detailed interpretation of the meteoritic isotopic anomalies involving p-nuclides still eludes us to a large extent. This concerns in particular the Xe-HL case. It is also concluded that no reliable p-process chronometry can be built yet on the p-radionuclides ^{92}Nb$^{\rm g}$ or ^{146}Sm in view of the uncertainties still affecting their production in each of the considered p-process sites and of the possible variations of their yields from one site to the other.

At this point, it is of interest to remind of some items the study of which would help putting the p-process modeling on safer grounds. On the astrophysics side, let us cite (1) a more reliable estimate of the s-nuclide seeds for the p-process in massive stars (including e.g. the effect of rotation) and in the material accreted by white dwarfs with (sub-) Chandrasekhar mass prior to their explosion. This question relates directly to a good knowledge of the rate of the key neutron production reaction $^{22}{\rm Ne}\,(\alpha\,,{\rm n})\,^{25}{\rm Mg}$.
(2) Some progress in the description of convection in advanced stages of the evolution of massive stars, and in particular in their O-Ne layers. The possibility of survival at the explosion stage of at least a fraction of pre-explosively produced p-nuclides remains an uncertain issue.
(3) The modeling of the p-process in massive stars with a quite broad range of metallicities. Yields from such stars are needed to predict the evolution of the p-nuclide content of galaxies.
(4) The exploration of the impact on the p-process (and on many other questions) of the multi-dimensional modeling of the pre-SNII and core-collapse SNII or SNIb/c phases.
(5) The calculation of p-process yields from a variety of Chandrasekhar mass Type I supernova models of the deflagration or delayed detonation types, as well as of sub-Chandrasekhar He-detonation models. In the latter case, a

study of the p-nuclide production in the framework of a 3D simulation has been conducted recently.

On the more nuclear physics side, and apart from the already mentioned $^{22}\mathrm{Ne}(\alpha, \mathrm{n})^{25}\mathrm{Mg}$ reaction, it is of substantial importance to predict reliably the rates of thousands of nucleon or α-particle radiative captures and of the inverse transformations. For a long time to come, almost all these data will have to be provided by theory, but experiments have to help constraining and improving the models as much as possible. Efforts have to be started in order to measure directly the cross sections of photoreactions near threshold. In addition, more experiments should have to be conducted at sub-Coulomb energies on radiative proton and α-particle captures.

I hope that this review of the p-process will convince astrophysicsts, nuclear physicists and cosmochemists hunting for isotopic anomalies that the nuclear astrophysics 'p-nuts' are quite appealing after all, and can be enjoyed within a broad interdisciplinary research gathering.

Acknowledgements It is a pleasure to thank very warmly the organisers of the School for having made my stay in Kodaikanal most enjoyable. I also express my appreciation to all the participants to the School for their essential contribution to the very nice atmosphere they have succeeded to create inside, as well as outside, the lecture room.

References

1. S. E. Woosley et al., Rev. Mod. Phys. **2002**, 1015 (2002)
2. C. A. Meakin, D. Arnett, ApJ **667**, 448 (2007)
3. D. Arnett et al., preprint [arXiv:0809.1625] (2008)
4. H.-Th. Janka et al., Phys. Rep. **442**, 38 (2007)
5. C. E. Rolfs, W. S. Rodney, *Cauldrons in the Cosmos* (Chicago: Univ. Chicago Press) (1988)
6. C. Angulo et al., (the NACRE Collaboration), Nucl. Phys. **A656**, 3 (1999)
7. C. Iliadis et al., ApJS **142**, 105 (2002)
8. M. Arnould et al., A&A **347**, 572 (1999)
9. J. A. Stoesz, F. Herwig, MNRAS **340**, 763 (2003)
10. R. G. Izzard et al., A&A **466**, 641 (2007)
11. L. R. Buchmann, C.A. Barnes, Nucl. Phy **A777**, 254 (2006
12. M. Katsuma, Phys. Rev. C **78**, 034606 (2008)
13. G. Imbriani et al., ApJ **558**, 903 (2001)
14. T. Spilane et al., in *Frontiers in Nuclear Structure, Astrophysics, and Reactions: FINUSTAR 2*, AIP Conf. Proc., Vol. **1012**, Eds. P. Demetriou, R. Julin, & S.V. Harrisopulos (New York: American Institue of Physics), p. 184 (2008)
15. A. Chieffi et al., ApJ **502**, 737 (1998)
16. S. E. Woosley et al., ApJ **175**, 731 (1972)

17. D.D. Clayton, *Principles of Stellar Evolution and Nucleosynthesis* (Chicago: Univ. Chicago Press) (1968)
18. W. D. Arnett, *Supernovae and Nucleosynthesis* (Princeton: Princeton Univ. Press) (1996)
19. K. Sumiyoshi et al., ApJ **629**, 922 (2005)
20. M. Liebendörfer, et al., ApJ **620**, 840 (2005)
21. K. Nomoto, ApJ **322**, 206 (1987)
22. F. S. Kitaura et al., A&A **450**, 345 (2006)
23. L. Siess, in *Stars and Nuclei: A Tribute to Manuel Forestini*, ed. by T. Montmerle & C. Kahane, EAS Pub. Series, (Les Ulis: EDP Sciences) **19**, p. 103 (2006)
24. M. Arnould et al., Phys. Rep. **450**, 97 (2006)
25. H.-Th.Janka et al., A&A **485**,199 (2008)
26. A. Mezzacappa, in *Supernovae as Cosmological Lighthouses*, ed. by M. Turetto et al., ASP Conf. Series **342**, 175 (2005)
27. A. Burrows et al., ApJ **640**, 878 (2006)
28. M. Limongi, A. Chieffi, ApJ **592**, 404 (2003)
29. E. M. Burbidge et al., Rev. Mod. Phys. **29**, 547 (1957)
30. K. Lodders, ApJ **591**, 1220 (2003)
31. M. Asplund et al., ASPC **336**, 25 (2005)
32. M. Arnould, S. Goriely, Phys. Rep. **384**, 1 (2003)
33. D. D. Clayton et al., Ann. Phys. **12**, 331 (1961)
34. F. Käppeler et al., Rep. Prog. Phys. **52**, 945 (1989)
35. C. Arlandini et al., ApJ **525**,886 (1999)
36. J. Simmerer et al., ApJ **617**, 1091 (2004)
37. S. Goriely, A&A **342**, 881 (1999)
38. S. Goriely, M. Arnould, A&A **312**, 327 (1996)
39. S. Goriely, A&A **327**, 845 (1997)
40. H. Palme, H. Beer, in *Landolt Börnstein*, New Series, Group VI, Astronomy and Astrophysics, Subvolume **3a** (Berlin: Springer), p. 196 (1993)
41. S. Goriely et al., A&A **444**, L1 (2005)
42. G. Cescutti et al., A&A **448**, 557 (2006)
43. J. W. Truran et al., PASP **114**, 1293 (2002)
44. P. S. Barklem, A&A **439**, 129 (2005)
45. L. Mashonkina, G. Zhao, A&A **456**, 313 (2006)
46. P. Magain, A&A **297**, 686 (1995)
47. D. L. Lambert, C. Allende Prieto, MNRAS **335**, 325 (2002)
48. L. Mashonkina et al., A&A **343**, 519 (1999)
49. S. Honda et al., ApJ **607**, 474 (2004)
50. B. Plez et al., A&A **428**, L9 (2004)
51. S. Van Eck et al., A&A **404**, 291 (2003)
52. M. Paul et al., ApJ **558**, L133 (2001)
53. D. E. Brownlee et al., ASPC **104**, 223(1996)
54. J.-P. Meyeret al., ApJ **487**, 182 (1997)
55. A. J. Westphal et al., Nat **396**, 50 (1998)
56. A. J. Westphal et al., *Adv. Space Res.*, **27**, 797 (2001)
57. B. A. Weaver, A.J. Westphal, *Adv. Space Res.* **35**, 167 (2005)
58. P. A. Seeger.et al., ApJS **11**, 121 (1965)
59. R. D. Hoffman et al., ApJ **482**, 951 (1997)

60. B. S. Meyer et al., ApJ **498**, 808 (1998)
61. M. D. Delano, A.G.W. Cameron, Astophys. Space Sci. **10**, 203 (1971)
62. S. E. Woosley, R.D. Hoffman, ApJ **395**, 202 (1992)
63. S. Tsuruta, A. G. W. Cameron, *Can. J. Phys.*, **43**, 2056 (1965)
64. S. E. Woosley et al., ApJ **433**, 229 (1994)
65. K. Takahashi et al., A&A **286**, 857 (1994)
66. Y.-Z. Qian, S.E. Woosley, ApJ **471**, 331 (1996)
67. S. Wanajo, Y. Ishimaru, Nucl.Phy. **A777**, 676c (2006)
68. D. Argast et al., A&A **416**, 997 (2004)
69. K. Yokoi et al., A&A **117**, 65 (1983)
70. W. A. Fowler, F. Hoyle, Ann. Phys. **10**, 280 (1960)
71. J. Audouze, A&A **8**, 436 (1970)
72. K. L. Hainebach et al., ApJ **205**, 920 (1976)
73. T. G. Harrison, ApJS **36**, 199 (1978)
74. J. Audouze, J. W. Truran, ApJ **202**, 204 (1975)
75. M. Arnould,*PhD thesis*, Université Libre de Bruxelles (unpublished) (1971)
76. M. Arnould, A&A **46**, 117 (1976)
77. S. E. Woosley, W. M. Howard, ApJS **36**, 285 (1978)
78. M. Rayet et al., A&A **298**, 517 (1995)
79. T. Rauscher et al., ApJ **576**, 323 (2002)
80. P. Kroup et al., MNRAS **262**, 545 (1993)
81. T. Weaver, S.E. Woosley, Phys. Rep. **227**, 65 (1993)
82. G.C. Jordan IV, B.S. Meyer, ApJ **617**, L131 (2004)
83. C. Fröhlich et al., PRL **96**, 142502 (2006)
84. Röpke, et al., in *Proc. 13th EPS Conf. 'Beyond Einstein - Physics of the 21st Century*, Eds. A.M. Cruise & L. Ouwehand, ESA-SP **637**, p. 161 (2006)
85. I. Jr. Iben, I. Jr.Tutukov, ApJ **370**, 615 (1991)
86. W. Hillebrandt, J.C. Niemeyer, ARAA **38**, 191 (2000)
87. S. Goriely et al., A&A **383**, L27 (2002)
88. S. Goriely et al., A&A **383**, L27 (2002)
89. S. Goriely, L. Siess, in *From Lithium to Uranium: Elemental tracers of early cosmic evolution* ed. by V. Hill et al., Proc. of IAU symposium Nr 228, (Cambridge: Cambridge University Press), p. 451 (2005)

Part III

Cosmochemistry and Solar System Abundances

Cosmochemistry

Bruce Fegley, Jr.[1] and Laura Schaefer[2]

[1] Planetary Chemistry Laboratory, Department of Earth and Planetary Sciences, Washington University, St. Louis, MO 63130-4899 bfegley@wustl.edu
[2] laura-s@wustl.edu

Keywords: cosmochemistry, condensation calculations, elements, solar nebula

1 Introduction

Cosmochemistry is the chemistry of the cosmos. This is a broad topic that ranges from the nucleosynthesis of elements in stars to their chemistry on the Earth today. In this chapter, we describe chemical equilibrium (or condensation) calculations of the cosmochemical behavior of the elements.

2 Computational Methods

Chemical equilibrium calculations predict the distribution of each element between its gaseous, solid, and liquid compounds as a function of temperature, pressure, and bulk elemental composition. These calculations are often called condensation calculations because they show the stable phases that condense out of a cooling gas with solar system elemental abundances. However, chemical equilibrium calculations are path independent because the Gibbs energy is a state function, i.e., its differential dG is an exact (or perfect) differential. Thus, the results of chemical equilibrium calculations apply equally well to heating or cooling of a solar composition system.

The inputs to the chemical equilibrium calculations are the temperature, pressure, bulk elemental composition, and thermodynamic data for all compounds included in the calculations. The temperatures and pressures used in the calculations depend on the system being studied, e.g., a protoplanetary accretion disk, the photospheric region of a cool star, the ejecta from a nova or supernova, a planetary atmosphere, and so on. The bulk elemental composition is the set of elemental abundances that are appropriate for the system

A. Goswami and B.E. Reddy (eds.), *Principles and Perspectives in Cosmochemistry*, Astrophysics and Space Science Proceedings,
DOI 10.1007/978-3-642-10352-0_7,

being studied. For example, solar elemental abundances are used for a protoplanetary accretion disk around a solar metallicity star. On the other hand, chemical equilibrium calculations for the atmospheres of cool carbon stars use bulk compositions that are richer in C, N, and s-process elements than solar abundances [1].

The thermodynamic data for each compound are either the standard Gibbs energy ΔG^o or equilibrium constant K_{eq} for forming the compound from its constituent elements in their reference states. The reference states are generally the stable form of an element at the ambient temperature. For example, the reaction

$$2H_2(gas) + O_2(gas) = 2H_2O(gas) \tag{1}$$

forms water vapour from its constituent elements. A second example is the reaction

$$4Al(metal) + 3O_2(gas) = 2Al_2O_3(corundum) \tag{2}$$

This forms Al_2O_3, which is the mineral corundum. The reference state of Al is metal below its melting point, liquid Al from the melting point (933.6 K) to the boiling point (2,798 K), and Al gas at higher temperatures. The coexisting phases at the transition points (i.e., solid and liquid at the melting point or liquid and gas at the boiling point) are in equilibrium with one another, and either can be used as the reference state for aluminum.

The standard Gibbs energy for the formation of water vapour, corundum, or any other compound, from its constituent elements is calculated from tabular data in literature compilations using polynomial equations. The equilibrium constant is related to the ΔG^o of a reaction by the equation

$$\Delta G^o = -RTlnK_{eq} \tag{3}$$

The thermodynamic data come from a number of literature compilations and from primary articles in the scientific literature. Fegley and Lodders [2] list the data sources for a large number of compounds. Lodders [3, 4] published some updates to this list. The other updates that have been made will be described in future publications.

The chemical equilibrium calculations are done by sophisticated computer codes, such as the CONDOR code [2]. This code simultaneously considers the dual constraints of mass balance and chemical equilibrium. The operation of the CONDOR code and the general principles of chemical equilibrium calculations are best illustrated using a simplified version of iron chemistry in solar composition material. We define the total elemental abundance of iron as A(Fe). This is the atomic abundance of Fe relative to 10^6 Si atoms and is 838,000 Fe atoms [5]. The mole fraction (X) of total iron (ΣFe) in all Fe-bearing compounds is

$$X_{\Sigma Fe} = \frac{A(Fe)}{A(H + H_2 + He)} \tag{4}$$

The term $A(H + H_2 + He)$ is the sum of the H and He abundances with the temperature-dependent H and H_2 equilibrium taken into account. Molecular H_2, H, and He are the three major gases in solar composition material over most temperatures and pressures where chemical compounds exist. In the actual computation, other gases such as H^+, CO, H_2O, N_2, Ne, other ions, etc., are also included in the denominator; however, these are neglected in the description here for clarity. Multiplying $X_{\Sigma Fe}$ by the total pressure P_{tot} gives the partial pressure sum for iron

$$P_{\Sigma Fe} = X_{\Sigma Fe} P_T = P_{Fe} + P_{FeS} + P_{FeH} + P_{FeO} + .. \tag{5}$$

Equation (5) is a mass balance equation that counts the total number of iron atoms in all Fe-bearing gases and equates this sum to the total amount of iron. The partial pressures of Fe-bearing gases with more than one iron atom are multiplied by the number of iron atoms in the gas to insure that all atoms are counted. In this example the Fe-bearing gases included in the mass balance equation are the most abundant ones at low pressures characteristic of cool stellar atmospheres, protoplanetary disks, and ejecta from nova and supernova. All of these Fe-bearing gases contain one Fe atom and their partial pressures are multiplied by unity. If any gases containing two Fe atoms, such as Fe_2Cl_4, were present, their partial pressures would be multiplied by two. We must now relate the partial pressures of the different Fe-bearing gases in the mass balance equation to their chemical equilibrium abundances.

This is done by rewriting equation (5) in terms of the thermodynamic activity of Fe (a_{Fe}), the equilibrium constants (K_i) for forming the Fe-bearing gases from the constituent elements in their reference states, and the thermodynamic activities and fugacities of all other elements combined with iron in the gases.

$$P_{\Sigma Fe} = X_{\Sigma Fe} P_T = a_{Fe}[K_{Fe} + K_{FeS} f_{S_2}^{1/2} + K_{FeH} f_{H_2}^{1/2} + K_{FeO} f_{O_2}^{1/2}] \tag{6}$$

The actual mass balance sum for iron in the CONDOR code includes about 21 Fe-bearing gases. The most important ones in a low-pressure, solar composition gas are Fe, FeS, FeH, and FeO. Analogous forms of equation (6) are written for each element in the code. The a_i and f_i terms in equation (6) are the elemental activities and fugacities of the respective elements, e.g., f_{H2} is the fugacity of hydrogen.

Equation (6) combines the mass balance and chemical equilibrium constraints for iron. It also shows that the chemistry of iron is coupled to that of other elements because the fugacities of sulfur, hydrogen, and oxygen are included in equation (6). In general, the chemistry of all the elements is coupled, and the mass balance equations form a set of coupled, nonlinear equations that are solved iteratively. An initial guess is assumed for the activity or fugacity of each element. These guesses can be optimized if the major gas for each element is known, but this is not necessary for the code to operate properly. The CONDOR code solves the set of mass balance equations and returns the thermodynamic activity or fugacity for each element, the abundances of all gases (molecules, radicals, atoms, and ions) included in the code, and information on the quality of the calculated results for each element. The convergence criterion requires that the calculated and input abundances for each element agree to within 1 part in 10^{15}. However, for all practical purposes a satisfactory solution is reached when the calculated and input abundances for each element agree to within 1 part in 10,000 (0.01%). At present the CONDOR code considers over 3,600 solid, liquid, and gaseous compounds for all naturally occurring elements in the periodic table.

The CONDOR code also takes possible condensates (both solid and liquid) into account. For example, Fe metal (or liquid Fe depending on the temperature) forms if the thermodynamic activity of Fe (a_{Fe}) is equal to or greater than unity. The code computes the temperature at which a_{Fe} reaches unity (the Fe metal condensation temperature), resets a_{Fe} to unity at all temperatures below this point, and adds a new term for the abundance of Fe metal to the mass balance equation.

The stabilities of condensates containing two or more elements are computed considering compound formation from the elements in their respective reference states.

For example, the reaction

$$Fe(metal) + S(refstate) = FeS(troilite) \tag{7}$$

is used for troilite, which is the most common Fe sulfide in meteorites. The reference state of sulfur is solid orthorhombic sulfur, solid monoclinic sulfur, liquid sulfur, or S_2 gas depending on the temperature. As temperature decreases, reaction (7) shifts to the right and the thermodynamic activity of troilite increases. Troilite condenses when its thermodynamic activity reaches unity and this is calculated from

$$a_{FeS} = a_{Fe} a_s K_{FeS} \tag{8}$$

where K_{FeS} is the temperature-dependent equilibrium constant for troilite formation from its constituent elements, and a_{Fe} and a_S are taken from the gas-phase equilibrium calculations described above. In general, a pure phase such as Fe-metal, corundum (Al_2O_3), or troilite (FeS) starts to condense from the gas at the temperature where its thermodynamic activity reaches unity. The thermodynamic activity is less than unity at higher temperatures where the condensate phase is unstable, and is fixed at unity at all temperatures where the condensate is stable.

Once troilite is stable, the fraction of Fe (α_{Fe}) condensed in troilite is calculated and the gas-phase abundance of total iron (P_{Fe}) is reduced by multiplying by (1 - α_{Fe}). Analogous corrections are made for all elements distributed between the gas and condensates. The gas-phase and gas-solid chemical equilibria are coupled and solved simultaneously using iterative techniques. Palme and Fegley [6] illustrate this procedure for condensation of the minerals enstatite $MgSiO_3$ and forsterite Mg_2SiO_4, which involve coupled equilibria for the elements Mg, O, and Si.

The total abundance of each condensate is limited by the abundance of the least abundant element in the condensate. For example, the mineral schreibersite Fe_3P forms by reaction of P-bearing gases with Fe metal at about 1300 K and 10^{-4} bar total pressure. Phosphorus has an atomic abundance of 8,373 atoms, which is about 1% of the atomic abundance of iron. There are 3 Fe atoms in each molecule of schreibersite. Thus Fe_3P formation consumes only 3% of the total iron abundance, while removing all phosphorus from the gas phase. Likewise, troilite formation removes all sulfur from the gas because its abundance is only 53% of that of iron, while unreacted Fe metal remains present at lower temperatures until it is consumed by formation of Fe-bearing oxides and silicates.

Several points are worth emphasizing. The first point is mass balance. The total amount of each element is conserved in the chemical equilibrium calculations. Thus the abundances of all gases and all condensed phases (solids and/or liquids) sum to the total elemental abundance - no less and no more. The second point is that chemical equilibrium is completely independent of the size, shape, and state of aggregation of condensed phases - a point demonstrated by Willard Gibbs over 130 years ago. Finally, the third point is that chemical equilibrium is path independent. Thus, the results of chemical equilibrium calculations are independent of any particular reaction. A particular chemical reaction does not need to be specified because all possible reactions give the same result at chemical equilibrium. This is completely different than chemical kinetic models where the results of the model are critically dependent on the reactions that are included. However, a chemical equilibrium calculation does not depend on kinetics, is independent of kinetics, and does not need a particular list of reactions. This point may seem obvious, but is often misunderstood.

3 Cosmochemical Behaviour of the Elements

This section summarizes the results of chemical equilibrium calculations for naturally occurring elements. We review the cosmochemical behavior of the elements in order of increasing volatility, i.e., decreasing condensation temperature using the cosmochemical classification scheme.

3.1 Refractory Elements

Refractory elements are the first elements to condense (or the last ones to evaporate) from solar composition material. The refractory elements are subdivided into refractory siderophiles (elements found in metal) and refractory lithophiles (elements found as oxides and/or silicates). The refractory siderophiles are the noble or platinum group metals (Ru, Os, Rh, Ir, and Pt, but not Pd), molybdenum Mo, tungsten W, and rhenium Re. All of the refractory siderophiles are trace elements with low abundances. In the 1970s micron-size metal nuggets composed of the refractory siderophile elements were discovered inside Ca, Al-rich inclusions (CAIs) in the Allende meteorite. This is a stony meteorite, which is a type CV3 carbonaceous chondrite. The compositions of the refractory metal nuggets in Allende and other carbonaceous chondrites are reproduced by chemical equilibrium calculations [7, 8].

The actinides (U, Th, Pu), alkaline earths (Be, Mg, Ca, Sr, Ba), lanthanides (elements La - Lu), Al, and the elements in groups 3b (Sc, Y), 4b (Ti, Zr, Hf), and 5b (V, Nb, Ta) of the periodic table are refractory lithophile elements. The refractory lithophiles are 5% of the total mass of the rock in solar composition material. Aluminum Al, calcium Ca, and titanium Ti are the three most abundant refractory lithophiles, and they form minerals that are the host phases for most of the less abundant refractory lithophile elements such as the actinides, lanthanides, and transition elements in group 5b of the periodic table. Some of the less abundant refractory lithophiles - the group 4b elements Zr, Hf, and the group 3b elements Y and Sc - condense as oxides before any Ca, Al, Ti-bearing minerals form [9]. But the rest condense into the more abundant host phases.

Figure 1 shows the percentage distribution of aluminum between its major gases and the condensed phases that are stable at chemical equilibrium in a solar composition gas as a function of temperature at a total pressure of 10^{-4} bar. This pressure is representative of that in the inner regions of protoplanetary accretion disks (such as the solar nebula) and photospheric regions of cool stars.

At the high temperatures that are shown on the right hand side of Figure 1, all Al is in the gas phase, mainly as monatomic Al and aluminum hydroxide AlOH gases. For example at 2000 K, about 92% of all aluminum is in Al gas, about 7.5% is in AlOH gas, and the remaining 0.5% is in other Al-bearing gases. Corundum Al_2O_3 condenses at 1640 K and is the first Al-bearing mineral (see the small wedge at the top right of Figure 1). However, Figure 1 shows that corundum is only stable for a few degrees (the wedge disappears). Grossite $CaAl_4O_7$ and hibonite $CaAl_{12}O_{19}$ form by reaction of corundum with the surrounding gas. These reactions destroy all the corundum and produce grossite and hibonite. At 1600 K, most of the Al is in these minerals. The remainder is mainly in gaseous AlOH and Al. In fact, Figure 2 shows that other Al-bearing gases are still present but that these are less abundant than Al or AlOH. In turn, other reactions form the mineral gehlenite $CaAl_2SiO_7$ from grossite and hibonite with decreasing temperature. The rest of Figure 1 shows the appearance, disappearance, and abundance of other Al-bearing condensates formed at chemical equilibrium at lower temperatures.

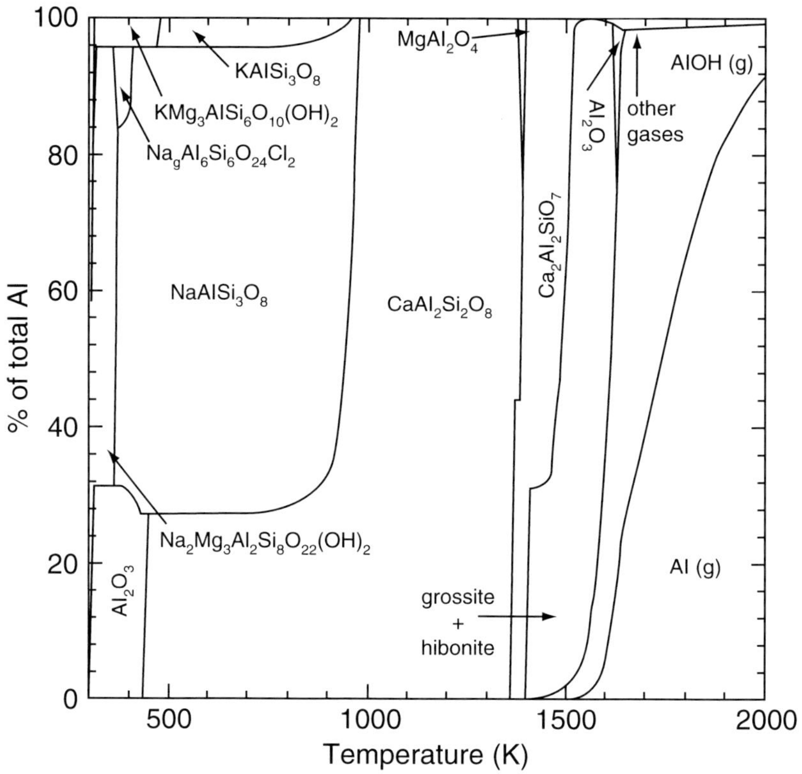

Fig. 1. Percentage distribution of aluminum between different phases in a solar composition gas as a function of temperature for a total pressure of 10^{-4} bar. Gaseous species are indicated with (g), all other species are condensed phases

Figure 2 shows the mole fractions of the major Al-bearing gases that are stable at chemical equilibrium in a solar composition gas as a function of temperature at a total pressure of 10^{-4} bar. The temperature range in Figure 2 is only from 1200 - 2000 K because the mole fractions of all the Al-bearing gases decrease strongly with decreasing temperature below the condensation temperature of corundum. Figures 1 and 2 are complementary to one another. Figure 1 gives the percentage distribution relative to total aluminum while Figure 2 gives the mole fractions, defined as the partial pressure of a gas relative to the total pressure. For example, the mole fraction of AlOH gas is

$$X_{AlOH} = \frac{P_{AlOH}}{P_T} \qquad (9)$$

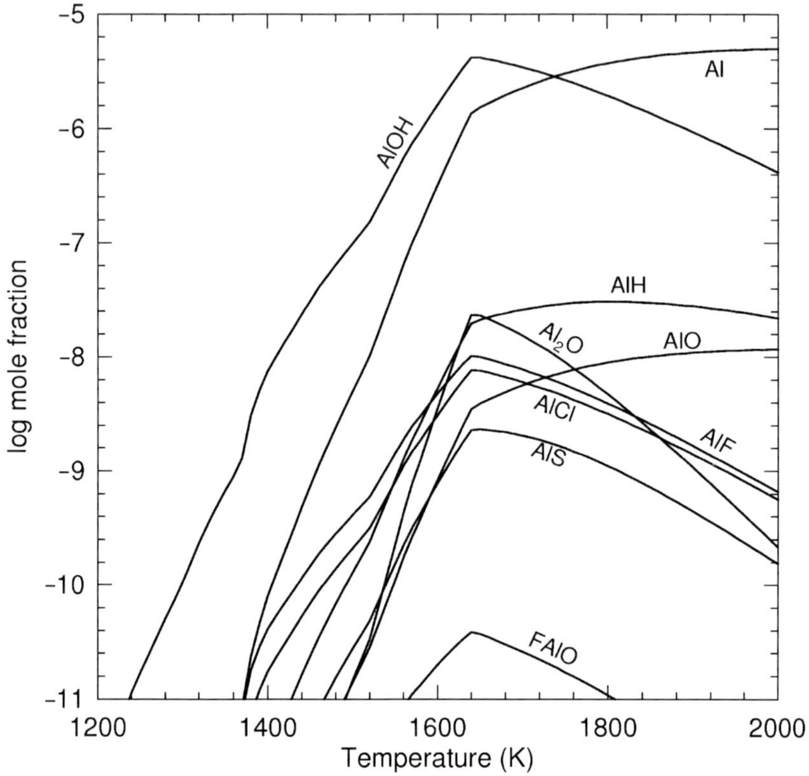

Fig. 2. Gas phase chemistry of aluminum in a solar composition gas as a function of temperature at a total pressure of 10^{-4} bar. The abundances of the gases are shown as mole fractions

This definition of mole fraction is identical to saying that the mole fraction is equal to the number of moles of a gas relative to the moles of all gases, or equivalently that the mole fraction of a gas is equal to its volume fraction. Figure 2, and similar figures for other elements, show gases that are not abundant enough to appear on a percentage distribution graph such as Figure 1. This information is important for astronomical observations of the photospheric regions of cool stars, of protoplanetary accretion disks, of circumstellar envelopes, and of the atmospheres of gas giant planets.

Table 1 lists the names and formulas of the minerals shown in Figure 1 and subsequent figures. For example, Table 1 shows that the mineral grossite has the formula $CaAl_4O_7$ and that $MgAl_2O_4$ is the mineral spinel. Table 2

Table 1. Minerals included in Figures

Mineral name	Formula
Alabandite	MnS
Albite	$NaAlSi_3O_8$
Anorthite	$CaAl_2Si_2O_8$
Bornite	Cu_5FeS_4
Chlorapatite	$Ca_5(PO_4)_3Cl$
Chromite	$FeCr_2O_4$
(Cobalt Sulfide)[a]	Co_9S_8
Corundum	Al_2O_3
Diopside	$CaMgSi_2O_6$
Enstatite	$MgSiO_3$
Eskolaite	Cr_2O_3
Fayalite	Fe_2SiO_4
Fluorapatite	$Ca_5(PO_4)_3F$
Forsterite	Mg_2SiO_4
Gehlenite	$Ca_2Al_2SiO_7$
Glaucophane	$Na_2Mg_3Al_2Si_8O_{22}(OH)_2$
Grossite	$CaAl_4O_7$
Halite	$NaCl$
Heazlewoodite	Ni_3S_2
Hibonite	$CaAl_{12}O_{19}$
Hydroxyapatite	$Ca_5(PO_4)_3OH$
Ilmenite	$FeTiO_3$
Iron metal	Fe
(Iron Selenide)[a]	$FeSe_{0.961}$
Karelianite	V_2O_3
Orthoclase	$KAlSi_3O_8$
Perovskite	$CaTiO_3$
Phlogopite	$KMg_3AlSi_3O_{10}(OH)_2$
Picrochromite	$MgCr_2O_4$
Rutile	TiO_2
Schreibersite	Fe_3P
Sodalite	$3NaAlSiO_4 \cdot NaCl$
(Sodiam Bromide)[a]	$NaBr$
Spinel	$MgAl_2O_4$
Talc	$Mg_3Si_4O_{10}(OH)_2$
Tephroite	Mn_2SiO_4
(Titanium Oxide)[a]	Ti_4O_7
Tremolite	$Ca_2Mg_5Si_8O_{22}(OH)_2$
Troilite	FeS
Zincochromite	$ZnCr_2O_4$

[a] No known mineral name

Table 2. Abundances and Condensation Temperatures of the Elements in the Solar Nebula

	Element	Symbol	abundance	T_c (K)	Initial Condensate {Dissolving Species}	50% T_c(K)	Major gases
1	Hydrogen	H	2.431×10^{10}	182	H_2O ice	-	H_2
2	Helium	He	2.343×10^{9}	<3	He ice	-	He
3	Oxygen	O	1.413×10^{7}	182	H_2O ice[a]	180	CO,H_2O
4	Carbon	C	7.079×10^{6}	78	$CH_4 \cdot 7H_2O$	40	CO, CH_4
5	Neon	Ne	2.148×10^{6}	9.3	Ne ice	9.1	Ne
6	Nitrogen	N	1.950×10^{6}	131	$NH_3{\cdot}H_2O$	123	N_2, NH_3
7	Magnesium	Mg	1.020×10^{6}	1390	Spinel	-	Mg
				1354	Forsterite[b]	1336	
8	Silicon	Si	$\equiv1.00\times10^{6}$	1529	Gehlenite	-	SiO, SiS
				1290	Enstatite	1265	
9	Iron	Fe	8.380×10^{5}	1358	Fe metal[c]	1325	Fe
10	Sulfur	S	4.449×10^{5}	704	FeS	655	H_2S, HS
11	Argon	Ar	9.032×10^{4}	48	Ar.6H_2O	47	Ar
12	Aluminum	Al	8.410×10^{4}	1640	Al_2O_3	1653	Al, AlOH, Al_2O, AlS,
				1620	$CaAl_{12}O_{19}$	1615	AlH, AlO
13	Calcium	Ca	6.287×10^{4}	1620	$CaAl_{12}O_{19}$	-	Ca
				1529	Gehlenite	1500	
14	Sodium	Na	5.751×10^{4}	970	{$NaAlSi_3O_8$}	960	Na,NaCl
15	Nickel	Ni	4.780×10^{4}	1280	{Ni}	1260	Ni
16	Chromium	Cr	1.286×10^{4}	870	Cr_2O_3		Cr
17	Manganese	Mn	9168	1050	MnS	-	Mn
18	Phosphorus	P	8373	1285	Fe_3P	1265	PO,P,PN,PS
19	Chlorine	Cl	5237	650	$Ca_5(PO_4)_3Cl$	-	HCl,NaCl,KCl
				-	$Na_4[Al_3Si_3O_{12}]Cl$	400	
20	Potassium	K	3692	950	{$KAlSi_3O_8$}	915	K,KCl,KOH
21	Titanium	Ti	2422	1593	$CaTiO_3$	1575	TiO,TiO_2
22	Cobalt	Co	2323	1190	{Co}	1170	Co
23	Zinc	Zn	1226	660	$ZnCr_2O_4$	650	Zn
24	Fluorine	F	841.1	710	$Ca_5[PO_4]_3F$	710	HF
25	Copper	Cu	527.0	960	{Cu}	940	Cu
26	Vanadium	V	288.4	-	{VO,V_2O_3}	1250	VO_2, VO
27	Selenium	Se	65.79	-	ZnSe	520	H_2Se,GeSe
28	Krpton	Kr	55.15	53	Kr·6H_2O	52	Kr
29	Bromine	Br	11.32	370	NaBr	360	Br,HBr

[a] 22.75% of oxygen is condensed into rock before water ice condensation. [b] Major condensed reservoir of element. [c] Condensation temperature of pure iron metal.

lists the elements discussed in this chapter in decreasing order of abundance. Table 2 also gives the major gases and initial condensates of these elements. For example, Table 2 shows that Al is the twelfth most abundant element in solar composition material with an abundance of 8.41×10^4 Al atoms (per 10^6 Si atoms), its initial condensates are corundum Al_2O_3 at 1640 K and hibonite $CaAl_{12}O_{19}$ at 1620 K, and its major gases are Al and AlOH.

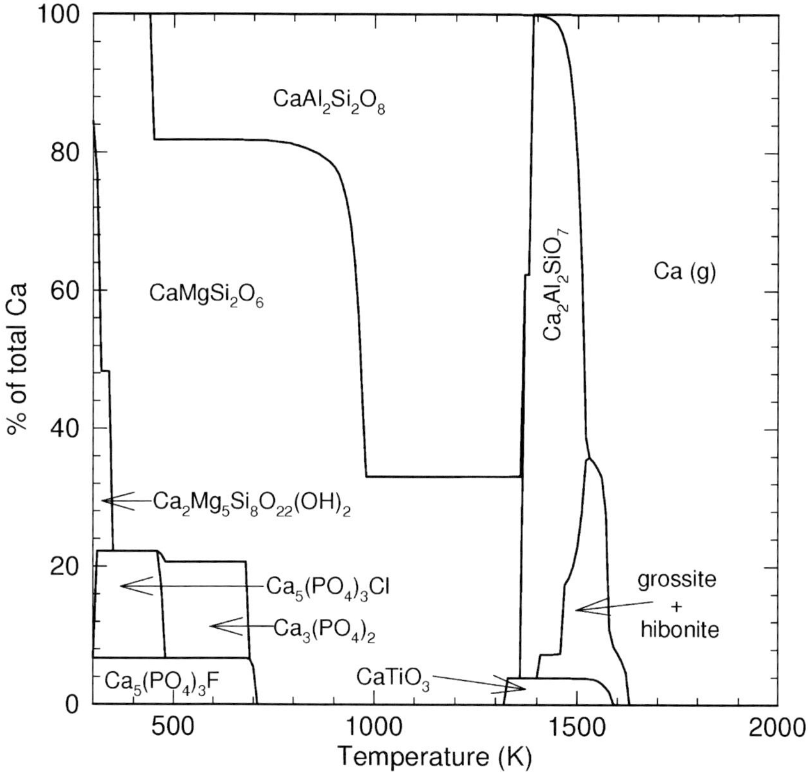

Fig. 3. Percentage distribution of calcium between different phases in a solar composition gas as a function of temperature for a total pressure of 10^{-4} bar. Gaseous species are indicated with (g), all other species are condensed phases. Hydroxyapatite ($Ca_5(PO_4)_3OH$) is not shown because it occurs only at 300 K (and lower) with an abundance of 15.5% of total Ca

Figures 3 and 4 for Ca and Ti, respectively, give similar information for the major gases and condensates of these two refractory lithophiles. For example,

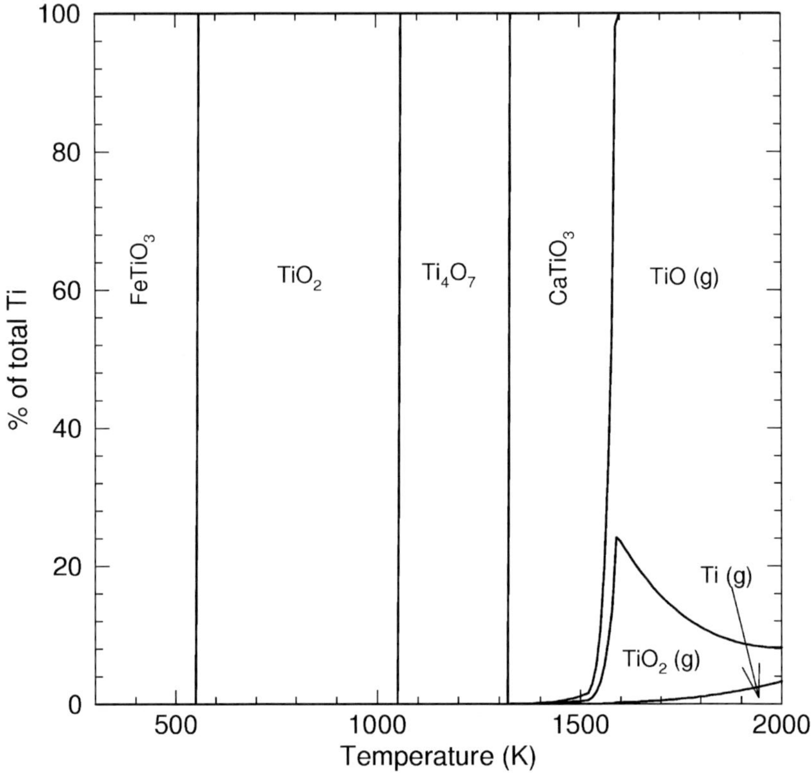

Fig. 4. Percentage distribution of titanium between different phases in a solar composition gas as a function of temperature for a total pressure of 10^{-4} bar. Gaseous species are indicated with (g), all other species are condensed phases

monatomic Ca gas is the major Ca-bearing gas and hibonite $CaAl_{12}O_{19}$ is the first Ca-bearing condensate. Likewise, titanium monoxide TiO, TiO_2, and monatomic Ti are the major Ti-bearing gases and perovskite $CaTiO_3$ is the first Ti-bearing condensate.

The results of the chemical equilibrium calculations for the refractory lithophiles are also confirmed by the mineralogy and chemistry of the Ca, Al-rich inclusions in Allende and other meteorites. The major minerals in CAIs are the same ones predicted by the calculations, namely melilite (a solid solution of gehlenite $CaAl_2SiO_7$ and åkermanite $Ca_2MgSi_2O_7$), spinel, corundum, grossite, hibonite, and perovskite. Chemical analyses of CAIs show that the refractory lithophiles are enriched by an average of 20 times solar elemental abundances, as expected for complete condensation of elements that constitute 5% of rocky material. The less abundant refractory lithophiles, such as the lanthanides (or rare earth elements REE) are found dissolved in perovskite and hibonite, as predicted by chemical equilibrium calculations. Kornacki and Fegley [9, 10] describe chemical equilibrium calculations for refractory lithophiles and their abundances in CAIs.

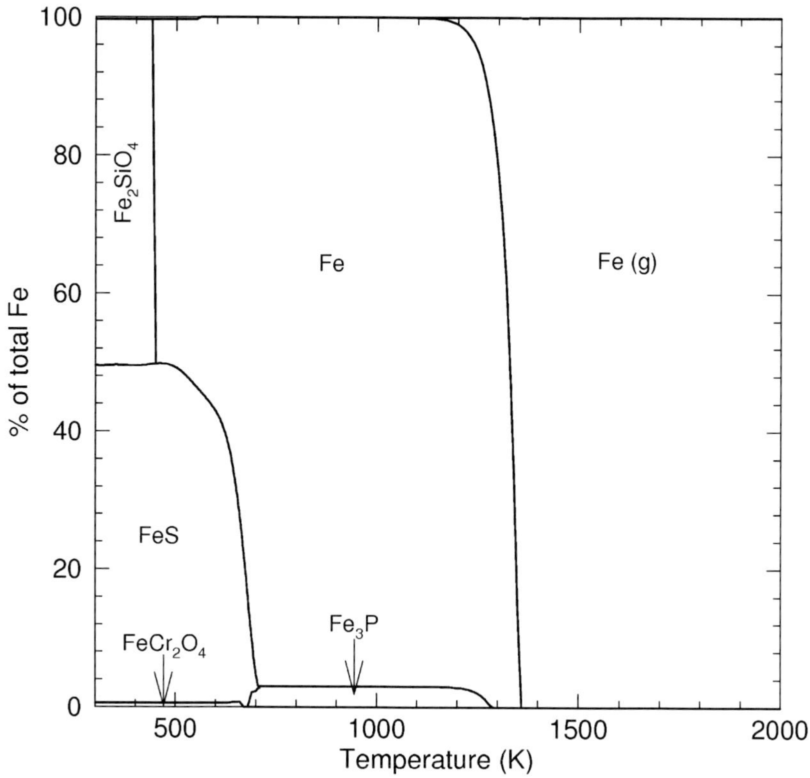

Fig. 5. Percentage distribution of iron between different phases in a solar composition gas as a function of temperature for a total pressure of 10^{-4} bar. Gaseous species are indicated with (g), all other species are condensed phases

3.2 Major elements

Iron Fe, magnesium Mg, and silicon Si are the next elements to condense and they constitute the major elements because they make up most of the rock in solar composition material. Figure 5 shows the percentage distribution of iron between its major gases and condensates. Iron metal condenses at 1357 K at 10^{-4} bar total pressure and is 50% condensed by 1325 K. The iron metal is not pure Fe, but is an Fe-rich metal alloy containing smaller amounts of nickel Ni, cobalt Co, and other siderophile elements. Iron metal is the only Fe-bearing condensate until schreibersite Fe_3P forms (1285 K, 10^{-4} bar). As mentioned earlier, phosphorus is about 1% as abundant as Fe, so formation of Fe_3P consumes only 3% of the total iron metal. The formation of troilite FeS at 704 K consumes about 55% of the remaining Fe metal. Hydrogen sulfide is the major S-bearing gas containing about 100% of all sulfur and troilite formation occurs via the net thermochemical reaction

$$H_2S(gas) + Fe(metal) = FeS(troilite) + H_2(gas) \tag{10}$$

This reaction has the same number of gas molecules on either side of the reaction and the FeS formation temperature of 704 K is thus independent of total pressure.

Chromite $FeCr_2O_4$ formation at 660 K consumes a tiny fraction of Fe metal, but the remainder of the metal remains present until it is oxidized to Fe^{2+}-bearing silicates such as fayalite Fe_2SiO_4, shown in Figure 5. This dissolves in solid solution with forsterite Mg_2SiO_4 forming olivine.

Figures 6 and 7 show the equilibrium chemistry of magnesium and silicon. Small amounts of magnesium and silicon condense at high temperatures as spinel $MgAl_2O_4$ (1390 K, shown in Figure 1), gehlenite $Ca_2Al_2SiO_7$ (1529 K), and *åkermanite* $Ca_2MgSi_2O_7$ (not shown). However the amounts of Mg and Si in these high temperature condensates are limited by Ca and Al, which are much less abundant than magnesium or silicon. Most Mg and Si condense as forsterite Mg_2SiO_4 (1354 K) and enstatite $MgSiO_3$ (1290 K). Forsterite condenses from the gas phase while enstatite forms by reaction of nebular gas with preexisting forsterite grains.

Figure 8 shows that at pressures greater than about 10^{-4} bar, Fe metal condenses at higher temperatures than forsterite Mg_2SiO_4. Conversely, at pressures less than about 10^{-4} bar, forsterite condenses at higher temperatures than Fe metal. The separation between the metal and forsterite condensation temperatures increases with increasing pressure from the crossover point. The planet Mercury is richer in Fe and poorer in silicates than expected from solar composition. This may indicate that the materials accreted by Mercury formed at pressures higher than the 10^{-4} bar crossover point for the iron and forsterite condensation curves.

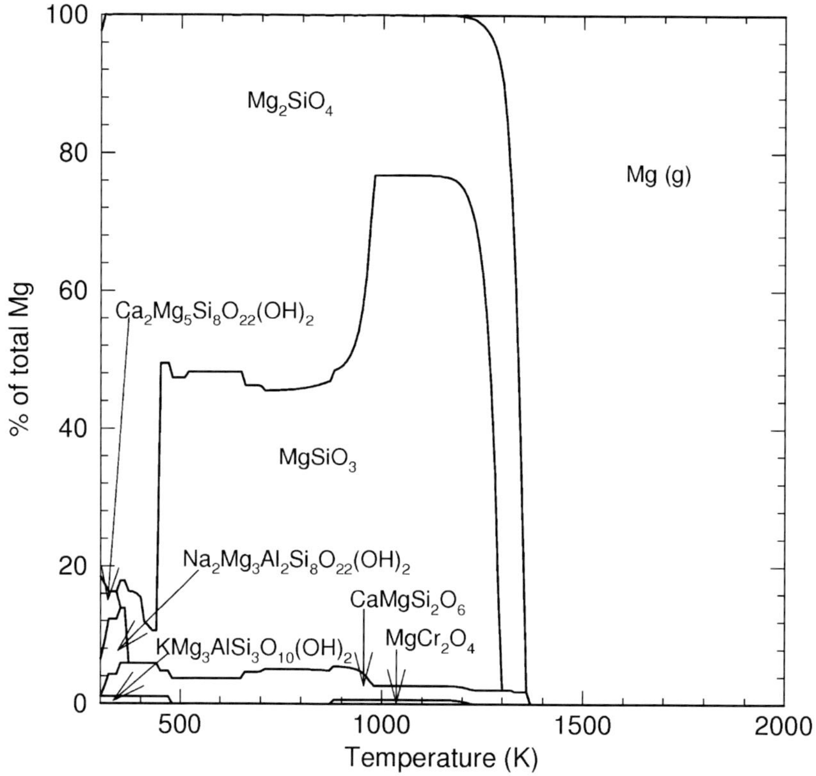

Fig. 6. Percentage distribution of magnesium between different phases in a solar composition gas as a function of temperature for a total pressure of 10^{-4} bar. Gaseous species are indicated with (g), all other species are condensed phases. Spinel ($MgAl_2O_4$) and talc ($Mg_3Si_4O_{10}(OH)_2$) are not shown due to their low abundance (1.4% and 0.9% total Mg at 300 K, respectively)

3.3 Moderately Volatile Elements

The moderately volatile elements are the next elements to condense from cooling solar composition material. The moderately volatile elements are a diverse set of elements including the alkali elements (Li, Na, K, Rb, Cs); transition metals (Cr, Co, Mn, Ni, Pd, Cu, Ag, Au, Zn); group 3 elements (B and Ga); group 4 elements (Ge and Sn); group 5 elements (P, As, Sb); group 6 elements (S, Se, Te); and two of the halogens (F and Cl). Moderately volatile elements have condensation temperatures between those of the major elements (Fe, Mg, Si) and troilite FeS. The moderately volatile elements with higher

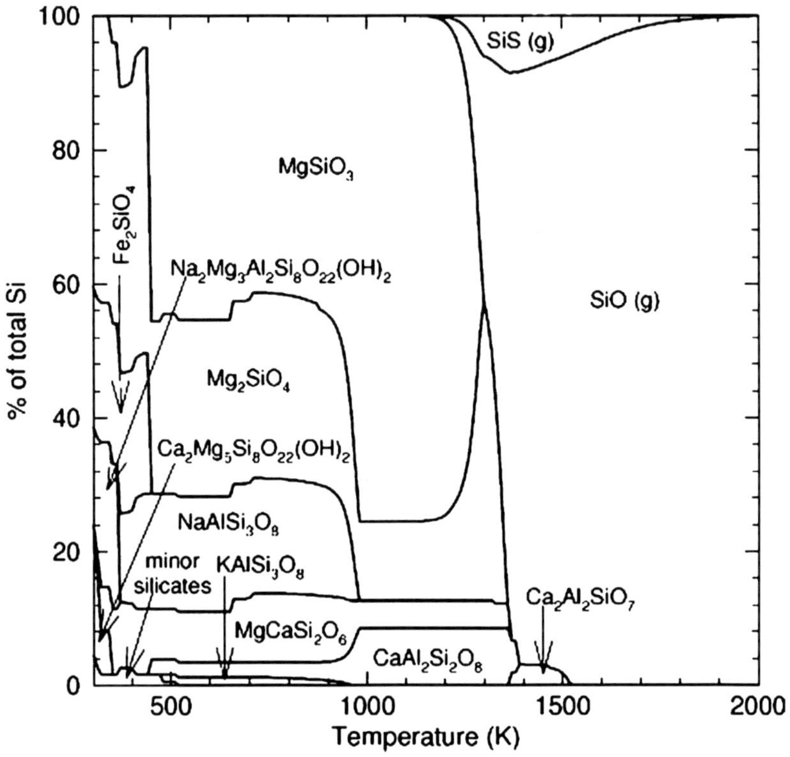

Fig. 7. Percentage distribution of silicon between different phases in a solar composition gas as a function of temperature for a total pressure of 10^{-4} bar. The abundances of the phases are shown as a cumulative percentage of the element. Gaseous species are indicated with (g), all other species are condensed phases. Not shown due to low abundance: $KMg_3AlSi_3O_{10}(OH)_2$ (1.1%), Mn_2SiO_4 (0.46%), and $Na_8Al_6Si_6O_{24}Cl_2$ (1.6%)

abundances condense as pure phases. For example, Figures 5 and 9 show that sulfur condenses as troilite FeS at 704 K. As mentioned above, H_2S is the dominant S-bearing gas at the troilite condensation temperature. Figure 9 shows that S, SH, and SiS are important and that PS and S_2 are minor S-bearing gases at higher temperatures. However, these gases are highly reactive and cannot survive at lower temperatures. For example, the condensation of forsterite Mg_2SiO_4 and enstatite $MgSiO_3$ removes all silicon, including SiS, from the gas. Condensation of schreibersite Fe_3P removes PS and all other

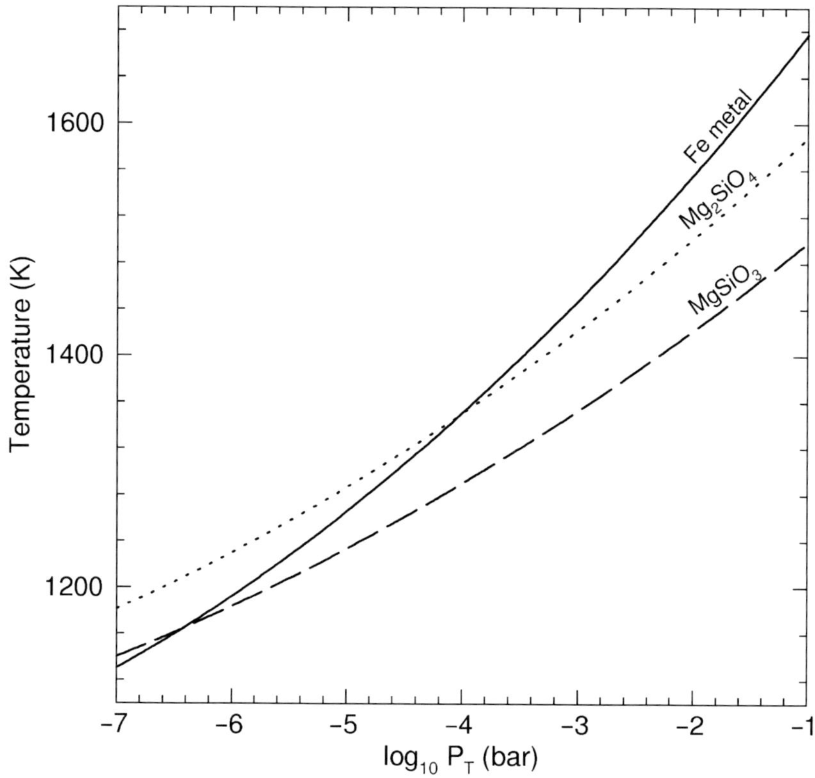

Fig. 8. Condensation temperatures of Fe metal, Mg_2SiO_4 forsterite, and $MgSiO_3$ enstatite in a solar composition gas as a function of total pressure

P-bearing gases. Finally, monatomic S, SH, and S_2 convert into H_2S with decreasing temperature.

Figures 5 and 10 show that phosphorus condenses as schreibersite Fe_3P at 1285 K. Schreibersite reacts with nebular gas at lower temperatures and forms fluorapatite $Ca_5(PO_4)_3F$ at 710 K and whitlockite $Ca_3(PO_4)_2$ at 680 K. The whitlockite reacts with HCl in the nebular gas to form chlorapatite $Ca_5(PO_4)_3Cl$ at 470 K. Chondrites contain all these phosphate minerals. Figures 11 and 12 show the equilibrium chemistry of fluorine and chlorine, which illustrate an interesting mass balance example. Fluorine is the least abundant element in fluorapatite and its abundance limits that of fluorapatite. The 3:1 ratio of P to F atoms in fluorapatite consumes about 30% of total phosphorus, leaving 70% of total phosphorus for chlorapatite formation.

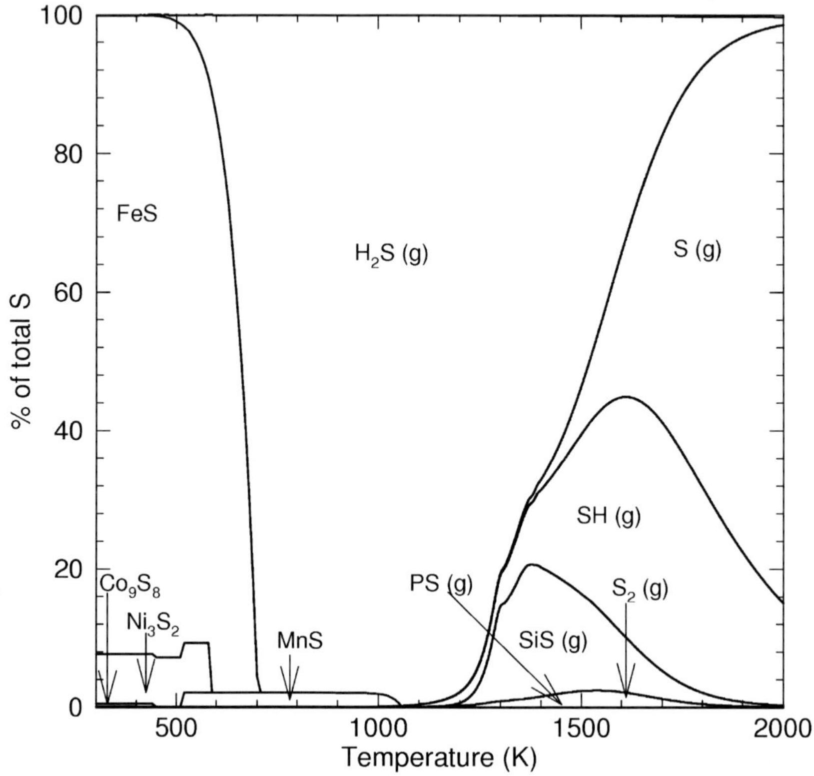

Fig. 9. Percentage distribution of sulfur between different phases in a solar composition gas as a function of temperature for a total pressure of 10^{-4} bar. Gaseous species are indicated with (g), all other species are condensed phases

However, this is not enough to condense all chlorine, which requires a 3:1 ratio of P to Cl atoms to condense all chlorine in chlorapatite. In fact, the total phosphorus abundance is too small to condense all Cl in chlorapatite. The residual chlorine, which is about 60% of total Cl, condenses as halite NaCl (420 K). However, this forms sodalite $Na_4(AlSiO_4)_3Cl$ (405 K, and again at 315 K), which decomposes back to halite plus nepheline (365 K), over the 405 - 315 K range. Both minerals are found in chondrites. Halite is probably present in many chondrites that contain water-soluble chlorine, even though it is not reported as one of the observed minerals.

Sodium and potassium condense initially as the feldspar minerals albite $NaAlSi_3O_8$ and orthoclase $KAlSi_3O_8$. Some of the albite reacts to form halite

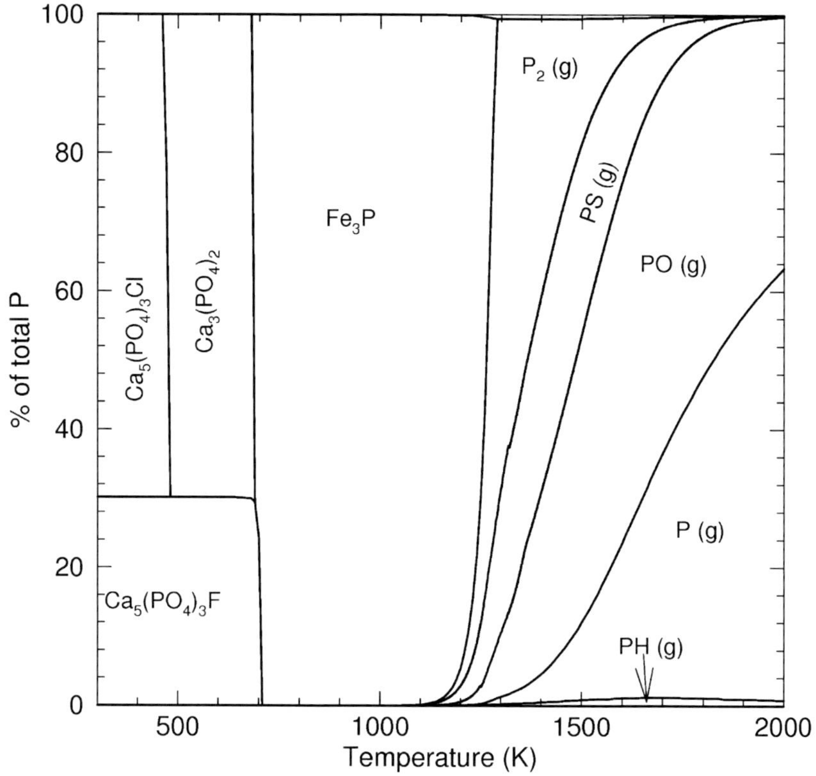

Fig. 10. Percentage distribution of phosphorus between different phases in a solar composition gas as a function of temperature for a total pressure of 10^{-4} bar. Gaseous species are indicated with (g), all other species are condensed phases. Hydroxyapatite ($Ca_5(PO_4)_3OH$) has an abundance of 70% of total P, but occurs only at 300 K (and lower), and is therefore not shown

at 420 K. Formation of glaucophane, a hydrous mineral called an amphibole, consumes the remaining albite at 360 K. Likewise, formation of phlogopite, a hydrous mineral called a mica, consumes all orthoclase at 470 K. Neither glaucophane nor phlogopite occurs in meteorites. Thus, although they become thermodynamically stable at low temperatures, they may not form because of slow chemical reaction rates.

Most of the moderately volatile elements have low abundances and condense in solid solutions in more abundant minerals. For example, selenium Se may dissolve in FeS as FeSe. Copper, Ag, and Au may dissolve in Fe alloy.

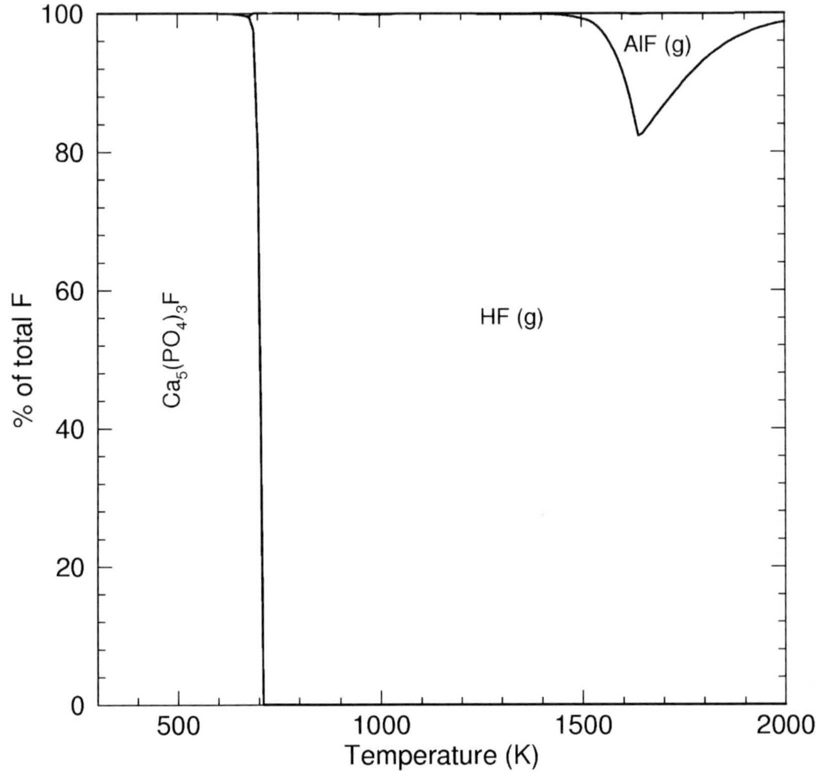

Fig. 11. Percentage distribution of fluorine between different phases in a solar composition gas as a function of temperature for a total pressure of 10^{-4} bar. Gaseous species are indicated with (g), all other species are condensed phases

Lithium, Mn, and Zn condense into forsterite and enstatite, e.g. as $LiSiO_3$, Mn_2SiO_4, $MnSiO_3$, Zn_2SiO_4, and $ZnSiO_3$. Boron and gallium substitute for Al in feldspar and Rb and Cs substitute for K in orthoclase. Some elements such as gallium are found in metal and silicates in meteorites, and gallium may dissolve in both phases during condensation.

3.4 Highly Volatile Elements

The highly volatile elements (Bi, Br, Cd, Hg, I, In, Pb, Tl) condense at temperatures between troilite and water ice. In general, the equilibrium chemistry of these elements is poorly known because thermodynamic data are unavailable

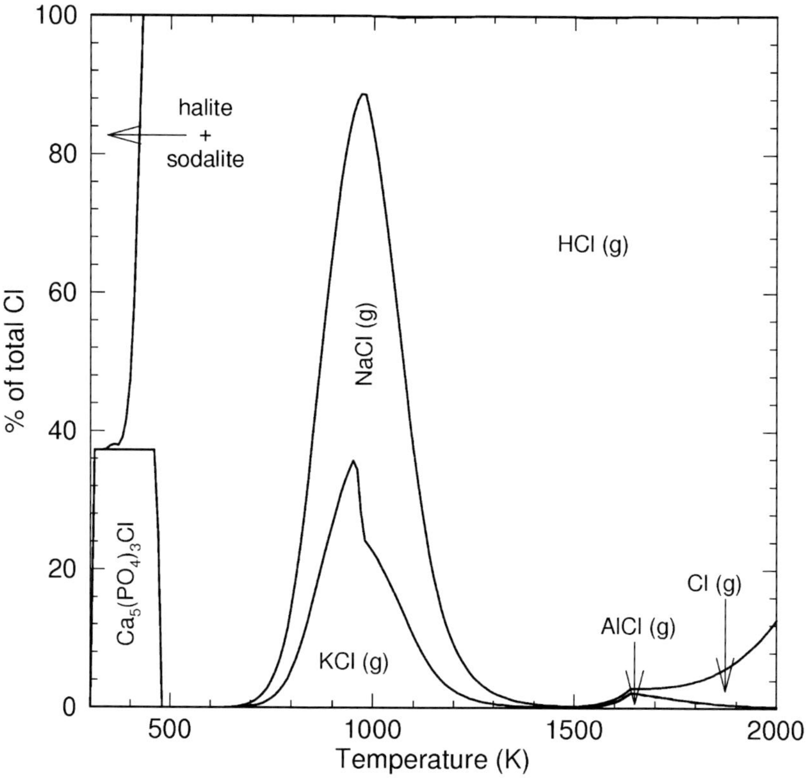

Fig. 12. Percentage distribution of chlorine between different phases in a solar composition gas as a function of temperature for a total pressure of 10^{-4} bar. Gaseous species are indicated with (g), all other species are condensed phases

for some pure phases and solid solutions. Most of the highly volatile elements condense in solid solution in more abundant minerals. For example, Lodders [5] calculated that mercury condenses as HgS, HgSe, and HgTe dissolved in solid solution in troilite FeS. Fifty percent of all Hg is condensed at 252 K. Her result predicts that the CI carbonaceous chondrites should contain the solar abundance of mercury. At present, the Hg solar abundance is estimated from r-process and s-process systematics of the elements. Bismuth Bi, and lead Pb condense in solid solution in iron metal alloy, while cadmium, In, and Tl condense in solid solution in troilite FeS. Bromine and iodine condense in solid solution in the apatite minerals. Lodders [5] gives a good discussion of condensation chemistry of the moderately and highly volatile elements in solar composition material.

3.5 Atmophile Elements

The atmophile elements (H, C, N, noble gases) condense at and below the condensation temperature of water (liquid or ice). Liquid water or water ice condenses at the temperature where the partial pressure of water vapour equals the vapour pressure over liquid water or water ice. The condensation curve for water (liquid or ice) is

$$\frac{10,000}{T_c(H_2O)} = 38.84 - 3.83 log_{10} P_T \tag{11}$$

Thus, water ice starts to condense at 185 K at 10^{-4} bar total pressure in solar composition material. Fifty percent of all water condenses by 180 K. Liquid water condenses instead of water ice at total pressures of 3.8 bar and above. Condensation of liquid water has important consequences such as the formation of aqueous NH_3 solutions. Aqueous NH_3 solutions are stable down to 173 K at one bar total pressure, which is the eutectic point, i.e., the lowest melting point, in the NH_3 - H_2O phase diagram [11].

The partial pressure of water vapour in solar composition material depends on several factors. Formation of anhydrous minerals consumes about 23% of total oxygen. The formation of hydrous minerals, such as glaucophane, tremolite, serpentine, and talc consumes more water vapour. For example, serpentine and brucite $Mg(OH)_2$ form by the hydration of forsterite Mg_2SiO_4 grains by water vapour

$$\begin{gathered} 2Mg_2SiO_4(forsterite) + 3H_2O(g) = \\ Mg_3Si_2O_5(OH)_4(serpentine) + Mg(OH)_2(brucite) \end{gathered} \tag{12}$$

An analogous reaction is formation of talc and brucite by hydration of enstatite grains

$$\begin{gathered} 4MgSiO_3(enstatite) + 2H_2O(g) = \\ Mg_3Si_4O_{10}(OH)_2(talc) + Mg(OH)_2(brucite) \end{gathered} \tag{13}$$

These hydration reactions provide hydrous minerals at low temperatures in solar composition material ($\sim$ 250 K for serpentine and $\sim$ 300 K for talc). By mass, serpentine contains 13.0% water, talc 4.8% water, and brucite 3.1% water. Theoretical models of gas - grain kinetics predict that hydration of forsterite and enstatite to serpentine and talc are too slow to proceed over the lifetime of the solar nebula [12]. These predictions are consistent with

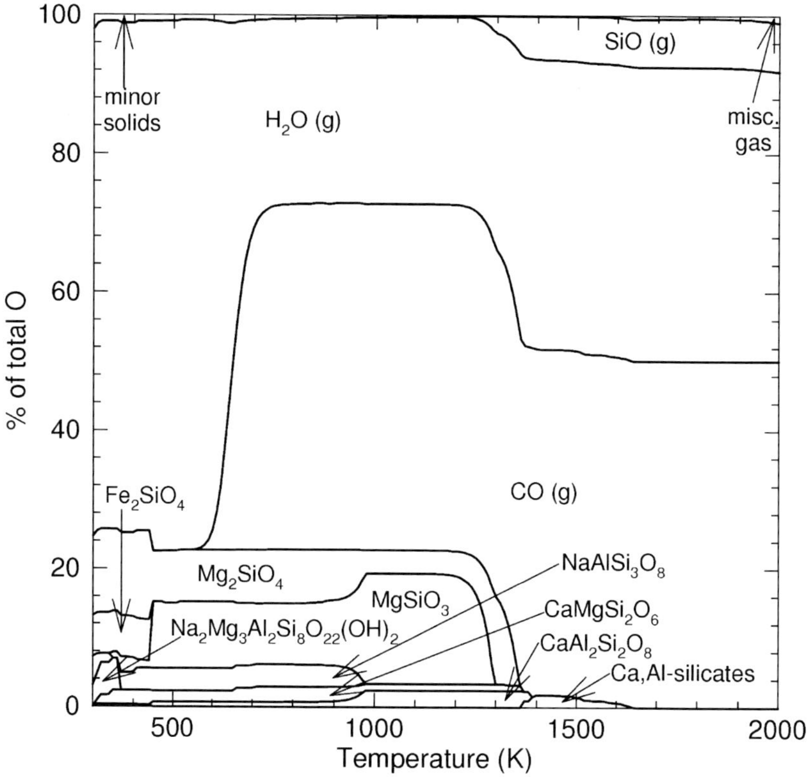

Fig. 13. Percentage distribution of oxygen between different phases in a solar composition gas as a function of temperature for a total pressure of 10^{-4} bar. Gaseous species are indicated with (g), all other species are condensed phases. The minor solids include: Al_2O_3 (0.28%), $Ca_2Mg_5Si_8O_{22}(OH)_2$ (4.15%), $Ca_5(PO_4)_3$(F,Cl,OH) (0.25%), $Ca_3(PO_4)_2$ (0.17%), $CaTiO_3$ (0.05%), Cr_2O_3 (0.14%), $FeCr_2O_4$ (0.15%), $FeTiO_3$ (0.05%), $KAlSi_3O_8$ (0.21%), $KMg_3AlSi_3O_{10}(OH)_2$ (0.31%), $MgAl_2O_4$ (0.40%), $MgCr_2O_4$ (0.18%), $Mg_3Si_4O_{10}(OH)_2$ (0.27%), Mn_2SiO_4 (0.13%), $Na_8Al_6Si_6O_{24}Cl_2$ (0.44%), TiO_2 (0.03%), Ti_4O_7 (0.03%), VO (0.002%), V_2O_3 (0.003%), $ZnCr_2O_4$ (0.03%)

petrological studies of meteorites containing hydrous silicates, which conclude that hydrous minerals formed on the meteorite parent bodies.

The partitioning of carbon between CO and CH_4 also affects the water vapour partial pressure. As discussed below, CO is the major C-bearing gas at high temperatures and low pressures while CH_4 is the major C-bearing gas at low temperatures and high pressures. Because the bulk C/O ratio is 0.5 in solar composition material, the water vapour abundance increases by about a factor of two when CO converts to methane. All these factors are summarized in Figure 13, which shows the percentage distribution of oxygen between the major O-bearing gases and minerals. At complete chemical equilibrium, the anhydrous and hydrous minerals consume about 26% of all oxygen and about 74% is left to condense as water or water ice.

The condensation chemistry of carbon and nitrogen depends on the major C- and N-bearing gases at low temperatures in solar composition material. At 10^{-4} bar total pressure, CH_4 and NH_3 are the predominant C- and N-bearing gases at temperatures below the water ice (or liquid water) condensation curve. Ammonia condenses as ammonia monohydrate $NH_3{\cdot}H_2O$ at 131 K, 10^{-4} bar total pressure via the net reaction

$$NH_3(g) + H_2O(ice) = NH_3 \cdot H_2O(ice) \tag{14}$$

Ammonia monohydrate (also called ammonium hydroxide NH_4OH) is a distinct compound and is not a clathrate hydrate. Ammonia ice and/or liquid NH_3 could form in the absence of liquid water. This happens in Jupiter's atmosphere where NH_3 ice clouds form because gravitational sedimentation keeps liquid water droplets in a cloud layer hundreds of kilometers below the NH_3 ice clouds. Thus, no water is present in the upper atmosphere where the NH_3 ice clouds form. Liquid NH_3 does not condense in the solar nebula or on the gas giant planets in our solar system (Jupiter, Saturn, Uranus, and Neptune) because the total pressure and/or NH_3 gas abundance is too low. The NH_3 gas pressure would have to be at least as high as the triple point pressure (0.06 bar at 195.5 K) for liquid NH_3 condensation. Assuming that all nitrogen is present as NH_3, the total pressure at low temperatures would have to be about 444 bar for liquid NH_3 to form (i.e., to reach 0.06 bar NH_3 pressure at the triple point) in solar composition material. Such high pressures are unreasonable for the solar nebula but they may occur in extrasolar gas giant planets or in the interiors of icy planets and satellites.

Methane condenses at lower temperatures as methane clathrate hydrate ($CH_4 \cdot 7H_2O$) at 78 K and CH_4 ice. A clathrate hydrate is a cage compound in which a gas molecule is trapped inside the water ice crystal lattice. One gas atom is trapped for every seven water molecules. Clathrate hydrates form via reactions exemplified by

$$CH_4(gas) + 7H_2O(ice) = CH_4 \cdot 7H_2O(clathrate\ hydrate) \tag{15}$$

However, there is not enough water ice to condense all CH_4 as a clathrate hydrate because the CH_4 to H_2O ratio is 1:7 in the clathrate hydrate while the solar C/O ratio is 0.5. Thus, the residual CH_4 condenses as CH_4 ice at lower temperatures. Hydrogen and helium do not condense because temperatures never get low enough for solid hydrogen or liquid helium to form. Neon ice forms at about 9 K, but it is doubtful if temperatures ever get this low in any astronomical environment where condensation occurs. The other noble gases (Ar, Kr, Xe) either condense as ices (solid Ar, Kr, Xe) or form clathrate hydrates (Ar 48 K, Kr 53 K, Xe 69 K). The condensation temperatures given for the noble gas clathrate hydrates assume that sufficient water ice is available for reaction and that all of each noble gas forms a clathrate hydrate. If no water ice is available, the noble gases condense as ices at lower temperatures than their clathrates form.

The condensation chemistry of carbon and nitrogen depends on whether or not CH_4 and NH_3 are the predominant C- and N-bearing gases. The two major C-bearing gases in solar composition material over a wide P, T range are CO and CH_4. These two gases are converted into one another by the net reaction

$$CO + 3H_2 = CH_4 + H_2O \tag{16}$$

$$logK_{16} = \frac{11,069.94}{T} - 1.17969logT - 8.96596 \tag{17}$$

Equation (17) gives the equilibrium constant for reaction (16) from 298 - 2500 K. Chemical equilibrium calculations using this data and the solar elemental abundances show that CO is stable at high temperatures and low pressures while CH_4 is stable at low temperatures and high pressures in solar composition material (e.g., see [13]). For example, at 10^{-4} bar total pressure, CO is the major C-bearing gas at temperatures greater than 625 K, CH_4 is the major C-bearing gas at temperatures less than 625 K, and the two gases have equal abundances at 625 K. The equal abundance point shifts to higher temperatures with higher pressures, as shown in Figure 14.

However, the kinetics of the gas phase conversion of CO to CH_4 become so slow that it may not happen unless grain catalyzed reactions occur. If all carbon remained as CO at the low temperatures in solar composition material, then the water ice abundance was decreased below the amount which could condense if CO were converted to CH_4. On the other hand, if CO were efficiently converted to CH_4 and/or other hydrocarbons, then a sizable fraction of the total O was released from CO and was available for formation of water ice. The water ice/rock mass ratios in "icy" bodies formed in the solar nebula, where CO was the dominant carbon gas, are predicted to be lower than the water ice/rock ratios in "icy" bodies formed in the subnebulae around Jupiter and Saturn, where CH_4 was the dominant carbon gas. (Planetary scientists think that the Galilean satellites (Io, Europa, Ganymede, and Callisto) of Jupiter and Titan and other regular satellites of Saturn formed in miniature versions of the solar nebula known as subnebulae. These existed around Jupiter and Saturn during their formation and were higher density regions with different chemistry than the surrounding solar nebula.) To first

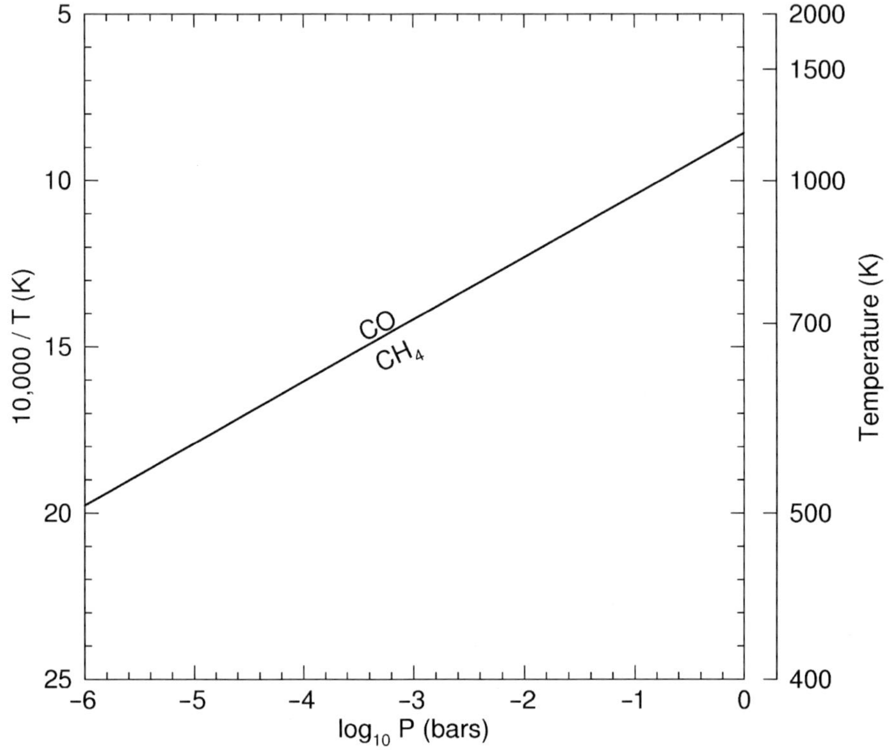

Fig. 14. Percentage distribution of carbon between CO and CH_4 in a solar composition gas as a function of temperature and pressure. The line shows where CO and CH_4 have equal abundances. Above the line, CO is more abundant, whereas below the line, CH_4 is more abundant

approximation, this distinction is observed and supports the chemical modeling. The two major N-bearing gases in solar composition material are N_2 and NH_3 (see Figure 15). They are converted into one another by the net thermochemical reaction

$$N_2 + 3H_2 = 2NH_3 \tag{18}$$

$$logK = (\frac{6051.59}{T}) - 1.21176logT - 7.89739 \tag{19}$$

Equation (19) gives the equilibrium constant from 298 - 2500 K for reaction (18). Chemical equilibrium calculations using this data and the solar elemental abundances show that N_2 is the dominant N-bearing gas at high temperatures and low pressures and that NH_3 is the major N-bearing gas at low temperatures and high pressures. Figure 15 shows the N_2/NH_3 equal abundance line over a wide P, T range. At higher temperatures, N_2 is the only nitrogen-bearing gas of any importance. Along the 10^{-4} bar isobar, NH_3 remains the second most abundant N-bearing gas until about 1670 K where monatomic N becomes the second most abundant gas. However, even at 2000 K, 10^{-4} bar the N_2/N molecular ratio is 100,000 and all other N-bearing gases are less abundant.

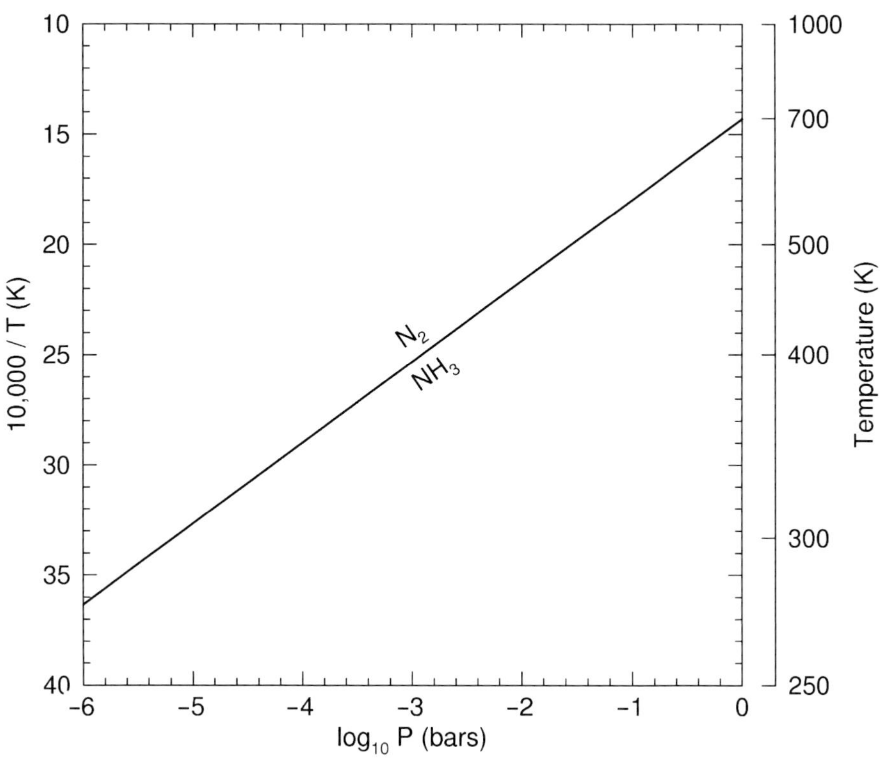

Fig. 15. Percentage distribution of nitrogen between N_2 and NH_3 in a solar composition gas as a function of temperature and pressure. The line shows where N_2 and NH_3 have equal abundances. Above the line, N_2 is more abundant, whereas below the line, NH_3 is more abundant

The N_2/NH_3 equal abundance line is slightly different from the CO/CH_4 equal abundance line because equal abundances of N_2 and NH_3 do not correspond to 50% of total nitrogen in each gas. Figure 15 shows that at 10^{-4} bar total pressure, N_2 and NH_3 have equal abundances at 345 K, but 50% of total nitrogen is in each gas at 320 K. At lower temperatures, NH_3 first condenses as ammonium carbonate NH_4HCO_3, or ammonium carbamate (NH_4COONH_2). The exact amount of NH_3 in these compounds depends upon the amount of CO_2 and is hard to quantify because the amount of CO_2 depends on the rate of the CO to CO_2 conversion in different astronomical environments such as the solar nebula or another protoplanetary accretion disk. The remaining NH_3 condenses as ammonia monohydrate ($NH_3 \cdot H_2O$) at 131 K, 10^{-4} bar.

However, the gas phase conversion of N_2 to NH_3 is a slow reaction and may not occur over long times, such as the lifetime of the solar nebula. Industrial production of NH_3 from N_2 (the Bosch - Haber process) uses Fe-based catalysts to speed up the reaction. Iron-rich metal grains and magnetite Fe_3O_4 grains are common in chondritic meteorites and it is likely that such grains catalyzed the N_2 to NH_3 conversion in the solar nebula and other solar composition systems, e.g. see [14]. If N_2 remains the major N-bearing gas due to kinetic limitations, N-bearing ices do not form until very low temperatures where either N_2 clathrate hydrate or N_2 ice condense. Temperatures of 20 - 40 K are required for this to happen.

The most important of the atmophile elements is hydrogen, which is the most abundant element in solar composition material. Hydrogen's dominance controls the chemistry of solar composition material. With the exception of helium, which is non-reactive, hydrogen is about 1,000 times as abundant as all other elements combined. Thus, hydrogen-bearing gases (hydrides) are major or important gases at chemical equilibrium for many elements. A few examples are H_2O, CH_4 (at low temperatures), NH_3 (at low temperatures), H_2S, HF, HCl, and HBr.

Most hydrogen remains in elemental form because no other element amounts to more than about 0.1% of the hydrogen elemental abundance. Figures 16 and 17 illustrate this point. They show the mole fractions of the major H-bearing gases. By definition, the mole fractions of all gases that are present add up to unity. Figure 17 shows that the major H-bearing gases are always elemental gases. With decreasing temperature the major H-bearing gas changes from ionized hydrogen H^+ (at temperatures higher than those shown on the graph), to monatomic hydrogen H, and finally to molecular hydrogen H_2. The hydrides and all other H-bearing gases are much less abundant than the different allotropes of hydrogen.

Figure 16 shows the percentage distribution (on a logarithmic scale) of hydrogen between the major H-bearing gases over the same temperature range. Figure 16 shows that H_2 is 50% dissociated to monatomic H at 2,230 K (i.e., the temperature where each gas has a mole fraction of 0.50). Monatomic H is 50% ionized to H^+ at 8,700 K, above the highest temperature on the graphs. Thermal ionization of monatomic H occurs via the reaction

$$H(gas) = H^{+}(gas) + e^{-}(gas) \qquad (20)$$

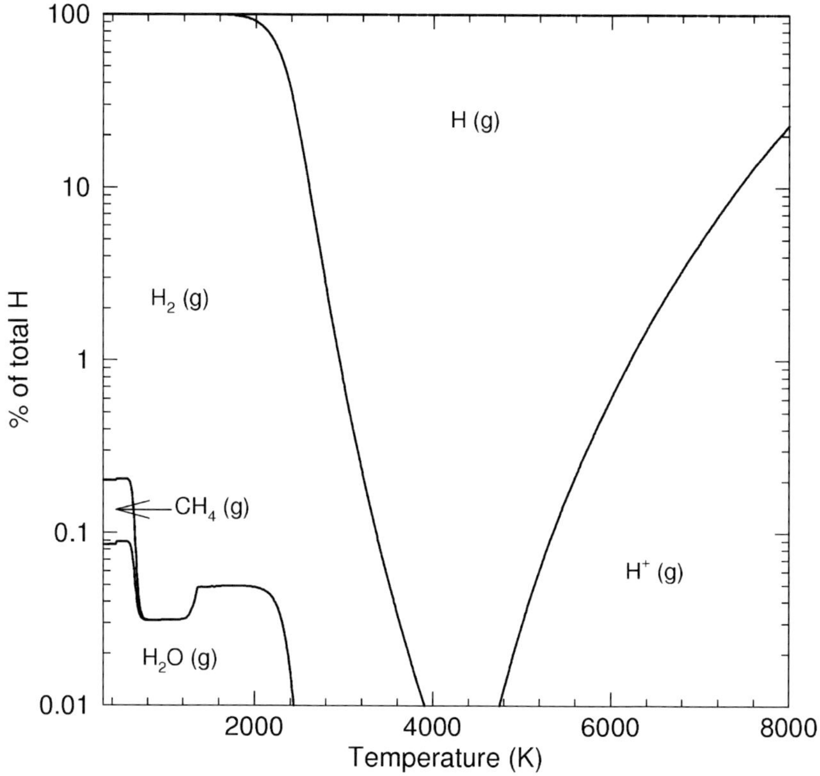

Fig. 16. Percentage distribution of hydrogen between different phases in a solar composition gas as a function of temperature for a total pressure of 10^{-4} bar. Gaseous species are indicated with (g). Solid phases ($Na_2Mg_3Al_2Si_8O_{22}(OH)_2$, $Ca_5(PO_4)_3OH$, $Ca_2Mg_5Si_8O_{22}(OH)_2$, $KMg_3AlSi_3O_{10}(OH)_2$, $Mg_3Si_4O_{10}(OH)_2$) are not shown, due to their low abundances ($1 < \times 10^{-4}\%$)

The electron pressure in this reaction is that produced by the ionization of all elements in a solar composition gas - not only the electron pressure due to hydrogen ionization. As the total pressure decreases, thermal dissociation and ionization of hydrogen become more important at lower temperatures. For example, hydrogen is 50% dissociated at 1,880 K and 50% ionized at 7,100 K at 10^{-6} bar total pressure.

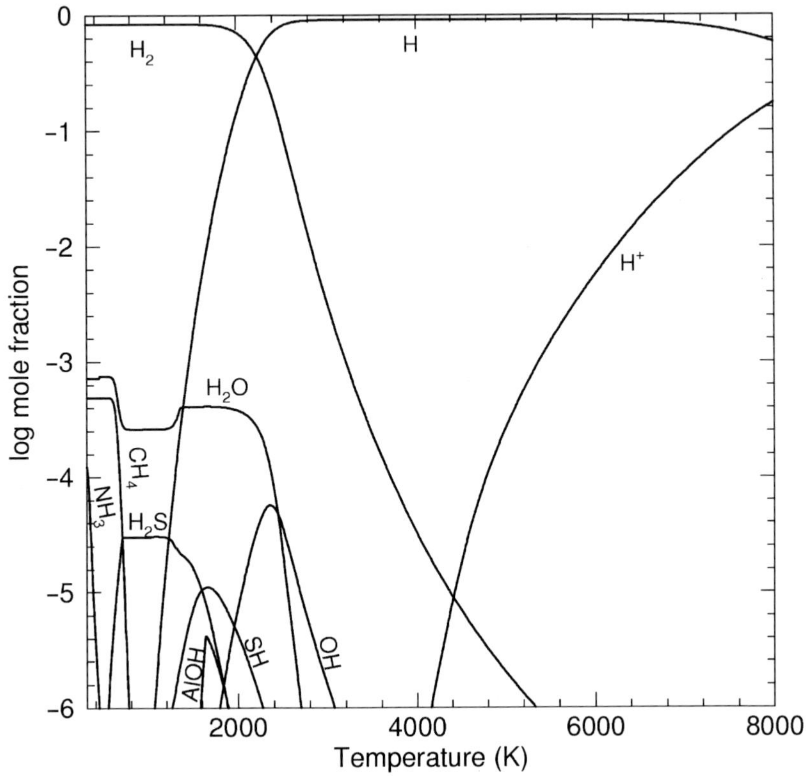

Fig. 17. Gas phase chemistry of hydrogen in a solar composition gas as a function of temperature at a total pressure of 10^{-4} bar. The abundances of the gases are shown in mole fractions

4 Summary

In this chapter, we reviewed the methods and results of chemical equilibrium calculations applied to solar composition material. These types of calculations are applicable to chemistry in a variety of astronomical environments including the atmospheres and circumstellar envelopes of cool stars, the solar nebula and protoplanetary accretion disks around other stars, planetary atmospheres, and the atmospheres of brown dwarfs. The results of chemical equilibrium calculations have guided studies of elemental abundances in meteorites and presolar grains and as a result have helped to refine nucleosynthetic models of element formation in stars.

Acknowledgment This work was supported by funding from the NASA Astrobiology and Planetary Atmospheres Program.

References

1. K. Lodders, B. Fegley, Jr.: Condensation Chemistry of Carbon Stars. In: *Astrophysical Implications of the Laboratory Study of Presolar Materials*, AIP Conference Proceedings v. 402, ed by T. J. Bernatowicz, E. Zinner (American Institute of Physics, Woodbury, NY 1997) pp 391-423
2. B. Fegley, Jr., K. Lodders: Icarus **110**, 117 (1994)
3. K. Lodders: J. Phys. Chem. Ref. Data **28**, 1705 (1999)
4. K. Lodders: J. Phys. Chem. Ref. Data **33**, 357 (2004)
5. K. Lodders: Astrophys. J. **591**, 1220 (2003)
6. H. Palme, B. Fegley, Jr.: Earth Planet. Sci. Lett. **101**, 180 (1990)
7. H. Palme, F. Wlotzka: Earth Planet. Sci. Lett. **33**, 45 (1976)
8. B. Fegley, Jr., H. Palme: Earth Planet. Sci. Lett. **72**, 311 (1985)
9. A. S. Kornacki, B. Fegley, Jr.: Earth Planet. Sci. Lett. **79**, 217 (1986)
10. A. S. Kornacki, B. Fegley, Jr.: Proc. 14th Lunar Planet. Sci. Conf. J. Geophys. Res. **89**, B588 (1984)
11. J. S. Lewis: Earth Planet. Sci. Lett. **15**, 286 (1972)
12. B. Fegley, Jr.: Trends of Volatile Elements in the Solar System. In: Workshop on the Origins of Solar Systems ed by J. A. Nuth, P. Sylvester (LPI, Houston 1988) Tech. Rep. No. **88-04**, p. 51.
13. K. Lodders, B. Fegley, Jr.: Icarus **155**, 393 (2002)
14. B. Fegley, Jr., R. G. Prinn: Solar Nebula Chemistry: Implications for Volatiles in the Solar System. In: The Formation and Evolution of Planetary Systems, ed by H. Weaver, L. Danly (Cambridge University Press, Cambridge, UK 1989) pp 171-211

Solar System Abundances of the Elements

Katharina Lodders

Planetary Chemistry Laboratory, Dept. of Earth & Planetary Sciences and McDonnell Center for the Space Sciences, Washington University, Campus Box 1169, One Brookings Drive, Saint Louis, MO 63130, USA

Summary. Representative abundances of the chemical elements for use as a solar abundance standard in astronomical and planetary studies are summarized. Updated abundance tables for solar system abundances based on meteorites and photospheric measurements are presented.

Keywords: solar system abundances, solar photosphere, meteorites, CI chondrites

1 Motivations to Study Solar System Elemental Abundances

The investigation of which chemical elements exist in nature and in what quantities has a long history. The determination of elemental abundances in various celestial objects is still a very active field in astronomy, planetary science, and meteoritics. There are multiple motivations for studying the solar system abundances of the chemical elements. One reason to study this overall composition of the solar system is to understand how the diversity of planetary compositions, including that of our home planet, can be explained, since all planets in the solar system share a common origin from the material of the protosolar disk (the solar nebula).

The composition of the sun determines how the sun works and evolves over time, as composition influences the interior structure of the sun. Although the Sun is mainly composed of H and He, other heavy elements such as C, N, O, Ne, Fe, etc., are important opacity sources that influence the energy transport out of the sun through radiation and convection. The sun is a typical main sequence dwarf star and its composition is a useful baseline for comparison to abundances in other dwarf stars and to changes that appear in advanced stages of stellar evolution. For example, relative to the suns composition, red giant stars show observable abundance variations that are the result of

A. Goswami and B.E. Reddy (eds.), *Principles and Perspectives in Cosmochemistry*, Astrophysics and Space Science Proceedings,
DOI 10.1007/978-3-642-10352-0_8,

nucleosynthesis operating in giant stars, and these products have been dredged up from stellar interiors to the stellar exteriors.

The solar system abundances are a useful local galactic abundance standard because many nearby dwarf stars are similar in composition; however, in detail there are some stochastic abundance variations (e.g [1,2,3,5]). The term cosmic abundances should be avoided because abundances generally decrease with galactocentric distance. There are also abundance differences between our galaxy and galaxies at high red-shift; hence there is no generic cosmic composition that applies to all cosmic systems.

Finally, solar abundances are a critical test of nucleosynthesis models and models of Galactic chemical evolution [6, 7, 8]. Ideally, such models should quantitatively explain the elemental and nuclide distributions of solar system matter.

The sun has the most mass (>99%) of the solar system objects and therefore it is the prime target for studying solar system abundances. Most elements can be measured in the sun's photosphere, but data from the solar chromosphere and corona, solar energetic particles, solar wind, and solar cosmic rays (from solar flares), help to evaluate abundances of elements that have weak absorption lines (because these elements are low in abundance or only have blended absorption lines in the photospheric spectrum).

Below we will see that meteorites, smaller rocks from asteroidal objects delivered to Earth, provide important information for solar system abundances of non-volatile elements. Other sources to refine solar system abundances are analysis of other solar system objects such as the gas-giant planets, comets and the interplanetary dust particles from comets. Outside the solar system, the compositions of hot B stars, planetary nebulae, Galactic cosmic rays (GCR), the nearby interstellar medium (ISM) and H II regions have been employed to amend the solar system abundances of some elements.

Solar or solar system abundance data derived from meteorites and the solar photosphere are reviewed periodically. References [9, 10, 11, 12, 13, 14, 15, 16, 17] give some compilations that summarize information on photosphere and meteoritic abundances used as solar system abundance standards since 1989.

2 Meteorites as Abundance Standards for Non-Volatile Solar System Matter

About a century ago, it became more evident that meteorites may contain chemical and mineralogical information about the earliest solid objects that existed in the solar system, and that they may carry resemblance to the materials that accreted to planets like the Earth. Several different meteorite groups are recognized. With respect to abundances, the chondritic meteorites are the most important ones. Meteorites that contain small silicate spheres are called chondrites (after Greek chondros for sphere). In addition to silicate minerals, most chondrites contain FeNi metal and iron sulfide (the mineral troilite, FeS, is most common), and a host of minor minerals. The major chondrite groups are ordinary, enstatite and carbonaceous chondrites; each group has further divisions based on the different proportions of their major minerals. The most common meteorites are, as the name implies, the ordinary chondrites. The enstatite chondrites, named after their most common silicate mineral enstatite, contain exotic sulfides and metal phases that are only stable under highly reducing conditions. The carbonaceous chondrites contain up to several percent of carbon, depending on sub-type. The carbonaceous meteorites of Ivuna type (abbreviated as CI chondrites for Carbonaceous and Ivuna) are the most important ones for solar system abundance determinations.

The CI chondrites experienced severe aqueous alteration on their parent asteroid, and if they ever contained chondrules, metal, and sulfides as the other carbonaceous chondrites do, these phases were largely erased. The CI chondrites mainly consist of fine grained, hydrous silicates, magnetite (Fe_3O_4) probably produced by oxidation of FeNi metal, Fe-Ni bearing sulfides other than troilite, and various salts. Despite the absence of chondrules, the CI chondrites are still chondrites because their overall elemental composition for many elements is much closer to chondritic meteorites than to iron meteorites or so-called achondrites, that mainly consist of silicates and experienced severe melting.

Up to the 1950s, the CI chondrites were not given too much attention for their potential role as abundance standards. The classical abundance papers by [18,19,20] used elemental analyses of silicates, metal, and sulfides from different chondrites and assumed some representative proportion of metal, silicate, and sulfides to come up with a meteoritic abundance standard. By the 1970s, the picture emerged that the CI chondrites as well as another carbonaceous chondrite group named CM after the Mighei meteorite may be suitable groups to represent the solar system abundances of elements that provide the cations in rock-forming minerals. However, the relative contents of volatile elements is lower in CM chondrites than in CI chondrites, and a correlation of CM- to CI chondrite abundance ratios with condensation temperatures implies that volatility-related fractions occurred in CM chondrites.

Figure 1 shows the concentration ratios for CM chondrites to CI chondrites as a function of condensation temperature. The symbols indicate the mineral phase hosting the elements. There is a smooth decrease in the concentration ratio with condensation temperatures which is independent of mineral host phase. This correlation indicates that the elemental abundances in CM chondrites are volatility controlled. The higher concentration ratio for refractory elements plots above unity and reflects that CM chondrites accumulated a higher proportion of refractory elements. The ratio below unity is for the volatile elements incomplete condensation or accumulation. This limits the use of CM chondrites as abundance standards for elements with condensation temperatures less than $\sim$ 1500 K. (see chapter by B. Fegley and L. Schaefer in this volume for condensation chemistry of the elements).

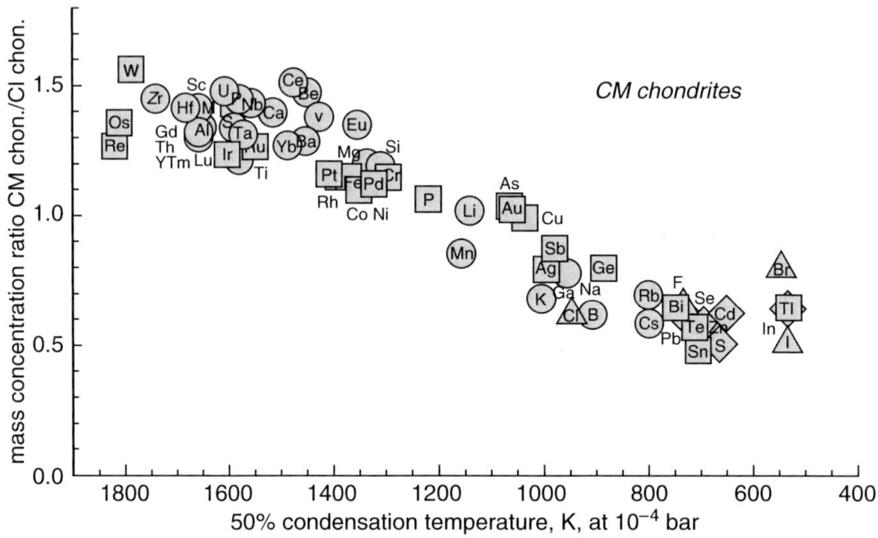

Fig. 1. The decreasing element concentration ratio of CM- over CI chondrites indicates volatility related fractionations in CM meteorites and makes them of limited use as an abundance standard. The different symbol shapes indicate the principal mineral host phase for the elements (circle: lithophile elements in silicate and oxides; box: siderophile elements in metal alloy; chalcophile: sulfides; triangle: halogen). Data sources for CM chondrites: [21] plus updates; CI chondrites: [17]

The EH enstatite chondrite group also has higher relative abundances of volatile elements, however, in detail chemical fractionations of volatile and non-volatile elements are apparent. In a comparison of the abundances of all chondrite groups to those in the solar photosphere, only the CI chondrites are found to have the closest match. Finally, the CI chondrites give the best agreement (of all chondrite groups) between observed and theoretically predicted nuclide abundances as a function of mass number (see e.g., [22] for arguments in favor of CI chondrites as standard).

Part of the reason why it took so long to recognize the significance of CI chondrites is simply that CI chondrites are very rare. Out of the $\sim$ 1000 recorded observed meteorite falls from which material is preserved, only 5 CI chondrites are known. Among the 40,000 or so meteorites collected in Antarctica, only a few are CI chondrites. The meteorites are very fragile and decompose easily, for example, if placed in water, CI chondrites immediately begin to disintegrate. Hence, CI chondrites that are found a long time after their fall are not useful for abundance studies, as chemical information is easily altered or lost.

Table 1 lists the 5 observed CI chondrite falls, which are named after the nearest town to their fall location. Sufficient mass for study is available only for three of them, notably the Orgueil meteorite, which is probably one of the most and best analyzed rocks on this planet.

Table 1. Observed CI chondrite meteorite falls

Meteorite	Date of Fall	Country	Preserved Mass
Alais	15 March 1806	France	6 kg
Orgueil	14 May 1868	France	14 kg
Tonk	22 January 1911	India	10 g
Ivuna	16 December 1938	Tanzania	0.7 kg
Revelstoke	31 March 1965	Canada	$\leq$1g

The other reason why it took so long to realize that CI chondrites are a chemically special meteorite group is that early chemical analysis methods were not able to analyze trace element abundances in small samples with sufficient precision. The advances came through the application of instrumental neutron activation analyses and mass-spectroscopic measurements, which revealed the chemical differences in minor element abundances. Compared to all other chondrites, the CI chondrites have the highest relative contents of volatile elements, such as C, O, alkalis, chalcogenides, and halogens. In contrast, the abundances of refractory elements such as Al, Ca, Si, Mg, Fe, alkaline earths, and rare earth elements are less variable among chondrites, but important abundance differences for these elements exist between the chondrite groups and provide lively discussions among meteoriticists.

At the same time that elemental abundance analyses improved for rock samples, spectroscopic abundance determinations for the solar photosphere advanced. A comparison of elemental abundances in the solar photosphere to CI chondrites then showed the best agreement for most elements; the exceptions being H, C, N, O, and the noble gases. These elements form extremely volatile compounds that may have never accreted to the CI chondrite parent asteroid or were easily lost from CI chondrite material while in space or in the terrestrial environment.

2.1 Composition of CI chondrites

The reference composition of CI chondrites in Table 2 is based on a new review by [17]; henceforth LPG09. The previous evaluations of the composition of CI chondrites were done by [13]; henceforth L03 and [14]; henceforth PJ03. These two evaluations used slightly different approaches to derive recommended CI chondrite compositions. PJ03 emphasized the use of the Orgueil meteorite as the CI standard rock, because it is the most massive of the 5 CI chondrite falls and therefore the most analyzed one (see Table 1). L03 used data from all CI chondrites and computed weighted average compositions. However, since most analytical data are for the Orgueil meteorite, these weighted averages are also dominated by this chondrite. A comparison of data from Orgueil and the other four CI chondrites in [13] shows that compositional differences among CI chondrites are relatively small.

In our new CI chondrite compilation [17], we included many new data on trace elements that have become available in recent years, primarily because of improvements in instrumentation. In particular, application of Inductively Coupled Plasma Mass Spectrometry (ICP-MS) led to many new high quality analyses. These data and resulting updates of CI chondrite compositions are discussed in [17], however, an extensive discussion of the database containing 200+ references on CI chondrite analyses is deferred to the future [23].

In addition to computing average concentrations from several reliable analyses for a given element, we employed the element ratio method to find standard values for element concentrations in CI chondrites. This method relies on the fact that concentration *ratios* of elements with similar cosmo- and geochemical properties are usually more constant than the absolute measured element concentration. The reason for using this method is that analytical data for CI meteorites can show variable absolute concentrations for some elements, which is caused by the variable water contents and/or massive alteration of fine grained matrix material. In addition, small scale heterogeneities naturally arise because these meteorites are breccias (see [24, 25]). If there are compositional heterogeneities in the meteorite, the measured concentration of elements residing in common mineral phases may vary from sample to sample, however, the concentration ratios of these elements relative to each other will remain constant. Therefore, it is sometimes practical to take the concentration ratio of a pair of geochemically similar elements from several samples and to use the absolute abundance of the accurately determined element of that pair to calculate the concentration of the other element. Since there are only three CI chondrites for which larger numbers of analysis exist, it is also useful to compare the element concentration ratios in CI chondrites to that of other carbonaceous chondrites. The inclusion of data, e.g., from CM chondrites, which are relatively closely related to the CI chondrites, increases the statistics for abundance determinations by the ratio method. However, in using other data than for CI chondrites, the explicit assumption is made that chemical fractionations of the elements are absent between CI chondrites and the other chondrite groups invoked. This assumption is usually justified as long as CM chondrites and refractory elements are involved. Figure 1 shows that the concentration ratio of refractory elements in CI and CM chondrites plots at a comparable constant level.

The element ratio method and the direct average method usually give consistent results within 3% for most elements, which is easily within the estimated uncertainty from the statistics or individual analyses. However, including recent new analyses, the data spread has become wider for the elements Y, Zr, Hf, Nb and Ta. This spread seems to indicate real heterogeneities for these elements in CI chondrite samples, and the ratio method may lead to more reliable results (see LPG09 for more details).

The selected concentrations for the Orgueil meteorite, considered as most representative for the CI chondrites, are listed in Table 2; details about the data sources are in LPG09. Concentrations are given as parts per million (ppm) by mass (10,000 ppm = 1 mass%). Corresponding atomic abundances normalized to 10^6 Si atoms (the 'cosmochemical abundance scale') are listed as well.

3 Photospheric abundances

Russell [26] reported the earliest comprehensive analysis of the solar photosphere for 56 elements. Since then, numerous element abundance data have been derived by spectroscopy of the solar photosphere.

Converting the absorption lines into abundances requires knowledge of line positions of neutral and ionized atoms, as well as their transition probabilities and lifetimes of the excited atomic states. In addition, a model of the solar atmosphere is needed. In the past years, atomic properties have seen many experimental updates, especially for the rare earth elements (see below). Older solar atmospheric models used local thermodynamic equilibrium (LTE) to describe the population of the quantum states of neutral and ionized atoms and molecules according to the Boltzmann and Saha equations. However, the ionization and excitation temperatures describing the state of the gas in a photospheric layer may not be identical as required for LTE. Models that include the deviations from LTE (=non-LTE) are used more frequently, and deviations from LTE are modeled by including treatments for radiative and collision processes (see, e.g., [27, 28]).

Solar atmospheric models have evolved from one dimension to more complicated 2D and 3D models designed to take into account effects of convection and granulation on radiative transfer in the solar atmosphere. Recent applications of 3D models instead of older 1D models leads to lower abundances of several elements, notably oxygen, from previously determined values. Significant reductions of photospheric abundances in other elements (e.g., Na, Al, Si) were also found (see, e.g., [15]). However, different 3D model assumptions lead to different results, see for example the discussion by [29] for silicon.

Table 2. CI chondrite composition

Z		ppm	σ	Si = 10^6	σ
1	H	19700	2000	5.13E+06	5.1E+05
2	He	0.00917		0.601	
3	Li	1.47	0.19	55.6	7.2
4	Be	0.0210	0.0015	0.612	0.043
5	B	0.775	0.078	18.8	1.9
6	C	34800	3500	7.60E+05	7.6E+04
7	N	2950	440	55300	8300
8	O	459000	46000	7.63E+06	7.6E+05
9	F	58.2	8.7	804	121
10	Ne	1.80E-04		0.00235	
11	Na	4990	250	5.70E+04	2.8E+03
12	Mg	95800	2900	1.03E+06	3E+04
13	Al	8500	260	8.27E+04	2.5E+03
14	Si	107000	3000	1.00E+06	3E+04
15	P	967	97	8190	820
16	S	53500	2700	4.38E+05	2.2E+04
17	Cl	698	105	5170	780
18	Ar	0.00133		0.00962	
19	K	544	27	3650	180
20	Ca	9220	461	60400	3000
21	Sc	5.9	0.3	34.4	1.7
22	Ti	451	36	2470	200
23	V	54.3	2.7	280	14
24	Cr	2650	80	13400	400
25	Mn	1930	58	9220	280
26	Fe	185000	6000	8.70E+05	2.6E+04
27	Co	506	15	2250	70
28	Ni	10800	300	4.83E+04	1.4E+03
29	Cu	131	13	541	54
30	Zn	323	32	1300	130
31	Ga	9.71	0.49	36.6	1.8
32	Ge	32.6	3.26	118	12
33	As	1.74	0.16	6.10	0.55
34	Se	20.3	1.42	67.5	4.7
35	Br	3.26	0.49	10.7	1.6
36	Kr	5.22E-05		1.64E-04	
37	Rb	2.31	0.16	7.10	0.50
38	Sr	7.81	0.55	23.4	1.6
39	Y	1.53	0.2	4.52	0.45
40	Zr	3.62	0.4	10.4	1.0

Table 2. CI chondrite composition – Continued

Z		ppm	σ	Si = 10^6	σ
41	Nb	0.279	0.028	0.788	0.079
42	Mo	0.973	0.097	2.66	0.27
44	Ru	0.686	0.041	1.78	0.11
45	Rh	0.139	0.014	0.355	0.035
46	Pd	0.558	0.028	1.38	0.07
47	Ag	0.201	0.010	0.489	0.024
48	Cd	0.674	0.047	1.57	0.11
49	In	0.0778	0.0054	0.178	0.012
50	Sn	1.63	0.24	3.60	0.54
51	Sb	0.145	0.021	0.313	0.047
52	Te	2.28	0.16	4.69	0.33
53	I	0.53	0.11	1.10	0.22
54	Xe	1.74E-04		3.48E-04	
55	Cs	0.188	0.009	0.371	0.019
56	Ba	2.41	0.14	4.61	0.28
57	La	0.242	0.012	0.457	0.023
58	Ce	0.622	0.031	1.17	0.06
59	Pr	0.0946	0.0066	0.176	0.012
60	Nd	0.471	0.024	0.857	0 .043
62	Sm	0.152	0.008	0.265	0.013
63	Eu	0.0578	0.0029	0.100	0.005
64	Gd	0.205	0.010	0.342	0.017
65	Tb	0.0384	0.0027	0.063	0.004
66	Dy	0.255	0.013	0.412	0.021
67	Ho	0.0572	0.0040	0.091	0.006
68	Er	0.163	0.008	0.256	0.013
69	Tm	0.0261	0.0018	0.041	0.003
70	Yb	0.169	0.008	0.256	0.013
71	Lu	0.0253	0.0013	0.038	0.002
72	Hf	0.106	0.005	0.156	0.008
73	Ta	0.0145	0.0015	0.0210	0.0021
74	W	0.0960	0.0096	0.137	0.014
75	Re	0.0393	0.0039	0.0554	0.0055
76	Os	0.493	0.039	0.680	0.054
77	Ir	0.469	0.023	0.640	0.032
78	Pt	0.947	0.076	1.27	0.10
79	Au	0.146	0.015	0.195	0.019
80	Hg	0.350	0.070	0.458	0 .092
81	Tl	0.142	0.011	0.182	0.015
82	Pb	2.63	0.18	3.33	0.23
83	Bi	0.110	0.010	0.138	0.012
90	Th	0.0310	0.0025	0.0351	0.0028
92	U	8.10E-03	6.5E-04	8.93E-03	7.1E-04

Hence abundances derived with 3D models still have to be regarded with some caution until model assumptions and details are sorted out.

Lower abundances for some elements derived from these models also cause problems for standard solar models that describe the evolution of the sun to its current radius and luminosity (see [30]). Another problem is that some of the 3D abundances compare worse to meteoritic data than before. In the following preference is given to elemental abundances derived with more conservative solar atmospheric models; however, for some elements (e.g., P, S, Eu) the results from 1D/2D and some 3D models produce consistent results.

The solar photospheric abundances from various literature sources are listed in Table 3. Several new measurements have become available since the compilation by L03 that lists references to data not described below.

Elemental abundances are normalized to 10^{12} atoms of hydrogen. The ratio of the number of atoms of an element N(X), relative to the number of hydrogen atoms, N(H) is given on a logarithmic scale, and frequently used notations are:

$A(X) = \log N(X)/N(H) + 12 = \log\epsilon X$

Uncertainties for the logarithmic scale are given in logarithmic units dex equivalent to an uncertainty factor on a linear scale. The uncertainty relation is $\sigma(in\%) = (10^{\sigma(in\ dex)} - 1) \times 100$.

As already mentioned, most elements are determined by absorption spectroscopy. Exceptions are rare gases and some elements that have no accessible or only heavily blended lines for use in quantitative spectroscopy of the photosphere (e.g., As, Se, Br, Te, I, Cs). Sunspot spectra are used for e.g., F, Cl, In, and Tl, however, the abundance uncertainty for these elements is rather large. For the noble gases, theoretical considerations or data from other objects must be used to obtain representative solar values (see below and [13]). The following describes updates for several elements.

Helium: The He discovery was made from spectral lines in the coronal spectrum during a solar eclipse in 1868; however, despite being first discovered in the sun, the He abundance cannot be determined spectroscopically in the solar photosphere. The He abundance is determined from results of helioseismic models, as described below in the section for the present-day mass fractions of H, He, and heavy elements.

Lithium: The value of A(Li) = 1.1 ± 0.1 from [31] used in [13] is kept. The analysis from [32] of A(Li)=1.0 ± 0.1 was amended with 3D models by [15] to give 1.05 ± 0.10. Considering the already large uncertainty and the uncertainties in the 3D models, the previous selection has been kept.

Beryllium: The photospheric and meteoritic Be abundance determinations are associated with difficulties [13]. By 2003, there were two conflicting photospheric determinations (A(Be)= 1.15 ± 0.10 and 1.40 ± 0.09). A subsequent re-analysis of the photospheric Be abundance by [33] yields A(Be) = 1.38 ± 0.09, in support of the higher photospheric value. The meteoritic value of A(Be) = 1.41 ± 0.08 in L03 was based on chemical element systematics in

CM and CV chondrites because there was only one previous determination of Be in CI chondrites. This situation has improved since [34] report 2 Be analyses for the Orgueil CI chondrite. The average of the 3 Be determinations is 0.021 ± 0.002 ppm (by mass), corresponding to a meteoritic A(Be)= 1.32 ± 0.03. Within the larger uncertainties, the photospheric and meteoritic Be abundances are in agreement, suggesting that no Be destruction has occurred over the Sun's lifetime. However, taken at face value, a lower Be abundance in CI chondrites than in the photosphere is not easily accounted for by any physical or chemical fractionation process.

Carbon: The C abundance of A(C) = 8.39 (± 0.04) from [35] was selected in [13]. This value was confirmed by [36], and derived from the analysis of CO by [37]. However, this C abundance is based on 3D models from one group and an independent confirmation of this value by another model is desirable.

Nitrogen: The N abundance is among the more uncertain elemental abundances. Asplund et al. [15] (henceforth also A05) derived N abundances from neutral N (NI) and the NH molecule. An LTE analysis of the NI lines gives A(N) = 7.88 ± 0.08, a NLTE analysis gives 7.85 ± 0.08. The N abundance from the NH molecule and 3D models gives a lower abundance of A(N) = 7.73 ± 0.05. A05 recommend A(N) = 7.78 ± 0.06 in their table. However, this seems quite low in comparison to previous estimates of e.g., 7.83 ± 0.11 in L03 that was based on [27]. A recent detailed study by [38] found 7.86 ± 0.12, which is recommended here.

Oxygen: The recommended O abundance of 8.73 ± 0.07 is an average from the recent determinations by [39, 40, 41].

Caffau et al. [39] recommend A(O)= 8.76 ± 0.07 from a detailed analyses of several oxygen lines and 3D atmospheric models. They include different model NLTE corrections, as well as treatments of collisions with H atoms on NLTE level populations of O. Within the (unfortunately still larger) uncertainties, their O abundance is closer to most of the recent low O abundance determinations than to the older values advocated by [9, 12]. Melendez et al. [41] derived the O abundances from several O lines and compared various model approaches, which lead to an average O abundance of A(O) = 8.71 ± 0.02, where the uncertainty does not really cover the real uncertainty of the *true* O abundance. Using different 3D models, [40] report A(O) = 8.72 ± 0.06 from modeling of several O I lines.

A review on the problem of the solar O and other light element abundances is given by [30]. The photospheric O abundance determination remains enigmatic, as the value obtained by [35, 33] appears to be erratically low, whereas the older abundances quoted in compilations such as [9] (AG89), [12] (GS98), as well as the recent study by [42] probably give values that are much too high.

The downward adjustment in solar O by [35] was substantial compared to the values given in the compilations by [12] (A(O)= 8.83) and [9] (A(O) = 8.93). The decrease was mainly due to the realization that the O line commonly used in the abundance analysis is blended with a Ni line. The value by Caffau et al. is 0.07 dex higher (factor 1.17) than the abundances derived by [35]; and it is also higher than the low values determined by Asplund and coworkers. [39] give these reasons why the O abundances was previously underestimated. In the studies by Allende Pietro et al. and Asplund et al., lower equivalent widths were used, and effects of collisions of H atoms in the calculations of the statistical equilibrium of O were not considered.

Ayres et al. ([42]) derived the O abundance from weak CO absorptions but did not find support for a lower O abundance. They recommend A(O) = 8.85, much closer to the value in [9] (A(O) = 8.93). However, to obtain the absolute O abundance using the CO molecule, Ayres et al. assumed that C/O = 0.5. Oxygen is about twice as abundant as carbon, and carbon is largely tied into the CO molecule. An analysis of the CO abundance thus provides only the lower limit to the total C abundance (as C is present also in other gases such as C, CH etc), and the O abundance can only be derived if the total C/O ratio and the C abundance is known. [42] find A(C) =8.54, and with an assumed C/O of 0.5, their corresponding O abundance is A(O) = 8.85. However, using the same CO lines [37] find a lower O abundance with their 3D models than reported by [42]. Given the problem that the O abundances from CO requires various assumptions about the distribution of C and the C/O ratio, these abundance determinations appear even more uncertain.

Adopting A(C) = 8.39 and A(O) = 8.73 $\pm$ 0.07 gives a C/O ratio of C/O = 0.457, less than the C/O ratio of 0.50 in more recent studies and compilations, but still somewhat higher than the value of C/O = 0.427 from AG89.

The O abundance has been steadily revised downward from [26] value over the years, likewise, the abundance values of C and N from more recent analysis tend to be smaller. Figure 2 shows historical abundance trends for the more abundant elements C, N, O, Si, and Fe from various sources starting with [26] and including data sources quoted in the compilations mentioned above. The variations on the graph appear small, however, one should not forget that the data are plotted on a log scale and a difference of 0.5 dex corresponds to about a factor of 3 change in abundance. The Fe abundance was a big problem until the early 1970s when improved transition probabilities and lifetimes confirmed that the photospheric Fe abundance had been underestimated by about a factor of 3-10, which did not match the meteoritic value. This issue has now been put to rest and the photospheric and meteoritic value are now in perfect agreement. However, the issue of the C, N, and O abundances cannot be aided with meteoritic abundances, and future analyses have to resolve this.

Neon Results by [43] from Ne I and Ne II lines in nearby, early type B stars yield log A(Ne) = 7.97 $\pm$ 0.07, which can be taken as characteristic of the present day ISM. If neon contributions from more massive AGB stars to the ISM over the past 4.6 Gyr are negligible, this value may be taken as representative for the Sun. Landi et al. ([44]) obtained A(Ne) = 8.11 $\pm$ 0.1 from solar flare measurements in the ultra-violet. This Ne abundance is derived independently of the O abundance, unlike other Ne abundance determinations that rely on Ne/O ratios and an adopted O abundance (see e.g., [13] and below).

The average from these studies gives A(Ne) = 8.05 ± 0.06, but considering the larger uncertainties in the determinations, an overall uncertainty of 0.10 dex (25%) is easily warranted for the recommended value here.

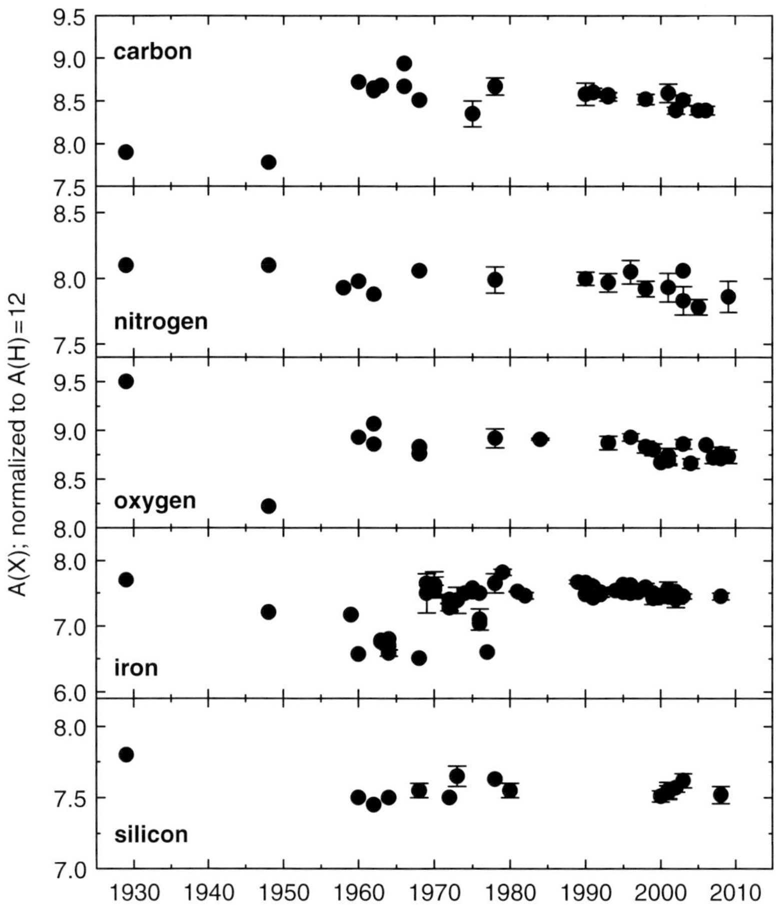

Fig. 2. Photospheric abundance determinations over time

This Ne value compares relatively well to the Ne abundance adopted in older compilations by [8] (8.09) or [11](8.08). However, the Ne values in more recent compilations (A(Ne) = 7.87 ± 0.1 in L03, and A(Ne) = 7.84 ± 0.06 in A05) are lower by 26% (0.10 dex). The Ne abundance in previous compilations was mainly based on the characteristic Ne/O ratio of 0.15 for the local ISM and solar energetic particles. The Ne abundance was then calculated from the adopted O abundances. Since the adopted O abundances have changed to lower values in these compilations, the Ne abundances dropped as well. Using Ne/O = 0.15 and our selected O abundance from above, the Ne abundance from the ratio method gives A(Ne) = 7.91, about 0.14 dex lower than the preferred Ne value which is independent of an adopted O abundance. More recently, [45] recommends Ne/O = 0.17 ± 0.05 for the photosphere from extreme UV measurements of supergranule cell center regions. This ratio and the selected O abundance (A(O)=8.73) yield A(Ne) = 7.96 ($\pm$ 0.15, uncertainty only from ratio). This value is lower than the recommended value, but agrees within uncertainties permitted by the given Ne/O ratio.

In principle, the Ne abundance can also be derived from solar wind data. In 2006, Bochsler [46] reported A(Ne) = 8.08 ± 0.12. Bochsler [47] further analyzed the solar wind data for fractionations, and finds A(Ne) =7.96 ± 0.13, and also A(O) = 8.87 ± 0.11, with relatively large uncertainties that do not help to resolve the issue of the uncertain O and Ne abundances.

Neon is the third most abundant heavy element after O and C, and the heavy elements are important opacity sources that influence radiative transfer in the sun. Good agreement of standard solar models and helioseismological observations existed until about the year 2000. Then more sophisticated photospheric modeling began to yield lower C, N, O, Ne and other heavier element abundances (see Figure 2, and compilations by L03, A05). A decrease in heavy element abundances led to solar model results that no longer stood the test from helioseismology. A detailed review of this problem is given by [30] . The recommended N, O, and Ne abundances here are larger than previously recommended in L03 and A05, and it will be interesting to see if these abundances can bring solar models again in closer agreement with helioseismological constraints.

Sodium: the previously selected Na value of 6.3 ± 0.03 in L03 is in agreement with A(Na) = 6.27 measured by [3]; [15] found A(Na) = 6.17 ± 0.04 from six Na lines and their 3D model atmospheres, which is $\sim$ 25% lower than the meteoritic value as well as previously determined photospheric Na abundances. The reason for this difference is not yet clear.

Aluminum: A 3D analysis by A05 gives A(Al) = 6.37 ± 0.06, which is lower than the previous value of 6.47 ± 0.07 from 1D models selected in L03. The older value is kept as it is much better in agreement with meteoritic data. The Al/Si from 3D models in A05 is 1.2 times that of CI chondrites, which is a large difference that still needs to be understood.

Silicon: The A(Si) = 7.52 ± 0.06 selected here is from [29], which is not that different from 7.51 ± 0.04 reported by [48] and 7.54 ± 0.05 by [27].

Phosphorus: The value of A(P) = 5.49 ± 0.04 in L03 is changed to the recent result of A(P) = 5.46 ± 0.04 by [49]. The new value is based on 3D atmospheric models. According to [49] the value for P with 1D models is not significantly different. A lower value, A(P) = 5.36 ± 0.04 from a different 3D analysis was given in the compilation by A05. Here the well documented analysis by [49] is taken for the photospheric abundance, which agrees well with the meteoritic value of 5.43 ± 0.04.

Sulfur: New results with 3D models by [50,51] lead to A(S) = 7.14 ± 0.01. They found that 3D models have no big effect on S abundances compared to 1D models.

Argon: The value of A(Ar) = 6.50 is derived from various independent sources since the Ar abundance cannot be determined spectroscopically in the photosphere (see [52]).

Potassium: [53] confirmed the K abundance of A(K) = 5.12 ± 0.03 used in L03. The value A(K) = 5.08 ± 0.07 proposed by A05 appears too low, a similar situation as for Na above.

Calcium: [3] find A(Ca) = 6.33 ± 0.07 in their LTE analysis of the solar spectrum (the same value as in [54]), which is lower than the previously selected value in L03. However, the uncertainty for the value from [3] is high. The result from 3D models by A05 is about 5% lower. A new Ca analysis for the photosphere using different model atmospheres is needed.

Scandium: The Sc abundance remains uncertain, [55] recommends a range of 3.07 < A(Sc) < 3.13. To cover the range of values reported in several recent papers A(Sc) = 3.10 with an appropriate uncertainty of 0.1 dex is recommended.

Titanium: The value from the LTE analysis by [3] is adopted.

Chromium: [56] found A(Cr) = 5.64 ± 0.1, which is identical with the photospheric value listed in L03 but with much smaller uncertainty.

Manganese: Two recent studies give A(Mn) = 5.37 ± 0.05 [57] and 5.36 ± 0.10 [58] and [3] found the same value as [57]. These values are only slightly lower than the value 5.39 ± 0.03 given in L03. The analyses of the photospheric Mn abundance seem to confirm that the photospheric Mn abundance is lower than the meteoritic value of 5.50 ± 0.01. Assuming that indeed both the photospheric and meteoritic data are reliable, the cause of this abundance difference must be found.

Nickel: The A(Ni) = 6.23 ± 0.04 from the LTE analysis by [3] is similar to the previous value and has a lower uncertainty.

Zirconium: [59] found A(Zr) = 2.58 ± 0.02 from a 3D analysis.

Palladium: A value of A(Pd) = 1.66 ± 0.04 was reported by [60]

Indium: Previous determinations of the photospheric In abundance lead to a value that is substantially higher (∼ 400%) than the CI chondritic value. [61] suggested that this large difference could be the result of the relatively high volatility of In. Incomplete condensation of any element into the materials assembled to the CI chondrite parent body would lead to a relative depletion of a volatile element in CI chondrites when compared to the Sun. At 10^{-4} bar total pressure, half of all In is condensed at 536 K, comparable to the 50% condensation temperatures of other volatile elements like Tl (532K), S (664K) and Pb (730K; see L03). However, the abundances of these and other volatile elements are in closer agreement for the photosphere and CI chondrites, so a fractionation due to volatility cannot explain the huge photospheric In abundance.

A high In abundance can also be ruled out considering the abundance distribution of the elements and nuclides as a function of atomic mass. A high In abundance would introduce spikes in the otherwise rather smooth abundances curve (see below). A recent investigation by [62] shows that the often used In line at 451.13 nm in the solar sun spot spectrum is blended by some line of a currently not identified element, which causes an apparently higher In abundance. Vitas et al. [62] obtained A(In) = 1.50 with no potential blends considered, and this value is adopted here as an upper limit for the photospheric In abundance. Their study of other In lines further indicates that the solar In abundance is unlikely to be higher than the meteoritic value of A(In) = 0.78, for which [62] also find support from nuclide distribution systematics from nucleosynthesis.

Rare earth elements: In the past years, many improvements have been made in abundance analyses of the REE through measurements of atomic lifetimes and transition probabilities, notably by the Wisconsin group. A recent paper by [63] summarizes the efforts and gives abundances for the REE. Now the REE abundances are among the best-known abundances for the sun. The following lists several papers on REE that appeared since 2003; the new values for Ce, Dy, Tm, Yb, and Lu are from [63].

Praseodymium: The value of A(Pr) = 0.76 ± 0.02 from [63] is preferred to the previously selected value of A(Pr) = 0.71 ± 0.08 in L03. The value based on [64] of A(Pr) = 0.58 ± 0.10 adopted by [16] is erroneously low, which is easily seen from a comparison to the well-established meteoritic value of 0.78 ± 0.03.

Neodymium: [65]; A(Nd) = 1.45 ± 0.05

Samarium: [66]; A(Sm) = 1.00 ± 0.03

Europium: [67] employ a 3D hydrodynamic model atmosphere and find A(Eu) = 0.52 ± 0.02 [67]. They find that 3D effects are negligible for the Eu determination in the Sun. This value is identical to the value reported by [68].

Gadolinium: [69]; A(Gd) = 1.11 ± 0.03.

Holmium: [70]; A(Ho) = 0.51 ± 0.1.

Erbium: [71]; A(Er) = 0.96 ± 0.03

Hafnium: The photospheric abundance of A(Hf) = 0.88 ± 0.08 selected in L03 is confirmed with a reanalysis by [72] who determined improved transition probabilities. Another recent re-determination of the photospheric Hf abundance gives A(Hf) = 0.87 ± 0.04 [73].

Osmium: [74] found A(Os) = 1.25 ± 0.11. This is significantly lower than the value of 1.45 ± 0.10 used in previous compilations. Both Os values seem to be problematic when compared to the meteoritic value of 1.37 ± 0.03. Assuming that the meteoritic value is reliable, the older photospheric value is 17% too low and the new one 30% too high. Grevesse et al. [16] selected the newer, smaller value but a conservative approach is to adopt the value with the smaller difference to the meteoritic value. The older value of A(Os) = 1.45 also appears more reasonable considering abundance systematics in the Pt-element region. Nucleosynthesis models predict that Os should be more abundant than Ir, as seen in CI chondrites. Overall, a new analysis of the photospheric Os abundance is needed.

Platinum: [75] find that the photospheric Pt abundance is not very reliable. The value selected in L03 is kept here, but is assigned a 0.3 dex uncertainty (factor of 2) to emphasize its low reliability.

Thallium: The Tl value of 0.95 ± 0.2 is from the linear average of the end-member composition of $0.72 \leq A(Tl) \leq 1.1$ selected in L03 that was found for sunspot spectra. The uncertainty quoted here is to indicate the derived range. There are no new measurements.

Thorium: The Th abundance is difficult to determine because the only accessible Th line in the photospheric spectrum is heavily blended with Ni I and Ni II. Caffau et al. [73] report a nominal Th abundance A(Th) = 0.08 ± 0.03, which should not be over-interpreted because of the line blends.

4 Recommended Present-Day Solar Abundances

4.1 Cosmochemical and Astronomical Abundance Scale Conversion

In order to compare the atomic silicon-normalized CI chondrite abundances in Table 2 (N(Si) = 10^6 atoms; cosmochemical abundance scale) with the photospheric abundances on the hydrogen-normalized scale (A(H) = 12; astronomical abundance scale) in Table 3, the data must be converted to a common scale. One cannot easily convert the meteoritic data to the H-normalized astronomical abundance scale because H is depleted in meteorites. However, a comparison can be done for the non-volatile rock-forming elements. The difference of the logarithmic Si-normalized abundances of CI chondrites to the abundances on the astronomical scale is more or less constant for many elements. This shows that the relative abundances in the photosphere and CI chondrites are similar.

The link for both abundance scales is an average conversion constant that is calculated by subtracting the logarithm of the Si-normalized meteoritic abundances (Table 2) from the logarithmic H-normalized photospheric abundances (Table 3) for all elements heavier than neon that have uncertainties < 0.1 dex, i.e., below ~25%, in their photospheric abundance determinations. There are 39 elements that qualify and Figure 3 shows a comparison of the photospheric and CI chondritic abundances for these elements on a linear scale (note that the conversion constant for the log scales is equivalent to a scale factor on linear abundance scales). The scale conversion constant is 1.533 ± 0.042; and the cosmochemical and astronomical scales are coupled as:

$A(X) = \log N(X) + 1.533$

Previously, the conversion constant was somewhat larger. For example, AG89 used 1.554 ± 0.020 from only 12 elements, which resulted in a smaller nominal uncertainty of the conversion constant. L03 found a value of 1.539

$\pm$ 0.046 for 35 elements for which photospheric abundances were determined with less than 25% uncertainty, and used a constant of 1.540, which is exactly the log of the ratio of Si in the astronomic to the meteoritic scale. The slightly lower conversion factor found here is the result of the systematic decrease of the reported photospheric abundance values.

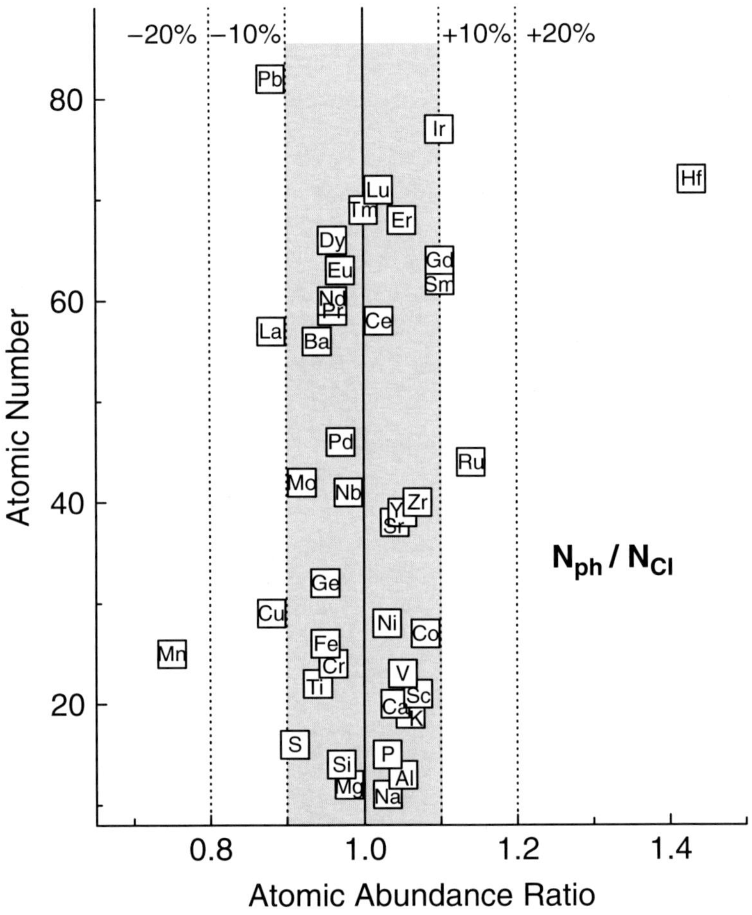

Fig. 3. The photospheric/CI chondritic abundance ratios for 39 elements that are well determined for the solar photosphere. The grey-shaded region shows agreement within 10%

It is notable that the conversion constant can be derived from a large range of elements with different properties (e.g., atomic number, mass, first ionization potential, condensation temperature). The premise in linking the meteoritic and solar data is that there are no chemical and physical fractionations of the elements (except for the obvious loss of highly volatile elements from meteorites). The small spread in the conversion factor indicates that there is basic agreement of solar and meteoritic abundances. There is no apparent dependence of the conversion factor on atomic number, mass or any other elemental property. In addition, the solar/meteoritic abundance ratios are independent of the geochemical character of an element, whether it is lithophile, siderophile or chalcophile, which indicates that any chemical and physical fractionation of silicates, metal, and sulfides did not affect CI chondrite abundances. A reasonable estimate for the uncertainty of the relative scale of solar and meteoritic abundances is about 10%.

4.2 Comparison of Photospheric and Meteoritic Abundances

The photospheric and CI chondrite abundances on the astronomical abundance scale are given in Table 3. Agreement within 10% for meteoritic and photospheric data exists for 40 elements (the 39 shown in Figure 3 plus the light element Be). This increase in elements that show good agreement is mainly due to the recent improvements in photospheric measurements.

The largest differences are for the highly volatile elements that form low-temperature ices and/or exist in gaseous form in the terrestrial atmosphere. The largest depletion is for the noble gases. The depletion sequence for N, C, H reflects the general lack of solid nitrogen compounds in meteorites and the predominance of oxides and silicates.

Only Li is clearly consumed in the interior of Sun by nuclear reactions, which explains the ~150 times smaller photospheric abundance. Nominally, B is depleted in the Sun by about 20% but within the stated uncertainties it is apparently not affected. Beryllium is another fragile element like Li and B, and may be subject to destruction in the Sun. However, the comparison of abundances for the photosphere and CI chondrites indicate that there was no Be loss in the Sun; indeed; the nominal Be abundance for the Sun is higher than in CI chondrites.

The difference between photospheric and CI chondrite abundances exceeds 10% for 21 other elements (see LPG09 for a detailed comparison). However, in most cases the combined uncertainties of the photospheric and meteoritic determinations are larger than the difference in abundance, and solar and meteoritic abundances agree within error limits. Elements with abundance differences larger than the combined error bars are W, Rb, Ga, Hf, and Mn. The abundances of Ga, Rb, and W need to be redetermined in the photosphere to resolve the differences. There are new photospheric analyses for Hf and Mn suggesting that the differences in photospheric and meteoritic values could be real since there are no plausible reasons to doubt the results. Line blending may not be the culprit as this usually leads to over-estimated abundances for the photosphere (e.g., like for Indium as noted above); however, the photospheric Mn value is ~1.4x lower than in CI chondrites. Manganese can be accurately measured in meteorites. Its concentration in Orgueil is similar to that in two other CI meteorites, Alais and Ivuna (L03), and it fits with the abundances of other elements of similar volatility. It seems that an unidentified problem in the photospheric abundance analysis may cause the discrepancy of the meteoritic and solar Mn abundances.

The problem is reversed for Hf and the meteoritic abundance of Hf is less than that of the photosphere. This could indicate a problem with the photospheric abundance determination and suggest that line blending is more severe than already corrected for in current models. However, two recent Hf analyses using different models essentially obtain the same abundance, and if there is a problem with the analysis, it remains elusive. The Hf concentration in CI chondrites has been accurately determined, because Hf is important for Lu-Hf and W-Hf dating. The very constant Lu/Hf ratio in meteorites closely ties Hf to other refractory elements, which do not show large differences in abundance to the sun as does Hf. This issue awaits resolution.

Overall, the agreement between photospheric and meteoritic abundances has improved further with new photospheric and meteoritic data.

4.3 Combined Solar Abundances from CI Chondrites and Photospheric Data

The CI chondritic and photospheric abundances can be combined to select a set of recommended present-day solar system abundances. Here the same procedure as in L03 is used to construct such an abundance set. The recommended data are from photospheric values for ultra-highly volatile elements like H, C, N, and O and from various sources and theory for the noble gases (see above for He, Ne, Ar, and L03 for Kr and Xe). The CI chondrite data are the obvious choice for elements that are only determined in CI chondrites but also for elements that have photospheric abundance determinations with high uncertainties. Several elements are equally well determined in CI chondrites and in the photosphere, and an average of their Si-normalized abundances is used. The recommended present day abundances are converted to the astronomical scale using the same conversion constant (1.533) between the astronomical and cosmochemical abundance scales as described before.

Figure 4 shows the recommended abundances of the elements as a function of atomic number. The large abundances of H and He are not shown to avoid scale compression in the diagram. Overall, abundances decrease relatively smoothly with increasing atomic number. The regular pattern that odd numbered elements are less abundant than their even-numbered neighbors holds from the lightest to the heaviest elements; originally Harkins established this observation for the lighter elements up to the Fe region. Notable exceptions are the low abundances of Li, Be, and B, that consist of fragile nuclei that are easily destroyed in stellar interiors. The elemental abundance distribution is not controlled by the chemical properties of the elements but instead by nuclear properties.

Table 3. Cl chondrite, solar (mainly photosphere), and recommended present day solar abundances

		Cl chondrites		Sun			recommended	
Z		A(X)		A(X)		Note	A(X)	
1	H	8.24	0.04	12.00		s	12.00	
2	He	1.31		10.925	0.02	s,t	10.925	0.02
3	Li	3.28	0.05	1.10	0.10	m	3.28	0.05
4	Be	1.32	0.03	1.38	0.09	m	1.32	0.03
5	B	2.81	0.04	2.70	0.17	m	2.81	0.04
6	C	7.41	0.04	8.39	0.04	s	8.39	0.04
7	N	6.28	0.06	7.86	0.12	s	7.86	0.12
8	O	8.42	0.04	8.73	0.07	s	8.73	0.07
9	F	4.44	0.06	4.56	0.30	m	4.44	0.06
10	Ne	-1.10		8.05	0.10	s,t	8.05	0.10
11	Na	6.29	0.02	6.30	0.03	a	6.29	0.04
12	Mg	7.55	0.01	7.54	0.06	a	7.54	0.06
13	Al	6.45	0.01	6.47	0.07	a	6.46	0.07
14	Si	7.53	0.01	7.52	0.06	m	7.53	0.06
15	P	5.45	0.04	5.46	0.04	a	5.45	0.05
16	S	7.17	0.02	7.14	0.01	a	7.16	0.02
17	Cl	5.25	0.06	5.50	0.30	m	5.25	0.06
18	Ar	-0.48		6.50	0.10	s,t	6.50	0.10
19	K	5.10	0.02	5.12	0.03	a	5.11	0.04
20	Ca	6.31	0.02	6.33	0.07	m	6.31	0.02
21	Sc	3.07	0.02	3.10	0.10	m	3.07	0.02
22	Ti	4.93	0.03	4.90	0.06	m	4.93	0.03
23	V	3.98	0.02	4.00	0.02	a	3.99	0.03
24	Cr	5.66	0.01	5.64	0.01	a	5.65	0.02
25	Mn	5.50	0.01	5.37	0.05	m	5.50	0.01
26	Fe	7.47	0.01	7.45	0.08	a	7.46	0.08
27	Co	4.89	0.01	4.92	0.08	a	4.90	0.08
28	Ni	6.22	0.01	6.23	0.04	a	6.22	0.04
29	Cu	4.27	0.04	4.21	0.04	m	4.27	0.04
30	Zn	4.65	0.04	4.62	0.15	m	4.65	0.04
31	Ga	3.10	0.02	2.88	0.10	m	3.10	0.02
32	Ge	3.60	0.04	3.58	0.05	a	3.59	0.06
33	As	2.32	0.04			m	2.32	0.04
34	Se	3.36	0.03			m	3.36	0.03
35	Br	2.56	0.06			m	2.56	0.06
36	Kr	-2.25		3.28	0.08	t	3.28	0.08
37	Rb	2.38	0.03	2.60	0.10	m	2.38	0.03
38	Sr	2.90	0.03	2.92	0.05	m	2.90	0.03
39	Y	2.19	0.04	2.21	0.02	a	2.20	0.04
40	Zr	2.55	0.04	2.58	0.02	a	2.57	0.04

Table 3. – Continued

		Cl chondrites		Sun			recommended	
Z		A(X)		A(X)		Note	A(X)	
41	Nb	1.43	0.04	1.42	0.06	a	1.42	0.07
42	Mo	1.96	0.04	1.92	0.05	a	1.94	0.06
44	Ru	1.78	0.03	1.84	0.07	m	1.78	0.03
45	Rh	1.08	0.04	1.12	0.12	a	1.10	0.13
46	Pd	1.67	0 .02	1.66	0.04	a	1.67	0.04
47	Ag	1.22	0.02	0.94	0.30	m	1.22	0.02
48	Cd	1.73	0.03	1.77	0.11	m	1.73	0.03
49	In	0.78	0.03	1.50	UL	m	0.78	0.03
50	Sn	2.09	0.06	2.00	0.30	m	2.09	0.06
51	Sb	1.03	0.06	1.00	0.30	m	1.03	0.06
52	Te	2.20	0.03			m	2.20	0.03
53	I	1.57	0.08			m	1.57	0.08
54	Xe	-1.93		2.27	0.08	t	2.27	0.08
55	Cs	1.10	0.02			m	1.10	0.02
56	Ba	2.20	0.03	2.17	0.07	a	2.18	0.07
57	La	1.19	0.02	1.14	0.03	m	1.19	0.02
58	Ce	1.60	0.02	1.61	0.06	a	1.60	0.06
59	Pr	0.78	0.03	0.76	0.04	a	0 .77	0.05
60	Nd	1.47	0.02	1.45	0.05	m	1.47	0.02
62	Sm	0.96	0.02	1.00	0.05	m	0.96	0.02
63	Eu	0.53	0.02	0.52	0.04	a	0.53	0.04
64	Gd	1.07	0.02	1.11	0.05	a	1.09	0.06
65	Tb	0.34	0.03	0.28	0.10	m	0.34	0.03
66	Dy	1.15	0.02	1.13	0.06	a	1.14	0.06
67	Ho	0.49	0.03	0.51	0.10	m	0.49	0.03
68	Er	0.94	0.02	0.96	0.06	a	0.95	0.06
69	Tm	0.14	0.03	0.14	0.04	m	0.14	0.03
70	Yb	0.94	0.02	0.86	0.10	m	0.94	0.02
71	Lu	0.11	0 .02	0.12	0.08	m	0.11	0.02
72	Hf	0.73	0.02	0.88	0.08	m	0.73	0.02
73	Ta	-0.14	0.04			m	-0.14	0.04
74	W	0 .67	0.04	1.11	0.15	m	0.67	0.04
75	Re	0.28	0.04			m	0.28	0.04
76	Os	1.37	0.03	1.45	0.11	m	1.37	0.03
77	Ir	1.34	0.02	1.38	0.05	a	1.36	0.06
78	Pt	1.64	0.03	1.74	0.30	m	1.64	0.03
79	Au	0.82	0.04	1.01	0.18	m	0.82	0.04
80	Hg	1.19	0.08			m	1 .19	0.08
81	Tl	0.79	0.03	0.95	0.20	m	0.79	0.03
82	Pb	2.06	0.03	2.00	0.06	m	2.06	0.03
83	Bi	0.67	0.04			m	0.67	0.04
90	Th	0.08	0.03	0.08	UL	m	0.08	0.03
92	U	-0.52	0.03	-0.47	UL	m	-0.52	0.03

Abundances on the astronomical scale with log N(H)=12.
Note: a = average of meteoritic and solar value; m = meteoritic value; t = theoretical and/or indirectly determined.
UL: upper limit

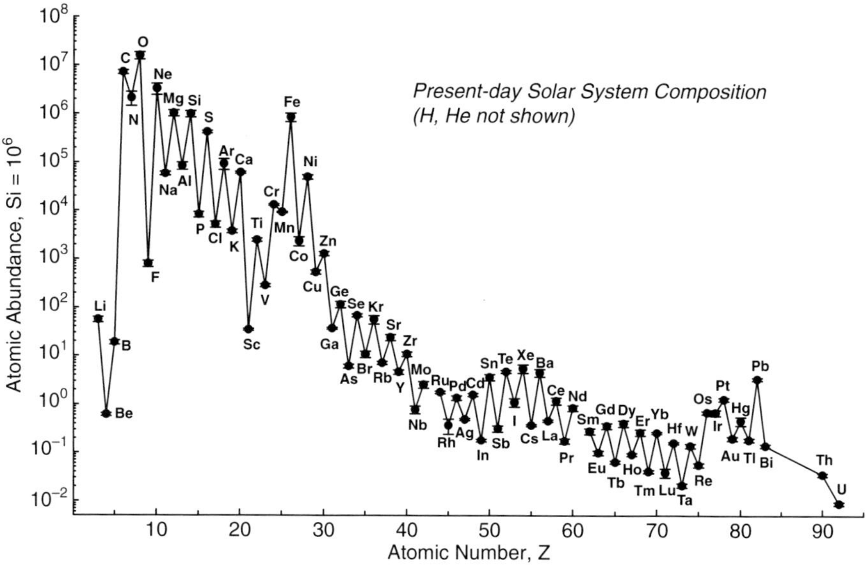

Fig. 4. Abundances of the elements as a function of atomic number

4.4 Mass Fractions X, Y, and Z in Present-Day Solar Material

Many applications in planetary sciences and astronomy use mass fractions of the elements rather than atomic abundances that we have dealt with so far. Mass fractions are also involved when the He abundance is to be derived. Although a value for He is listed in the tables above, the He abundance cannot be derived from the meteoritic nor the photospheric analyses. Thus, at this point one only has the atomic abundances of all elements relative to H or Si, except for He. Using the atomic weights of the elements, the relative atomic elemental abundances are converted to mass concentration ratios. Then the ratio of the mass sum of all heavy elements relative to the mass of H is obtained, which is needed to derive the He mass fraction from different constraints, and from that we finally obtain the atomic He abundance.

The mass fraction of H is usually abbreviated as X, that of He as Y, and the sum of the mass fractions of all other heavy elements as Z. The overall sum of these mass fractions is X+Y+Z=1. Absolute mass fractions of X, Y, and Z can be derived if the ratio of Z/X is known from atomic abundance analysis (the Z/X ratio can always be computed without knowing the He abundance), and if either the mass fraction of H or He is known independently.

The mass fraction of He can be inferred from inversion of helioseismic data by matching the sound speeds of H and He dominated mixtures under physical conditions appropriate for the solar convection zone. The mass fraction Z, which combines all other heavy elements is also important as it governs opacities and thus the density structure of the Sun's outer convection zone. The depth of the solar convection zone derived from helioseismic data poses constraints on the permissible fraction of heavy elements (a detailed review on helioseismology and the He abundance problem is given by [30]). The helioseismic inversion models require Z/X ratios and heavy element abundances (for opacities) as inputs. Therefore the He mass fraction and the He abundance from such models is not independent of X/Z. Ideally, one would use the abundance data and X/Z from the new compilation here to find the corresponding He abundance from helioseismic models and fits to solar data. One should not necessarily adopt a He abundance that is based on models that are calibrated to different Z/X than found for the new compilation of elemental abundances.

One can also obtain the absolute fractions of X, Y, and Z if the Z/X and the X (hydrogen mass fraction) are known. [76,30] have shown that the estimated mass fraction of H from helioseismic models is relatively independent of Z/X ratios in the range of $0.0171 < Z/X < 0.0245$. Their models calibrated to Z/X=0.0171 and 0.0218 yield an average $X=0.7389 \pm 0.0034$ [76]. If the H mass fraction is indeed independent of the Z/X ratio, and if compositional variations within Z (mainly governed by the mass fractions of O, C, Ne, see below) also do not alter this conclusion much, we can use this X to estimate the He mass fraction.

With $Z/X = 0.0191$ found from the abundances in Table 3, and assuming $X = 0.739$, one obtains $Z = 0.0141$, and from this, $Y = 1-X-Z = 0.2469$. This mass fraction of He corresponds to an atomic He abundance of A(He) = 10.925.

The mass fraction of heavy elements (Z=0.014) is intermediate to those in the compilations by GS98 (Z=0.017) and G07 (Z=0.012); see Table 4 for a comparison for present-day solar values. The He mass fraction Y is smaller than that in previous compilations, but the (rounded) He abundance is the same for the three compilations in Table 4. The hydrogen mass fraction of X=0.739 [76] adopted here is essentially the same as in [15, 16], and the smaller value for [12] seems to be due to the different model assumptions for deriving the He abundance there.

Table 4. Protosolar mass fractions and He abundance

Present-Day	Z/X	X	Y	Z	A(He)
this work	0.0191	0.7390	0.2469	0.0141	10.925
A05, G07	0.0165	0.7383	0.2495	0.0122	10.93
GS98	0.0231	0.7347	0.2483	0.0169	10.93

Note: AG89 is not included because they do not give the present-day He abundance or Z/X. For protosolar values see below.

Table 5. Concentration of present-day solar composition (mass %)

	this work	A05,G07	GS98
H (=X)	73.90	73.92	73.47
He (=Y)	24.69	24.86	24.83
O	0.63	0.54	0.79
C	0.22	0.22	0.29
Ne	0.17	0.10	0.18
Fe	0.12	0.12	0.13
N	0.07	0.06	0.08
Si	0.07	0.07	0.07
Mg	0.06	0.06	0.07
S	0.03	0.03	0.05
all other elements	0.04	0.02	0.04
total heavy elements (=Z)	1.41	1.22	1.69

Note: Elements in order of decreasing concentration by mass.

Overall, the mass fractions derived here are closer to the ~10 year old GS98 compilation than to the most recent ones by L03, A05 or G07. Ideally, the He abundance proposed here needs to be evaluated with results from helioseismic models calibrated to the abundances of the elements (other than He) and the X/Z ratio found here.

The fraction for Z obtained here is higher than in the compilations by e.g., L03, A05, G07. The mass increase in Z should help to resurrect the standard solar models, which agreed with helioseismic constraints when the GS98 abundances were used, but crumbled under the too low Z values that were suggested in more recent compilations (see review by [30]). Table 5 compares the mass fractions of the most abundant elements in present-day solar material (note that the order by mass is different from that by atomic abundances). About half of the mass fraction of Z is from O, followed by C, Ne, and Fe. The higher Z here comes mainly from increased O and Ne abundances, which may help to eradicate the problem with the incompatibility of standard solar models and recommended present-day solar abundances.

5 Solar System Abundances 4.56 Gyr Ago

The data discussed above are for present-day abundances in the photosphere and meteorites. However, two processes affected the solar abundances over time. The first is element settling from the solar photosphere into the Sun's interior; the second is decay of radioactive isotopes that contribute to the overall atomic abundance of an element. The first, discussed in the following, is more important for the sun and large-scale modeling; the changes in isotopic compositions and their effects on abundances are comparably minor but important for radiometric dating. The isotopic effects are considered in the solar system abundance table in this section, but are not described at length here.

Settling or diffusion of heavy elements from the photosphere to the interior boundary layer of the convection zone and beyond lowered the elemental abundances (relative to H) from protosolar values 4.56 Gyr ago (see [30]). Over the Sun's lifetime, diffusion decreased abundances of elements heavier than He by ~13% from original protosolar values, whereas that of He dropped a little more by about ~15%; modeling these depletions also depends on opacities, hence abundances. With these estimates, the proto-solar abundances (subscript 0) are calculated from the present-day data for the astronomical scales as

$$A(He)_0 = \mathrm{A(He)} + 0.061,$$

and for all elements heavier than He it is

$$A(X)_0 = \mathrm{A(X)} + 0.053 = \log \mathrm{N(X)} + 1.586$$

The atomic abundances of the elements on both abundance scales are given in Table 6. Note that on the cosmochemical abundance scale ($\mathrm{N(Si)}=10^6$), the relative abundances of the heavy elements do not change from the data in Table 3 because the scale is normalized to Si, one of the heavy elements. Only H and He change from present-day values: the relative H abundance is less, and the He abundance is slightly higher (because of the higher diffusive loss).

For completeness, the protosolar mass fractions X, Y, and Z are summarized in Table 7. However, as noted before for the present day solar mass fractions, it is up to the solar models and helioseismology to derive the best-fitting current and proto solar He mass fractions from the given abundances of the other elements.

6 Abundance of the Nuclides

The abundance of an element is determined by the number and abundances of its stable isotopes, which in turn depends on the stability of the nuclei during thermonuclear reactions in stellar interiors. Already in the 1910s, Harkins

Table 6. Solar system abundances 4.56 Gyr ago

Z		N(Si)=10^6	σ	log N(H)=12	σ
1	H	2.59E+10		12.00	
2	He	2.51E+09	1.2E+08	10.986	0.02
3	Li	55.6	7.2	3.33	0.05
4	Be	0.612	0.043	1.37	0.03
5	B	18.8	1.9	2.86	0.04
6	C	7.19E+06	6.9E+05	8.44	0.04
7	N	2.12E+06	6.8E+05	7.91	0.12
8	O	1.57E+07	2.8E+06	8.78	0.07
9	F	804	121	4.49	0.06
10	Ne	3.29E+06	8.5E+05	8.10	0.10
11	Na	57700	5100	6.35	0.04
12	Mg	1.03E+06	1.5E+05	7.60	0.06
13	Al	84600	15300	6.51	0.07
14	Si	1.00E+06	2E+04	7.59	0.08
15	P	8300	1100	5.51	0.05
16	S	4.21E+05	2.4E+04	7.21	0.02
17	Cl	5170	780	5.30	0.06
18	Ar	92700	24000	6.55	0.10
19	K	3760	330	5.16	0.04
20	Ca	60400	3000	6.37	0.02
21	Sc	34.4	1.7	3.12	0.02
22	Ti	2470	200	4.98	0.03
23	V	286	20	4.04	0.03
24	Cr	13100	500	5.70	0.02
25	Mn	9220	280	5.55	0.01
26	Fe	8.48E+05	1.69E+05	7.51	0.08
27	Co	2350	500	4.96	0.08
28	Ni	49000	5000	6.28	0.04
29	Cu	541	54	4.32	0.04
30	Zn	1300	130	4.70	0.04
31	Ga	36.6	1.8	3.15	0.02
32	Ge	115	18	3.65	0.06
33	As	6.10	0.55	2.37	0.04
34	Se	67.5	4.7	3.42	0.03
35	Br	10.7	1.6	2.62	0.06
36	Kr	55.8	11.3	3.33	0.08
37	Rb	7.23	0.51	2.45	0.03
38	Sr	23.3	1.6	2.95	0.03
39	Y	4.63	0.50	2.25	0.04
40	Zr	10.8	1.2	2.62	0.04

Table 6. – Continued

Z		N(Si)=10^6	σ	log N(H)=12	σ
41	Nb	0.780	0.139	1.48	0.07
42	Mo	2.55	0.40	1.99	0.06
44	Ru	1.78	0.11	1.84	0.03
45	Rh	0.370	0.128	1.15	0.13
46	Pd	1.36	0.15	1.72	0.04
47	Ag	0.489	0.024	1.28	0.02
48	Cd	1.57	0.11	1.78	0.03
49	In	0.178	0.012	0.84	0.03
50	Sn	3.60	0.54	2.14	0.06
51	Sb	0.313	0.047	1.08	0.06
52	Te	4.69	0.33	2.26	0.03
53	I	1.10	0.22	1.63	0.08
54	Xe	5.46	1.10	2.32	0.08
55	Cs	0.371	0.019	1.16	0.02
56	Ba	4.47	0.81	2.24	0.07
57	La	0.457	0.023	1.25	0.02
58	Ce	1.18	0.19	1.66	0.06
59	Pr	0.172	0.020	0.82	0.05
60	Nd	0.856	0.043	1.52	0.02
62	Sm	0.267	0.013	1.01	0.02
63	Eu	0.10	0.01	0.58	0.04
64	Gd	0.360	0.049	1.14	0.06
65	Tb	0.06	0.00	0.39	0.03
66	Dy	0.404	0.062	1.19	0.06
67	Ho	0.09	0.01	0.55	0.03
68	Er	0.262	0.042	1.00	0.06
69	Tm	0.04	0.00	0.19	0.03
70	Yb	0.256	0.013	0.99	0.02
71	Lu	0.0380	0.0019	0.17	0.02
72	Hf	0.156	0.008	0.78	0.02
73	Ta	0.0210	0.0021	-0.09	0.04
74	W	0.137	0.014	0.72	0.04
75	Re	0.0581	0.0058	0.35	0.04
76	Os	0.678	0.054	1.42	0.03
77	Ir	0.672	0.092	1.41	0.06
78	Pt	1.27	0.10	1.69	0.03
79	Au	0.195	0.019	0.88	0.04
80	Hg	0.458	0.092	1.25	0.08
81	Tl	0.182	0.015	0.85	0.03
82	Pb	3.31	0.23	2.11	0.03
83	Bi	0.138	0.012	0.73	0.04
90	Th	0.0440	0.0035	0.23	0.03
92	U	0.0238	0.0019	-0.04	0.03

Table 7. Protosolar mass fractions and He abundance

	Z_0/X_0	X_0	Y_0	Z_0	$A(He)_0$
this work[(a)]	0.0215	0.7112	0.2735	0.0153	10.986
A05, G07[(b)]	0.0185	0.7133	0.2735	0.0132	10.985
GS98[(c)]	0.0231	0.7086	0.2750	0.0163	10.99
AG89[(d)]	0.0267	0.7068	0.2743	0.0189	10.99

[(a)] present-day to protosolar values are converted using ($A(He)_0$ = A(He) + 0.061, all other elements (except H) from $A(X)_0$ = A(X) + 0.053

[(b)] G07 suggest $A(He)_0$ = A(He)+0.057, all other elements (except H) $A(X)_0$ = A(X) + 0.05.

[(c)] GS98: Changes in Z due to diffusion were not assumed (Z/X = Z_0/X_0); a ∼ 10% loss of He from the photosphere was considered.

[(d)] Changes in Z due to diffusion were not assumed (Z/X = Z_0/X_0)

and Oddo made the observation that elements with even atomic numbers are more abundant than their odd-numbered neighbors, which finds its explanation in the nuclear properties of the elements (see also Figure 4). An element is defined by its atomic number (Z), which is the number of positively charged nucleons (=protons) in its atoms. Atoms belonging to the same element may have different atomic masses due to a different number of neutral nucleons (=neutrons, neutron number N). In 1913, Soddy coined the term isotope for atoms with the same proton number but different neutron numbers after the Greek isos-topos meaning at the same place in the periodic table of the elements. The term isotope is in specific reference to a given element; whereas in a discussion of properties of atomic nuclei of different elements (with different Z and N) the generic term nuclide is usually more appropriate. However, often the terms nuclide and isotope are used as inter-exchangeable.

The sum of the number of protons (Z) and neutrons (N) is referred to as mass number A = Z+N. The mass number A is usually used when the nuclide abundance distributions are discussed, which is analogous to using the proton or atomic number Z for discussing elemental distributions.

There are 280 naturally occurring nuclides that make up the 83 stable and long-lived elements. These are all the elements up to Bi with Z = 83, except for unstable Tc (Z = 43) and Pm (Z = 61) that only have short-lived isotopes, but the long-lived Th and U bring the total back to 83. Here long-lived or short-lived is with respect to the half-life of an isotope against radioactive decay and the age of the solar system. Long-lived means then an element is still present in measurable quantities since the solar system formed 4.6 Gyr ago, and radioactive isotopes with half-lives above ∼0.6 Gyr usually qualify for this. Of the 280 nuclides, 266 are stable, and 14 have large half-lives such as ^{40}K, ^{232}Th, ^{235}U, ^{238}U, that find practical use in radiometric age dating of terrestrial and extraterrestrial samples, nuclear power reactors, and weaponry.

Considering only the atomic number, one finds that of the 83 elements, 43 have even Z, and 40 odd Z (note that Tc and Pm with only short-lived isotopes have odd atomic numbers, but Th and U have even ones), which reflects the higher stability of an atomic nucleus with even number of protons. This extends further to nuclei that also have an even number of neutrons. The proton and neutron numbers in the nuclei of the 266 stable nuclides lead to the following groupings:

Z even, N even: 159 nuclides
Z even, N odd: 53 nuclides
Z odd, N even: 50 nuclides
Z odd, N odd: 4 nuclides (^{2}H, ^{6}Li, ^{10}B, ^{14}N)

Since an element's abundance is the sum of the abundances of the element's isotopes, a lower number of odd-Z numbered nuclides (50+4) than even-Z (159+53) means that there is a lower abundance of odd-Z elements. This is a simple explanation for the odd-even abundance distribution noted by [77].

The mass numbers of the stable and long-lived nuclides range from A=1 (^{1}H) to 209 (^{209}Bi) except for gaps at A=5 and 8. After ^{209}Bi we only have longer-lived nuclides of the actinides Th and U with the mass numbers 232, 235, and 238. Several nuclides have the same mass number but are isotopes of different elements, simply because A is given by Z+N. In comparisons of the nuclide distributions as function of mass number, the nuclides with the same A (isobaric nuclides, or isobars) are often summed up.

Table 8 summarizes the nuclide abundances and Figure 5 shows the abundance distribution of the nuclides as a function of mass number at the time of solar system formation 4.56 Ga ago. Figure 5 shows that nuclides with even A (shown as closed symbols) have usually higher abundances than the odd numbered nuclides (open symbols). Further, the odd numbered nuclides plot parallel to the even numbered A in a somewhat smoother distribution curve. This behavior of nuclide distribution with mass number compares well to the behavior of elemental distribution with atomic number (Figure 4).

Abundances peak at mass numbers for closed proton and neutron shells. These nuclear shells are analogous to the closed electron shells that characterize atomic properties. The magic numbers for nuclear stability are 2, 8, 20, 28, 50, 82, and 126; and nuclides with Z and/or N equal to these magic numbers are the ones that show large abundances in the diagram of abundance versus mass number (A=Z+N). This is particularly notable for the light doubly-magic nuclei with equal magic Z and N, e.g., ^{4}He (Z=N=2), ^{16}O (Z=N=8), and ^{40}Ca (Z=N=20). Beyond the region of nuclides with mass numbers of 56 (the Fe-peak region), abundances decline more or less smoothly and spike at certain mass number regions. The nuclides beyond the Fe peak are products from neutron-capture processes. The peaks in the distribution correspond to regions where either nuclides are preferentially made by the slow-neutron capture (s-) process operating in red giant stars (e.g, Y and Ba regions) or by the rapid-neutron capture (r-) process probably operating in supernovae (e.g., Pt region); see, e.g., [78, 79, 80] for reviews on stellar nucleosynthesis. Here the *slow* and *rapid* are in reference to beta-decay timescales of the intermediate, unstable nuclei produced during the neutron-capture processes. The nuclide yields from these processes depend on the neutron energies and flux, but also on the abundance and stability of the target nuclei against neutron-capture which in turn depends on Z and N. Hence the abundance distribution becomes controlled by the more stable magic nuclides that serve as bottlenecks for the overall yields in the neutron capture processes.

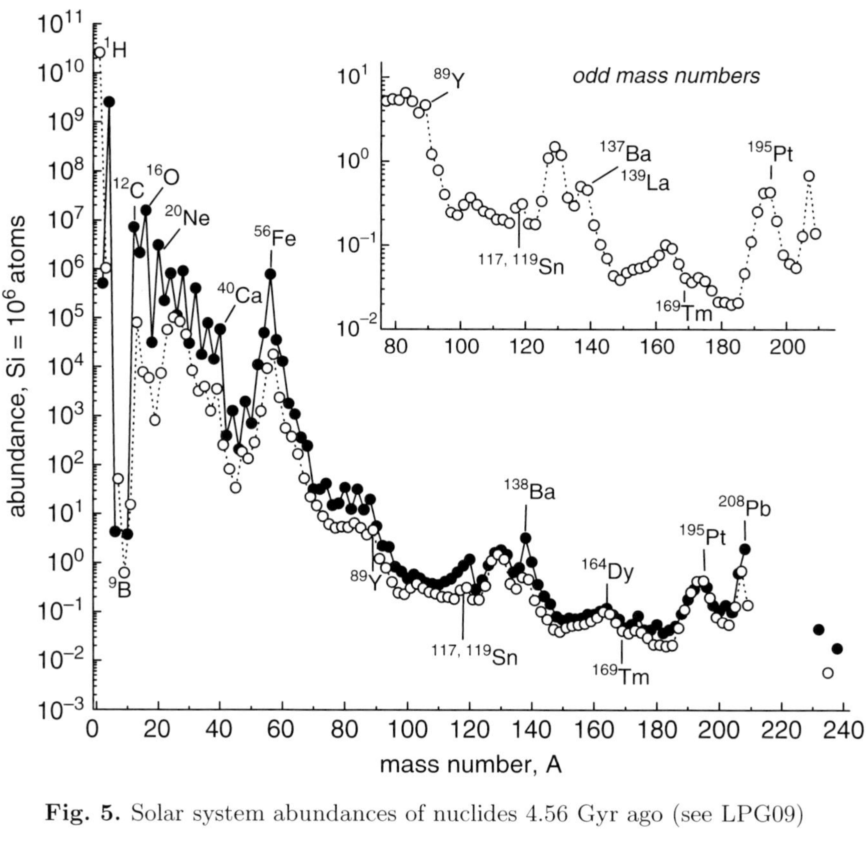

Fig. 5. Solar system abundances of nuclides 4.56 Gyr ago (see LPG09)

Table 8. Solar system nuclide abundances 4.56 Gyr ago

Z		A	atom%	N	Z		A	atom%	N
1	H	1	99.9981	2.59E+10	15	P	31	100	8300
1	H	2	0.00194	5.03E+05					
			100	2.59E+10	16	S	32	95.018	400258
2	He	3	0.0166	1.03E+06	16	S	33	0.75	3160
2	He	4	99.9834	2.51E+09	16	S	34	4.215	17800
			100	2.51E+09	16	S	36	0.017	72
3	Li	6	7.589	4.2				100	421245
3	Li	7	92.411	51.4	17	Cl	35	75.771	3920
			100	55.6	17	Cl	37	24.229	1250
4	Be	9	100	0.612				100	5170
					18	Ar	36	84.595	78400
5	B	10	19.820	3.7	18	Ar	38	15.381	14300
5	B	11	80.180	15.1	18	Ar	40	0.024	22
			100	18.8				100	92700
6	C	12	98.889	7.11E+06	19	K	39	93.132	3500
6	C	13	1.111	7.99E+04	19	K*	40	0.147	6
			100	7.19E+06	19	K	41	6.721	253
7	N	14	99.634	2.12E+06				100	3760
7	N	15	0.366	7.78E+03	20	Ca	40	96.941	58500
			100	2.12E+06	20	Ca	42	0.647	391
8	O	16	99.763	1.57E+07	20	Ca	43	0.135	82
8	O	17	0.037	5.90E+03	20	Ca	44	2.086	1260
8	O	18	0.200	3.15E+04	20	Ca	46	0.004	2
			100	1.57E+07	20	Ca	48	0.187	113
9	F	19	100	804				100	60400
					21	Sc	45	100	34.4
10	Ne	20	92.9431	3.06E+06					
10	Ne	21	0.2228	7.33E+03	22	Ti	46	8.249	204
10	Ne	22	6.8341	2.25E+05	22	Ti	47	7.437	184
			100	3.29E+06	22	Ti	48	73.72	1820
11	Na	23	100	57700	22	Ti	49	5.409	134
					22	Ti	50	5.185	128
12	Mg	24	78.992	8.10E+05				100	2470
12	Mg	25	10.003	1.03E+05	23	V	50	0.2497	0.7
12	Mg	26	11.005	1.13E+05	23	V	51	99.7503	285.7
			100	1.03E+06				100	286.4
13	Al	27	100	8.46E+04	24	Cr	50	4.3452	569
14	Si	28	92.230	9.22E+05	24	Cr	52	83.7895	11000
14	Si	29	4.683	4.68E+04	24	Cr	53	9.5006	1240
14	Si	30	3.087	3.09E+04	24	Cr	54	2.3647	309
			100	1.00E+06				100	13100

Table 8. – Continued

Z		A	atom%	N	Z		A	atom%	N
25	Mn	55	100	9220	35	Br	79	50.686	5.43
					35	Br	81	49.314	5.28
26	Fe	54	5.845	49600				100	10.7
26	Fe	56	91.754	7.78E+05	36	Kr	78	0.362	0.20
26	Fe	57	2.1191	18000	36	Kr	80	2.326	1.30
26	Fe	58	0.2819	2390	36	Kr	82	11.655	6.51
			100	8.48E+05	36	Kr	83	11.546	6.45
27	Co	59	100	2350	36	Kr	84	56.903	31.78
					36	Kr	86	17.208	9.61
28	Ni	58	68.0769	33400				100	55.8
28	Ni	60	26.2231	12900	37	Rb	85	70.844	5.121
28	Ni	61	1.1399	559	37	Rb*	87	29.156	2.108
28	Ni	62	3.6345	1780				100	7.23
28	Ni	64	0.9256	454	38	Sr	84	0.5580	0.13
			100	49000	38	Sr	86	9.8678	2.30
29	Cu	63	69.174	374	38	Sr	87	6.8961	1.60
29	Cu	65	30.826	167	38	Sr	88	82.6781	19.2
			100	541				100	23.3
30	Zn	64	48.63	630	39	Y	89	100	4.63
30	Zn	66	27.9	362					
30	Zn	67	4.1	53	40	Zr	90	51.452	5.546
30	Zn	68	18.75	243	40	Zr	91	11.223	1.210
30	Zn	70	0.62	8	40	Zr	92	17.146	1.848
			100	1300	40	Zr	94	17.38	1.873
31	Ga	69	60.108	22.0	40	Zr	96	2.799	0.302
31	Ga	71	39.892	14.6				100	10.78
			100	36.6	41	Nb	93	100	0.780
32	Ge	70	21.234	24.3					
32	Ge	72	27.662	31.7	42	Mo	92	14.525	0.370
32	Ge	73	7.717	8.8	42	Mo	94	9.151	0.233
32	Ge	74	35.943	41.2	42	Mo	95	15.838	0.404
32	Ge	76	7.444	8.5	42	Mo	96	16.672	0.425
			100	115	42	Mo	97	9.599	0.245
33	As	75	100	6.10	42	Mo	98	24.391	0.622
					42	Mo	100	9.824	0.250
34	Se	74	0.89	0.60				100	2.55
34	Se	76	9.37	6.32	44	Ru	96	5.542	0.099
34	Se	77	7.64	5.15	44	Ru	98	1.869	0.033
34	Se	78	23.77	16.04	44	Ru	99	12.758	0.227
34	Se	80	49.61	33.48	44	Ru	100	12.599	0.224
34	Se	82	8.73	5.89	44	Ru	101	17.060	0.304

Table 8. – Continued

Z		A	atom%	N	Z		A	atom%	N
44	Ru	102	31.552	0.562	52	Te	120	0.096	0.005
44	Ru	104	18.621	0.332	52	Te	122	2.603	0.122
			100	1.78	52	Te	123	0.908	0.043
45	Rh	103	100	0.370	52	Te	124	4.816	0.226
					52	Te	125	7.139	0.335
46	Pd	102	1.02	0.0139	52	Te	126	18.952	0.889
46	Pd	104	11.14	0.1513	52	Te	128	31.687	1.486
46	Pd	105	22.33	0.3032	52	Te	130	33.799	1.585
46	Pd	106	27.33	0.371				100	4.69
					53	I	127	100	1.10
46	Pd	108	26.46	0.359					
46	Pd	110	11.72	0.159	54	Xe	124	0.129	0.007
			100	1.36	54	Xe	126	0.112	0.006
47	Ag	107	51.839	0.254	54	Xe	128	2.23	0.122
47	Ag	109	48.161	0.236	54	Xe	129	27.46	1.499
			100	0.489	54	Xe	130	4.38	0.239
48	Cd	106	1.25	0.020	54	Xe	131	21.80	1.190
48	Cd	108	0.89	0.014	54	Xe	132	26.36	1.438
48	Cd	110	12.49	0.197	54	Xe	134	9.66	0.527
48	Cd	111	12.8	0.201	54	Xe	136	7.87	0.429
48	Cd	112	24.13	0.380				100	5.46
48	Cd	113	12.22	0.192	55	Cs	133	100	0.371
48	Cd	114	28.73	0.452					
48	Cd	116	7.49	0.118	56	Ba	130	0.106	0.005
			100	1.57	56	Ba	132	0.101	0.005
49	In	113	4.288	0.008	56	Ba	134	2.417	0.108
49	In	115	95.712	0.170	56	Ba	135	6.592	0.295
			100	0.178	56	Ba	136	7.853	0.351
50	Sn	112	0.971	0.035	56	Ba	137	11.232	0.502
50	Sn	114	0.659	0.024	56	Ba	138	71.699	3.205
50	Sn	115	0.339	0.012				100	4.471
50	Sn	116	14.536	0.524	57	La*	138	0.091	0.000
50	Sn	117	7.676	0.277	57	La	139	99.909	0.457
50	Sn	118	24.223	0.873				100	0.457
50	Sn	119	8.585	0.309	58	Ce	136	0.186	0.002
50	Sn	120	32.593	1.175	58	Ce	138	0.250	0.003
50	Sn	122	4.629	0.167	58	Ce	140	88.450	1.043
50	Sn	124	5.789	0.209	58	Ce	142	11.114	0.131
			100	3.60				100	1.180
51	Sb	121	57.213	0.179	59	Pr	141	100	0.172
51	Sb	123	42.787	0.134					
			100	0.313					

Table 8. – Continued

Z		A	atom%	N	Z		A	atom%	N
60	Nd	142	27.044	0.231	68	Er	166	33.503	0.088
60	Nd	143	12.023	0.103	68	Er	167	22.869	0.060
60	Nd	144	23.729	0.203	68	Er	168	26.978	0.071
60	Nd	145	8.763	0.075	68	Er	170	14.910	0.039
60	Nd	146	17.130	0.147				100	0.262
60	Nd	148	5.716	0.049	69	Tm	169	100	0.0406
60	Nd	150	5.596	0.048					
			100	0.856	70	Yb	168	0.12	0.0003
62	Sm	144	3.073	0.008	70	Yb	170	2.98	0.0076
62	Sm*	147	14.993	0.041	70	Yb	171	14.09	0.0361
62	Sm*	148	11.241	0.030	70	Yb	172	21.69	0.0556
62	Sm	149	13.819	0.037	70	Yb	173	16.10	0.0413
62	Sm	150	7.380	0.020	70	Yb	174	32.03	0.0821
62	Sm	152	26.742	0.071	70	Yb	176	13.00	0.0333
62	Sm	154	22.752	0.060				100	0.256
			100	0.267	71	Lu	175	97.1795	0.0370
63	Eu	151	47.81	0.0471	71	Lu*	176	2.8205	0.0011
63	Eu	153	52.19	0.0514				100	0.0380
			100	0.0984	72	Hf	174	0.162	0.0003
64	Gd	152	0.203	0.0007	72	Hf	176	5.206	0.0081
64	Gd	154	2.181	0.0078	72	Hf	177	18.606	0.0290
64	Gd	155	14.800	0.0533	72	Hf	178	27.297	0.0425
64	Gd	156	20.466	0.0736	72	Hf	179	13.629	0.0212
64	Gd	157	15.652	0.0563	72	Hf	180	35.100	0.0547
64	Gd	158	24.835	0.0894				100	0.156
64	Gd	160	21.864	0.0787	73	Ta	180	0.0123	2.6E-06
			100	0.360	73	Ta	181	99.9877	0.0210
65	Tb	159	100	0.0634				100	0.0210
					74	W	180	0.120	0.0002
66	Dy	156	0.056	0.0002	74	W	182	26.499	0.0363
66	Dy	158	0.095	0.0004	74	W	183	14.314	0.0196
66	Dy	160	2.329	0.0094	74	W	184	30.642	0.0420
66	Dy	161	18.889	0.0762	74	W	186	28.426	0.0390
66	Dy	162	25.475	0.1028				100	0.137
66	Dy	163	24.896	0.1005	75	Re	185	35.662	0.0207
66	Dy	164	28.260	0.1141	75	Re*	187	64.338	0.0374
			100	0.404				100	0.0581
67	Ho	165	100	0.0910	76	Os	184	0.020	0.0001
					76	Os	186	1.598	0.0108
68	Er	162	0.139	0.0004	76	Os	187	1.271	0.0086
68	Er	164	1.601	0.0042	76	Os	188	13.337	0.0904

Table 8. – Continued

Z		A	atom%	N	Z		A	atom%	N
76	Os	189	16.261	0.110	80	Hg	200	23.10	0.106
76	Os	190	26.444	0.179	80	Hg	201	13.18	0.060
76	Os	192	41.070	0.278	80	Hg	202	29.86	0.137
			100	0.678	80	Hg	204	6.87	0.031
77	Ir	191	37.272	0.250				100	0.458
77	Ir	193	62.728	0.421	81	Tl	203	29.524	0.054
			100	0.672	81	Tl	205	70.476	0.129
78	Pt*	190	0.014	0.0002				100	0.182
78	Pt	192	0.783	0.010	82	Pb	204	1.997	0.066
78	Pt	194	32.967	0.420	82	Pb	206	18.582	0.614
78	Pt	195	33.832	0.431	82	Pb	207	20.563	0.680
78	Pt	196	25.242	0.322	82	Pb	208	58.858	1.946
78	Pt	198	7.163	0.091				100	3.306
			100	1.27	83	Bi	209	100	0.1382
79	Au	197	100	0.195	90	Th*	232	100	0.0440
80	Hg	196	0.15	0.001	92	U*	234	0.002	4.9E-07
80	Hg	198	9.97	0.046	92	U*	235	24.286	0.0058
80	Hg	199	16.87	0.077	92	U*	238	75.712	0.0180
								100	0.0238

Only a few nuclides beyond the Fe-group are exclusively produced by the s- or r- process; most nuclides have varying abundance contributions from both processes. If the contribution from each process for each isotope is known, the overall contribution of the r- and s- process to the elemental abundance can be estimated. The review on heavy element synthesis by [80] includes a recent table on the r- and s- contributions to each element. A small number of proton-rich nuclides cannot be produced by the r- and s- process and are produced instead by the p-process, which probably involves neutrino induced disintegration of heavier nuclides. Like the r- process, the p- process is not yet completely understood. However, except for Mo, where p-process isotopes contribute about 25% to the elemental abundance, the contribution from p-process nuclides to overall elemental abundances is usually quite small.

Table 8 lists the percent contribution of the isotope(s) for each element, and the atomic abundance relative to 10^6 silicon atoms at the time of solar system formation.

The abundances of radioactive isotopes (indicated by a star next to the element symbol) are adjusted accordingly. Table 8 is an update to the Table in L03, and includes several revisions of isotopic compositions, e.g., for Mo [81],Dy [82, 83],Er [84],Yb [85],and Lu [86].

Acknowledgements I thank Bruce Fegley and Herbert Palme for discussions and comments.

References

1. B. Edvardsson, J. Andersen, B. Gustafsson, D. L. Lambert, P. E. Nissen, J. Tomkin, A&A **275**, 101 (1993)
2. B. Nordstrom, M. Mayor, J. Andersen, J. Holmberg, F. Pont, B. R. Jorgensen, E. H. Olsen, S. Udry, N. Mowlavi, A&A **419**, 989 (2004)
3. B. E. Reddy, J. Tomkin, D. L. Lambert, C. Allende Prieto, MNRAS **340**, 304 (2003)
4. B. R. Olsen, E. H. Udry, S. Mowlavi, A&A **419**, 989 (2003)
5. B. E. Reddy, D. L. Lambert, C. Allende Prieto, MNRAS **367**, 1329 (2006)
6. F. X. Timmes, S. E. Woosley, T. A. Weaver, ApJS **98**, 617 (1995)
7. M. Samland, ApJ **496**, 155 (1998)
8. A. Goswami, N. Prantzos, A&A **359** 191 (2000)
9. A. Anders, N. Grevesse, Geochim. Cosmochim. Acta **53**, 197 (1989)
10. H. Palme, H. Beer in *Landolt-Bernstein* Group VI, A&A ed. by H. H. Voigt, Springer, Berlin, 3a, 196 (1993)
11. N. Grevesse, A. Noels, A. J. Sauval, ASPC **99**, 117(1996)
12. N. Grevesse, A. J. Sauval, Space Sci. Rev. **85**, 161 (1998); GS98
13. K. Lodders, ApJ **691**, 1220 (2003); L03
14. H. Palme, A. Jones, in *Meteorites, Comets, and Planets* ed. by A. M. Davis; Vol. 1 *Treatise on Geochemistry* eds. H.D. Holland & K. K. Turekian Elsevier-Pergamon, Oxford, 41 (2003); PJ03
15. M. Asplund, N. Grevesse, A. J. Sauval, ASPC **336**, 25 (2005); A05
16. N. Grevesse, M. Asplund, A. J. Sauval, Space Sci. Rev. **130**, 105 (2007); G07
17. K. Lodders, H. Palme, H. P. Gail, in *Landolt-Böernstein New Series, Group IV/4B*, chapter 4.4, ed. by Trümper, J. E. (ed.), Berlin, Heidelberg, New York: Springer-Verlag, p. 560 (2009); LPG09
18. V. M. Goldschmidt, *Geochemische Verteilungsgesetzte der Elemente. IX. Die Mengenverhltnisse der Elemente und der Atom-Arten, Skrifter Norske Videnskaps Akademi I. Mathematisk-Naturvidenskapelig klasse*, No 4, Oslo 1937/1938, pp. 148. see also V.M. Goldschmidt, J. Amer. Chem. Soc., 1937, 655 (1937)
19. H. Brown, Rev. Mod. Phys **21**, 625 (1949)
20. H. E. Suess, H. C. Urey, Rev. Mod. Phys **28**, 53 (1956)
21. K. Lodders, B. Fegley, *The Planetary Scientist's Companion*, Oxford Univ. Press (1998)
22. E. Anders, Geochim. Cosmochim. Acta **35**, 516 (1971); AG89
23. K. Lodders, H. Palme, in preparation (2009)
24. A. Greshake, W. Kloeck, P. Arndt, M., Maetz, G. J. Flynn, S. Bajt, S., A. Bischoff, Meteoritics Planet. Sci. **33**, 267 (1998)
25. A. Morlok, A. Bischoff, T. Stephan, C. Floss, E. Zinner, E. K. Jessberger, Geochim. Cosmochim. Acta **70**, 5371 (2006)
26. H. N. Russell, ApJ **70**, 11 (1929)
27. H. Holweger, in *Solar and galactic composition* ed. by R.F. Wimmer-Schweinsgruber, AIP Conf. Proc. **598**, 23 (2001)
28. M. Steffen, & H. Holweger, A&A **387**, 258 (2002)
29. J. R. Shi, T. Gehren, K. Butler, L. I. Mashonkina, G. Zhao, A&A **486**, 303 (2008)

30. S. Basu, H. M. Anita, Phys. Rep. **457**, 217 (2008)
31. M. Carlsson, R. J. Rutten, J. H. M. J. Brules, N. G. Shchukina, A&A **288**, 860 (1994)
32. E.A. Müller, E. Peytremann, R. de la Reza, Solar Physics **41**, 53 (1975)
33. M. Asplund, A&A **417**, 769 (2004)
34. A. Makishima, E. Nakamura, Geostandard and Geoanalytical Research **30**, 245 (2006)
35. C. Allende Pietro, D. L. Lambert, M. Asplund, ApJ **573**, L137 (2002)
36. M. Asplund, N. Grevesse, A. J. Sauval, C. Allende Prieto, R. Blomme, A&A **431**, 693 (2005)
37. P. C. Scott, M. Asplund, N. Grevesse, A. J. Sauval, A&A **456**, 675 (2006)
38. E. Caffau, E. Maiorca, P. Bonifacio, R. Faraggiana et al., A&A **498**, 877 (2009)
39. E. A. Caffau, H. G. Ludwig, M. Steffen, T. R. Ayres, P. Bonifacio, R. Cyarel, B.Freytag, B. Plez, A&A **488**, 1031 (2008)
40. H. G. Ludwig, M. Steffen in *Precision Spectroscopy in Astrophysics*, Proc. of the ESO/Lisbon/Aveiro conference in Aveiro, Portugal, 11-15 Sep 2006, ed. by N.C. Santos, L. Pasquini, A.C.M. Correia, and M. Romanielleo, Springer-Verlag Berlin Heidelberg, Germany, 133 (2008)
41. J. Melendez, M. Asplund, A&A **490**, 817 (2008)
42. T. R. Ayres, C. Plymate, C. U. Keller, ApJS **165**, 618 (2006)
43. T. Morel, K. Butler, A&A **487**, 307 (2008)
44. E. Landi, U.Feldman, & G. A. Doschek, ApJ **659**, 743 (2007)
45. P. R. Young, A&A **444**, L45 (2005)
46. P. Bochsler, F. Auchere, R. M. Skoug, in *Proc. SOHO 17*, ed. by H Lacoste, Ouwehand, L., ESA SP-616, 28 (2006)
47. P. Bochsler, A&A **471**, 315 (2007)
48. M. Asplund, A&A **359**, 755 (2000)
49. E. Caffau, M. Steffen, L. Sbordone, H. G. Ludwig, P.Bonifacio, A&A **473**, L9 (2007)
50. E. Caffau, H. G. Ludwig A&A **467**, L11 (2007)
51. E. Caffau, R. Faraggiana, P. Bonifacio, H. G. Ludwig, M. Steffen, A&A **699**, 699 (2007)
52. K. Lodders, ApJ **674**, 607 (2008)
53. H. W. Zhang, K. Butler, T. Gehren, J. R. Shi, G. Zhao, A&A **457**, 645 (2006)
54. D. L. Lambert, B. Warner, MNRAS **138**, 181 (1968)
55. H. W. Zhang, T. Gehren, G. Zhao, A&A **48**, 489 (2008)
56. J. S. Sobeck, J. R. Lawler, C. Sneden, ApJ **667**, 1267 (2007)
57. M. Bergemann, T. Gehren, A&A **473**, 291 (2007)
58. R. Blackwell-Whitehead, M. Bergemann, A&A **472**, L43 (2007)
59. G. Ljung, H. Nilsson, M. Asplund, S. Johansson, A&A **456**, 1181 (2006)
60. H. L. Xu, Z. W. Sun, Z. W. Dai, Z. K. Jiang, P. Palmeri, P. Quinet, E. Biemont, A&A **452**, 357 (2006)
61. G. Gonzales, MNRAS **371**, 781 (2006)
62. N. Vitas, I. Vince, M. Lugaro, O. Andriyenko, M. Gosic, R. J. Rutten, MNRAS **384**, 370 (2008)
63. C. Sneden, J. E. Lawler, J. J. Cowan, I. I. Ivans, E. A. Den Hartog, ApJS **182**, 80 (2009)

64. S. Ivarsson, G. M., Wahlgren, H. G. Ludwig, Bull. A&AS **35**, 1451 (2003)
65. E. A. Den Hartog, J. E. Lawler, C. Sneden, J. J. Cowan, ApJS **148**, 543 (2003)
66. J. E. Lawler, E. A., Den Hartog, C. Sneden, J. J. Cowan, ApJS **162**, 227 (2006)
67. A. Mucciarelli, E. Caffau, B. Freytag, H. G. Ludwig, P. Bonifacio, A&A **484**, 841 (2008)
68. J. E. Lawler, M. E. Wickliffe, E. A. Den Hartog, C. Sneden, ApJ **563**, 1675 (2001)
69. E. A. Den Hartoog, J. E. Lawler, C. Sneden, J. J. Cowan, ApJS **167**, 292 (2006)
70. J. E. Lawler, C. Sneden, J. J. Cowan, ApJ **604**, 850 (2004)
71. J. E. Lawler, C. Sneden, J. J. Cowan, J. F. Wyart, I. I. Ivans, J. S. Sobeck, M. H.Stockett, E. A. Den Hartog, ApJS **178**, 71 (2008)
72. J. E. Lawler, E. A. D. Hartog, Z. E. Labby, C. Sneden, J. J. Cowan, I. I. Ivans, ApJS **169**, 120 (2007)
73. E. Caffau, E. Maiorca, P. Bonifacio, R. Faraggiana, et al., A&A **483**, 591 (2008)
74. P. Quinet, P. Palmari, E. Biemont, A. Jorrison, et al., A&A **448**, 1207 (2006)
75. E. A. Den Hartog, M. T. Herd, J. E. Lawler, C. Sneden, J. J. Cowan, T. C. Beers, ApJ **619**, 639 (2005)
76. S. Basu, H. M. Anita, 2004, ApJ **606**, L85 (2004)
77. W.D. Harkins, J. Am. Chem. Soc. **39**, 856 (1917)
78. G. Wallerstein, I. Iben, P. Parker, A. M. Boesgaard, G. M. Hale, A. E. Champagne, et al., Rev. Mod. Phys.**69**, 995 (1997)
79. S. E. Woosley, A. Heger, T. A. Weaver, Rev. Mod. Phys. **74**, 1015 (2002)
80. C. Sneden, J. J. Cowan, R, Gallino, ARAA **46**, 145 (2008)
81. M. E. Wieser, J. R. DeLaeter, Phys. Rev. **C75**, 055802 (2007)
82. I. Segal, L. Halicz, I. T. Platzner, Int. J. Mass Spec. **216**, 177 (2002)
83. T. L. Chang, W. J. Li, M. T. Zhao, J. Wang, Q. Y. Qian, Int. J. Mass Spec. **207**, 13 (2001)
84. T. L. Chang, M. T. Zhao, W. J. Li, J. Wang, Q. Y. Qian, Int. J. Mass Spec. **177**, 131 (1998)
85. J. R. De Laeter, N. Bukilic, Int. J. Mass. Spec. **252**, 222 (2006)
86. J. R. Laeter, N. Bukilic, Phys. Rev. **C73**, 045806 (2006)

Cosmochemistry: A Perspective

Aruna Goswami

Indian Institute of Astrophysics, Koramangala, Bangalore 560034
aruna@iiap.res.in

Cosmochemistry can be said to aim at constructing a quantitative and continuous history of chemical evolution of our universe starting from the appearance of hydrogenic matter immediately after the Big Bang. While the main events of this history are embodied in stellar nucleosynthesis, the scope of the complete history is much larger. Many important aspects of cosmochemistry still remain unclear, although improved observations and steadily increasing computing power hold exciting promises. The Kodai school on 'Synthesis of Elements in Stars' provided glimpses of status and challenges of diverse issues in this emerging discipline; the following paragraphs provide a synthesis of the various contributions to arrive at a perspective.

The non-linear and multi-scale physics that govern the life of a star are not amenable to linear or simplified calculations. However, much insight about stellar nuclear reactions, stellar evolution as well as stellar explosions have come from computer simulations. As also noted by Ray (this volume) astronomical observations across the entire spectrum of electromagnetic radiation and other indicators like neutrinos have significantly aided theoretical developments. A major contribution has come from the wider availability of powerful telescopes from different platforms. Similarly, high speed computers had given important feedback to theoretical developments in many areas. Another important and encouraging trend in cosmochemistry is the increasing feedback from laboratories. Crucial inputs to the field of nuclear astrophysics have come from laboratory experiments involving radioactive ion beams and intense high-energy particle beams. Since short lived nuclei as found in SN ejecta can only be studied close to their sites of formation/in the laboratory before they decay, such studies can further contribute to the advancement of nuclear astrophysics. With growing and affordable and increasing computing power, therefore, computational astrophysics in general, and computational cosmochemistry in particular, is an area that holds many promises.

Many concepts and techniques of cosmochemistry have a generic nature. The methods and results of chemical equilibrium calculations applied to solar composition are applicable to chemistry in a variety of astronomical

A. Goswami and B.E. Reddy (eds.), *Principles and Perspectives in Cosmochemistry*, Astrophysics and Space Science Proceedings,
DOI 10.1007/978-3-642-10352-0_9,

environments including the atmospheres and circumstellar envelopes of cool stars, the solar nebula and protoplanetary accretion disks around other stars, planetary atmospheres, and the atmospheres of brown dwarfs. Similarly, the results of chemical equilibrium calculations have guided studies of elemental abundances in meteorites and presolar grains and as a result have helped to refine and constrain models of nucleosynthesis in stars.

Recent estimates indicate that as much as half of all elements heavier than iron are made by the *s*-process in low-mass AGB stars. However, while our understanding of the inner workings of AGB stars has significantly improved in the recent past, there are many unknowns that render predictions uncertain. As emphasized by Karakas (this volumn) our lack of knowledge about convective mixing processes in stars is one of the greatest sources of uncertainty. Other modeling uncertainties such as mass-loss rates, opacities, and reaction rates are also important and need careful investigation.

Many challenges also remain in understanding the structure and evolution of low and intermediate-mass stars prior to the AGB phase, such as the mechanisms of formation of ^{13}C pockets and the extent of extra-mixing processes during the first dredge-up and asymptotic giant branch phases. The processes driving mass loss in AGB stars remain a major challenge, an important requirement is to develop techniques to accurately measure many of the reaction rates of interest for AGB nucleosynthesis. Better understanding of nucleosynthesis in low-metallicity AGB stars is another essential requirement to solve the puzzle surrounding the abundances of many carbon-rich metal-poor stars in the Galaxy. The mass-transfer processes that occur in binary star systems also remain poorly understood; AGB stars thus remain a challenging and critical gap in our understanding of stellar evolution.

Supernova events as well as pre-supernova stages of stars play important roles in cosmochemistry. However, a number of aspects of astrophysics of both pre-supernova and supernova stages of evolution of massive stars remain unclear, as emphasized by Marcel Arnould (this volume). A major challenge comes from the uncertainties in the rates of some nuclear reactions and of weak interaction processes. Multi-dimensional simulations, especially of the late pre-supernova phases, are expected to be particularly helpful to understand these processes. Similarly, considerable efforts are still required for successful simulations of supernova explosions based on less uncertain pre-supernova models. Here again, the multi-dimensional treatment of a variety of physical effects, including rotation, magnetic fields, instabilities of different origins, and the transport of neutrinos needs careful inclusion.

Among the most vexing nucleosynthetic puzzles is the r-process. Some success has been achieved with regard to p-process elements for which calculations conducted with 'realistic' models for the O-Ne-rich layers of SNeII broadly indicate that they may all provide yields enriched with a suite of p-nuclides at a level that is compatible (within a factor of 3 or so) with the solar values. As highlighted by Marcel Arnould this relative success, however, does not remove a host of puzzling problems. It is still not clear, for example, if the underproduction of the p-isotopes of Mo and Ru in SNII explosions is in fact due to some misrepresentation of the production in the He-burning core of massive stars of the s-nuclide seeds for the p-process. The modelling of the s-process in these conditions indeed faces uncertainties from both nuclear physics and astrophysics which are highly non-trivial.

The arena of cosmochemistry is emerging as a broad interdisciplinary area of research. It is one area where astrophysical processes at different scales, both spatial and temporal, combine to provide an integrative view of the chemical history of the Galaxy/galaxies, their structure and origin. The primary goal of this collection of lectures is to put together important components of cosmochemistry which we hope will serve as a contemporary and useful reference for researchers in this exciting and challenging area.

MIX
Papier aus verantwortungsvollen Quellen
Paper from responsible sources
FSC® C105338

Printed by Books on Demand, Germany